Photocatalytic Nanomaterials for Environmental Applications

Edited by
Rajesh J. Tayade
Vimal Gandhi

Photocatalytic nanomaterials have a great potential in such applications as reduction of carbon dioxide and degradation of various pollutants. They are equally important in the production and storage of energy, e.g. in the conversion of solar energy to electricity, and the production of hydrogen in photoelectrochemical cells.

Research on synthesis, characterization and specific applications is reported for titanium oxide and a number of other promising catalysts, such as silver phosphate, cerium oxide, zinc oxide and zinc sulfide.

Keywords: Photocatalytic Nanomaterials, Nanocomposites, Solar Energy Conversion, Carbon Dioxide Reduction, Hydrogen Generation, Degradation of Pollutants, Titanium Oxide, Silver Phosphate, Cerium Oxide, Zinc Oxide, Zinc Sulfide

Photocatalytic Nanomaterials for Environmental Applications

Edited by

Rajesh J. Tayade
Inorganic Materials & Catalysis Division,
Central Salt & Marine Chemicals Research Institute,
Bhavnagar-364002, Gujarat
INDIA

Vimal Gandhi
Department of Chemical Engineering,
Dharmsinh Desai University,
Nadiad- 387 001, Gujarat,
India

Published by **Materials Research Forum LLC**
Millersville, PA 17551, USA

Published as part of the book series
Materials Research Foundations
Volume 27 (2018)
ISSN 2471-8890 (Print)
ISSN 2471-8904 (Online)

Print ISBN 978-1-945291-58-6
ePDF ISBN 978-1-945291-59-3

Distributed worldwide by

Materials Research Forum LLC
105 Springdale Lane
Millersville, PA 17551
USA
http://www.mrforum.com

Manufactured in the United States of America
10 9 8 7 6 5 4 3 2 1

Table of Contents

Preface

Concerns related to environmental problems and the energy crisis have created great challenges for scientist and technologist at a global level. For the last 3-4 decades, major attention has been given for photocatalysis to overcome these problems. Photocatalysis is an effective catalytic process with potential applications in solving above mentioned problems by degrading pollutants, reducing CO_2 levels and water splitting, respectively. Till today, various photocatalytic materials, photocatalytic reactor, degradation of various organic compound present in air and water have been studied and demonstrated. In spite of significant research work carried out in the field of photocatalysis, there are several challenges needing to be addressed such as enhancement in photocatalytic activity, development of visible light activated photocatalysts, search of alternative photocatalytic materials, design and development of photocatalytic reactors based on the irradiation source. In this present special topic book entitled as "Photocatalytic Nanomaterials for Environmental Applications" we have tried to report the latest development and original applications and theoretical research in the area of photocatalysis.

This special topic book is a result from contributions experts from the international scientific community in the photocatalysis and nanomaterials development field. It thoroughly covers future prospects of various photocatalytic materials, their synthesis methods, and modifications for enhancement in photocatalytic activity as well as their application in various fields like photocatalytic degradation of environment pollutant, energy production and conversion of solar energy to electrical energy, energy storage, chemical transformation etc. It gives a comprehensive picture of photocatalytic materials and has posed several scientific and technological challenges in this area. This book will provide latest and in-depth coverage to the various photocatalytic materials used in the last decade for various applications.

The first five chapters of the book mainly focus on synthesis, characterization and various applications of titanium dioxide as photocatalyst. Chapter 1 discusses on various method of synthesis, morphological variety and application of titanium dioxide nanomaterials. To enhance the photocatalytic activity of the titanium dioxide, different TiO2-high surface area materials based composite photocatalysts are reviewed in detail in Chapter 2. A variety of visible light active photocatalysts with their synthetic methods and applications have been discussed in Chapter 3 to address the issue of UV light source in the photocatalysis. The three review chapters are supported by experimental work on synthesis of TiO_2-carbon composites, its characterization and application for degradation of isoniazid explained in Chapter 4. In Chapter 5, experimental work related to synthesis of TiO_2 nanotubes by using a low-temperature hydrothermal method without templates and investigated the effect of annealing temperature and dissolved oxygen for the degradation of carcinogenic phenol has been discussed.

The next four chapters (Chapter 6-9) enlighten the relatively newer field of application of photocatalysis related to photoelectrochemical reduction of carbon dioxide to solar fuel and photoelectrochemical cell for the production of hydrogen. Chapter 6 mainly focused on the understanding of reaction mechanism in photon-assisted reduction of carbon dioxide including the fundamental review on photophysics and photochemistry of semiconductor materials along with possible suggestions for the design of new materials. Chapter 7 is a review article on photo-electrochemical reduction of carbon dioxide to solar fuel with appropriate case studies. In Chapter 8, Cadmium Sulphide based photoelectrochemical cells are discussed in detail to channelize solar energy directly into chemical energy and provide hydrogen fuel. Experimental work related to enhancement in hydrogen storage capacity using synthesized TiO_2 -multiwall carbon nanotube with its characterization discussed in Chapter 9.

To provide alternate photocatalytic materials of titanium dioxide, researchers are trying to synthesize/develop variety of photocatalytic materials like silver phosphate, cerium oxide, zinc oxide, zinc sulfide etc. for different purposes. Silver phosphate based photocatalytic materials review with its synthesis methods and different applications in Chapter 10. Shape-control synthesis of bare/doped ceria with its photocatalytic applications has been discussed in detail in Chapter 11. Chapter 12 focused on experimental work related to synthesis of Sm^{+3} doped ceria, its detailed characterization and photocatalytic degradation of methylene blue. Experimental work of Chapter 13 discusses about enhancement in photocatalytic transformation of diphenylamine into N-phenyo-p-benzoquinonimine in the presence of Zinc sulfide particles mixed with different photocatalytic materials due to intra particle charge transfer. A review on recent advances in Copper Zinc Tin Sulfide -CZTS (Cu2ZnSnS4) film preparation, its application in the field of solar cell development and as photocatalysts is covered in Chapter 14.

Chapter 15 focuses on modelling and optimization of photocatalytic degradation of 4-Chlorophenol (4-CP) using Response Surface Methodology (RSM) and Artificial Neural Network (ANN) using Titanium nanotube synthesized by hydrothermal method. Chapter 16 focuses on the effect of ultra sound on the synthesis of nanoparticles as well as in the removal of pollutants from the effluents.

This book is indeed the result of remarkable cooperation of many distinguished experts, who came together to contribute their research work and comprehensive, in-depth and up-to-date review chapters. We are very thankful to all contributing authors who, in spite of their busy life in research and teaching, willingly accepted the call to contribute and sent their manuscript in time. We would also like to express our gratitude to all the publishers and authors and others for granting us the copyright permissions to use their illustrations. Although sincere efforts were made to obtain the copyright permissions from the respective owners to include the citation with the reproduced materials, we would like to offer our honest apologies to any copyright holder if unknowingly, their right is being infringed.

We would like to take this opportunity to express our sincere gratitude and also like to acknowledge the sincere efforts of Mr. Thomas Wohlbier of Materials Research Forum LLC, publishing Authority and his team, for in evolving this Special Topic Book into its final shape.

Rajesh J. Tayade
Inorganic Materials & Catalysis Division, Central Salt & Marine Chemicals Research Institute, Bhavnagar-364002, Gujarat, India

Vimal Gandhi
Department of Chemical Engineering, Dharmsinh Desai University, Nadiad- 387 001, Gujarat, India

Chapter 1

TiO_2 Nanomaterials a Future Prospect

Sajith N.V. [1], Soumya B. Narendranath[1], Sheethu Jose[1], Pradeepan Periyat[2]*

[1]Department of Chemistry, Central University of Kerala, Kasaragod, India. 671314
[2]Department of Chemistry, University of Calicut, 673635, India
Email :pperiyat@uoc.ac.in

Abstract

As a potential candidate for various applications such as self- cleaning coatings, electrode material for Li ion battery, dye sensitized solar cells, photo catalytic hydrogen generation, water purification etc. TiO_2 nanomaterials becomes an interesting topic for research all over the world. This review focuses on recent progresses in structure, methods of synthesis, morphological variety and applications of TiO_2 nanomaterials.

Keywords

Nanomaterials, TiO_2, Photocatalyst, Self-cleaning Coatings, Hydrogen Generation, Water Purification

Contents

1. Introduction

Semiconductor photocatalysis is the most widely explored area in the current era of increased fossil fuel depletion and the resulted environmental pollution [1-5]. Photocatalysis over a semiconductor utilizes solar energy which is ultimately free of cost and highly abundant in nature. Earth absorbs approximately 51% of the total incoming solar radiation [6]. The remaining energy is reflected back into outer space and absorbed by the atmosphere which is approximately 3,850,000 EJ per year (EJ-exajoule=1018 J). Out of this huge amount of energy, we use only 500 EJ per year, i.e. earth absorbs a massive amount of solar radiation in one hour compared to the whole world uses in one year [7]. It has been a constant challenge to exploit the huge amount of sun light falling on earth. Recently technologies to harvest solar energy using nano semiconducting materials have been designed and used to meet the present and future energy crisis [8-11]. Major efforts were taken in designing metal oxide nanoparticles for superior solar energy harvesting.

Metal oxide nanomaterials have proven their efficiency in the fields of heterogeneous catalysis [12-15], photocatalysis [16-19], energy conversion and storage [20-26], sensors

[27-32], pigments [33-36], protective coatings [37-40], electronic devices [41-45] etc. When a bulk material becomes nano, the nanomaterials retain structural and mechanical stability owing to a low surface free energy [46, 47]. Also, the bulk to nano transition accompanies the modifications in size, shape, crystal structure, morphology, optical, electronic and properties and surface chemistry [48-54]. Among the various metal oxide semiconducting nanomaterials, TiO_2 garnered special attention due to its thermal and chemical stability, fine tunable optical and electronic band gap energy, non-toxicity [55-61]. After the discovery of Honda Fujishima effect in 1972 [62], extensive research works have been focused on solar light harvesting via TiO_2 based nano semiconducting systems. TiO_2 and its various modifications proved to be excellent candidate for various technological applications such as photo catalytic water splitting [63], dye sensitized solar cells [64], self-cleaning coatings [65], sensors and as electrode material for Li ion battery [66]. However, the photo conversion efficiency of TiO_2 is limited because of its large band gap energy (3.0-3.2 eV) [56]. The broad band gap energy inhibits the utilization of visible (43%) and IR (52%) region of solar spectra which comprise the major part of it. Pure TiO_2 could absorb only the remaining 5% UV radiation [67] which is a major bottle neck for TiO_2 photocatalysis. Researchers all over the world are working on extending the absorption of TiO_2 in the visible region by reducing the band gap energy and thereby attain maximum photo catalytic activity.

A variety of modification strategies have been proposed and implemented to improve the photo absorption and consequent photocatalytic activities. Metal and non-metal doping [8, 68-80], dye sensitization [81-87], semiconductor nanocomposites [88-90], reduction [91] etc. paved pathways for the fine tuning of bandgap energy of TiO_2. This review focuses mainly on the crystal and electronic structure modifications of TiO_2 and the resulting enhancement in the photo sensitized activities. The structural aspects, basic electronic structure, morphologies and their variations upon modifications were discussed in detail.

2. Synthesis of TiO_2 nanoparticles

2.1 Sol-gel method

The sol-gel method is a widely accepted one for the synthesis of ceramic materials and metal oxide nanoparticles [92-94]. In a sol-gel process, a sol in the form of colloidal suspension is formed from the hydrolysis and polymerization reactions of the precursors. Typical precursors used in TiO_2 synthesis are of two types. Inorganic metal salts such as Titanium tetrachloride ($TiCl_4$), Titanium oxysulphate ($TiOSO_4$) and metal alkoxide compounds like Titanium isopropoxide, Titanium butoxide etc. [95]. During polymerization, solvent is escaped from the reaction medium and the sol is transformed to a solid gel. Finally, the gel is calcined to form the required phase. The gel can be fabricated into film or sheets by dip coating or spin coating. The sol-gel method has potential applications due to the easy fabrication of powders, films, fibers, monoliths, composites and porous media. These can be transforms into ceramic material upon heat treatment. The material obtained after this processing will have high purity, homogeneity and controlled porosity [96]. The reactions involved in a sol-gel process are represented through equations (1) and (2)

$$\text{Hydrolysis: -T-OR} + H_2O \rightarrow \text{Ti-(OH)}_4 + \text{R-OH} \quad (1)$$

$$\text{Condensation: -Ti-OH} + \text{H-OTi-} \rightarrow \text{-Ti-O-Ti-} + H_2O \quad (2)$$

Where, R = alkyl group.

The various factors affecting the sol-gel process are reaction temperature, reactivity of metal alkoxide, pH of the reaction medium and nature of the solvent and additive [97]. During the condensation reaction (equation (2)), Ti-O-Ti chains were developed. Hydrolysis is favoured by the presence of excess of water and condensation is favoured by low water content in the reaction medium [98].

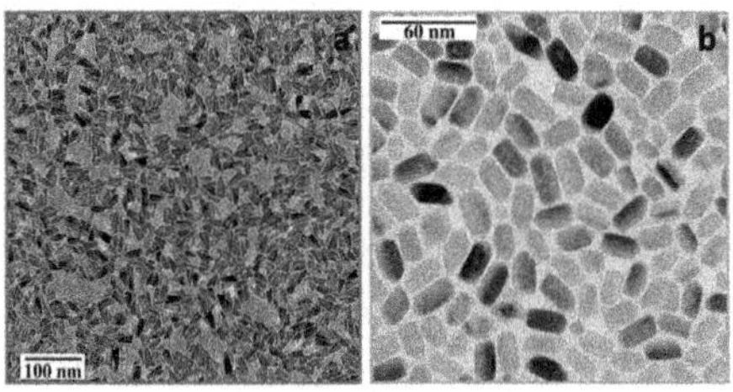

Figure 1. TEM images of a) Tetramethyl ammonium hydroxide capped TiO_2 and b) autoclaving the Tetramethyl ammonium hydroxide capped TiO_2 solution. Reprinted with permission from John Wiley and Sons (1999) [99].

TiO_2 nanomaterials were synthesized with various morphological varieties through the sol-gel method using titanium alkoxide and tetramethyl ammonium hydroxide [98, 99]. Here titanium alkoxide is added to tetramethyl ammonium hydroxide in alcoholic medium and heated at 100 °C for 6 h [99]. The TEM images of tetramethyl ammonium hydroxide capped TiO_2 and the TiO_2 nanoparticles obtained after autoclaving the said solution were given in Fig. 1. Shape controlling agents can be used in the sol-gel method which will fine tune the morphology of the resulting nanoparticles [100, 101]. Sugimoto *et al.* added Triethanolamine (TEOA) as shape controlling agent to an aqueous solution of titanium tetraisopropoxide (TTIP). They could get ellipsoidal nanoparticles instead cuboid at a pH of 11 [102, 103].

Nano TiO_2 was synthesized via the sol-gel method by P. Periyat et al. and his co-workers from titanium isopropoxide, glacial acetic acid and water mixture in the ratio 1:10:100. In their experiment, de-ionized water was added to a mixture of titanium isopropoxide and glacial acetic acid with constant stirring for 3 h followed by drying at 100 °C on water bath for 12 h [104]. SEM analysis indicated that the particles undergo agglomeration and TEM images shows a crystallite size of 16-28 nm at 700 °C.

2.2 Hydrothermal method

Hydrothermal method widely used for the production of crystalline materials and the synthesis is usually carried out in a special type of steel pressure vessel called autoclave with or without Teflon lining inside. The reaction is carried out in aqueous solution under controlled temperature and pressure. The temperature can be higher than the boiling point of water which produces the pressure of vapor saturation. The developed internal pressure depends on the temperature and volume of the solution in the autoclave. Many researchers found this method as a suitable method for the synthesis of TiO_2 nanoparticles [105-111]. For e.g., alcoholic solution of titanium tetraisopropoxide in aqueous medium was heated at 240 °C for 4 h produced TiO_2 nanoparticles with a controlled size of 7-25 nm [110]. The TEM images of the as prepared TiO_2 nanomaterials with varying size on varying the composition of solvent system and the concentration of TTIP are shown in Fig.2. Andersson et al. [105] reported a low temperature synthetic method in which a reverse micro emulsion system with Triton X-100, n-hexanol, cyclohexane and Tetrabutyltitanate dissolved in HCl or HNO_3 was prepared and then the microemulsion was transferred into an autoclave. The autoclave was heated at 120 °C for 13 h, and the precipitate obtained after the hydrothermal process was washed with ethanol and dried in a desiccator. In a different strategy, the sol formed by the hydrolysis of titaniumalkoxide on prolonged hydrothermal treatment could yield monodispersed TiO_2 nanoparticles. $TiCl_4$ and $TiCl_3$ can also be used as precursors for hydrothermal synthesis of TiO_2 nanoparticles which probably have rod morphology. TiO_2 white powder on hydrothermal heating gives TiO_2 nanowires. Here bulk powder was dispersed in NaOH solution around 200 °C for 3 days.

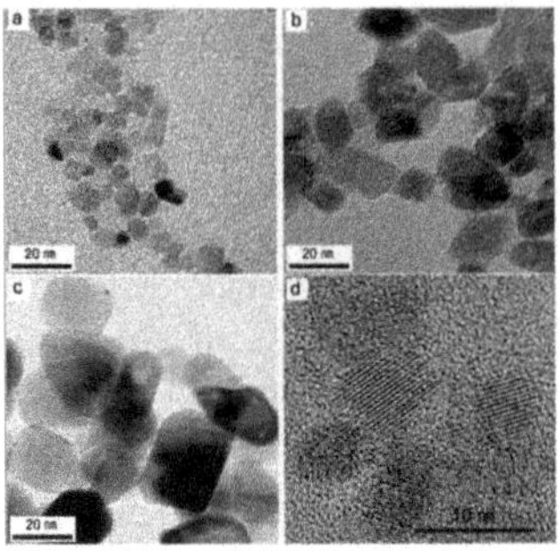

Figure 2. TEM images of hydrothermally synthesized a) 7nm sized TiO_2 nanoparticles from 0.10M TTIP in 4:1 ethanol/water b) 15nm particles from 0.04M TTIP in 1:2 ethanol/water c) 25nm particles from 0.02M TTIP in 1:8 ethanol/water and d) HRTEM image for a 7nm particle. Reprinted with permission from the American Chemical Society (2003) [110].

In addition to TiO_2 nanoparticles, TiO_2 nanowires have also been synthesized hydrothermally by adding 1g anatase TiO_2 powder into 10M NaOH solution in a Teflon-lined autoclave of 50 mL capacity at 200°C for 24 h [112] (SEM image is shown in Fig.3a). Jianming Li *et al.* found that hydrothermally synthesized TiO_2 (B) nanowires with ultrahigh surface area (TEM image is shown in Fig.3b) act as a good anode material for lithium-ion batteries, especially on fast charging and discharging performance [66].

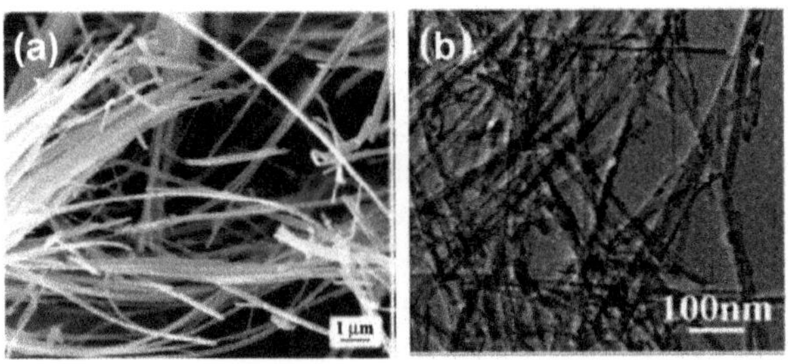

Figure 3.a) SEM image of anatase TiO_2 nanowire b) TEM image of TiO_2 (B) nanowire. Reprinted with permission from Elsevier (2002) and the Royal Society of Chemistry (2011) [112, 66].

Wei and co-workers reported a simple hydrothermal soft chemical process for the synthesis of TiO_2 nanowires. In this procedure, they reported treatment of a layered material $Na_2Ti_3O_7$ with 0.1 M HCl and then transferred into an autoclave at a temperature of 140-170°C for 3-7 days. The resultant product was filtered, washed with water and dried at 60°C for 4h produced TiO_2 nanowires with diameters 20-100 nm [113].

2.3 Solvothermal method

The solvothermal method is similar to hydrothermal method with the only difference being the solvent used here is non-aqueous, and in effect the temperature can be raised to much higher values because a variety of high boiling organic solvents can be used [114]. This method has better control in the crystallinity, size and morphology of TiO_2 nanoparticles in comparison with the hydrothermal method. This method involves the use of a solvent in a state of moderate to high pressure (1-10000 atm) and temperature (100-1000 °C) [115]. Using this method various nanoparticle having narrow size distribution and dispersity can be prepared [116]. Solvothermal methods are favoured for reducing the overgrowth of nuclei and regulated growth process leading to the high viscosity, low polarity, surface tension and high boiling point of organic solvents [117]. The current state of art gives an idea about the synthesis of TiO_2 nanoparticles *via* the solvothermal method with and without the help of surfactants [118-121]. Using titanic acid nanobelts (TAN) as precursor Yuhui Cao *et al.* prepared anatase TiO_2 nanocrystals exposed with (001) facets by solvothermal method in HF-C_4H_9OH mixed solution [122]. The percentage of exposure of (001) facets can be varied by adjusting the amount of HF. Photocatalytic measurements reveals that the degradation of methyl orange, methylene blue and rhodamine B become higher when TiO_2 nanocrystal having 77% (001) facet. TiO_2 with varied morphologies synthesized from hydrolysis of tetrabutyl titanate in a mixture of ethanol and glycerol by solvothermal preparation combined with post annealing [123]. Wen *et al.* reported the synthesis of ultralong single crystalline TiO_2 nanowires by solvothermal method using commercial Degussa P25, NaOH and absolute ethanol are the precursors at a temperature of 170-200 °C [118]. The same group reported the solvothermal synthesis of bamboo shaped Ag-doped TiO_2 nanowire heterojunctions by mixing 0.2 M titanium butoxide in ethanol, 10 M NaOH solution and silver nitrate at a temperature of 200 °C/24 h [119]. The TEM images of the synthesized bamboo shaped Ag-doped nanowires are shown in Fig. 4.

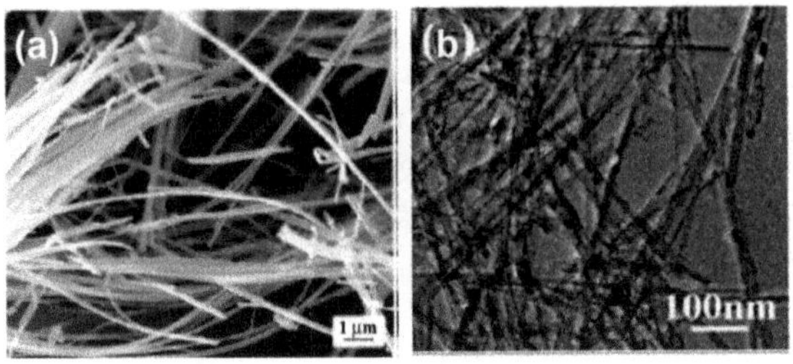

Figure 3.a) SEM image of anatase TiO_2 nanowire b) TEM image of TiO_2 (B) nanowire. Reprinted with permission from Elsevier (2002) and the Royal Society of Chemistry (2011) [112, 66].

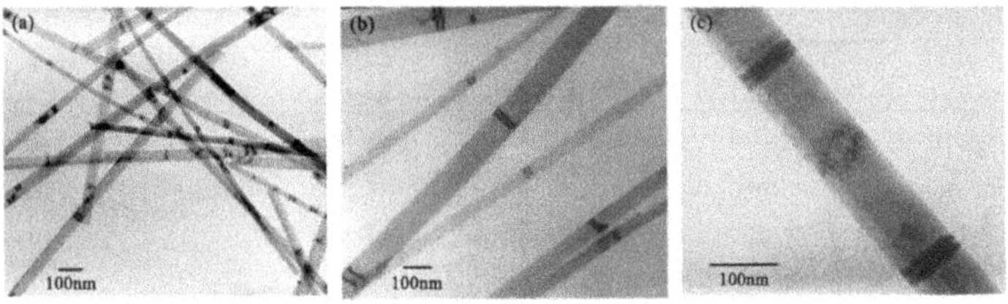

Figure 4. TEM images of the synthesized bamboo-shaped Ag-doped TiO_2 nanowires: (a) low magnification; (b) higher magnification; (c) a typical individual nanowire with two knots. Reprinted with permission from the American Chemical Society (2005) [119].

2.4 Chemical vapour deposition method

Chemical vapour deposition (CVD) is an important method for synthesizing advanced semiconductor materials [124]. In the CVD method materials in vapor state are condensed to form solid state material on reaction with a heated substrate [125]. When deposition proceeds, a chemical reaction takes place on the surface. This method is usually used to form coatings to improve the electrical, mechanical, optical, thermal, corrosion resistance and wear resistance of various materials [116].

Many precursors such as titanium tetra-isopropoxide (TTIP) [126], titanium tetrachloride [127], tetra-nitratotitanium [128] *etc.* can be used to make TiO_2 coating on a surface. For the preparation of TiO_2, Z. Ding *et al.* choose TTIP as precursor, because it is less reactive with water and results in easy handling [129]. Seifried *et al.* reported the preparation of nanocrystalline TiO_2 with particle sizes below 10 nm and thick crystalline

TiO_2 films having grain sizes below 30 nm by pyrolysis of TTIP under a gaseous atmosphere of helium and oxygen using liquid precursor delivery [130]. The preparation of single-phase rutile TiO_2 thin films has been carried out on $Pt/Ti/SiO_2/Si$ substrate by the Laser Chemical Vapour Deposition (LCVD) method and the developed TiO_2 thin films possess three different morphology including powder, Wulff- shaped and granules [131].

2.5 Electrodeposition method

Electro deposition is an important method for synthesizing nanomaterials and is used to produce a metallic coating on a surface by the process of reduction at the cathode. The surface to be coated is taken as cathode and the metal to be deposited as anode. The electrolyte used should be a salt solution of the metal to be deposited. Upon passing electricity the metal to be deposited, attracted towards the cathode and reduced to metallic state to form a coating on the surface of the cathode. Based on the mode of applying external current to the electro deposition system, it can be either direct or pulsed electro deposition. Compared to direct current electro deposition pulse electro deposition has the advantage of process controllable parameters such as control over structure, composition and properties [132].

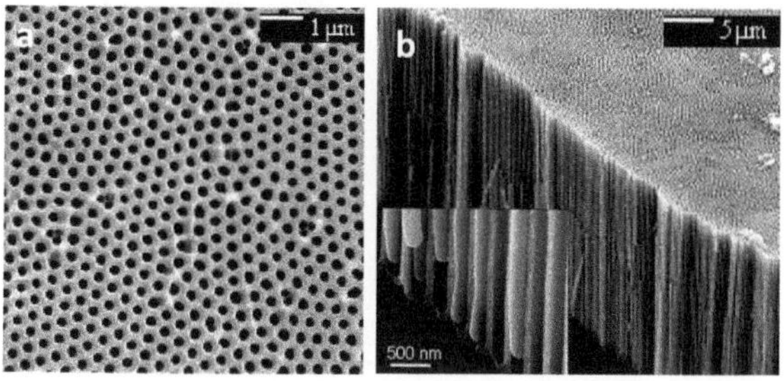

Figure 5. Top view and cross-sectional SEM image of TiO_2 electrodeposited on AAM pores. Reprinted with permission from Elsevier (2005) [133].

Y. Lei *et al.* synthesized TiO_2 nanowire with the use of the template of anodic alumina membrane (AAM). SEM image was depicted in Fig. 5 [133]. The experiment involves pulse electrodeposition of 0.2M $TiCl_3$ solution of pH 2 as titanium and/or its compounds into the pores of AAMs [98].

3. Structural features and stability of TiO_2 nanomaterials

TiO_2 mainly exists in four distinct polymorphs, *viz*, anatase, rutile, brookite and TiO_2 (B) [1, 134, 135]. As shown in Fig. 6, all four crystalline forms contain TiO_6 octahedra differing in their shared edges and corners leading to distortion in the polyhedral units [136]. Anatase and rutile belong to the tetragonal crystal system, while former has a zig-zag chains of TiO_6 octahedra connected through four edge sharing and the later composed of linear chains of TiO_6 octahedra connected by sharing two edges [137]. Rutile is the most thermodynamically stable form of TiO_2 [138]. Brookite crystallizes in an orthorhombic system in which both edges and corners are connected [139]. Brookite commonly occurs as a minor secondary phase along with anatase and rutile, therefore it is difficult to produce Brookite as pure phase [140]. TiO_2 (B) belongs to the monoclinic crystal system in which both edges and corners are shared by the TiO_6 octahedra having a perovskite-like layered structure [135]. Among the different phases of TiO_2 (mainly anatase, rutile, brookite and TiO_2 (B)) [141], rutile is thermodynamically stable while anatase, brookite and TiO_2 (B) are metastable transform to rutile at a particular temperature range (400-700 °C) [142] depending upon the methods and conditions of preparation [143].

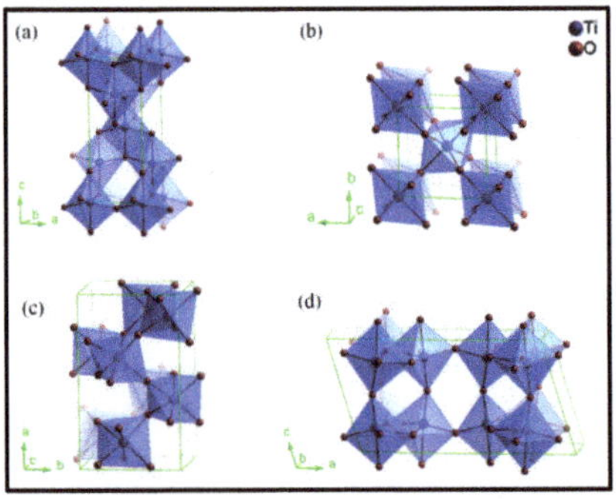

Figure 6. Crystalline structures of TiO_2 in different phases: (a)anatase, (b) rutile, (c) brookite, and (d) TiO2(B). Reprinted with permission from the American Chemical Society (2014) [135].

Anatase is the most widely studied crystal system among the other existing phases of TiO_2. Rutile is the high temperature stable phase. On the other hand, anatase and brookite are stabilized with particle size ranges in the nanoscale. The heat treatment on anatase and brookite phases result in the formation of rutile-TiO_2. More precisely, anatase to rutile, brookite to rutile, anatase to brookite to rutile and brookite to anatase to rutile transformations occur with heating. These transformations can be explained on the basis of surface energy which depends on the particle size. Surface free energy is defined as the free energy change associated with a solid when it is separated into two parts at a large distance or to increase the surface area. The thermodynamic stability is controlled by the surface energy which varies among the three polymorphs of TiO_2. The subsequent coarsening *via* heat treatment resulted in increased particle size and this phenomenon stabilizes the rutile at higher temperature. On the other hand, anatase and/or brookite stabilized with smaller particle size compared to rutile at moderate temperature. Significant studies are in harmony with the anatase/brookite to rutile transformations except some anomalous observations. Zang and Banfield found that the thermodynamic stability of anatase was below 11 nm, for brookite, it was between 11 to 35 nm and finally rutile was stable above 35 nm. Banfield *et al.* observed that after attaining rutile structure from anatase and brookite, it grew much faster compared to the rest. A similar conclusion was made by Li *et al.* which describe the different growth rates of anatase and rutile. As the temperature increases the anatase and rutile particle size increases, but rutile has a higher growth rate after nucleation compared to anatase. Interestingly, anatase phase have maximum photo catalytic activity due to the extent of minimum electron-hole recombination and its greater affinity towards the adsorption of organic compounds [144]. Therefore, the attainment of anatase phase stability favors photo degradation of pollutants.

4. Electronic structure and optical properties of TiO_2 nanoparticles

Theoretical calculations and experimental observations suggest that TiO_2 has an electronic band structure in which the Ti 3d orbitals form the conduction band levels and O 2p orbitals form the valance band levels. Asahi *et al.* studied the optical and electronic properties of anatase TiO_2 using first-principles calculations with full-potential linearized augmented plane-wave (FLAPW) method [145]. Outer most electrons of free Ti atoms are filled in 3d and 4s orbitals. While forming TiO_2, the 3d orbitals or conduction bands split into t_{2g} and e_g levels, while 4s are being unaffected. d_{xy} orbitals contribute to the bottom of the conduction band. Valance band composed of orbitals of oxygen, *i.e.*, σ, π and non-bonding p orbitals. The top of the valence is associated with the non-bonding p orbitals. Fig. 7 shows the molecular orbital bonding diagram for anatase TiO_2.

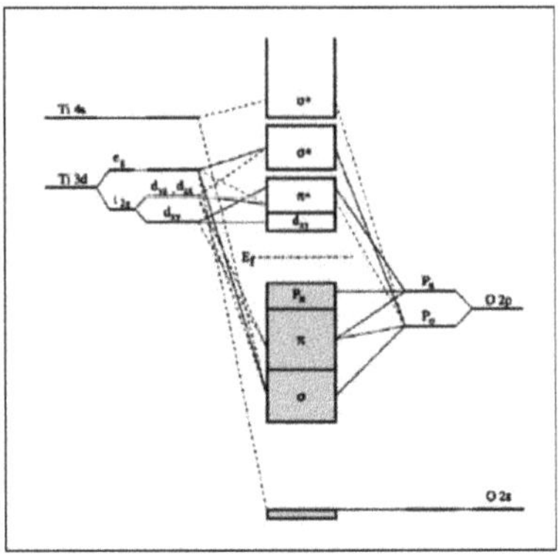

Figure 7. Molecular-orbital bonding structure for anatase TiO_2: (a) atomic levels; (b) crystal-field split levels; (c) final interaction states. The thin-solid and dashed lines represent large and small contributions, respectively. Reprinted with permission from the American Physical Society (2000) [143].

5. Modifications on TiO_2

Modifications of TiO_2 nanomaterials are primarily aimed to enhance its optical properties. The wide band gap of TiO_2 limits the utility in photo assisted applications. Extensive studies have been done by the scientific community to improve the performance of TiO_2 and novels designs were contributed. One of the significant pathways is band gap engineering through doping [98, 146]. Doping with metal and/or non-metal has resulted in efficient materials for photocatalysis as well as other functional applications. Other remarkable strategies are sensitization of TiO_2 with organic (dyes) [147] or inorganic (metal or low band gap semiconductors) [137], making solid solutions [148], design of metal TiO_2 nanocomposites [149] and reduction of pure TiO_2 [91].

5.1 Doping

As structural integrity plays an important role in designing the photocatalytic activities of a semiconductor nanomaterial, doping in TiO_2 is a great challenge to the research community [150]. Doping inserts foreign elements into the crystal lattice of TiO_2 which changes the chemical composition [151, 152]. The optical properties of TiO_2 is directly related to its electronic structure where chemical composition determines the electronic structure [153]. On the other hand, changing the chemical composition through doping has significant impacts on the optical properties of TiO_2 *via* modifying the electronic structure.

5.1.1 Metal ion doping

Various methods are employed for doping of metal ion in TiO_2, *viz*, sol-gel, hydrothermal, solid state method *etc*. Doping with alkali [154], alkaline earth [155-157], transition [158-164] and rare earth metals [165-170] were reported for TiO_2 nanomaterials. Li, Na and K doping were attempted by Bessekhouad *et al.* [154]. S. Liu *et al.* reported the enhanced absorption properties of various concentrations of Ca doped TiO_2 nanoparticles [157]. TiO_2 nano materials doped with different amounts of cerium synthesized by Yan *et al.* via the sol-gel method creates additional electronic states above the valence band of TiO_2 capture the photo formed holes and decreases the recombination rate of photo generated electron and hole results enhanced visible light absorption. The UV-vis absorption spectra of Ce doped TiO_2 is shown in Fig.8. This will enhance the photo degradation of methylene blue compared to pure TiO_2 [171].

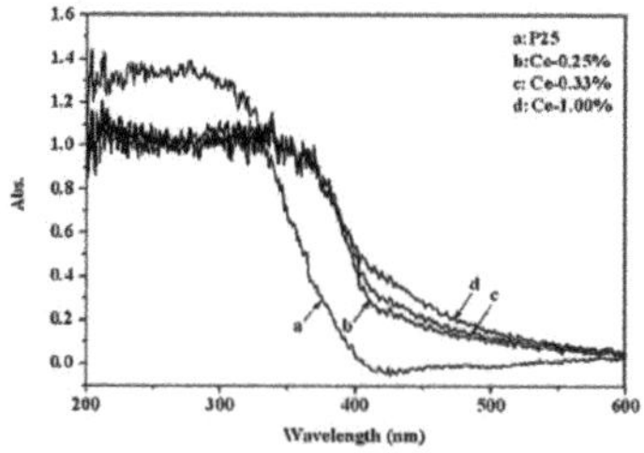

Figure 8. UV-vis absorption spectra of Ce doped TiO_2 with different concentration of cerium. Reprinted with permission from Elsevier (2012) [169].

C. Malengreux *et al.* studied screening of the photocatalytic activity of undoped and Fe^{3+}, Cr^{3+}, La^{3+} or Eu^{3+}doped TiO_2 by evaluating the degradation of 4-nitrophenol under UV-visible light. The dopant nature and content with an optimal content significantly influences the photocatalytic activity of Fe^{3+} or La^{3+} single doped as well as La^{3+}- Fe^{3+} and Eu^{3+}- Fe^{3+} co-doped catalysts whereas Cr^{3+}doped TiO_2 shows an impairment in the photodegradation of 4-nitrophenol. While no significant effect of dopant has been observed in the case of Eu^{3+} doped TiO_2 [172].

Higher anatase phase stability of Gd^{3+} doped TiO_2 reported by Zhang *et al.* using titanium alkoxide and gadolinium nitrate as precursor shows anatase phase stability up to 800 °C[173]. Recently Mg reduction of white TiO_2 nanoparticles followed by removal of excess Mg with HCl and distilled water changes colour from white to grey to blue – grey to black and is applicable for solar water evaporation [174]. Increasing amount of Mg produce more oxygen vacancy in TiO_2 which is responsible for its enhanced light absorption.

La^{3+} doped anatase TiO_2 synthesized by Li *et al.* through sol-gel process inhibits the phase transformation of TiO_2, reduce the crystallite size, enhance thermal stability and increase oxygen vacancy on the surface of TiO_2 [165]. The XRD patterns of La^{3+}-TiO_2 photo catalysts in different ratios were represented in Fig. 9.

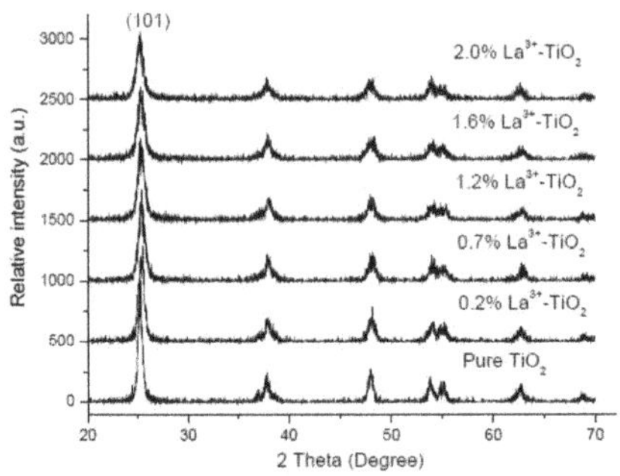

Figure 9. XRD plots of La^{3+}-TiO_2 photo catalysts in different ratios. Reprinted with permission from Elsevier (2004) [163].

Cao *et al.* developed Sn^{4+} doped TiO_2 nanofilm by the plasma enhanced CVD technique resulting in more surface defects on the TiO_2 surface [175]. Recently Mn doping TiO_2 was found to be beneficial for the extended absorption of TiO_2 [176]. The resulted material had a black color and showed higher photocatalytic activity compared to P25 under sunlight.

Even though metal ion doping has positive effects on TiO_2, higher cost of metal ion precursor, poor thermal stability and generation of secondary impurity phases which affects the purity, in fact diminishes the photocatalytic activity are some drawbacks of metal ion doping [177].

5.1.2 Non-metal ion doping

Taking into consideration the poor thermal stability and possibility for charge carrier recombination of metal ion doped TiO_2, in recent years non-metal ion doping is a promising way to avoid these problems. For instance, Yin *et al.* prepared N doped TiO_2

by solvothermal route showed violet in colour upon calcinations at 20-800 °C for 1 h changes to weak violet, bright yellow, weak yellow and gray. Although the colour did not change to white denoting the thermal stability of Ti-N bonding in $TiO_{2-x}N_x$ [178]. Suresh *et al.* found that incorporation of N into TiO_2 using urea tends to retains 11% anatase phase stability at 900 °C [179].

Visible light responsible S-doped TiO_2 nanocrystal powder was successfully synthesized by G. Yang *et al.* through a simple low-temperature solvothermal method using thiourea as the sulphur source. It shows higher photocatalytic activity in the visible region for the degradation of methylene blue and phenol [180] (Fig. 10). Unlike metal ion doping non-metal ion doping is less likely to form recombination centers and hence enhances the photocatalytic activity. Yu *et al.* prepared phosphorus doped TiO_2 by sol-gel method on calcinations at 400 and 800 °C showed higher photocatalytic activity for the degradation of methylene blue compared to Degussa P25 under UV light, suggesting that P^{5+} restrained photocatalytic activity even at higher calcinations temperature.

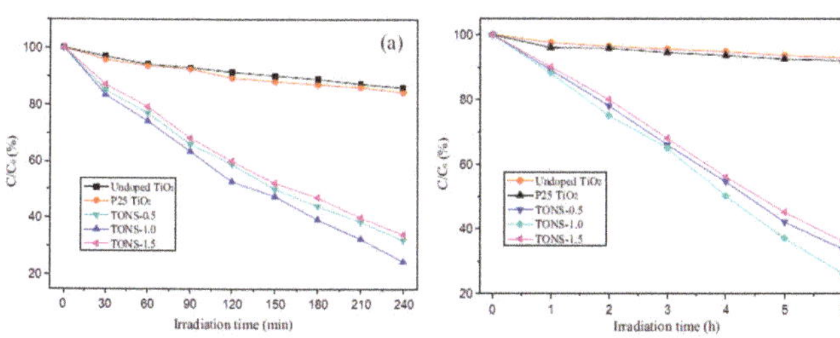

Figure 10. Photocatalytic degradation Methylene Blue (MB) over TONS-1.0 and pure TiO_2 samples under visible light irradiation (λ>420nm.) Reprinted with permission from Elsevier (2012) [177].

However, for higher concentration of P^{5+} the activity diminishes due to the formation of secondary oxides like TiP_2O_7 and $(TiO)_2P_2O_7$ [181]. Fluorine doping on TiO_2 creates several advantageous effects found by Li *et al.* It includes formation of surface oxygen vacancies, the enhancement of surface acidity and the increase of defect states *ie.*, Ti^{3+} ions [182].

5.1.3 Effect of doping on electronic structure of TiO_2

As the conduction band minimum (CBM) of TiO_2 is contributed by the Ti 3d orbitals, it can be expected that d orbitals from dopant could influence the CBM [145]. Effect of transition metal doping on the band gap energy of TiO_2 was widely studied [145, 183-185]. 3d transition elements, notably, V, Cr, Mn and Fe were successfully doped into the TiO_2 lattice and found the formation of impurity levels within the band gap. Umebayashi *et al.* used full-potential linearized augmented plane wave (FLAPW) method to study the doping effects of V, Cr, Mn, Fe, Co and Ni [186]. They observed a consistent shift in the energy levels created by the dopant (Fig. 11). Metal doping creates oxygen vacancies (in case of V doping) or impurity levels in the band gap which in turn red shift the absorption edge of the metal doped TiO_2 [187]. Conversely, these factors may reduce the thermal stability and increase the chance for charge carrier recombination.

Energy states of doped anions (non-metal ions) hybridize well with the valence band states of the semiconductor oxide material, and either shift the valence band upward or broaden, resulting in the reduction of the overall band gap. Chen and Burda compared the red shift in the absorption edges of bare and C-, N-, and S-doped TiO_2 nanomaterials *via* X-ray photoelectron spectroscopy [188]. Using the FLAPW method, Asahi *et al.* analysed the electronic structures of C, N, F, P, or S doped TiO_2 [77]. Anion doping with N^{3+}, C^{4+} or S^{2-} is expected to make a photocatalyst absorb in the visible light regime [189].

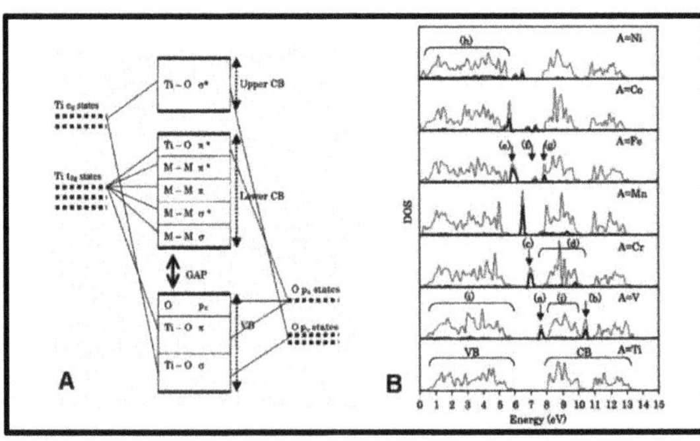

Figure 11. (A) Bonding diagram of TiO_2 (B) DOS of the metal-doped TiO_2 ($Ti_{1-x}A_xO_2$: A) V, Cr, Mn, Fe, Co, or Ni). Gray solid lines:total DOS. Black solid lines: dopant's DOS. The states are labeled (a) to (j). Reprinted with permission from Elsevier (2002) [183].

Among these, nitrogen doping can be expected to give promising materials because of the ample overlapping of O 2p and N 2p states which reduces the effective band gap [190]. Doping with sulphur will also lead to the same effect but the large ionic radius of sulphur will hinder the proper incorporation of it into the oxygen lattice [146]. However, C and P doping is found to lead to recombination centers [146]. The mixing of C with O 2p states was too weak to produce a significant band gap narrowing [191].

5.2 Sensitized TiO_2

Coupling of TiO_2 with a visible light active semiconductor forms an inorganic semiconductor sensitized TiO_2. Sulphides of some p block elements and transition metals act as sensitizer to TiO_2 [192]. CdS, PbS, Bi_2S_3 and Sb_2S_3 have appropriate energy level corresponds to the TiO_2 conduction band. Here the sensitizer absorbs the visible light part of solar energy which in turn generates electron-hole pairs [193-195]. The energetically favourable positions of the photogenerated electrons and the conduction band minimum of TiO_2 makes the possible electron transfer from the sensitizer to TiO_2 [196]. Hensel *et al.* synthesized nanocomposite of N-doped TiO_2 and CdSe, which possesses visible light absorption features [197]. They demonstrated that the synergistic effect of N-doping and CdSe sensitization makes the modified TiO_2 a better candidate of photoelectrochemical activities (Fig. 12). Shen *et al.* successfully implemented coupling of p-type $CuInS_2$ quantum dots with TiO_2 nanoparticles [198]. It formed a p-n heterojunction through which better electron-hole separation could be achieved. A composite of MoS_2/graphene and TiO_2 was found to be efficient for H_2 generation under visible light irradiation [199].

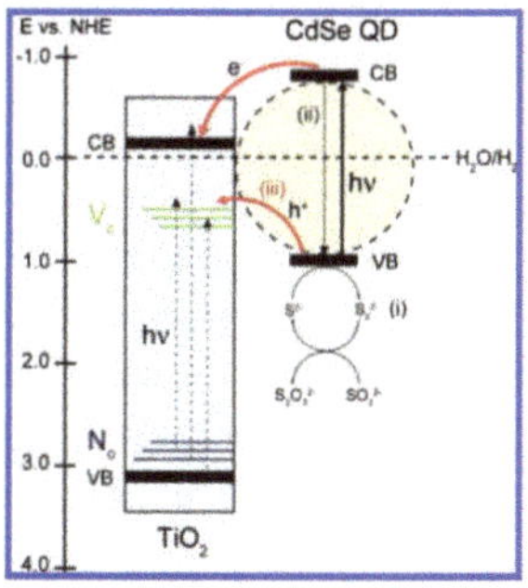

Figure 12. Schematic representation of N-doped TiO_2/CdSe nanocomposite. Reprinted with permission from the American Chemical Society (2010) [194].

Kang *et al.* prepared PbS sensitized TiO_2 nanoarrays for electrochemical applications [200]. Metal nanoparticle deposition on the surface of TiO_2 nanoparticles introduce new surface-active sites [201]. Redox reaction can happen on the active sites. Noble metals such as Au, Pd, Pt and Ag are deposited on the surface of the TiO_2 nanoparticles to achieve H_2 production as these noble metals possess low H_2 overpotential [202]. Pan and Xu synthesized noble metal nanoparticles deposited TiO_2 [203]. Noble metals used in their study were Ag, Au and Pd and these nanocomposites were used in the reduction of Cr(IV) and oxidation of benzyl alcohol. Zhou *et al.* derived a very interesting system containing ordered mesoporous anatase TiO_2 channels in which Ag clusters are confined [204]. It has remarkable photocatalytic activity towards the degradation of phenol. TiO_2 nanotubes with Ag nanoparticles also found to act as anode material in dye sensitized solar cells [205]. Au@TiO_2 yolk shell nanoparticles were developed by Sun *et al.* The oxygen vacancies in TiO_2 and noble metal deposition play the key role in improving the visible light response and enhanced charge separation respectively [206]. Z. Lin and co-workers developed palladium quantum dots deposited on highly ordered TiO_2 nanotube arrays [207]. The charge transfer was promoted by the symbiosis of TiO_2 nanotubular morphology and the fine dispersion of Pd quantum dots on the surface of TiO_2. Bimetallic deposition is also reported in TiO_2 surfaces. Luna *et al.* demonstrated the enhanced H_2 generation activity of Ni-Pd deposition on TiO_2 [208]. Singhal and Kumar also attempted to develop bimetallic-TiO_2 nanocomposites [209]. They synthesized various bimetal/TiO_2 nanocomposites, *viz*, Pt-Pd/TiO_2, Au-Pd/TiO_2 and Ag-Pd/TiO_2 for selective conversion of CO_2 to hydrocarbons [209]. Among these, $Ag_{1\%}Pd_{1\%}/TiO_2$ was found to be the best catalyst.

5.3 Reduction

The reduction of pure TiO_2 under hydrogen atmosphere resulted in black TiO_2 nanomaterials. The first report on black TiO_2 nanomaterial was published by Chen *et al.* in 2011 [91]. This black TiO_2 nanomaterial possesses a band gap of approximately 1.5 eV. Black TiO_2 nanomaterials possess inherent self-structural modification including oxygen vacancies, self-doped Ti^{3+}*etc.* which makes it absorb almost entire region of solar spectra [91, 210, 211]. These features make it as a superior candidate for various functional applications. Chen *et al.* synthesized black TiO_2 by heating white TiO_2 nanoparticles at 200 °C under 20.0 bar hydrogen atmosphere for 5 days [91, 211]. Reduction in presence of $H_2/N_2/Ar$ at low/high pressures were reported by many groups [98, 211]. All TiO_2 nanoparticles were characterized by reduced band gap energies. Fig.13 shows the structure, photographs and HRTEM images of white and black TiO_2.

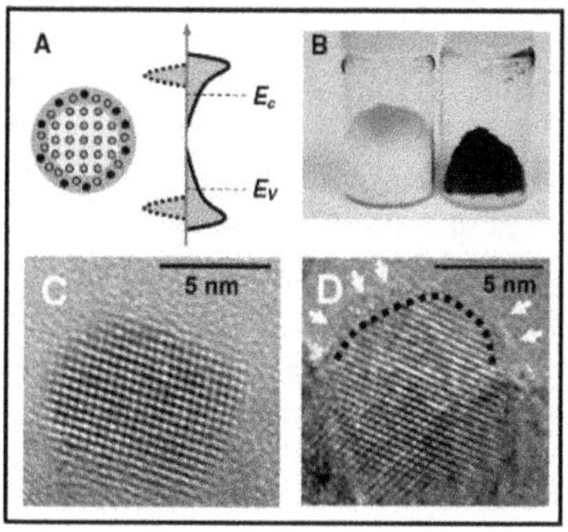

Figure 13.A) Structure and electronic DOS of black TiO_2 B) photographs of white and black TiO_2 C) HRTEM of TiO_2 and D) disordered black TiO_2. Reprinted with permission from the Royal Society of Chemistry (2015) [207].

6. Applications of TiO_2 nanomaterials

6.1 Self-cleaning coatings

The area of self-cleaning coating is divided into hydrophilic and hydrophobic. Both perform clean-up activity through the action of water. Hydrophilic coating acts through the formation of a film of water *via* water drops spread over the surface, which removes the contaminants away along with this spreading. And for hydrophobic coating water drops roll off from the surface quickly due to the water repellent and low adhesive nature of the surface [212]. Hydrophilic coating has an additional advantage of break down dirt particles chemically under sunlight [65].

Hydrophobic surface- A hydrophobic self-cleaning coating require very high static water contact angle *i.e.* usually, Θ_s>160°C and a very low roll-off angle [213]. If a surface has these two properties it is known as either super hydrophobic or ultra-phobic. The waxy surface of lotus leaves along with the microscopic structures contributes towards this kind of surface [214]. I. Kartini *et al.* developed a super hydrophobic film with water contact angle 155.5° using hybrid layers of TiO_2 and dodecylamine. The film restrained hydrophobicity for 4 weeks under high relative humidity (>90%) in outdoor applications, because of the outward orientation of the hydrocarbon chains in the hybrid film [215].

Hydrophilic surface- Self-cleaning windows are coated with a transparent layer of TiO_2 which perform clean up activity by the combination of two properties, that are photo induced chemical break down of organic dirt adsorbed onto the window and the generation of water film causes hydrophilic surface, thereby reduces the water contact angle and dirt is washed away [214]. TiO_2 has become the material of preference in self-cleaning activity, since titania is highly efficient for the photocatalytic degradation of dirt particles under direct sunlight and also it is inexpensive, nontoxic and easily available chemical [216]. The super hydrophilic character of TiO_2 was first found by Fujishima and co- workers, they reported that if the TiO_2 film is exposed to ultra violet light, water contact angle reduces to 0°causes spreading of water droplets on the surface, thereby attains removal of dirt materials from the surface [217].

6.2 Li ion battery

TiO_2 is preferentially considered as the negative electrode for Li ion battery, because of its ability to being reduced on lithium insertion at lower potential. In TiO_2 based Li ion batteries TiO_2 nanocrystals on the surface of its anode instead of carbon has the anode surface area of about 100 m^2/g compared to 3 m^2/g for carbon. This facilitates electrons to enter and leave the anode easily and thus bring about fast recharging, thereby provides high current when required [218].

Porous TiO_2 urchins having superior electrochemical performance as anode for lithium ion battery including high capacity (206.2 mA h g^{-1} at 0.5 C), superior rate performance (94.4 mA h g^{-1} at 20 C) and stable cycling stability (94.3% capacity retention over 1000 cycles at 10 C *Vs* third cycle) reported by Yi. Cai *et al.* hydrothermally *via* in situ self-sacrificing template method using TiO_2/oleylamine as precursors followed by ion-exchange and calcination [219]. The improved electrochemical performance of TiO_2 nanomaterial can be ascribed to their characteristic pore structures and high structural stability. The two CV curves in Fig.14 roughly overlap with each other indicates the high redox reversibility of the as prepared TiO_2 material.

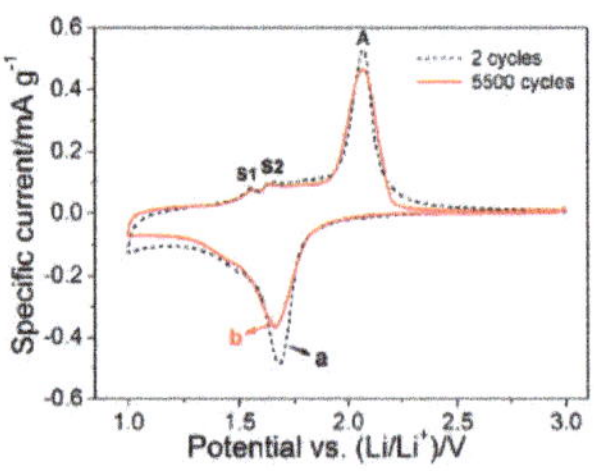

Figure 14. CV curves of a fresh electrode made of TiO_2 urchins recorded in the second sweep (a) and after galvanostatic cycling at 10 C over 5500 cycles (b). (Scan rate: 0.2 mV s_1). Reprinted with permission from Elsevier (2016) [216].

A TiO_2-CNT composite as anode material for lithium ion battery showed a superior cycling stability up to 50 cycles with a capacity of 230 mA hg^{-1}. This was synthesized by a magnetron sputtering technique onto a free-standing CNT film with paper like morphology, known as Bucky paper. The greater stability of the as prepared anode material can be attributed to the strong adhesion of TiO_2 nanocrystals on the CNT support, which already possess high electronic conductivity and good flexibility [218, 219].

TiO_2 (B) hollow nanostructure shows excellent electrochemical performance as anode material for lithium ion batteries [220]. This high performance is due to the following reasons: i) More open TiO_2 (B) structure with significant voids and continuous channels makes the material an excellent host structure for intercalation, ii) hollow structure furnish enough space to accommodate Li^+ ion and thereby enhances the specific capacity of the battery, iii) hollow nanostructures provide large contact area between electrode and electrolyte in such a way that diffusion distance for Li^+ transport reduced and, iv) the void space in hollow structure lessen the local volume change and the problem of pulverization and aggregation of electrode material, consequently enhances the cycling performance [221].

6.3 Dye sensitized solar cells (DSSC)

In the present scenario DSSC have great attention in the field of academic research and industrial applications because it is a potential low cost technically and economically credible alternative to traditional silicon solar cells [82]. In general, DSSC consists of mesoporous TiO_2 nanofilm covered with dye molecule on a fluorine doped tin oxide (FTO) glass plate and a platinum 15 (Pt)-coated counter electrode are in contact with a liquid electrolyte (I^-/I^{3-}).

Nowadays many studies have been conducted to alleviate the problem of charge recombination. Recently M.A. Hossain *et al.* reported both ends opened TiO_2 nanotube and TiO_2 nanoparticles to form heterostructure photoanode having excellent light scattering and harvesting, long excited electron life time, unidirectional electron pathway and desirable great photo electric conversion efficiency. The hetero-nanostructure photoanode showed superior solar cell activity with higher Jsc (20.01 mA 364 cm^{-2}) and PCE (8.56%) when the both-ends-opened TiO_2 nano- tubes were sandwiched between the TiO_2 nanoparticles layers [222].

At present many researchers are involved in developing and modifying photoanode by using nanoparticles, nanorods, nanotubes and nanofluids to enhance the light absorption [223], scattering [224], charge transport [225] and minimising charge recombination [226], in fact the (010) faceted leaf like anatase TiO_2 nanocrystals synthesized by *in situ* topochemical process using exfoliated layered titanate nanosheets exhibits better performance as the light scattering layer for DSSC and can increase 28% of energy conversion efficiency by improving 33% of J_{sc} because of the enhanced light absorption [227].

6.4 Electrochromic devices

An electrochromic device (ECD) consists of a liquid or solid or gel electrolyte sandwiched between an electrochromic electrode and a counter electrode. Colour change happens in it as a result of charging/discharging this cell on the application of electric potential [228]. The principle set up of an electrochromic device is represented in Fig. 15.

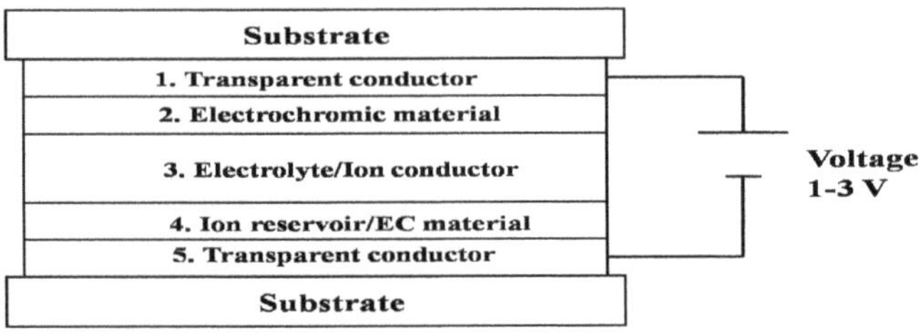

Figure 15. The principle set up of an electrochromic device. Adapted from [225].

The problem of using expensive metal oxides like WO_3, NiO, MnO_3 or IrO_3 as electrochromic electrode can be solved by the low cost, non-toxic and easily available TiO_2 nanomaterials, in fact. Diasanayake *et al.* reported a novel design of an ECD using a transparent nanoparticulate TiO_2 film as the electrochromic material and SnO_2 as the counter electrode with polymethyl methacrylate (PMMA) based gel as the electrolyte. An attractive reversible colour change between blue and colourless noticed on using this ECD [229].

Electrochromic anatase TiO_2 thin film on F-doped tin oxide (FTO) substrate was successfully developed by N.N. Dinh *et al.* through the doctor blade method using colloidal TiO_2 solution with particle size 15 nm exhibits good reversible coloration and bleaching process in a solution of 1M $LiClO_4$ in polypropylene carbonate. The potential application of this porous nano TiO_2 film is in large area electrochromic windows [230].

S. Berger *et al.* reported an anodization process through which a thin Ti metal layer on a conductive glass fully transforms into a TiO_2 nanotubular array. The obtained layer demonstrates very good contrast behaviour and a high cycling stability, thus an electrochromic device prototype was constructed using the as prepared TiO_2/ITO/glass electrode with 1 M $LiClO_4$ in propylene carbonate as electrolyte shown in Fig 15 [231]. TiO_2-WO_3 composite nanotube fabricated using TiW alloy anodization. This TiO_2-WO_3 composite nanotubes showed better ion insertion and electrochromic properties only when small amount of WO_3 (0.2 at %) are present [232].

6.5 Water splitting

Hydrogen is considered as the future fuel; because it is pollution - free renewable energy source can be storable with high energy density. 95% of the H_2 used for commercial purpose is derived through water splitting *via* electrolysis [233]. Among the various semiconducting metal oxides TiO_2 is the most extensively studied one, because in addition to strong catalytic activity and long-life time of electron/hole pairs it has strong chemical stability against photo corrosion and abundant reserves on the earth. Although, the energy conversion efficiency from solar to hydrogen by TiO_2 in photocatalytic water splitting is still low due to recombination of photogenerated electron – hole pairs, fast backward reaction, inability to utilize visible region (band gap energy 3.2 eV, only UV light can be utilized) of the sunlight. This difficulty can be resolved by the addition of hole scavengers, noble metal loading, ion doping (both cation and anion), composite semiconductors, dye sensitization, metal ion implantation *etc.* [233].

The fundamental requirements for the photocatalytic water splitting are, firstly a bandgap of more than 1.23 eV is the thermodynamically required energy condition whereas 2.0-2.4 eV is the energy of photon in the visible region [234]. Secondly, the bottom level of the CB of the photocatalyst should be more negative than the reduction potential of hydrogen (E_{H^+/H_2}) and the top level of the VB must be more positive than the oxidation potential of oxygen (E_{O_2}/H_2O), which facilitates the efficient transportation of electrons and holes [233, 235]. A schematic representation of mechanism of photocatalytic water splitting is depicted in Fig. 16.

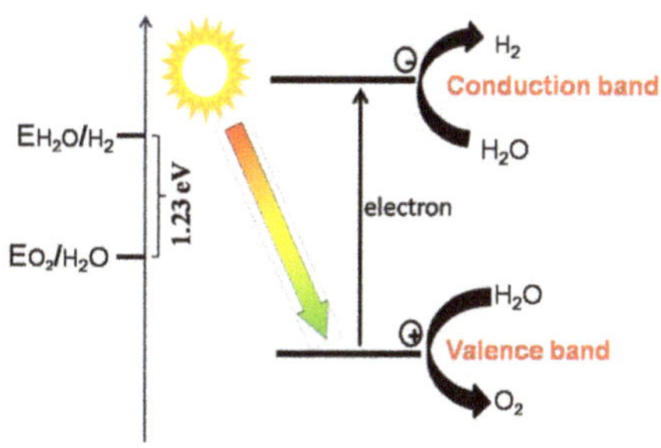

Figure 16. Mechanism of photocatalytic water splitting by TiO_2.

Tandem systems with separate semiconductors for water reduction and oxidation on TiO_2 based photocatalytic water-splitting are reported by Abe *et al.* [236] in which Pt-TiO_2 (anatase) preferentially catalyze water reduction using iodide as the sacrificial electron donor and TiO_2 (rutile) as the water oxidation catalyst with IO_3 as the electron acceptor under UV irradiation.

Apart from these, various other applications are also possible with TiO_2 nanomaterials with and without modifications, these includes cancer therapy [237], fuel cell [238], supercapacitor [239], water purification [240], printing ink industry [241], field emission [242], microwave absorption [243], sensor applications[30]*etc.*

7. Conclusions

Over the past decades, great efforts on TiO_2 nanomaterials have attracted extensive scientific interest for their synthesis, modifications, morphology and applications. Tremendous efforts have been made to synthesize TiO_2 nanomaterials through different methods such as hydrothermal, solvothermal, chemical vapour deposition, sol-gel and electrodeposition are briefly explained with recent research findings along with changes in properties on modification by metals and non-metals. This chemical modification results disordered surface, oxygen vacancies and Ti^{3+} to produce changes in surface, electrical, optical and electronic properties. Changes in electronic property imply bandgap narrowing or midgap states which enhances the visible light absorption and thus high photocatalytic activity. TiO_2 nanomaterials with various morphologies including nanotube, nanorod, nanowire and nanosphere are described shortly with modern synthesis routes. The most promising photocatalytic activity of TiO_2 found many industrial applications in the field of water purification, water splitting, self-cleaning coatings, dye sensitized solar cells *etc.* and apart from these other applications such as electrochromic devices, Li ion batteries and fuel cell were also pointed out here. Moreover, both theoretical and experimental studies based on modifications with different ions including band gap engineering and optical absorption analysis are under progress to develop this multifaceted material towards various practical applications.

Acknowledgement

The author SBN and SJ acknowledge DST-SERB, India for financial assistance.

References

[1] O. Carp, C.L. Huisman, A. Reller. Photoinduced reactivity of titanium dioxide, Prog. Solid State Chem. 32 (2004) 33-177. https://doi.org/10.1016/j.progsolidstchem.2004.08.001

[2] P.T. Anastas, M.M. Kirchhoff. Origins, current status, and future challenges of green chemistry, Acc. Chem. Res. 35 (2002) 686-94. https://doi.org/10.1021/ar010065m

[3] M. Momirlan, T. Veziroglu. Current status of hydrogen energy, Renew. Sustain. Energ. Rev. 6 (2002) 141-79. https://doi.org/10.1016/S1364-0321(02)00004-7

[4] X. Hu, G. Li, J.C. Yu. Design, fabrication, and modification of nanostructured semiconductor materials for environmental and energy applications, Langmuir. 26 (2009) 3031-9. https://doi.org/10.1021/la902142b

[5] S. Chaturvedi, P.N. Dave, N. Shah. Applications of nano-catalyst in new era, J. Saudi Chem. Soc. 16 (2012) 307-25.

[6] G. Rottman. Measurement of total and spectral solar irradiance. Solar Variability and Planetary Climates: Springer; 2006. p. 39-51. https://doi.org/10.1016/j.jscs.2011.01.015

[7] http://solarcellcentral.com/solar_page.html.

[8] P.V. Kamat. Meeting the clean energy demand: nanostructure architectures for solar energy conversion, J. Phys. Chem. C. 111 (2007) 2834-60. https://doi.org/10.1021/jp066952u

[9] Z.L. Wang, W. Wu. Nanotechnology-enabled energy harvesting for self-powered micro-/nanosystems, Angew. Chem. Int. Ed. 51 (2012) 11700-21. https://doi.org/10.1002/anie.201201656

[10] A.K. Hussein. Applications of nanotechnology in renewable energies—A comprehensive overview and understanding, Renew. Sustain. Energ. Rev. 42 (2015) 460-76. https://doi.org/10.1016/j.rser.2014.10.027

[11] N. Armaroli, V. Balzani. Solar electricity and solar fuels: status and perspectives in the context of the energy transition, Chem. A Eur. J. 22 (2016) 32-57. https://doi.org/10.1002/chem.201503580

[12] J.C. Colmenares, R. Luque. Heterogeneous photocatalytic nanomaterials: prospects and challenges in selective transformations of biomass-derived compounds, Chem. Soc. Rev. 43 (2014) 765-78. https://doi.org/10.1039/C3CS60262A

[13] W.Y. Teoh, J.A. Scott, R. Amal. Progress in heterogeneous photocatalysis: from classical radical chemistry to engineering nanomaterials and solar reactors, J. Phys. Chem. Lett. 3 (2012) 629-39. https://doi.org/10.1021/jz3000646

[14] K. An, G.A. Somorjai. Nanocatalysis I: Synthesis of Metal and Bimetallic Nanoparticles and Porous Oxides and Their Catalytic Reaction Studies, Catal. Lett. 145 (2015) 233-48. https://doi.org/10.1007/s10562-014-1399-x

[15] E.W. McFarland, H. Metiu. Catalysis by doped oxides, Chem. Rev. 113 (2013) 4391-427. https://doi.org/10.1021/cr300418s

[16] M. Miyauchi, A. Nakajima, T. Watanabe, K. Hashimoto. Photocatalysis and photoinduced hydrophilicity of various metal oxide thin films, Chem. Mater. 14 (2002) 2812-6. https://doi.org/10.1021/cm020076p

[17] S.G. Kumar, K.K. Rao. Comparison of modification strategies towards enhanced charge carrier separation and photocatalytic degradation activity of metal oxide semiconductors (TiO_2, WO_3 and ZnO), Appl. Surf. Sci. 391 (2017) 124-48. https://doi.org/10.1021/cm020076p

[18] X. Liu, J. Iocozzia, Y. Wang, X. Cui, Y. Chen, S. Zhao, *et al.* Noble metal–metal oxide nanohybrids with tailored nanostructures for efficient solar energy conversion, photocatalysis and environmental remediation, Energ. Environ. Sci. 10 (2017) 402-34. https://doi.org/10.1039/C6EE02265K

[19] R. Gusain, P. Kumar, O. P. Sharma, S.L. Jain, O.P. Khatri. Reduced graphene oxide–CuO nanocomposites for photocatalytic conversion of CO_2 into methanol under visible light irradiation, Appl. Catal. B- Environ. 181 (2016) 352-62. https://doi.org/10.1039/C6EE02265K

[20] A.S. Aricò, P. Bruce, B. Scrosati, J.-M. Tarascon, W. Van Schalkwijk. Nanostructured materials for advanced energy conversion and storage devices, Nat. Mater. 4 (2005) 366-77. https://doi.org/10.1038/nmat1368

[21] B.J. Eliasson, G. Moddel. Metal-oxide electron tunneling device for solar energy conversion. Google Patents; 2003.

[22] F.-t. Li, J. Ran, M. Jaroniec, S. Z. Qiao. Solution combustion synthesis of metal oxide nanomaterials for energy storage and conversion, Nanoscale. 7 (2015) 17590-610. https://doi.org/10.1039/C5NR05299H

[23] Z. Wang, C.-J. Liu. Preparation and application of iron oxide/graphene based composites for electrochemical energy storage and energy conversion devices: current status and perspective, Nano Energ. 11 (2015) 277-93. https://doi.org/10.1016/j.nanoen.2014.10.022

[24] X. Xia, J. Tu, Y. Zhang, X. Wang, C. Gu, X.-b. Zhao, *et al.* High-quality metal oxide core/shell nanowire arrays on conductive substrates for electrochemical energy storage, ACS nano. 6 (2012) 5531-8. https://doi.org/10.1021/nn301454q

[25] J. Wang, Q. Zhang, X. Li, B. Zhang, L. Mai, K. Zhang. Smart construction of three-dimensional hierarchical tubular transition metal oxide core/shell heterostructures with high-capacity and long-cycle-life lithium storage, Nano Energ. 12 (2015) 437-46. https://doi.org/10.1016/j.nanoen.2015.01.003

[26] H. Pan, Y. Shao, P. Yan, Y. Cheng, K.S. Han, Z. Nie, *et al.* Reversible aqueous zinc/manganese oxide energy storage from conversion reactions, Nat. Energy. 1 (2016) 16039. https://doi.org/10.1038/nenergy.2016.39

[27] D.R. Miller, S.A. Akbar, P.A. Morris Nanoscale metal oxide-based heterojunctions for gas sensing: a review, Sensors and Actuat B- Chem. 204 (2014) 250-72. https://doi.org/10.1016/j.snb.2014.07.074

[28] S.G. Chatterjee, S. Chatterjee, A.K. Ray, A.K. Chakraborty. Graphene–metal oxide nanohybrids for toxic gas sensor: A review. Sensors and Actuat. B- Chem. 221 (2015) 1170-81. https://doi.org/10.1016/j.snb.2015.07.070

[29] A. Mirzaei, S. Leonardi, G. Neri. Detection of hazardous volatile organic compounds (VOCs) by metal oxide nanostructures-based gas sensors: A review, Ceram. Int. 42 (2016) 15119-41. https://doi.org/10.1016/j.ceramint.2016.06.145

[30] J. Bai, B. Zhou. Titanium dioxide nanomaterials for sensor applications, Chem. Rev. 114 (2014) 10131-76. https://doi.org/10.1021/cr400625j

[31] S.J. Patil, A.V. Patil, C.G. Dighavkar, K.S. Thakare, R.Y. Borase, S.J. Nandre, *et al.* Semiconductor metal oxide compounds based gas sensors: A literature review, Front. Mater. Sci. 9 (2015) 14-37. https://doi.org/10.1007/s11706-015-0279-7

[32] N. Masson, R. Piedrahita, M. Hannigan, Approach for quantification of metal oxide type semiconductor gas sensors used for ambient air quality monitoring, Sensors and Actuat. B- Chem. 208 (2015) 339-45. https://doi.org/10.1016/j.snb.2014.11.032

[33] S. Roy, S. Kar, B. Bagchi, S. Das, Development of transition metal oxide–kaolin composite pigments for potential application in paint systems, Appl. Clay Sci. 107 (2015) 205-12. https://doi.org/10.1016/j.clay.2015.01.029

[34] D.R. Swiler, T.J. Detrie, E.A. Axtell, Rare earth-transition metal oxide pigments. Google Patents; 2003.

[35] S. Forestier, I. Hansenne, Cosmetic composition containing a mixture of metal oxide nanopigments and melanine pigments. Google Patents; 1997.

[36] P. Jeevanandam, R. Mulukutla, M. Phillips, S. Chaudhuri, L. Erickson, K.K. Klabunde, Near infrared reflectance properties of metal oxide nanoparticles, J. Phys. Chem. C. 111 (2007) 1912-8. https://doi.org/10.1021/jp066363o

[37] A.I. Foster, M.L. Sims, D. Young, Protective metal oxide films on metal or alloy substrate surfaces susceptible to coking, corrosion or catalytic activity. Google Patents; 1981.

[38] D. Wang, G.P. Bierwagen, Sol–gel coatings on metals for corrosion protection, Prog. Org. Coat. 64 (2009) 327-38. https://doi.org/10.1016/j.porgcoat.2008.08.010

[39] Y.-S. Kwon, A.A. Gromov, J. I. Strokova, Passivation of the surface of aluminum nanopowders by protective coatings of the different chemical origin, Appl. Surf. Sci. 253 (2007) 5558-64. https://doi.org/10.1016/j.apsusc.2006.12.124

[40] A. Mostafaei, F. Nasirpouri, Epoxy/polyaniline–ZnO nanorods hybrid nanocomposite coatings: Synthesis, characterization and corrosion protection performance of conducting paints, Prog. Org. Coat. 77 (2014) 146-59. https://doi.org/10.1016/j.porgcoat.2013.08.015

[41] C. Clavero. Plasmon-induced hot-electron generation at nanoparticle/metal-oxide interfaces for photovoltaic and photocatalytic devices, Nat. Photonics. 8 (2014) 95-103. https://doi.org/10.1038/nphoton.2013.238

[42] M. Prezioso, F. Merrikh-Bayat, B. Hoskins, G. Adam, K.K. Likharev, D.B. Strukov, Training and operation of an integrated neuromorphic network based on metal-oxide memristors, Nature. 521 (2015) 61-4. https://doi.org/10.1038/nature14441

[43] Z. Sun, T. Liao, Y. Dou, S. M. Hwang, M-S. Park, L. Jiang *et al.* Generalized self-assembly of scalable two-dimensional transition metal oxide nanosheets, Nat. Comm. 5 (2014) 3813. https://doi.org/10.1038/ncomms4813

[44] S. Park, K. . Kim, J.W. Jo, S. Sung, K.T. Kim, W.J. Lee *et al.* In-Depth Studies on Rapid Photochemical Activation of Various Sol–Gel Metal Oxide Films for Flexible Transparent Electronics, Adv. Funct. Mat. 25 (2015) 2807-15. https://doi.org/10.1002/adfm.201500545

[45] L. Petti, N. Münzenrieder, C. Vogt, H. Faber, L. Büthe, G. Cantarella *et al.* Metal oxide semiconductor thin-film transistors for flexible electronics, Appl. Phys. Rev. 3 (2016) 021303. https://doi.org/10.1063/1.4953034

[46] S.A. Corr. Metal oxide nanoparticles. Nanoscience 2012. p. 180-207. https://doi.org/10.1039/9781849734844-00180

[47] J. Pelegrina, F. Gennari, A. Condó, A.F. Guillermet, Predictive Gibbs-energy approach to crystalline/amorphous relative stability of nanoparticles: Size-effect calculations and experimental test, J. Alloy Compd. 689 (2016) 161-8. https://doi.org/10.1016/j.jallcom.2016.07.284

[48] A. Khalajhedayati, Z. Pan, T.J. Rupert, Manipulating the interfacial structure of nanomaterials to achieve a unique combination of strength and ductility, Nat. Comm.. 7 (2016).

[49] A.P. Alivisatos. Perspectives on the physical chemistry of semiconductor nanocrystals, J. Phys. Chem. 100 (1996) 13226-39. https://doi.org/10.1021/jp9535506

[50] A.M. Smith, S. Nie, Semiconductor nanocrystals: structure, properties, and band gap engineering, Acc Chem Res. 43 (2010) 190-200. https://doi.org/10.1021/ar9001069

[51] G. Eda, T. Fujita, H. Yamaguchi, D. Voiry, M. Chen, M. Chhowalla, Coherent atomic and electronic heterostructures of single-layer MoS_2, Acs Nano. 6 (2012) 7311-7. https://doi.org/10.1021/nn302422x

[52] V. N. Mochalin, O. Shenderova, D. Ho, Y. Gogotsi, The properties and applications of nanodiamonds, Nat. Nanotechnol. 7 (2012) 11-23. https://doi.org/10.1038/nnano.2011.209

[53] L. Yang, L. Li, M. Zhao, G. Li, Size-induced variations in bulk/surface structures and their impact on photoluminescence properties of GdVO 4: Eu^{3+} nanoparticles, Phys. Chem. Chem. Phys. 14 (2012) 9956-65. https://doi.org/10.1039/c2cp41136a

[54] R. Buonsanti, A. Llordes, S. Aloni, B.A. Helms, D.J. Milliron, Tunable infrared absorption and visible transparency of colloidal aluminum-doped zinc oxide nanocrystals, Nano Lett. 11 (2011) 4706-10. https://doi.org/10.1021/nl203030f

[55] M. Ma, F.W. Mont, X. Yan, J. Cho, E.F. Schubert, G. B. Kim *et al.* Effects of the refractive index of the encapsulant on the light-extraction efficiency of light-emitting diodes, Opt. Express. 19 (2011) A1135-A40. https://doi.org/10.1364/OE.19.0A1135

[56] V. Etacheri, C. Di Valentin, J. Schneider, D. Bahnemann, S.C. Pillai, Visible-light activation of TiO_2 photocatalysts: advances in theory and experiments, J. Photoch. Photobio. C. 25 (2015) 1-29. https://doi.org/10.1016/j.jphotochemrev.2015.08.003

[57] A. Barnard, P. Zapol, L. Curtiss, Modeling the morphology and phase stability of TiO_2 nanocrystals in water, J. Chem. Theory and Computation. 1 (2005) 107-16. https://doi.org/10.1021/ct0499635

[58] H. Shi, Magaye R., Castranova V., Zhao J. Titanium dioxide nanoparticles: a review of current toxicological data, Particle and fibre toxicology. 10 (2013) 15. https://doi.org/10.1186/1743-8977-10-15

[59] P. Tao, Y. Li, A. Rungta, A. Viswanath, J. Gao, B. C. Benicewicz *et al.* TiO_2 nanocomposites with high refractive index and transparency, J. Mater. Chem.. 21 (2011) 18623-9. https://doi.org/10.1039/c1jm13093e

[60] D.O. Scanlon, C.W. Dunnill, J. Buckeridge, S.A. Shevlin, A.J. Logsdail, S.M. Woodley *et al.* Band alignment of rutile and anatase TiO_2, Nat. Mater.. 12 (2013) 798-801. https://doi.org/10.1038/nmat3697

[61] N. Serpone. Is the band gap of pristine TiO_2 narrowed by anion-and cation-doping of titanium dioxide in second-generation photocatalysts? : ACS Publications; 2006.

[62] A. Fujishima, K. Honda, Electrochemical photolysis of water at a semiconductor electrode, Nature. 238 (1972) 37-8. https://doi.org/10.1038/238037a0

[63] R. Li, Y. Weng, X. Zhou, X. Wang, Y. Mi, R. Chong *et al.* Achieving overall water splitting using titanium dioxide-based photocatalysts of different phases, Energy Environ. Sci. 8 (2015) 2377-82. https://doi.org/10.1039/C5EE01398D

[64] J. Yu, J. Fan, L. Kangle. Anatase TiO_2 nanosheets with exposed (001) facets: improved photoelectric conversion efficiency in dye-sensitized solar cells, Nanoscale. 2 (2010) 2144-9. https://doi.org/10.1039/c0nr00427h

[65] I.P. Parkin, R.G. Palgrave, Self-cleaning coatings, J. Mater. Chem. 15 (2005) 1689-95. https://doi.org/10.1039/b412803f

[66] J. Li, W. Wan, H. Zhou, J. Li, D. Xu, Hydrothermal synthesis of TiO_2 (B) nanowires with ultrahigh surface area and their fast charging and discharging properties in Li-ion batteries, Chem. Comm. 47 (2011) 3439-41. https://doi.org/10.1039/c0cc04634e

[67] R. Levinson, P. Berdahl, H. Akbari, Solar spectral optical properties of pigments—Part I: model for deriving scattering and absorption coefficients from transmittance and reflectance measurements, Sol. Energy Mater. Sol. Cells. 89 (2005) 319-49. https://doi.org/10.1016/j.solmat.2004.11.012

[68] P. Moriarty, D. Honnery, What is the global potential for renewable energy?, Renew. Sustain. Energ. Rev. 16 (2012) 244-52. https://doi.org/10.1016/j.rser.2011.07.151

[69] N. Lior. Sustainable energy development: the present (2009) situation and possible paths to the future, Energy. 35 (2010) 3976-94. https://doi.org/10.1016/j.energy.2010.03.034

[70] A. Blakers. Sustainable energy options, Asian Perspective. 39 (2015) 559-90.

[71] V. Balzani, A. Credi, M. Venturi, Photochemical conversion of solar energy, ChemSusChem. 1 (2008) 26-58. https://doi.org/10.1002/cssc.200700087

[72] S. Chu, A. Majumdar, Opportunities and challenges for a sustainable energy future, Nature. 488 (2012) 294. https://doi.org/10.1038/nature11475

[73] J. Schneider, M. Matsuoka, M. Takeuchi, J. Zhang, Y. Horiuchi, M. Anpo *et al.* Understanding TiO_2 photocatalysis: mechanisms and materials, Chem. Rev. 114 (2014) 9919-86. https://doi.org/10.1021/cr5001892

[74] F. Zhang, J. Zhao, T. Shen, H. Hidaka, E. Pelizzetti, N. Serpone, TiO_2-assisted photodegradation of dye pollutants II. Adsorption and degradation kinetics of eosin in TiO_2 dispersions under visible light irradiation, Appl. Catal. B-Environ. 15 (1998) 147-56. https://doi.org/10.1016/S0926-3373(97)00043-X

[75] C. Chen, X. Li, W. Ma, J. Zhao, H. Hidaka, N. Serpone, Effect of transition metal ions on the TiO2-assisted photodegradation of dyes under visible irradiation: a probe for the interfacial electron transfer process and reaction mechanism, J. Phys. Chem. B. 106 (2002) 318-24. https://doi.org/10.1021/jp0119025

[76] W. Li, D. Li, Y. Lin, P. Wang, W. Chen, X. Fu *et al.* Evidence for the active species involved in the photodegradation process of methyl orange on TiO_2, J. Phys. Chem. C. 116 (2012) 3552-60. https://doi.org/10.1021/jp209661d

[77] R. Asahi, Morikawa T., Irie H., Ohwaki T. Nitrogen-doped titanium dioxide as visible-light-sensitive photocatalyst: designs, developments, and prospects, Chem. Rev. 114 (2014) 9824-52. https://doi.org/10.1021/cr5000738

[78] S.A. Ansari, M. M. Khan, M.O. Ansari, M.H. Cho, Nitrogen-doped titanium dioxide (N-doped TiO_2) for visible light photocatalysis, New J. Chem. 40 (2016) 3000-9. https://doi.org/10.1039/C5NJ03478G

[79] G. Li, L. Chen, N.M. Dimitrijevic, K.A. Gray. Visible light photocatalytic properties of anion-doped TiO_2 materials prepared from a molecular titanium

precursor, Chem. Phys. Lett. 451 (2008) 75-9. https://doi.org/10.1016/j.cplett.2007.11.071

[80] H. Wang, J. Lewis, Second-generation photocatalytic materials: anion-doped TiO_2, J. Phys. Conden. Matter. 18 (2005) 421. https://doi.org/10.1088/0953-8984/18/2/006

[81] S. Anderson, E.C. Constable, MP. Dare-Edwards, J.B. Goodenough, A. Hamnett, K. R. Seddon *et al.* Chemical modification of a titanium (IV) oxide electrode to give stable dye sensitisation without a supersensitiser, Nature. 280 (1979) 571-3. https://doi.org/10.1038/280571a0

[82] B. O'regan, M. Grfitzeli, A low-cost, high-efficiency solar cell based on dye-sensitized, Nature. 353 (1991) 737-40. https://doi.org/10.1038/353737a0

[83] M. Zhang, C. Chen, W. Ma, J. Zhao, Visible-Light-Induced Aerobic Oxidation of Alcohols in a Coupled Photocatalytic System of Dye-Sensitized TiO_2 and TEMPO, Angew. Chem. 120 (2008) 9876-9. https://doi.org/10.1002/ange.200803630

[84] Z. Wang, H. Wang, B. Liu, W. Qiu, J. Zhang, S. Ran *et al.* Transferable and flexible nanorod-assembled TiO_2 cloths for dye-sensitized solar cells, photodetectors, and photocatalysts, ACS nano. 5 (2011) 8412-9. https://doi.org/10.1021/nn203315k

[85] Y. Shengyuan, Z. Peining, A.S. Nair, S. Ramakrishna, Rice grain-shaped TiO_2 mesostructures—synthesis, characterization and applications in dye-sensitized solar cells and photocatalysis, J. Mater. Chem. 21 (2011) 6541-8. https://doi.org/10.1039/c0jm04512h

[86] P. Chowdhury, J. Moreira, H. Gomaa, A.K. Ray, Visible-solar-light-driven photocatalytic degradation of phenol with dye-sensitized TiO_2: parametric and kinetic study, Ind. Eng. Chem. Res. 51 (2012) 4523-32. https://doi.org/10.1021/ie2025213

[87] Z. Li, Y. Fang, S. Xu, Squaraine dye sensitized TiO_2 nanocomposites with enhanced visible-light photocatalytic activity, Mater. Lett. 93 (2013) 345-8. https://doi.org/10.1016/j.matlet.2012.11.135

[88] D. Y. Leung, X. Fu, C. Wang, M. Ni, M. K. Leung, X. Wang *et al.* Hydrogen Production over Titania-Based Photocatalysts, ChemSusChem. 3 (2010) 681-94. https://doi.org/10.1002/cssc.201000014

[89] J. Cao, J-Z. Sun, H-Y. Li, J. Hong, M. Wang, A facile room-temperature chemical reduction method to TiO_2@ CdS core/sheath heterostructure nanowires, J. Mater. Chem. 14 (2004) 1203-6. https://doi.org/10.1039/b313541a

[90] H. Tada, T. Kiyonaga, S. Naya, Rational design and applications of highly efficient reaction systems photocatalyzed by noble metal nanoparticle-loaded titanium (IV) dioxide, Chem. Soc. Rev. 38 (2009) 1849-58. https://doi.org/10.1039/b822385h

[91] X. Chen, L. Liu, Y.Y. Peter, S.S. Mao, Increasing solar absorption for photocatalysis with black hydrogenated titanium dioxide nanocrystals, Science. 331 (2011) 746-50. https://doi.org/10.1126/science.1200448

[92] Z-L. Lu, E. Lindner, H. A. Mayer, Applications of sol– gel-processed interphase catalysts, Chem. Rev. 102 (2002) 3543-78. https://doi.org/10.1021/cr010358t

[93] R.A. Caruso, M. Antonietti, Sol– gel nanocoating: an approach to the preparation of structured materials, Chem. Mater. 13 (2001) 3272-82. https://doi.org/10.1021/cm001257z

[94] S. Mann, S.L. Burkett, S.A. Davis, C.E. Fowler, N.H. Mendelson, S.D. Sims *et al.* Sol– gel synthesis of organized matter, Chemistry of Materials. 9 (1997) 2300-10. https://doi.org/10.1021/cm970274u

[95] B. Liu, E.S. Aydil, Growth of oriented single-crystalline rutile TiO_2 nanorods on transparent conducting substrates for dye-sensitized solar cells, J Am. Chem. Soc. 131 (2009) 3985-90. https://doi.org/10.1021/ja8078972

[96] C.J. Brinker, G.W. Scherer, SOL-GEL SCIENCE. In: George W. Scherer CJB, editor. Boston San Diego New York: Academic Press, INC; 1990.

[97] C.-C. Wang, J.Y. Ying, Sol– gel synthesis and hydrothermal processing of anatase and rutile titania nanocrystals, Chem. Mater. 11 (1999) 3113-20. https://doi.org/10.1021/cm990180f

[98] X. Chen, S.S. Mao, Titanium dioxide nanomaterials: synthesis, properties, modifications, and applications, Chem. Rev. 107 (2007) 2891-959. https://doi.org/10.1021/cm990180f

[99] A. Chemseddine, T. Moritz, Nanostructuring titania: control over nanocrystal structure, size, shape, and organization, Eur. J. Inorg. Chem. 1999 (1999) 235-45. https://doi.org/10.1002/(SICI)1099-0682(19990202)1999:2%3C235::AID-EJIC235%3E3.0.CO;2-N

[100] T. Sugimoto, K. Okada, H. Itoh, Synthetic of uniform spindle-type titania particles by the gel–sol method. J. Colloid. Interface Sci. 1997. https://doi.org/10.1006/jcis.1997.5037

[101] T. Sugimoto, X. Zhou, A. Muramatsu, Synthesis of uniform anatase TiO_2 nanoparticles by gel–sol method: 4. Shape control, J. Colloid Interface Sci. 259 (2003) 53-61. https://doi.org/10.1016/S0021-9797(03)00035-3

[102] T. Sugimoto, X. Zhou, Synthesis of uniform anatase TiO_2 nanoparticles by the Gel–Sol method: 2. adsorption of OH^- ions to Ti $(OH)_4$ gel and TiO_2 particles, J. Colloid Interface Sci. 252 (2002) 347-53. https://doi.org/10.1006/jcis.2002.8480

[103] T. Sugimoto, X. Zhou, A. Muramatsu, Synthesis of Uniform Anatase TiO_2 Nanoparticles by Gel–Sol Method, J. Colloid Interface Sci. 252 (2002) 347-53. https://doi.org/10.1006/jcis.2002.8480

[104] B. Naufal, P. Jaseela, P. Periyat, Direct Sunlight Active Sm^{3+} Doped TiO_2 Photocatalyst. Proc Materials Science Forum: Trans Tech Publ; 2016. p. 33-44.

[105] M. Andersson, L. Österlund, S. Ljungström, A. Palmqvist, Preparation of nanosize anatase and rutile TiO_2 by hydrothermal treatment of microemulsions and their activity for photocatalytic wet oxidation of phenol, J. Phys. Chem. B. 106 (2002) 10674-9. https://doi.org/10.1021/jp025715y

[106] J. Yang, S. Mei, J.M. Ferreira, Hydrothermal synthesis of nanosized titania powders: influence of peptization and peptizing agents on the crystalline phases and phase transitions, J. Am. Ceram. Soc. 83 (2000) 1361-8. https://doi.org/10.1111/j.1151-2916.2000.tb01394.x

[107] J. Yang, S. Mei, J.M. Ferreira, Hydrothermal synthesis of nanosized titania powders: influence of tetraalkyl ammonium hydroxides on particle characteristics, J. Am. Ceram. Soc. 84 (2001) 1696-702. https://doi.org/10.1111/j.1151-2916.2001.tb00901.x

[108] J. Yang, S. Mei, J.M. Ferreira, Hydrothermal synthesis of well-dispersed TiO_2 nano-crystals, J. Mater. Res. 17 (2002) 2197-200. https://doi.org/10.1557/JMR.2002.0323

[109] J. Yang, S. Mei, J.M. Ferreira, In situ preparation of weakly flocculated aqueous anatase suspensions by a hydrothermal technique, J. Colloid Interface Sci. 260 (2003) 82-8. https://doi.org/10.1016/S0021-9797(02)00190-X

[110] S.Y. Chae, M.K. Park, S.K. Lee, T.Y. Kim, S.K. Kim, W.I. Lee, Preparation of size-controlled TiO_2 nanoparticles and derivation of optically transparent

photocatalytic films, Chem. Mater. 15 (2003) 3326-31. https://doi.org/10.1021/cm030171d

[111] F. Cot, A. Larbot, G. Nabias, L. Cot, Preparation and characterization of colloidal solution derived crystallized titania powder, J. Eur. Ceram. Soc. 18 (1998) 2175-81. https://doi.org/10.1016/S0955-2219(98)00143-5

[112] Y. Zhang, G. Li, Y. Jin, Y. Zhang, J. Zhang, L. Zhang, Hydrothermal synthesis and photoluminescence of TiO_2 nanowires, Chem. Phys. Lett. 365 (2002) 300-4. https://doi.org/10.1016/S0009-2614(02)01499-9

[113] M. Wei, Y. Konishi, H. Zhou, H. Sugihara, H. Arakawa A simple method to synthesize nanowires titanium dioxide from layered titanate particles, Chem. Phys. Lett. 400 (2004) 231-4. https://doi.org/10.1016/j.cplett.2004.10.114

[114] J. Xu, J. P. Ge, Y. D. Li, Solvothermal synthesis of monodisperse PbSe nanocrystals, J. Phys. Chem. B. 110 (2006) 2497-501. https://doi.org/10.1021/jp056521w

[115] B. Gersten, Solvothermal synthesis of nanoparticles, Chemfiles. 5 (2005) 11-2.

[116] M.M. Byranvand, A.N. Kharat, L. Fatholahi, Z.M. Beiranvand, A review on synthesis of nano-TiO_2 via different methods, J. Nanostruct. 3 (2013) 1-9.

[117] C. Xu, D. Nie, H. Chen, G. Zhao, Y. Liu, Fabrication of cobalt hollow microspheres via a PVP-assisted solvothermal process, Mater. Lett. 110 (2013) 87-90. https://doi.org/10.1016/j.matlet.2013.08.010

[118] B.M. Wen, C. Y. Liu, Y. Liu, Solvothermal synthesis of ultralong single-crystalline TiO_2 nanowires, New J. Chem. 29 (2005) 969-71. https://doi.org/10.1039/b502604k

[119] B. Wen, C. Liu, Y. Liu, Bamboo-shaped Ag-doped TiO_2 nanowires with heterojunctions, Inorg. Chem. 44 (2005) 6503-5.

[120] X. Wang, J. Zhuang, Q. Peng, Y. Li, A general strategy for nanocrystal synthesis, Nature. 437 (2005) 121. https://doi.org/10.1038/nature03968

[121] C.S. Kim, B. K. Moon, J. H. Park, B. C. Choi, H. J. Seo, Solvothermal synthesis of nanocrystalline TiO_2 in toluene with surfactant, J.Cryst. Growth. 257 (2003) 309-15. https://doi.org/10.1016/S0022-0248(03)01468-4

[122] Y. Cao, L. Zong, Q. Li, C. Li, J. Li, J. Yang, Solvothermal synthesis of TiO_2 nanocrystals with {001} facets using titanic acid nanobelts for superior

photocatalytic activity, Appl. Surf. Sci. 391 (2017) 311-7. https://doi.org/10.1016/j.apsusc.2016.06.198

[123] J. Li, C. Wang, P. Zheng, L. Zhang, G. Chen, C. Tang, *et al* Solvothermal preparation of micro/nanostructured TiO_2 with enhanced lithium storage capability, Mater. Chem. Phys. 190 (2017) 202-8. https://doi.org/10.1016/j.matchemphys.2016.12.049

[124] A. Sherman. Chemical vapor deposition for microelectronics: principles, technology, and applications, (1987).

[125] G.E.J. Poinern. A laboratory course in nanoscience and nanotechnology: CRC Press, 2014.

[126] L. Lei, H.P. Chu, X. Hu, P.L. Yue, Preparation of heterogeneous photocatalyst (TiO_2/alumina) by metallo-organic chemical vapor deposition, Ind. Eng. Chem. Res. 38 (1999) 3381-5. https://doi.org/10.1021/ie980677j

[127] K. Schrijnemakers, N. Impens, E. Vansant, Deposition of a titania coating on silica by means of the chemical surface coating, Langmuir. 15 (1999) 5807-13. https://doi.org/10.1021/la9812469

[128] D.C. Gilmer, D.G. Colombo, C.J. Taylor, J. Roberts, G. Haugstad, S.A. Campbell, *et al*. Low Temperature CVD of Crystalline Titanium Dioxide Films Using Tetranitratotitanium (IV), Chem. Vap. Deposition. 4 (1998) 9-11.

[129] Z. Ding, X. Hu, P.L. Yue, G.Q. Lu, P.F. Greenfield, Synthesis of anatase TiO_2 supported on porous solids by chemical vapor deposition, Catal. Today. 68 (2001) 173-82. https://doi.org/10.1016/S0920-5861(01)00298-X

[130] S. Seifried, M. Winterer, H. Hahn, Nanocrystalline titania films and particles by chemical vapor synthesis, Chem. Vap. Deposition. 6 (2000) 239-44.

[131] D. Guo, A. Ito, T. Goto, R. Tu, C. Wang, Q. Shen, *et al.* Preparation of rutile TiO_2 thin films by laser chemical vapor deposition method, J. Adv. Ceram. 2 (2013) 162-6. https://doi.org/10.1007/s40145-013-0056-y

[132] D.S. Jayakrishnan. Electrodeposition: The versatile technique for nanomaterials 2012.

[133] S. Liu, K. Huang, Straightforward fabrication of highly ordered TiO_2 nanowire arrays in AAM on aluminum substrate, Sol. Energ. Mater. Sol. Cells. 85 (2005) 125-31.

[134] M. Gopal, W. M. Chan, L. De Jonghe, Room temperature synthesis of crystalline metal oxides, J. Mater. Sci. 32 (1997) 6001-8. https://doi.org/10.1023/A:1018671212890

[135] T.P. Feist, P.K. Davies, The soft chemical synthesis of TiO_2 (B) from layered titanates, J.Solid State Chem. 101 (1992) 275-95. https://doi.org/10.1016/0022-4596(92)90184-W

[136] Y. Ma, X. Wang, Y. Jia, X. Chen, H. Han, C. Li, Titanium dioxide-based nanomaterials for photocatalytic fuel generations, Chem. Rev. 114 (2014) 9987-10043. https://doi.org/10.1021/cr500008u

[137] A.L. Linsebigler, G. Lu, J.T. Yates Jr, Photocatalysis on TiO_2 surfaces: principles, mechanisms, and selected results, Chem. Rev. 95 (1995) 735-58. https://doi.org/10.1021/cr00035a013

[138] H. Zhang, J.F. Banfield, Understanding polymorphic phase transformation behavior during growth of nanocrystalline aggregates: insights from TiO_2, J. Phys. Chem. B. 104 (2000) 3481-7. https://doi.org/10.1021/jp000499j

[139] W.H. Baur. Atomabstände und bindungswinkel im brookit, TiO_2, Acta Crystallogr. 14 (1961) 214-6. https://doi.org/10.1107/S0365110X61000747

[140] Z. Yanqing, S. Erwei, C. Suxian, L. Wenjun, H. Xingfang, Hydrothermal preparation and characterization of brookite-type TiO_2 nanocrystallites, J. Mater. Sci. Lett. 19 (2000) 1445-8. https://doi.org/10.1023/A:1011010306699

[141] P. Periyat, D.E. McCormack, S.J. Hinder, S.C. Pillai, One-pot synthesis of anionic (nitrogen) and cationic (sulfur) codoped high-temperature stable, visible light active, anatase photocatalysts, J. Phys. Chem. C. 113 (2009) 3246-53. https://doi.org/10.1021/jp808444y

[142] P. Periyat, B. Naufal, S.G. Ullattil, A review on high temperature stable anatase TiO_2 photocatalysts. Proc Materials Science Forum: Trans Tech Publ; 2016. p. 78-93.

[143] P.I. Gouma, M.J. Mills, Anatase-to-Rutile Transformation in Titania Powders, J. Am. Ceram.Soc. 84 (2001) 619-22. https://doi.org/10.1111/j.1151-2916.2001.tb00709.x

[144] G. Riegel, J. R. Bolton, Photocatalytic efficiency variability in TiO_2 particles, J. Phys. Chem. 99 (1995) 4215-24. https://doi.org/10.1021/j100012a050

[145] R. Asahi, Y. Taga, W. Mannstadt, A. J. Freeman, Electronic and optical properties of anatase TiO_2, Phys. Rev. B. 61 (2000) 7459. https://doi.org/10.1103/PhysRevB.61.7459

[146] R. Asahi, T. Morikawa, T. Ohwaki, K. Aoki, Y. Taga, Visible-light photocatalysis in nitrogen-doped titanium oxides, Science. 293 (2001) 269-71. https://doi.org/10.1126/science.1061051

[147] N.G. Park, J. Van de Lagemaat, A. Frank, Comparison of dye-sensitized rutile-and anatase-based TiO_2 solar cells, J. Phys. Chem. B. 104 (2000) 8989-94. https://doi.org/10.1021/jp994365l

[148] J.C. Yu, J. Lin, R.W. Kwok, $Ti_{1-x}Zr_xO_2$ Solid Solutions for the Photocatalytic Degradation of Acetone in Air, J. Phys. Chem. B. 102 (1998) 5094-8. https://doi.org/10.1021/jp980332e

[149] M. Dahl, Y. Liu, Y. Yin, Composite titanium dioxide nanomaterials, Chem. Rev. 114 (2014) 9853-89. https://doi.org/10.1021/cr400634p

[150] W. Li, Z. Wu, J. Wang, A. A. Elzatahry, D. Zhao, A perspective on mesoporous TiO_2 materials, Chem. Mater. 26 (2013) 287-98. https://doi.org/10.1021/cm4014859

[151] N. Serpone, D. Lawless, R. Khairutdinov, Size effects on the photophysical properties of colloidal anatase TiO_2 particles: size quantization versus direct transitions in this indirect semiconductor, J. Phys. Chem. 99 (1995) 16646-54. https://doi.org/10.1021/j100045a026

[152] H. Tong, S. Ouyang, Y. Bi, N. Umezawa, M. Oshikiri, J. Ye, Nano-photocatalytic materials: possibilities and challenges, Adv. Mater. 24 (2012) 229-51. https://doi.org/10.1002/adma.201102752

[153] T. Sasaki, M. Watanabe, Semiconductor nanosheet crystallites of quasi-TiO_2 and their optical properties, J. Phys. Chem. B. 101 (1997) 10159-61. https://doi.org/10.1021/jp9727658

[154] Y. Bessekhouad, D. Robert, J.V. Weber, N. Chaoui, Effect of alkaline-doped TiO_2 on photocatalytic efficiency, J. Photochem. Photobiol. A-Chem. 167 (2004) 49-57. https://doi.org/10.1016/j.jphotochem.2003.12.001

[155] J.G. Ma, C.R. Zhang, J.J. Gong, Y.Z. Wu, S.Z. Kou, H. Yang, *et al.* The Electronic Structures and Optical Properties of Alkaline-Earth Metals Doped Anatase TiO_2: A Comparative Study of Screened Hybrid Functional and

Generalized Gradient Approximation, Materials. 8 (2015) 5508-25. https://doi.org/10.3390/ma8085257

[156] J. Son, J. Chattopadhyay, D. Pak, Electrocatalytic performance of Ba-doped TiO_2 hollow spheres in water electrolysis, Int. J. Hydrogen Energy. 35 (2010) 420-7. https://doi.org/10.1016/j.ijhydene.2009.10.096

[157] S. Liu, Z. Min, D. Hu, Y. Liu, Synthesis of calcium doped TiO_2 nanomaterials and their visible light degradation property. Proc International Conference on Material and Environmental Engineering (ICMAEE 2014): Atlantis Press; 2014. https://doi.org/10.2991/icmaee-14.2014.12

[158] K. Nagaveni, M. Hegde, G. Madras, Structure and Photocatalytic Activity of $Ti_{1-x}M_xO_{2\pm\delta}$ (M= W, V, Ce, Zr, Fe, and Cu) Synthesized by Solution Combustion Method, J. Phys. Chem. B. 108 (2004) 20204-12. https://doi.org/10.1021/jp047917v

[159] F. Gracia, J. Holgado, A. Caballero, A. Gonzalez-Elipe, Structural, Optical, and Photoelectrochemical Properties of Mn^{+}–TiO_2 Model Thin Film Photocatalysts, J. Phys. Chem. B. 108 (2004) 17466-76. https://doi.org/10.1021/jp0484938

[160] X. Wang, J. . Li, H. Kamiyama, M. Katada, N. Ohashi, Y. Moriyoshi, *et al.* Pyrogenic iron (III)-doped TiO_2 nanopowders synthesized in RF thermal plasma: phase formation, defect structure, band gap, and magnetic properties, J. Am. Chem. Soc. 127 (2005) 10982-90. https://doi.org/10.1021/ja051240n

[161] W. Sangchay. Photocatalytic and antibacterial activity of Ag-doped TiO_2 nanoparticles, Asia-Pac. J. Sci. Tech. 18 (2017) 731-8.

[162] N. Shetti, D. Nayak, S. . Malode, R. Kulkarni, Electrochemical sensor based upon ruthenium doped TiO_2 nanoparticles for the determination of flufenamic acid, J. Electrochem. Soc. 164 (2017) B3036-B42. https://doi.org/10.1149/2.0031705jes

[163] B. Liu, H. Chen, C. Liu, S. Andrews, C. Hahn, P. Yang, Large-scale synthesis of transition-metal-doped TiO_2 nanowires with controllable overpotential, J. Am. Chem. Soc. 135 (2013) 9995-8. https://doi.org/10.1021/ja403761s

[164] M. García-Mota, A. Vojvodic, F. Abild-Pedersen, J.K. Nørskov Electronic origin of the surface reactivity of transition-metal-doped TiO_2 (110), J. Phys. Chem. C. 117 (2012) 460-5. https://doi.org/10.1021/jp310667r

[165] F. Li, X. Li, M. Hou, Photocatalytic degradation of 2-mercaptobenzothiazole in aqueous La^{3+}–TiO_2 suspension for odor control, Appl. Catal. B-Environ. 48 (2004) 185-94. https://doi.org/10.1016/j.apcatb.2003.10.003

[166] A.W. Xu, Y. Gao, H.Q. Liu, The preparation, characterization, and their photocatalytic activities of rare-earth-doped TiO_2 nanoparticles, J. Catal. 207 (2002) 151-7. https://doi.org/10.1006/jcat.2002.3539

[167] X.P. Zhao, J.B. Yin, Preparation and electrorheological characteristics of rare-earth-doped TiO_2 suspensions, Chem. Mater. 14 (2002) 2258-63. https://doi.org/10.1021/cm011522w

[168] J. Zhao, J. Yu, Y. Xie, Z. Le, X. Hong, S. Ci, *et al.* Lanthanum and neodymium doped barium ferrite-TiO_2/MCNTs/poly (3-methyl thiophene) composites with nest structures: preparation, characterization and electromagnetic microwave absorption properties, Sci. Rep. 6 (2016). https://doi.org/10.1038/srep20496

[169] P. Mazierski, W. Lisowski, T. Grzyb, M. J. Winiarski, T. Klimczuk, A. Mikołajczyk, *et al.* Enhanced photocatalytic properties of lanthanide-TiO_2 nanotubes: An experimental and theoretical study, Appl. Catal. B- Environ. 205 (2017) 376-85. https://doi.org/10.1016/j.apcatb.2016.12.044

[170] M. Chang, Y. Song, Y. Sheng, J. Chen, H. Guan, Z. Shi, *et al.* Photoluminescence and Photocatalysis Properties of Dual-Functional Eu^{3+}-Doped Anatase Nanocrystals, J. Phys. Chem. C. 121 (2017) 2369-79. https://doi.org/10.1021/acs.jpcc.6b11013

[171] N. Yan, Z. Zhu, J. Zhang, Z. Zhao, Q. Liu. Preparation and properties of Ce-doped TiO_2 photocatalyst, Mater. Res. Bul. 47 (2012) 1869-73. https://doi.org/10.1016/j.materresbull.2012.04.077

[172] [172] C.M. Malengreaux, S.L. Pirard, G. Léonard, J.G. Mahy, M. Herlitschke, B. Klobes, *et al.* Study of the photocatalytic activity of Fe^{3+}, Cr^{3+}, La^{3+} and Eu^{3+} single-doped and co-doped TiO_2 catalysts produced by aqueous sol-gel processing, J. Alloy. Compd. 691 (2017) 726-38. https://doi.org/10.1016/j.jallcom.2016.08.211

[173] Y. Zhang, H. Zhang, Y. Xu, Y. Wang, Significant effect of lanthanide doping on the texture and properties of nanocrystalline mesoporous TiO_2, J. Solid State Chem. 177 (2004) 3490-8. https://doi.org/10.1016/j.jssc.2004.05.026

[174] M. Ye, J. Jia, Z. Wu, C. Qian, R. Chen, P.G. O'Brien *et al.* Synthesis of Black TiOx Nanoparticles by Mg Reduction of TiO_2 Nanocrystals and their Application for Solar Water Evaporation, Adv. Energ. Mater. 7 (2017).

[175] J.D. Bryan, S.M. Heald, S A. Chambers, D.R. Gamelin, Strong room-temperature ferromagnetism in Co2+-doped TiO_2 made from colloidal nanocrystals, J Am. Chem. Soc. 126 (2004) 11640-7. https://doi.org/10.1021/ja047381r

[176] S.G. Ullattil, A.V. Thelappurath, S.N. Tadka, J. Kavil, B.K. Vijayan, P. Periyat, A Sol-solvothermal Processed 'Black TiO_2'as Photoanode Material in Dye Sensitized Solar Cells, Solar Energy. 155 (2017) 490-5. https://doi.org/10.1016/j.solener.2017.06.059

[177] D.J. Reidy, J.D. Holmes, C. Nagle, M.A. Morris, A highly thermally stable anatase phase prepared by doping with zirconia and silica coupled to a mesoporous type synthesis technique, J. Mater. Chem. 15 (2005) 3494-500. https://doi.org/10.1039/b503395k

[178] S. Yin,Y. Aita, M. Komatsu, T. Sato, Visible-light-induced photocatalytic activity of $TiO_{2-x}N_y$ prepared by solvothermal process in urea–alcohol system, J Eur. Ceram. Soc. 26 (2006) 2735-42. https://doi.org/10.1016/j.jeurceramsoc.2005.05.012

[179] S.C. Pillai, P. Periyat, R. George, D. E. McCormack, M.K. Seery, H. Hayden et al. Synthesis of high-temperature stable anatase TiO_2 photocatalyst, J. Phys. Chem. C. 111 (2007) 1605-11. https://doi.org/10.1021/jp065933h

[180] G. Yang, Z. Yan, T. Xiao, Low-temperature solvothermal synthesis of visible-light-responsive S-doped TiO 2 nanocrystal, Appl. Surf. Sci. 258 (2012) 4016-22. https://doi.org/10.1016/j.apsusc.2011.12.092

[181] H.-F. Yu, Phase development and photocatalytic ability of gel-derived P-doped TiO 2, J. Mater. Res. 22 (2007) 2565-72. https://doi.org/10.1557/jmr.2007.0316

[182] D. Li, N. Ohashi, S. Hishita, T. Kolodiazhnyi, H. Haneda, Origin of visible-light-driven photocatalysis: a comparative study on N/F-doped and N–F-codoped TiO_2 powders by means of experimental characterizations and theoretical calculations, J. Solid State Chem.. 178 (2005) 3293-302. https://doi.org/10.1016/j.jssc.2005.08.008

[183] W. Choi, A. Termin, M. R. Hoffmann, The role of metal ion dopants in quantum-sized TiO_2: correlation between photoreactivity and charge carrier recombination dynamics, J. Phys. Chem. 98 (1994) 13669-79. https://doi.org/10.1021/j100102a038

[184] A. Naldoni, M. Allieta, S. Santangelo, M. Marelli, F. Fabbri, S. Cappelli *et al.* Effect of nature and location of defects on bandgap narrowing in black TiO_2

nanoparticles, J. Am. Chem. Soc. 134 (2012) 7600-3. https://doi.org/10.1021/ja3012676

[185] M. Kapilashrami, Y. Zhang, Y-S. Liu, A. Hagfeldt, J. Guo, Probing the optical property and electronic structure of TiO_2 nanomaterials for renewable energy applications, Chem. Rev. 114 (2014) 9662-707. https://doi.org/10.1021/cr5000893

[186] T. Umebayashi, T. Yamaki, H. Itoh, K. Asai, Analysis of electronic structures of 3d transition metal-doped TiO 2 based on band calculations, J. Phys. Chem. Solids. 63 (2002) 1909-20. https://doi.org/10.1016/S0022-3697(02)00177-4

[187] T. Nishikawa, Y. Shinohara, T. Nakajima, M. Fujita, S. Mishima, Prospect of activating a photocatalyst by sunlight-A quantum chemical study of isomorphically substituted titania, Chem. Lett. 28 (1999) 1133-4. https://doi.org/10.1246/cl.1999.1133

[188] X. Chen, C. Burda, The electronic origin of the visible-light absorption properties of C-, N-and S-doped TiO_2 nanomaterials, J. Am. Chem. Soc. 130 (2008) 5018-9. https://doi.org/10.1021/ja711023z

[189] X. Chen, S. Shen, L. Guo, S.S. Mao, Semiconductor-based photocatalytic hydrogen generation, Chemical reviews. 110 (2010) 6503-70. https://doi.org/10.1021/cr1001645

[190] J. Zhang, W.Dang, Z. Ao, S.K.Cushing, N. Wu, Band gap narrowing in nitrogen-doped $La_2Ti_2O_7$ predicted by density-functional theory calculations, Phys. Chem. Chem. Phys. 17 (2015) 8994-9000. https://doi.org/10.1039/C5CP00157A

[191] J.-Y. Lee, J. Park, J-H. Cho, Electronic properties of N-and C-doped TiO_2, Appl. Phys. Lett. 87 (2005) 011904. https://doi.org/10.1063/1.1991982

[192] W. Ho, J. C. Yu, J. Lin, J. Yu, P. Li, Preparation and photocatalytic behavior of MoS_2 and WS_2 nanocluster sensitized TiO_2, Langmuir. 20 (2004) 5865-9. https://doi.org/10.1021/la049838g

[193] D. Fitzmaurice, H. Frei, J. Rabani, Time-Resolved Optical Study on the Charge Carrier Dynamics in a TiO_2/AgI Sandwich Colloid, J. Phys. Chem. 99 (1995) 9176-81. https://doi.org/10.1021/j100022a034

[194] P. Hoyer, R. Könenkamp, Photoconduction in porous TiO_2 sensitized by PbS quantum dots, Appl. Phys. Lett. 66 (1995) 349-51. https://doi.org/10.1063/1.114209

[195] R. Vogel, P. Hoyer, H. Weller, Quantum-sized PbS, CdS, Ag2S, Sb2S3, and Bi2S3 particles as sensitizers for various nanoporous wide-bandgap

semiconductors, J. Phys. Chem. 98 (1994) 3183-8. https://doi.org/10.1021/j100063a022

[196] M.K. Nazeeruddin, P. Pechy, T. Renouard, S.M. Zakeeruddin, R. Humphry-Baker, P. Comte *et al.* Engineering of efficient panchromatic sensitizers for nanocrystalline TiO_2-based solar cells, J. Am. Chem. Soc. 123 (2001) 1613-24. https://doi.org/10.1021/ja003299u

[197] J. Hensel, G. Wang, Y. Li, J.Z. Zhang, Synergistic effect of CdSe quantum dot sensitization and nitrogen doping of TiO_2 nanostructures for photoelectrochemical solar hydrogen generation, Nano Lett. 10 (2010) 478-83. https://doi.org/10.1021/nl903217w

[198] F. Shen, W. Que, Y. Liao, X. Yin, Photocatalytic activity of TiO_2 nanoparticles sensitized by $CuInS_2$ quantum dots, Ind. Eng. Chem. Res. 50 (2011) 9131-7. https://doi.org/10.1021/ie2007467

[199] Q. Xiang, J. Yu, M. Jaroniec, Synergetic effect of MoS_2 and graphene as cocatalysts for enhanced photocatalytic H_2 production activity of TiO_2 nanoparticles, J. Am. Chem. Soc. 134 (2012) 6575-8. https://doi.org/10.1021/ja302846n

[200] Q. Kang, S. Liu, L. Yang, Q. Cai, C. A. Grimes, Fabrication of PbS nanoparticle-sensitized TiO_2 nanotube arrays and their photoelectrochemical properties, ACS Appl. Mater. Interfaces. 3 (2011) 746-9. https://doi.org/10.1021/am101086t

[201] M. Murdoch, G. Waterhouse, M. Nadeem, J. Metson, M. Keane, R. Howe *et al.* The effect of gold loading and particle size on photocatalytic hydrogen production from ethanol over Au/TiO_2 nanoparticles, Nat. Chem. 3 (2011) 489-92. https://doi.org/10.1038/nchem.1048

[202] S. Min, G. Lu, Sites for high efficient photocatalytic hydrogen evolution on a limited-layered MoS_2 cocatalyst confined on graphene sheets—the role of graphene, The J. Phys. Chem. C. 116 (2012) 25415-24. https://doi.org/10.1021/jp3093786

[203] X. Pan, Y-J. Xu, Defect-mediated growth of noble-metal (Ag, Pt, and Pd) nanoparticles on TiO_2 with oxygen vacancies for photocatalytic redox reactions under visible light, J. Phys. Chem. C. 117 (2013) 17996-8005. https://doi.org/10.1021/jp4064802

[204] W. Zhou, T. Li, J. Wang, Y. Qu, K. Pan, Y. Xie *et al.* Composites of small Ag clusters confined in the channels of well-ordered mesoporous anatase TiO_2 and

their excellent solar-light-driven photocatalytic performance, Nano Res. 7 (2014) 731. https://doi.org/10.1007/s12274-014-0434-y

[205] X. Wei, P.S. Nbelayim, G. Kawamura, H. Muto, A. Matsuda, Ag nanoparticle-filled TiO_2 nanotube arrays prepared by anodization and electrophoretic deposition for dye-sensitized solar cells, Nanotechnology. 28 (2017) 135207. https://doi.org/10.1088/1361-6528/aa5f11

[206] H. Sun, Q. He, P. She, S. Zeng, K. Xu, J. Li *et al.* One-pot synthesis of Au@ TiO_2 yolk-shell nanoparticles with enhanced photocatalytic activity under visible light, J. Colloid Interface Sci. 505 (2017) 884-91. https://doi.org/10.1016/j.jcis.2017.06.072

[207] M. Ye, Gong J., Lai Y., Lin C., Lin Z. High-efficiency photoelectrocatalytic hydrogen generation enabled by palladium quantum dots-sensitized TiO_2 nanotube arrays, Journal of the American Chemical Society. 134 (2012) 15720-3. https://doi.org/10.1021/ja307449z

[208] A. L. Luna, D. Dragoe, K. Wang, P. Beaunier, E. K. Kowalska, B. Ohtani *et al.* Photocatalytic Hydrogen Evolution Using Ni-Pd/TiO_2: Correlation of Light Absorption, Charge-Carrier Dynamics and Quantum Efficiency, J. Phys. Chem. C. (2017). https://doi.org/10.1021/acs.jpcc.7b01167

[209] N. Singhal, U. Kumar, Noble metal modified TiO_2: selective photoreduction of CO 2 to hydrocarbons, Mol. Cat. 439 (2017) 91-9. https://doi.org/10.1016/j.mcat.2017.06.031

[210] X. Chen, L. Liu, Huang F. Black titanium dioxide (TiO_2) nanomaterials, Chem. Soc. Rev. 44 (2015) 1861-85. https://doi.org/10.1039/C4CS00330F

[211] X. Liu, G. Zhu, X. Wang, X. Yuan, T. Lin, F. Huang, Progress in black titania: a new material for advanced photocatalysis, Adv. Energ. Mater. 6 (2016).

[212] S. Banerjee, D. D. Dionysiou, S. C. Pillai, Self-cleaning applications of TiO_2 by photo-induced hydrophilicity and photocatalysis, Appl. Catal. B-Environ. 176 (2015) 396-428. https://doi.org/10.1016/j.apcatb.2015.03.058

[213] A. Marmur. The lotus effect: superhydrophobicity and metastability, Langmuir. 20 (2004) 3517-9. https://doi.org/10.1021/la036369u

[214] Y. Zhang, H. Wu, X. Yu, F. Chen, J. Wu, Microscopic observations of the lotus leaf for explaining the outstanding mechanical properties, J. Bionic Eng. 9 (2012) 84-90. https://doi.org/10.1016/S1672-6529(11)60100-5

[215] I. Kartini, S. J. Santosa, E. Febriyanti, O. R. Nugroho, H. Yu, L. Wang, Hybrid assembly of nanosol titania and dodecylamine for superhydrophobic self-cleaning glass, J. Nanopart. Res. 16 (2014) 2514. https://doi.org/10.1007/s11051-014-2514-z

[216] X. Wang, Z. Li, J. Shi, Y. Yu. One-dimensional titanium dioxide nanomaterials: nanowires, nanorods, and nanobelts, Chem. Rev. 114 (2014) 9346-84. https://doi.org/10.1021/cr400633s

[217] R. Wang, K. Hashimoto, A. Fujishima, M. Chikuni, E. Kojima, A. Kitamura *et al.* Light-induced amphiphilic surfaces, Nature. 388 (1997) 431. https://doi.org/10.1038/41233

[218] M.O. Guler, O. Cevher, T. Cetinkaya, U. Tocoglu, H. Akbulut, High capacity TiO_2 anode materials for Li-ion batteries, Energy Convers. Management. 72 (2013) 111-6. https://doi.org/10.1016/j.enconman.2012.11.026

[219] Y. Cai, H-E. Wang, S-Z. Huang, M. F. Yuen, H-H. Cai, C. Wang *et al.* Porous TiO_2 urchins for high performance Li-ion battery electrode: facile synthesis, characterization and structural evolution, Electrochim. Acta. 210 (2016) 206-14. https://doi.org/10.1016/j.electacta.2016.05.140

[220] A.G. Dylla, G. Henkelman, K.J. Stevenson, Lithium insertion in nanostructured TiO_2 (B) architectures, Acc. Chem. Res. 46 (2013) 1104-12. https://doi.org/10.1021/ar300176y

[221] J. Zhong, Q. Wang, Y. Yu, Solvothermal preparation of Ag nanoparticles sensitized TiO_2 nanotube arrays with enhanced photoelectrochemical performance, J. Alloy Compd. 620 (2015) 168-71. https://doi.org/10.1016/j.jallcom.2014.08.205

[222] M. A. Hossain, S. Oh, S. Lim, Fabrication of dye-sensitized solar cells using a both-ends-opened TiO_2 nanotube/nanoparticle hetero-nanostructure, J. Ind. and Eng. Chem. 51 (2017) 122-8. https://doi.org/10.1016/j.jiec.2017.02.022

[223] B.-X. Lei, J-Y. Liao, R. Zhang, J. Wang, C-Y. Su, D-B. Kuang. Ordered crystalline TiO_2 nanotube arrays on transparent FTO glass for efficient dye-sensitized solar cells, J. Phys. Chem. C. 114 (2010) 15228-33. https://doi.org/10.1021/jp105780v

[224] A. G. Niaki, A. Bakhshayesh, M. Mohammadi, Double-layer dye-sensitized solar cells based on Zn-doped TiO_2 transparent and light scattering layers: improving electron injection and light scattering effect, Sol. Energy. 103 (2014) 210-22. https://doi.org/10.1016/j.solener.2014.01.041

[225] X. C. Lau, Z. Wang, S. A. Mitra, C70-carbon nanotube complex for bulk heterojunction photovoltaic cells, Appl. Phys. Lett. 103 (2013) 243108. https://doi.org/10.1063/1.4847376

[226] J. Xia, N. Masaki, K. Jiang, S. Yanagida, Sputtered Nb_2O_5 as a novel blocking layer at conducting glass/TiO_2 interfaces in dye-sensitized ionic liquid solar cells, The J. Phys. Chem. C. 111 (2007) 8092-7. https://doi.org/10.1021/jp0707384

[227] J. Yang, G. Wang, D. Wang, C. Liu, Z. Zhang, A self-cleaning coating material of TiO_2 porous microspheres/cement composite with high-efficient photocatalytic depollution performance, Mater. Lett. 200 (2017) 1-5. https://doi.org/10.1016/j.matlet.2017.04.090

[228] C. G. Granqvist. Handbook of inorganic electrochromic materials: Elsevier, 1995.

[229] M. Diasanayake, G. Senadeera, H. Sarangika, P. Ekanayake, C. Thotawattage, H. Divarathne *et al.* TiO_2 as a Low Cost, Multi Functional Material, Mater. Today: Proc. 3 (2016) S40-S7. https://doi.org/10.1016/j.matpr.2016.01.006

[230] N.N. Dinh, N.M. Quyen, D.N. Chung, M. Zikova, V-V. Truong, Highly-efficient electrochromic performance of nanostructured TiO_2 films made by doctor blade technique, Sol. Energ. Mater. Sol. C. 95 (2011) 618-23. https://doi.org/10.1016/j.solmat.2010.09.028

[231] S. Berger, A. Ghicov, Y-C. Nah, P. Schmuki, Transparent TiO_2 nanotube electrodes *via* thin layer anodization: fabrication and use in electrochromic devices, Langmuir. 25 (2009) 4841-4. https://doi.org/10.1021/la9004399

[232] Y.-C. Nah, A. Ghicov, D. Kim, S. Berger, P. Schmuki, TiO_2– WO_3 composite nanotubes by alloy anodization: growth and enhanced electrochromic properties, J. Am. Chem. Soc. 130 (2008) 16154-5. https://doi.org/10.1021/ja807106y

[233] M. Ni, M. K. Leung, D. Y. Leung, K. Sumathy, A review and recent developments in photocatalytic water-splitting using TiO_2 for hydrogen production, Renew. Sustain. Energ. Rev. 11 (2007) 401-25. https://doi.org/10.1016/j.rser.2005.01.009

[234] H. Pan. Principles on design and fabrication of nanomaterials as photocatalysts for water-splitting, Renew. Sustain. Energ. Rev. 57 (2016) 584-601. https://doi.org/10.1016/j.rser.2015.12.117

[235] A. Kudo, Y. Miseki, Heterogeneous photocatalyst materials for water splitting, Chem. Soc. Rev. 38 (2009) 253-78. https://doi.org/10.1039/B800489G

[236] R. Abe, K. Sayama, K. Domen, H. Arakawa, A new type of water splitting system composed of two different TiO_2 photocatalysts (anatase, rutile) and a IO^{3-}/I^-

shuttle redox mediator, Chem. Phys. Lett.. 344 (2001) 339-44. https://doi.org/10.1016/S0009-2614(01)00790-4

[237] W. Ren, Y. Yan, L. Zeng, Z. Shi, A. Gong, P. Schaaf *et al.* A near infrared light triggered hydrogenated black TiO_2 for cancer photothermal therapy, Adv. healthcare mater. 4 (2015) 1526-36. https://doi.org/10.1002/adhm.201500273

[238] C. Zhang, H. Yu, Y. Li, Y. Gao, Y. Zhao, W. Song *et al.* Supported Noble Metals on Hydrogen-Treated TiO2 Nanotube Arrays as Highly Ordered Electrodes for Fuel Cells, ChemSusChem. 6 (2013) 659-66. https://doi.org/10.1002/cssc.201200828

[239] Q. Wang, Z. Wen, J. Li, A hybrid supercapacitor fabricated with a carbon nanotube cathode and a TiO_2–B nanowire anode, Adv. Funct. Mater.. 16 (2006) 2141-6. https://doi.org/10.1002/adfm.200500937

[240] L. Zhang, T. Kanki, N. Sano, A. Toyoda, Development of TiO_2 photocatalyst reaction for water purification, Sep. Purif. Techn. 31 (2003) 105-10. https://doi.org/10.1016/S1383-5866(02)00157-0

[241] S. Nishimoto, A. Kubo, K. Nohara, X. Zhang, N. Taneichi, T. Okui *et al.* TiO_2-based superhydrophobic–superhydrophilic patterns: Fabrication via an ink-jet technique and application in offset printing, Appl. Surf. Sc. 255 (2009) 6221-5. https://doi.org/10.1016/j.apsusc.2009.01.084

[242] W.-D. Zhu, C-W. Wang, J-B. Chen, D-S. Li, F. Zhou, H-L. Zhang. Enhanced field emission from hydrogenated TiO_2 nanotube arrays, Nanotechnology. 23 (2012) 455204. https://doi.org/10.1088/0957-4484/23/45/455204

[243] T. Xia, C. Zhang, N. A. Oyler, X. Chen, Hydrogenated TiO_2 nanocrystals: a novel microwave absorbing material, Adv. Mater. 25 (2013) 6905-10. https://doi.org/10.1002/adma.201303088

Chapter 2

TiO_2-High Surface Area Materials Based Composite Photocatalytic Nanomaterials for Degradation of Pollutants: A Review

Molly Thomas[1] and Thillai Sivakumar Natarajan*[2]

[1]Nanomaterials Discovery Laboratory, Department of Chemistry Dr. H. S. Gour Central University Sagar, Madhya Pradesh 470 003, India.

[2]School of Chemical and Bioprocess Engineering, University College Dublin (UCD), Belfield, Dublin 4, Ireland.

Email (tsknatarajan@gmail.com; thillai.natarajan@ucd.ie)

Abstract

Heterogeneous TiO_2 semiconductor based photocatalytic process is widely recognized oxidation technology for degradation of pollutants which completely converts the pollutants into water, carbon dioxide and inorganic compounds as compared to other conventional treatment technologies. The photocatalyst properties, reaction operational parameters and the lifetime of photogenerated electron hole pairs possess considerable influence in the degradation efficiency. The degradation reaction comprises of adsorption of pollutants onto the catalyst surface followed by reaction on the catalyst surface and desorption of degraded pollutants from the surface. Thus, the adsorption of pollutants is one of the important parameters for enriching the degradation efficiency which associates with the surface area of the photocatalysts. However, in some cases surface area did not favor the degradation efficiency in which the lifetime of the charge carriers improves the degradation efficiency. Herein, we aim to review on different TiO_2-high surface area materials based composite photocatalysts, different synthesis methodologies and their degradation efficiency.

Keywords

Photocatalysis, TiO_2, High Surface Area, Activated Carbon, Carbon Nanotube, Graphene Oxide, Zeolite, Silica, Irradiation, Pollutants Degradation

Contents

1. Introduction

The rapid growth of industrialization with increase in the population discharges several types of pollutants such as dyes, pharmaceuticals, phenols, herbicides, pesticides, other organics and volatile organic compounds into the environment (water and air), causes sincere concern to our earth eco-system and decline the fresh water resources. So, the most important challenges for the scientific community are to develop efficient purification methods for controlling the environmental pollution and provide good quality water and air. Consequently, different physical, chemical, thermal and biological treatment methodologies have been comprehensively utilized; however, they produce hazardous intermediates which create secondary pollution and require expensive method for disposal and regeneration of the solid adsorbents [1-6]. To overcome these problems, one of the advanced oxidation processes (AOP) such as heterogeneous semiconductor photocatalysis has been widely recognized as an alternative route for the degradation of pollutants into less harmful products at ambient pressure and temperatures [7,8]. Specifically, the discovery of photoelectrochemical splitting of water using TiO_2 semiconductor electrode by Fujishima and Honda enriched the interest of using TiO_2

semiconductor as a photocatalyst for the degradation of pollutants in the presence of light irradiation [9]. Moreover, sustainable solar light can also be used as an irradiation source. In addition to TiO_2, various semiconductors such as ZnO, WO_3, Sb_2O_3, SnO_2, Cu_2O, CeO_2, V_2O_5 and ZrO_2, etc., have been comprehensively studied for the degradation of several types of pollutants. Among these, TiO_2 has been broadly used potential photocatalyst due to its higher degradation ability, no photo-corrosion, cost-effective, eco-friendly nature and stable under chemical and biological conditions. Subsequently, series of investigation have been performed using TiO_2 as a photocatalyst; developed different geometry of TiO_2 based slurry and immobilized photocatalytic reactors [10-20]. Moreover, the effect of different operational parameters such as TiO_2 phase composition (anatase, rutile and brookite), surface area, shape, particle size, morphology, band positions, preparation method, initial concentration of pollutants, amount of catalysts, pH of the reaction medium, intensity of light irradiation, coating thickness, co-presence of other ions, presence of quenchers and temperature have been studied on the photocatalytic degradation efficiency of TiO_2 [21-25].

The TiO_2 based photocatalytic systems possess significant issues such as poor visible light response (wide band gap energy, 3.2 eV, uses ~5% of solar spectrum), lower surface area and the rapid recombination of photogenerated electron and holes which limits their commercial applicability. To overcome these issues, TiO_2 have been doped with metals, non-metals, anions, sensitized with dye molecules, and prepared composites with other semiconductors [13, 15, 19, 26-30]. However, the low adsorption capacity of TiO_2, poor stability of modified catalyst under long run reactions, leaching of dopants during reaction, regeneration of the catalyst and reproducibility of the degradation results problems decreases the commercial expectation of TiO_2. Therefore, the supporting of TiO_2 on high surface area materials meets these criteria and enhances the degradation efficiency through higher adsorption of pollutants, absorption of visible light and efficient separation and transportation of photogenerated electron hole pairs. Furthermore, the suitable supporting materials should fulfill the criteria such as (1) It should be transparent or at least allow some UV radiation to pass through it and be chemically inert or non-reactive to the adsorbed pollutants, intermediate products and the aqueous reaction medium, (2) The high surface area supporting material should have a strong adsorption affinity towards the pollutants to be degraded, (3) The supporting material should appropriately bond to the TiO_2 either through physically or chemically without reducing its reactivity, (4) The supporting material should allow for easy recovery of photocatalyst. Based on these criteria, the high surface area materials such as carbon (activated carbon (AC), graphene (GO), carbon nanotubes (CNTs), zeolite, silica, clay, and metal organic frameworks (MOFs) are extensively used as a potential supporting materials for TiO_2

semiconductor for the degradation of various pollutants [31-39]. Among these, AC, GO, CNTs, zeolite and silica have been majorly used because of their unique properties. Therefore, in the following sections, we summarized the degradation efficiency of TiO_2/carbon (AC, GO, CNTs), TiO_2/Zeolite and TiO_2/SiO_2 based composite photocatalysts.

2. Principles of semiconductor photocatalysis

Initially, photocatalyst (example: TiO_2) produces electron hole pairs by absorbing radiation energy (photons) equals or higher than their band-gap energy (E_{bg}), excites the electrons from valence band (VB) to conduction band (CB) and leaves holes in VB. Afterwards the photogenerated electrons reduce the surface adsorbed oxygen or oxygen dissolved in water to superoxide radical anions ($O_2^{\bullet -}$). Similarly, holes oxidize the surface adsorbed H_2O or hydroxyl (-OH) groups to hydroxyl radicals ($^{\bullet}OH$). Subsequently, these reactive radicals react with the pollutant molecules, degrade them into range of lower intermediates and further mineralized into water, carbon dioxide and other inorganic anions. Moreover, holes in the VB possess strong oxidizing power, thus these could directly oxidize the pollutants adsorbed on the photocatalyst surface. In addition, electrons in the CB possess strong reducing power which indirectly degrade the pollutants using $^{\bullet}OH$ radicals formed by photocleavage of *in-situ* produced hydrogen peroxide (H_2O_2), which is produced by reaction of $O_2^{\bullet -}$ with the proton (H^+). Moreover, in the case of dye degradation, the surface adsorbed dye also excites under light irradiation and migrates an electron into the CB of the photocatalyst and reduce the surface adsorbed O_2 or dissolved O_2 to $O_2^{\bullet -}$, which reacts with H^+ to form a H_2O_2 and further photocleaved into $^{\bullet}OH$ radicals under light irradiation. The reactions can be expressed as follows.

$$\text{Photocatalyst} + h\upsilon\ (\lambda \geq \text{band gap}) \rightarrow \text{Photocatalyst}\ (h^+_{VB}) + \text{Photocatalyst}\ (e^-_{CB})$$

$$\text{Photocatalyst}\ (h^+_{VB}) + H_2O \rightarrow \text{Photocatalyst} + H^+ + \bullet OH$$

$$\text{Photocatalyst}\ (h^+_{VB}) + \text{-}OH \rightarrow \text{Photocatalyst} + \bullet OH$$

$$\text{Photocatalyst}\ (e^-_{CB}) + O_2 \rightarrow \text{Photocatalyst} + O_2\bullet -$$

$$2O_2^{\bullet -} + 2H^+ \rightarrow H_2O_2$$

$$O_2^{\bullet -} + H^+ \rightarrow HO_2^{\bullet}$$

$$HO_2^{\bullet} + HO_2^{\bullet} \rightarrow H_2O_2$$

$H_2O_2 + h\upsilon \rightarrow 2\ \bullet OH$

$HO_2^{\bullet} + H^{+} +$ Photocatalyst (e^{-}_{CB}) $\rightarrow$ Photocatalyst + $OH^{-} + \bullet OH$

Photocatalyst (h^{+}_{VB}) + pollutants $\rightarrow$ oxidation products

$\bullet OH + O_2^{\bullet} -$ + water or air pollutants $\rightarrow$ intermediate products $\rightarrow CO_2 + H_2O$ + inorganic ions

3. Methodologies for photocatalysts preparation

As described above, the degradation efficiency of TiO_2 semiconductor photocatalysts are depending upon the catalyst's structural, morphological and electronic properties and catalytic reaction operational parameters. Various synthesis methodologies such as sol-gel, hydrothermal, solvothermal, precipitation, direct oxidation, sonochemical, microwave, chemical vapor deposition, electrodeposition, electrospinning template assisted route, combustion and wet impregnation have been applied for the preparation of TiO_2-high surface area materials based composite photocatalysts [15, 31-36, 40]. Herein, we briefly described some of these synthetic methodologies.

3.1 Sol-Gel Method

The sol-gel method is the most widely used wet-chemical technique for synthesizing semiconductor (typically metal oxide) photocatalysts. Using this method morphology of the photocatalyst materials can be altered by tailoring the parameters such as pH, temperature, time, concentration of reagent, amount of water, organic additives and nature of the precursors. The sol-gel process consists of following process to prepare the desire semiconductors (a) preparation of homogeneous solution of precursors (mainly metal alkoxides and metal chlorides), (b) conversion of the homogeneous solution into a sol by hydrolyzing them with suitable reagent (generally water with or without any acid/base), (c) conversion of formed sol into gel by aging (polycondenstation reaction), (d) drying the gel, and (e) thermal treatment/sintering of the produced gel. Titanium alkoxides, titanium tetrachloride, or titanium halogenide are extensively used as TiO_2 precursors, these are hydrolyzed and condensed to form inorganic polymers composed of M-O-M or M-OH-M bonds; further condensation leads to formation of gels. These gels can be dried and thermally treated to yield the desired TiO_2 semiconductor material [41, 42] and these reactions are expressed as follows.

(1) Hydrolysis

$$Ti(OR)_4 + H_2O \rightarrow HO\text{-}Ti\text{-}(OR)_3 + R\text{-}OH$$

$$Ti(OR)_4 + 4H_2O \rightarrow HO\text{-}Ti\text{-}(OH)_3 + 4R\text{-}OH$$

(2) Condensation

$$(RO)_3\text{-}Ti\text{-}OH + HO\text{-}Ti\text{-}(OR)_3 \rightarrow (RO)_3\text{-}Ti\text{-}O\text{-}Ti\text{-}(OR)_3 + H_2O$$

$$(HO)_3\text{-}Ti\text{-}OH + HO\text{-}Ti\text{-}(OH)_3 \rightarrow (HO)_3\text{-}Ti\text{-}O\text{-}Ti\text{-}(OH)_3 + H_2O$$

3.2 Hydrothermal and solvothermal method

Hydrothermal synthesis is also a widely employed method for producing TiO_2 semiconductor in aqueous solutions under controlled temperature and pressure, and normally performed in steel autoclaves with or without Teflon liners. Moreover, the crystallography, morphology and other physical-chemical properties of semiconductor highly depend on the hydrothermal conditions such as choice of TiO_2 precursors, the temperature, the concentration of reactants, hydrothermal duration, the subsequent post washing procedure and thermal treatment/sintering temperature and duration. The solvothermal method is similar to hydrothermal technique whereas the solvent is non-aqueous, and the reaction usually performed at higher temperature than the hydrothermal method due to the organic solvents with high boiling points. In addition, the solvents influence the solubility, reactivity, coordinating ability and diffusion behavior of the reactants which better controls the particle size, crystallinity and morphology of the final semiconductors as compared to hydrothermal method [15, 43].

3.3 Sonochemical method

The application of ultrasound to chemical reaction is a potential technique for synthesizing numerous nanostructured materials and it is a sustainable alternative route for saving energy during synthesis of nanomaterials. It arises from acoustic cavitation in liquid media which involves formation, growth and implosive collapse of bubbles in a liquid and the nanostructured materials can be effectively synthesized with required particle size distribution under atmospheric pressure and low temperature [44].

3.4 Microwave method

Microwave method is activating the energy to promote the reaction and highly depending on the ability of the reaction mixture to efficiently absorb the microwave energy. Photocatalytic materials with varied sizes, shapes and morphologies can be synthesized and physicochemical properties of the materials can be improved with slight modifications of the reaction components using microwaves as heating source to attain formation or crystallization of the material. So, the microwave assisted chemical reactions depend on the choice of solvents and their ability to convert the microwave energy into heat which is determined by loss tangent. Furthermore, the microwave hydrothermal method is superior to conventional hydrothermal methods in following ways: (1) the temperature of reaction attains rapidly and maintains them throughout the reaction; (2) the kinetics of crystallization is increased by one-to-two orders of magnitude comparison with conventional hydrothermal treatment; and (3) formation of new phases [45].

4. TiO_2 -high surface area materials composite photocatalysts

Various TiO_2-high surface area materials based composite photocatalysts have been investigated for the degradation of various pollutants. Among those, we will discuss TiO_2 and activated carbon, carbon nanotubes, graphene, zeolite, and silica based composite photocatalytic materials with relevant examples.

4.1 TiO_2-activated carbon based photocatalyst

As described above photocatalytic degradation efficiency of TiO_2 has been improved by metal and non-metal doping, metal nanoparticle loading, dye sensitization and coupling with semiconductors, however, commercialization of these system are not up to the mark and yet to be comprehended. To overcome these TiO_2 has been supported on varied materials, among those supporting on carbon materials (activated carbon, fullerene, CNTs and graphene) has received significant attention, which could tune the band gap, extend the visible light absorption, provide higher active sites for enriched adsorption of pollutants, improves the charge carrier separation and transportation, decreases the recombination rate of charge carriers and enhances the production of reactive radicals. The following section described about TiO_2-activated carbon based composite photocatalytic materials for the degradation of pollutants.

Activated carbon (AC) is also known as activated charcoal or active carbon, is highly porous and offer high surface area for distributing and immobilizing the TiO_2 nanoparticle on its surface and enhances the photocatalytic degradation efficiency. AC mainly prepared from carbonaceous materials such as coal (bituminous, lignite) or plant (lignocellulosic) based materials by carbonization and activation processes. The carbonization process includes drying and heating the raw material, and then it is pyrolyzed by heating them at temperature in the range of 400 to 600 °C under oxygen-free inert gas atmosphere to prevent the combustion of the carbonizing materials. Then these carbonized materials are activated by exposing them into different physical and chemical activating agents such as CO_2, steam, air, potassium hydroxide (KOH), sodium hydroxide (NaOH), zinc chloride ($ZnCl_2$), potassium carbonate (K_2CO_3), phosphoric acid

(H_3PO_4), and sulphuric acid (H_2SO_4) etc., to create and develop porosity in the carbon materials and thereby increase their surface area and adsorptive capacity [34, 46, 47].

Matos et al., studied the synergy effect of suspended mixture of AC (Merck, 775 m^2/g, 10 mg) and TiO_2 (50 mg) in the photocatalytic degradation of phenol. Pure TiO_2 completely degraded the phenol in 6 h of light irradiation whereas it was less than 3h for AC/TiO_2 composites. They also studied the degradation of phenol, 4-chlorophenol (4-CP) and herbicide 2, 4-dichlorophenoxyacetic acid (2,4-D) using suspended mixture of TiO_2 and two different commercial activated carbon, Purocarbon (1240 m^2/g, AC_{PC}) and Merck AC (775 m^2/g, AC_M). The results demonstrated that complete degradation of phenol, 4-CP and 2,4-D were observed using TiO_2 catalyst in 6, 13 and 3h of UV irradiation, whereas in the case of TiO_2/AC_M and TiO_2/AC_{PC} complete degradation was obtained in 3, 9 and 2h and 11, 13 and 5 h of UV irradiation. Overall TiO_2/AC_M composite showed 2.5 times higher activity than the pure TiO_2, attributed to the enriched adsorption of pollutants on AC followed by a mass transfer to photoactive TiO_2 through a common interface between AC and TiO_2. The low degradation efficiency of TiO_2/AC_{PC} is ascribed to the formation of higher concentration and distribution of intermediates (hydroquinone, benzoquinone, cathecol, resorcinol, 4-chlorocathecol, 4-chlororesorcinol, 2-chlorohydroquinone, and 2, 4-DCP) were observed in AC_{PC} as compared to TiO_2/AC_M and TiO_2 photocatalyst [48, 49]. Ao et al., immobilized the TiO_2 on AC filter for the degradation of indoor air pollutants such as nitrogen oxide (NO), benzene, toluene, ethylbenzene and o-xylene (BTEX). The results confirmed that TiO_2/AC filter notably enhanced the degradation of NO and BTEX at short residence time and high humidity levels; attributed to the high adsorption capacity of AC filter [50]. Liu et al., synthesized highly reusable and easily separable TiO_2/AC composite with different weight percentage of AC by hydrothermal method for the degradation of phenol, methyl orange (MO) and Cr(VI) pollutants. The results demonstrated that 76.3, 52.8 and 74.5% of phenol, MO and Cr (VI) were degraded using TiO_2 while these were significantly enhanced to 99.5, 100 and 100% using TiO_2/5%AC composite. This is may be attributed to the higher surface area of TiO_2/5%AC (96.97 m^2/g), leads to enriched adsorption of pollutants on their surface as compared to bare TiO_2 (58.37 m^2/g) and transferred into decomposition centre, illuminated TiO_2 [51].

Velasco et al., prepared P25 TiO_2/AC composite for phenol degradation, revealed that 78% of phenol was degraded using P25 TiO_2 whereas it was significantly enhanced to 98% using P25 TiO_2/AC composite. This is attributed to the high surface area of P25 TiO_2/AC (876 m^2/g) as compared to P25 TiO_2 (53 m^2/g) which led to enriched adsorption of phenol on P25 TiO_2/AC surface (54%) under dark condition than the P25 TiO_2 (3%). Subsequently, the spontaneous transfer of adsorbed pollutants to P25 TiO_2 surface rapidly decomposed the phenol. However, they observed that AC support itself has self-photoactivity and capable of degrading the phenol (98%) under UV irradiation and outperformed the degradation efficiency of AC/TiO_2 composite (degradation efficiency is similar (98%) but degradation rate is faster in AC support). So, immobilization of P25 TiO_2 on AC support decreasing the self-photoactivity of AC (antagonist effect), may be attributed to the drop in the porosity of the composite and blockage of photoactive centers in the AC support after immobilization of P25 TiO_2 [52]. In the same way, diverse research articles were reported on the preparation of TiO_2/AC composites using commercially available AC for decomposition of various pollutants [53-72] and summarized in Table 1.

The biomass derived AC based TiO_2 composite material synthesis has also received greater attention. Doi et al., synthesized wood derived AC and TiO_2 based composites by sol-gel method for the degradation of formaldehyde (HCHO) under UV irradiation. The results demonstrated that complete degradation of HCHO was observed in 100 min of UV irradiation; attributed to the synergic effect of AC and TiO_2 crystallites [73]. Chuang et al., carbonized moso bamboo (*Phyllostachys pubescens*) at 600 °C followed by activated at 800 °C under CO_2 atmosphere, and then combined with TiO_2 nanoparticle for the degradation of benzene and toluene. Carbonized bamboo with TiO_2 (1:2) showed higher percentage degradation of benzene (72%) and toluene (71%) than the other composite (1:1) and bare TiO_2, attributed to the higher surface area which leads to enhanced adsorption of

Table 1. TiO_2-activated carbon based composite photocatalyst for degradation of pollutants.

Catalyst	SA (m^2/g)	Method of preparation	P	C (mg/L or M)	D (%)	Ref
TiO_2/AC	NA	Solution mixing; thermal treat-ment	Phenol	94 mg/L	100 (3h) 100 (6h, TiO_2)	48
TiO_2/AC	232-463	Sol-gel	HCHO	NA	~100	53
TiO_2/AC	498-533	Hydrolytic precipitation	Phenol	50 mg/L	88	54
TiO_2 /AC	526	Sol-gel	MO	5×10^{-3} M	100	55
TiO_2 /AC	704	Sol-gel	Phenol	40 mg/L	NA	56
TiO_2 /AC	271-478	Hydrolytic precipitation	MO	(1-10) $\times10^{-3}$ M	89	57
TiO_2 -AC	800.92	Dip-coating	Phenol	NA	87	58
Pitch/activated carbon/TiO_2	728-811, 125 (TiO_2)	CCl_4 solvent mixing	MB	1.0×10^{-4} M	~70	59

TiO_2 -AC	138.10	Dip-coating	RhB	30 mg/L	93.1	60
TiO_2/AC	964-1060	Dip-hydrothermal; Pyrolysis	MO	50 mg/l	96.6	61
TiO_2 /SAC	391-1427	NA	HA	20 mg/L	NA	62
meso- TiO_2 /SAC	1649	Ion-exchange method and heat-treatment process	HA	20 mg/L	NA	63
TiO_2/ASAC	580-730	Impregnation and MOCVD	IPS	TOC_0 (578×10^{-6})	92	64
TiO_2/AC	136-848 (TiO_2/CK), 99-1033 (TiO_2/CW) ,	Sol-gel	oxidation of propene	100 mg/L	60	65
AC-TiO_2	1,012 (GAC–TiO_2), 845 (PAC–TiO_2)	Wet chemical	Waste water	19 L	100	66
TiO_2/porous carbon	370-569	Electrospinning technique	Organic dyes	10 mg/L	94	67
TiO_2-activated carbon	565.8-592.0	Carbonization-activation blending method	desulphurization	3000 ± 30 mg/L	-	68
TiO_2 /AC	288.70	Sol-gel	Dye samples	25-200 mg/L	100	69
TiO_2 /AC	849.2	Impregnation	AMX, AMP, DIC, PCM	50 mg/L	100	70
TiO_2-AC	100.337, 48 (TiO_2)	Sol-gel	RR 198	100 mg/L	97	71
TiO_2 /AC	129, 96 (TiO_2)	Sol-gel	TC	50 mg/L	NA	72

*NA-Data not available, P-Pollutants, C- concentration, D-Degradation

benzene and toluene on the surface of composites [74]. Le et al., combined coconut shell derived AC with TiO_2 anatase synthesized by chemical vapor deposition method or P25 TiO_2 for the degradation of methylene blue (MB) dye. AC/TiO_2 anatase (2:1) composite is exhibited higher percentage degradation of MB dye than the AC/TiO_2 anatase (1:1), AC/P25 TiO_2 (1:1 and 1:2), bare TiO_2 anatase and P25 TiO_2. This may be attributed to the higher surface area of AC (> 750 m^2/g) which led to enhanced adsorption of MB dye, concentrates the MB around the deposited TiO_2 for high photocatalytic activity [75].

We also prepared AC from Palm tree fruit waste material (peel of Palmyra tuber) by carbonization and KOH activation, subsequently AC/anatase TiO_2 nanotube (TNT) composite was synthesized by hydrothermal method for rhodamine 6G (RhB-6G) dye degradation. The results revealed that 10%-AC/ATNT composite displayed higher degradation performance (85%) as compared to bare TNT (78%), P25 TiO_2 (60%), and TiO_2 nanoparticle (56%). The incorporation of AC enhanced the surface area of TNT which led to higher percentage adsorption of RhB-6G dye, improved the visible light response and suppressed the recombination of photogenerated electron–hole pairs resulting in enhanced photocatalytic degradation performance of AC/TNT composite [38]. Furthermore, different waste materials such as sawdust of *Tabebuia Pentaphyla* wood, pine sawdust, coconut shell, bamboo (Phyllostachys pubescens Mazel), peach stones, Jordanian olive stones, cotton balls and rick husk derived AC based TiO_2 composite photocatalytic materials have been reported for the degradation of dyes and organic compounds [76-85, Table 1].

4.2 TiO_2- carbon nanotube based photocatalyst

In addition to activated carbon, carbon nanotube (CNTs) is another allotrope of carbon discovered by Ijima in 1991 that exhibit extraordinary strength, thermal stability and unique chemical and physical properties [86]. Moreover, CNTs possess distinctive tubular structure, large specific surface area, strong adsorption capacities, and conduct electrons. CNTs can be classified into single-walled and multiwalled CNTs which can be viewed as wrapped from one layer and multi layers of carbon atoms sheet. CNTs act as an electron sink which separate the photogenerated electrons from holes and enhances the efficiency of redox reactions. Moreover, the CNTs could also acts as photosensitizer which extends the visible light range of TiO_2 and transfer the electrons to the TiO_2 and produce superoxide radicals by reaction with adsorbed O_2. Due to the high surface area, the CNTs are also used in the field of gas storage application. Therefore, the high surface area of CNTs along with the strong mechanical properties and electrically conductive nature received greater attention to use them as supporting materials for TiO_2 which enhances the surface area of TiO_2, improves the charge carriers separation and transportation; and the degradation efficiency of TiO_2 [35, 87-89].

Gao et al., synthesized MWCNT/TiO_2 nanocomposite film by surfactant wrapping sol-gel method, revealed that surface area (196.72 m^2/g) and pore volume (0.136 cm^3/g) of TiO_2 was enhanced to 230.23 m^2/g and 0.337 cm^3/g after loading with CNTs. Photo-electro-catalytic activity study demonstrated that CNT/TiO_2 composite showed higher MB dye degradation as compared to TiO_2 alone [90]. Yen et al., prepared MWCNT/TiO_2 composite with different weight ratio of MWCNT by sol-gel and hydrothermal methods and studied the influence of synthesis methodologies on their morphology and degradation efficiency. The results demonstrated that surface area of TiO_2 were different for sol-gel (146.21 m^2/g) and hydrothermal method (193.67 m^2/g).

Similarly, surface area of MWCNT/TiO_2 which is synthesized by sol-gel method increased with increase in MWCNT weight ratio, obtained higher value of 260.81 m^2/g (weight ratio:8), whereas, it was decreased with increase in MWCNT weight ratio (74.11 m^2/g, weight ratio:8) in composite synthesized by hydrothermal method. This may be due to the larger particle size of TiO_2 in composite synthesized by hydrothermal method. The degradation results revealed that TiO_2 synthesized by sol-gel and hydrothermal methods showed 20.52 and 22.58% degradation of NOx; it was increased to 32.14 (sol-gel) and 26.51% (hydrothermal) using MWCNT/TiO_2 composite. This may be ascribed to the increased surface area, crystallite size and lifetime of photogenerated charge carriers [91]. Subsequently, different nanostructure of TiO_2 was combined with CNTs; An et al., loaded MWCNT on TiO_2 sub-micrometer sized sphere by hydrothermal method for the degradation of gaseous styrene and aqueous MO dye solution. The results demonstrated that 18.9% of MWCNT loaded TiO_2 microsphere showed higher percentage degradation of gaseous styrene (54%) and MO dye (91.6%) as compared to other percentage of MWCNT loaded composite, pure TiO_2 and P25 TiO_2, respectively. This may be ascribed to the synergetic roles of CNTs in composite photocatalysts which enhanced adsorption capacity, improved the charge carrier separation and transport and suppressed the charge carrier recombination [92].

Jung et al., prepared CNT/TiO_2 nanofibers by electrospinning method for MB dye degradation; 62% and 35% of MB dye was degraded using CNT/TiO_2 nanofiber (5%) and bare TiO_2 nanofiber. This may be due to the higher surface area of composite (52 m^2/g) than the bare TiO_2 nanofiber (10 m^2/g) and the effective separation of charge carriers in the composites [93]. Lu et al., prepared fluffy-ball shaped CNT/TiO_2 nanorod composite by solvothermal method, degraded 93% of MB dye in 70 min irradiation whereas it was less than 36% for pure TiO_2 nanorod. This is attributed to the higher surface area (26.56 m^2/g) than the pure TiO_2 nanorod (2.93 m^2/g) and efficient retardation of electron-hole recombination by added CNTs [94]. Moreover, to further enhance the surface area of TiO_2, charge carrier lifetime and their degradation efficiency, TiO_2 nanotubular (TNT) structure was synthesized followed by composites were made by merging with CNTs [95, 96]. Jiang et al., developed CNTs/TNTs (x) (x-mass ratio of CNTs to P25 TiO_2) composite by hydrothermal process followed by CNTs addition during nitric acid washing process. The surface area analysis revealed that uncalcined CNTs/TNT (0.10) composite showed higher surface area (172.26 m^2/g) than the bare TNT (76.52 m^2/g). However, the calcined CNTs/TNTs composite (350, 450, 550°C) displayed low surface area as compared to uncalcined counterpart but higher than the bare TNTs, this may be because of the deterioration of tubular structure of composites.

On the other hand, calcining the CNTs/TNTs (x) at 450 °C with different mass ratio of CNTs (x-0.05, 0.10 and 0.15) resulted in increase in the surface area of composite was obtained (106.34 m^2/g (x-0.05), 112.24 m^2/g (x-0.10) and 121.98 m^2/g (x-0.15)). The CNTs/TNTs (0.10)-450 composite showed higher degradation rate of MO (94%) than the other composites, bare TNTs (86.9%), P25 TiO_2 (85%) and CNTs (28%) [97].

Table 2. TiO_2-CNT based composite photocatalyst for degradation of pollutants.

Catalyst	**SA (m^2/g)**	**Method of preparation**	**P**	**C (mg/L or M)**	**D (%)**	**Ref**
MWNT/TiO_2	114-163, 107 (TiO_2)	Sol-gel	Phenol	50 mg/L	95	98
MWNT/TiO_2	111-163	Acid-catalyzed sol-gel	Phenol	50 mg/L	97.3, 44.2 (TiO_2)	99
TiO_2 /CNTs	137-168, 127 (TNT)	Sonochemical	Acetone	400 mg/L	NA	100
TiO_2/MWCNT	NA	Deposition	Phenol	50 mg/L	90, 20 (TiO_2)	101
TiO_2/CNTs	116	Liquid phase deposition	MO	100 mg/L	NA	102
MWCNTs-TiO_2	NA	Sol-gel	DNPC	35.8 mg/L	100, 70 (TiO_2)	103
CNT-TiO_2	70-192, 107 (TiO_2)	Modified acid catalyzed sol-gel method	BZD	50 mg/L	99	104
TiO_2/SWCNT	171 50 (P25)	Sol-solvothermal	RhB	10 mg/L	80 40 (P25)	105
CNT/TiO_2 core–shell	295-381 (uncalcined), 83-105 (calcined)	Surfactant wrapping sol-gel	MB	10 mg/L	~60	106
TNT-CNT	379	Hydrothermal	BZ	250 mg/L	90 (mineralization)	95
CNTs/P-TiO_2	156-165, 62 (TiO_2)	Hydrothermal	MO	20 mg/L	NA	107
MWCNT/TiO_2	NA	Solvothermal sol-gel	MB	10 mg/L	NA	108
MWCNT/ TiO_2	190	Sol-gel; reflux	MO	50×10^{-3} M	93	109
CdS/CNT-	67.9-103.6	NA	MB	1.0×10^{-5}	98.6	110

TiO_2				M		
CNTs/TNTs	86-172, 50 (P25)	CVD	MO	20 mg/L	94	97
MWCNT/ TiO_2	120-194, 68 (P25)	Sol-gel	MO	1.0×10^{-5} M	NA	111
MWCNT/TiO_2	NA	Hydrolysis followed by calcination	MB	10 mg/L	76 30-40 (TiO_2)	112
Fluffy-ball-shaped CNT/TiO_2 nanorod	26.56	Hydrothermal	MB	10 mg/L	93	94
TiO_2 /CNT/Cu composites	110	Sol-gel	MO	15 mg/L	NA	113
TiO_2/MWCNT s	101-133, 54 (P25)	Layer by doctor-blading technique	4-CP	0.1×10^{-3} M	NA	114
TiO_2-MWCNTs	NA	Sono-chemical	MB	5 mg/L	90 ~66 (TiO_2-P25)	115
MWCNT/TNT	158-355, 196.5 (TNT)	Hydrothermal	RhB-6G	50 mg/L	89 55 (P25)	39
Mesoporous TiO_2 spheres/MWC NTs	104-190	Sol-gel and solvothermal	RhB	10 mg/L	90.7	116
TC-NCs	NA	Sol-gel	MO	20 mg/L	100 (30 min) 100 (TiO_2, 80 min)	117

*NA-Data not available, P-Pollutants, C- concentration, D-Degradation

We also synthesized MWCNT/TNT composites with different weight percentage of MWCNT by hydrothermal method, revealed that surface area and pore volume of TNT (196.5 m^2/g, 0.58 cm^3/g) was significantly enhanced (275 m^2/g and 1.03 cm^3/g, 10%MWCNT/TNT). Degradation results demonstrated that 78 % of rhodamine 6G (RhB-6G) dye was degraded using TNT, whereas it was 89% for 10%MWCNT/TNT. This may be due to the higher surface area of composite which leads to enriched adsorption of RhB-6G dye on composite (15%) as compared to pristine TNT (6%). Moreover, this may also be attributed to the enhanced separation of photogenerated electron holes pairs on composites led to generation of higher concentration of reactive radicals for RhB-6G dye degradation [39]. Similarly, different studies were reported on CNTs/TiO_2 composites for degradation of dyes, organic compounds, pharmaceuticals, herbicides and volatile organic compounds, summarized in Table 2 [98-117].

Furthermore, the photocatalytic degradation efficiency of CNTs/TiO_2 composites was improved by enhancing the lifetime of photogenerated charge carriers on their surfaces through loading of metals (Ag, Cu, Au, Pd, Ce), non-metals (N, C, P) and other semiconductor materials (Cr_2O_3, Fe_3O_4, MgO, V_2O_5, WO_3, ZnO and CdS), respectively [107, 110, 113, 118-122].

4.3 TiO_2-graphene based photocatalyst

Graphene is a carbon allotrope, the single planar sheet of sp^2-bonded carbon atoms that are densely packed in a honeycomb crystal lattice. Moreover, it possesses high specific surface area (~2600 m^2/g), thermal conductivity (~5000 W m^{-1} K^{-1}), and excellent mobility of charge carriers at room temperature (200 000 cm^2 V^{-1} s^{-1}) which makes them ideal support for semiconductor materials to enhance the adsorption capacity of the catalyst and suppressing the recombination of the charge carriers. In addition to these, it is also tuning the band gap of parent semiconductors, acting as a photosensitiser and co-catalyst to improve the reaction efficiency. Thus, significant research efforts have been employed to develop graphene oxide (GO) or reduced graphene oxide (RGO) based semiconductor composite photocatalyst for environmental and energy application [123-125]. Specifically, making of graphene/TiO_2 composite greatly extended the visible light absorption of TiO_2, enhanced the surface area of TiO_2 and absorptivity of pollutants, and reduced the band gap of TiO_2 and recombination of electron hole pairs which resulted in enriched degradation efficiency. Zhang et al., hydrothermally synthesized P25 TiO_2/GO nanocomposite for MB dye degradation under UV and visible light irradiation and compared with P25 TiO_2/CNT composite. Under UV light irradiation, ~85, ~70 and 25% of MB were degraded by P25 TiO_2/GO, P25 TiO_2/CNT and P25 TiO_2. In case of visible light irradiation, ~12% of MB dye was degraded using P25 TiO_2 and it was significantly enhanced to 65 and 55% for P25 TiO_2/GO and P25 TiO_2/CNT composite. This may be ascribed to the enhanced adsorption of pollutants, however, no significant difference in the surface area of P25TiO_2, P25TiO_2/GO and P25TiO_2/TNT (47.572, 54.226, and 51.034 m^2/g) composites. Therefore, the enhanced adsorption of pollutants is accredited to the selective adsorption of the aromatic dye on the catalyst such as π- π stacking between MB and aromatic regions of the graphene. Moreover, the enhanced degradation of composite may also be attributed to the extended light absorption and effective separation and transportation of charge carriers which generates higher concentration of reactive radicals for dye decomposition. They also synthesized same P25 TiO_2/GO composite with different percentage of GO for the degradation of benzene, a volatile organic compound. P25 TiO_2/0.5%GO exhibited higher conversion of benzene, amount of CO_2 production and mineralization rate is higher than the other percentage of GO loaded composite and bare P25 TiO_2 [36, 126]. Jiang et al., assembled the TiO_2

nanoparticle on GO sheets by in-situ liquid phase deposition followed by calcination at 200 °C for the degradation of MO and Cr(VI) reduction. The surface area of GO/TiO_2 and bare TiO_2 are 80 m^2/g and 57 m^2/g, respectively. Degradation results revealed that about 95% of MO was degraded within 9 min whereas it was 40% for bare TiO_2 in 18 min; kinetic data showed 12 fold increase in rate constant value (0.317 min^{-1}) was observed as compared to bare TiO_2 (0.0253 min^{-1}). Likewise, for Cr(VI) reduction, GO/TiO_2 composite showed 3.9 and 5.4 times higher activity than the bare TiO_2 (0.0174 min^{-1}) and P25 (0.0127 min^{-1}). The enhanced degradation efficiency is may be ascribed to the enhanced adsorption capability of GO/TiO_2, efficient separation of charge carriers and produces high concentration of reactive radical species [127]. Moreover, Perera, et al., hydrothermally prepared the TiO_2 nanotube (TNT)/GO composite for malachite green oxalate dye degradation; revealed that 10%GO/TNT showed higher degradation efficiency (~89%) than the other percentage of GO loaded TNT composite and bare TNT. The enhanced degradation efficiency is because of the enhanced adsorption of pollutants, maximum interface contact between the TNT and GO surface and charge transfer reactions to create radical species [128]. Furthermore, photocatalytic degradation efficiency of TiO_2/GO composites was further improved by loading of metal and non-metals and other semiconductors. Gu et al., prepared visible light active N doped TiO_2/RGO and N, V co-doped TiO_2/RGO composite by hydrothermal method and studied their catalytic efficiency by degradation of acid orange 7 (AO7) azo dye. Results demonstrated that 64.9% and 74.6% of AO7 were decomposed by N–TiO_2/RGO and N, V–TiO_2/RGO composites whereas bare TiO_2 showed negligible effect on the photodegradation of AO7. The enhanced photocatalytic degradation efficiency of the nanocomposites under visible light irradiation is attributed to the enriched adsorption of pollutants, light absorption intensity and reduced recombination of electrons and holes [129]. Sun et al., prepared GO/ rutile TiO_2 nanorod nanocomposites by hydrothermal method followed by deposited the Cu_2O on the surface of GO/ rutile TiO_2 nanorod composite through chemical bath deposition process. Ternary Cu_2O/GO/rutile TiO_2 nanorod composite exhibited 2.8 and 1.5 times higher activity than the bare TiO_2 and GO/ rutile TiO_2 nanorod in MB dye degradation [130]. Recently, Appavu and Thiripuranthagan synthesized N, S co-doped TiO_2(NST)/GO nanocomposite with different weight percentage of GO by hydrothermal method for the decolorization of congo red (CR), MB and reactive orange 16 (RO16) dyes under visible light irradiation. The results demonstrated that NST/5%GO displayed the highest photocatalytic activity in the decolorization of the dyes (93% for CR, 95% for MB and 96% for RO16) than the other percentage of GO loaded composites and NST. This may be attributed to the higher surface area of NST/5%GO composite (142 m^2/g) than the other composite (NST/2.5%GO-125 m^2/g, NST/7.5%GO-136 m^2/g, NST/10%GO-129 m^2/g) and bare

NST (118 m^2/g) which enhances the percentage adsorption of dyes on NST/x%GO (19-28%) than the bare NST (5-10%) [131]. Likewise, various research works have been reported [132-145, Table 3] and still the research efforts are going on the designing of GO/TiO_2 based nanocomposite for the degradation of pollutants.

Table 3. TiO_2-graphene based composite photocatalyst for degradation of pollutants.

Catalyst	SA (m^2/g)	Method of preparation	P	C mg/L or M)	D (%)	Ref
P25-Graphene	51 (P25-GR) 47.57 (P25)	Hydrothermal	MB	10 mg/L	85, 20 (P25)	36
graphene/TiO_2 nanocrystals	190, 58 (P25)	Hydrolysis followed by hydrothermal	RhB	0.03×10^{-3} M	NA	132
TiO_2/graphene	175.7	Liquid phase deposition	MO	25mg/L	85, 40 (P25)	133
GO/ TiO_2	80, 57 (Neat TiO_2)	Liquid phase deposition	MO	10 mg/L	95, 40 (Neat TiO_2)	127
TiO_2-graphene	132-407	Thermal hydrolysis	But-ane	NA	NA	134
TO/GS	176, 90 (TiO_2)	Template-free self- assembly	RhB	4×10^{-5} M	100 (30 min), 75 (70 min, TiO_2)	135
hGO-TNT	NA	Hydrothermal	MG	13.1 mg/L	80	128
TiO_2 - RGO	99-169, 81.8(TiO_2)	Hydrothermal	RhB	2×10^{-5} M	98.8	136
graphene-TiO_2 nanowire	NA	Hydrothermal	MB	2 mg/L	NA	137
TiO_2/Graphene	NA	Solvothermal	HC-HCO	200mg/L	NA	138
N,V–TiO_2–RGO	NA	Hydrothermal	AO7	10 mg/L	74.6	129
Graphitic carbon-TiO_2	11-17, 26 (TiO_2)	Hydrothermal	NB	50 mg/L	96, 55 (TiO_2)	139
Cu_2O/graphene/ TiO_2 nanorod	59, 64 (TiO_2)	Chemical bath deposition	MB	5 mg/L	NA	130
Mesoporous N-RGO/TiO_2	84.3-100.2, 91 (N-TiO_2)	Hydrothermal	CR MB	3.12×10^{-5} M	CR 96 (72; N-TiO_2), MB 88 (55; N-TiO_2)	140

Nano TiO_2 - GO	55.66, 40.46 (NT)	Solvothermal	MB	5 mg/L	93	141
Porous TiO_2/graphene	NA	Template assisted	MB	10 mg/L	NA	142
TiO_2/rGO	129-142	Hydrothermal	MB CR RO-16	3.12×1 0^{-5} M	94 (MB; 120 min), 93 (CR; 50 min), 96 (RO16; 120 min)	143
TiO_2/RGO-X (X=C,S,T)	247 (TiO_2/ RGO-C), 277 (TiO_2/ RGO-S), 213 (TiO_2/ RGO-T)	Surfactant-assisted hydrothermal	MB	12 mg/L	97.5 (TiO_2 /RGO-C), 90 (TiO_2 /RGO-S), 90 (TiO_2 /RGO-T)	131
TNR-rGO	176-187 (a-TNR+rGO), 381 (a-TNR) 101-118 (TNR+rGO), 102 (TNR)	Hydrothermal	BPA	10 mg/L	NA	144
TPt-rGO	NA	Chemical-thermal method	AMR, SY, TAZ	2×10^{-5}M	99.56 (AMR), 99.15 (SY), 96.23 (TAZ)	145

*NA-Data not available, P-Pollutants, C- concentration, D-Degradation

4.4 TiO_2-Zeolites based photocatalyst

Zeolites are aluminosilicates with well-defined interconnected channels or cavities possess large surface area, internal pore volumes, unique uniform pores and channel sizes (3-8 Å). Moreover, zeolites exhibit several other characteristics which make them suitable hosts for semiconductor materials, such as (1) high thermal stability, chemical and thermal inertness, (2) transparency to UV-vis radiation above 240 nm which allows the light penetration and reach the guest material surface in zeolite framework, (3) actively participate in electron transfer processes, either as electron acceptor or electron donor, (4) shows high absorptivity for organic compounds in solution and (5) Modulation of the polarizing strength of the zeolite interior by changing the nature of internal charge balancing cations and the size of the channels which modifies the electronic states and conformational mobility of the guests within zeolites [146-148]. Xu and Langford, studied the photocatalytic efficiency of TiO_2 supported ZSM5 and zeolite A with different TiO_2 percentage by 4-chlorophenol and acetophenone degradation; 5%TiO_2/ZSM5 exhibited higher degradation efficiency than the TiO_2/zeolite A and bare TiO_2. This may be attributed to the higher surface area of 5%TiO_2/ZSM5 (392 m^2/g) as compared to 5.1%TiO_2/zeolite A (177 m^2/g) and bare TiO_2 (35 m^2/g) which enhances the concentration of adsorbed pollutants molecules and then adsorbed molecules were efficiently reached to light activated sites [149]. Zhu et al., studied the degradation performance of TiO_2 supported zeolites (13-x, Na-Y and 4A) with different TiO_2 loading by direct fast scarlet 4BS degradation. The adsorption study revealed that no detectable adsorption of direct fast scarlet 4BS dye on all catalysts, attributed that adsorption capacity is may not be the factor for 4BS dye degradation. The maximum diameter of direct fast scarlet 4BS dye is > 1 nm due to the naphthalene ring with the large substituents of -SO_3Na and -NHCONH- which is higher than the aperture pore size of 13X, Y and 4A type zeolites (< 1 nm). So, the big direct fast scarlet 4BS dye molecules cannot access the inner surface of zeolites and most of them adsorbed on the outer surface of zeolites and TiO_2, however, at the same time, the large dye molecules inhibit the surface adsorption of dye molecules because of the steric hindrance between the dye molecules. The degradation results reveal that 13X/TiO_2 type catalysts showed higher activity than the Y and 4A/TiO_2 zeolites, Y exhibit higher activity than the 4A/TiO_2 zeolites. The higher activity of 13X/TiO_2 type catalysts is due to the high Si/Al ratio, higher surface area and bigger aperture size of zeolite which favors the adsorption of more water molecules, hydroxyl ions and dye molecules. They also validated the catalytic efficiency by degradation of another dye, acid red 3B dye [150]. Chen et al., supported TiO_2 on ZSM-5 zeolite and studied the influence of pH on the degradation of phenol and benzene. The photocatalytic activity of TiO_2/ZSM-5 is higher in acidic condition; however the degradation efficiency of TiO_2/ZSM-5 and P25 TiO_2 is almost similar [151]. Tayade et al., incorporated TiO_2 onto NaY and HY zeolites for MB dye degradation, results revealed that 1% TiO_2 loaded NaY and HY zeolites exhibited higher photocatalytic activity (100 %, < 2h time) than the bare TiO_2 (82%, 4h). This is due to the higher surface area of TiO_2 coated zeolites (698 m^2/g-1%TiO_2/NaY, 740 m^2/g-1%TiO_2/HY) as compared to bare TiO_2 (124 m^2/g) which leads to higher percentage adsorption of MB dye on TiO_2 coated NaY and HY zeolites (23-25%-TiO_2/NaY, 32-39%-TiO_2/HY). Moreover, the fine dispersion of TiO_2 in cavities of the zeolite and separation of photogenerated electron hole pairs are also responsible for the higher photocatalytic activity of TiO_2 coated NaY and HY zeolites. They also synthesized Ag ion exchanged TiO_2 coated/NaY zeolite photocatalyst by sol-gel and ion-exchange methods for degradation of dyes (MB and malachite green) and organic compounds (nitrobenzene and acetophenone) [152, 153]. Wang et al., supported TiO_2 on Na-Y zeolite with different contents of TiO_2 by impregnation method and studied their catalytic activity by C.I. Basic Violet 10 dye degradation. The surface area of zeolite (750 m^2/g) was decreased with increasing the TiO_2 and reached to (300 m^2/g) for 50%TiO_2/Na-Y zeolite. All the catalysts showed no obvious adsorption of dye, because the diameter of dye (> 1 nm) is larger than the pore size of zeolite (< 1 nm) which did not allow the dye to access the inner surface of the zeolite. Moreover, the large dye molecules may have steric hindrance with another dye molecule and decrease the adsorption of dye molecules on the surface of the catalyst. So, the surface area of catalysts and followed by dye adsorption has insignificant in the photocatalytic degradation activity of the catalysts. Degradation studies reveal that 20%TiO_2/NA-Y zeolite exhibited higher activity than the other TiO_2 content loaded zeolite, however all the composite showed lower degradation performance than the P25 TiO_2 [154]. Zhang et al., encapsulated the TiO_2 into cavities of FAU-type zeolites by ion-exchange method and examined their degradation efficiency by degradation of charged substrates such as rhodamine B (cationic), reactive blue 19 (anionic), aniline (anionic), benzoate (anionic), tetramethylammonium (cationic) and dichloroacteate (anionic) in water. The negative charge of the zeolite framework promotes selective photocatalytic degradation of charged substrates. Therefore, degradation rate of composite photocatalysts is higher for cationic substrates as compared to anionic substrates [155]. Zhang et al., synthesized ternary TiO_2/MoS_2@zeolite composite photocatalysts by ultrasonic-hydrothermal synthesis method for the degradation of MO dye. The surface area of TiO_2/MoS_2@zeolite, TiO_2@zeolite, and zeolite are 139.0 m^2/g, 261.0 m^2/g and 18.5 m^2/g, respectively. Degradation results revealed that 95%, 77%, 55% and 1.5% of MO was degraded using TiO_2/MoS_2@zeolite, TiO_2@zeolite, P25 TiO_2 and zeolite. Though the surface area of ternary composite is lower, but it exhibited higher degradation efficiency and described that surface area not having major influence on the degradation efficiency. The enhanced degradation efficiency of ternary composite is attributed to the efficient separation of electron hole pairs under visible light and generates higher concentration of reactive radical species [156]. Recently, Setthaya et al., synthesized zeolites (faujasite and zeolite P1) and TiO_2-zeolite composite by solvothermal method using rice husk ash and metakaolin as sources of silica and alumina and titanium tetrabutoxide as the titanium source. TiO_2-zeolite composite is also synthesized by impregnation method. The surface area of zeolite, TiO_2-zeolite (solvothermal) and TiO_2-zeolite (impregnation) composite are 487.1 m^2/g, 452.1 m^2/g and 437.6 m^2/g. The degradation of MB dye study reveals that TiO_2-zeolite composite synthesized by both methods degraded 97.32% (solvothermal) and 99.43% (impregnation) of the MB as compared to pristine TiO_2 (82%). This is attributed to the enriched adsorption of MB dye (95.2%, 92.0%) than the bare TiO_2; the molecular diameter of MB dye (0.77 nm) is lower than the average pore size of synthesized zeolite (1.36 nm), so the MB dye could be stored in the large internal surface of zeolite.

Moreover, composite synthesized by impregnation method showed slightly higher degradation efficiency than the solvothermal method. This is due to the incorporation of TiO_2 on the external surface of zeolite than the solvothermal method [157]. Similarly, various research efforts have been performed [Table 4, 158-175] and still the research is ongoing to prepare TiO_2/zeolite based composite photocatalytic materials.

Table 4. TiO_2-Zeolite based composite photocatalyst for degradation of pollutants.

Catalyst	SA (m^2/g)	Method of preparation	P	C (mg/L or M)	D (%)	Ref
TiO_2 supported on ZSM5, zeolite A, silica, alumina	177 (zeolite A), 157-178 (alumina), 308-392 (ZSM5), 243-289 (silica), 35 (TiO_2 calcined)	Acid-catalyzed sol-gel	4-CP ACP	50 mg/L	NA	149
TiO_2/zeolite	557-625, 50 (TiO_2)	Ion-exchange	2,4-D	200 mg/L	100	158
Pt-TiO_2 /zeolites	31-118	Sol-gel	MO	20 mg/l	87.8	159
TiO_2/zeolite	NA	Sol-gel	TLE	42.5 mg/L	95, 55 (P25)	160
RH/MCM-41	590-760, 53 (TiO_2)	Sol-gel	TMA	1×10^{-3} M	100	161
Hβ zeolite- TiO_2	425-545, 50 (TiO_2)	Ion-exchange	PRX	200 mg/L	NA	162
TiO_2-SiMgOX	109-136	NA	SO_2	15 mg/L	NA	163
TiO_2/HZSM-5	115.7-242.9, 84 (TiO_2)	Sol-gel	MO	10 mg/L	99.5	164
Ag/ TiO_2-zeolite	59.45	Microwave-assisted hydrothermal	HA	50 mg/L	32 (minerali zation)	165
TiO_2/zeolite	103-267	Sol-gel	RhB	2 mg/L	NA	166
Cr/TiO_2/zeolite	92-205	Sol-gel	MO	10 mg/L	41.73	167
zeolite-supported TiO_2	551-751	Impregnation	2-POH	2.6×10^{-3} M	NA	168
TiO_2 /HZSM-11 zeolite	293-309	Hydrotherma l crystallizatio n	DDV P	1×10^{-4} M	93	169
zeolite/TiO_2	16	Sol-gel	HA	10 mg/L	80	170
TiO_2/zeolite	NA	Sol–gel; photo-reductive deposition	AMX	30 mg/L	88, 95 (P25)	171

TiO_2-zeolite	38-324	Sol-gel method	RB5	NA	NA	172
TiO_2/MoS_2@zeolite	139	Ultrasonic-hydrothermal	MO	20 mg/L	95, 55 (P25)	156
TiO_2/Zeolite	10-13	Impregnation	MB	20 mg/L	80	173
TiO_2/CLI	194-210 39 (TiO_2 NPs)	Hydrolysis	TPA	20 mg/L	94, ~80 (TiO_2 NPs)	174
ZTi-Y (Y= I - Impregnation; M-mechanical mixing)	338 (ZTi-I) 253(ZTi-M) 152 (TiO_2)	Impregnation; mechanical mixing	HCHO, TCE	15-25 mg/L	100, 70 (HCHO, TCE) 75, 62 (TiO_2)	175

*NA-Data not available, P-Pollutants, C- concentration, D-Degradation

4.5 TiO_2-SiO_2 based photocatalysts

Like zeolite material, the discovery of mesoporous silica (SiO_2) has broadened the application of SiO_2 as a support for catalytic materials and enhances their adsorption capacity and catalytic efficiency. The well-ordered hexagonal SiO_2 possess many advantages such as high surface area, large pore volume, pore size, thick pore wall, and greater stability. Moreover, the pore size and the thickness of the SiO_2 wall can be altered by varying the reaction parameters; therefore, these properties makes them as an ideal catalytic support and allow easier diffusion of molecules before and after the reactions [176-180]. Torimoto et al., evaluated the photocatalytic activity of TiO_2 loaded zeolite (mordenite), SiO_2, and activated carbon (AC) composite by degradation of aqueous solution of propyzamide. The adsorption study reveal that TiO_2 loaded mordenite, SiO_2, and AC exhibited higher adsorption capacity than the pure TiO_2 and the order is 70 wt % TiO_2/AC > 70 wt % TiO_2/SiO_2 > 70 wt % TiO_2/mordenite > TiO_2; however, the degradation of propyzamide is completely opposite to the adsorption order. Nevertheless, the mineralization study (amount of CO_2 production) demonstrated that TiO_2 loaded mordenite, SiO_2, and AC exhibited higher mineralization rate than the TiO_2 and consistent with the adsorption strength order (70 wt % TiO_2/AC > 70 wt % TiO_2/SiO_2 > 70 wt % TiO_2/mordenite > TiO_2), which proved that adsorption is most important factor and significantly improved the rate of mineralization of the propyzamide [181]. Dong et al., synthesized 2-D hexagonal mesoporous anatase TiO_2-SiO_2 nanocomposites with various Ti/Si ratios (90/10, 80/20, 70/30 and 60/40), studied their synchronous role of coupled adsorption and photocatalytic oxidation. The results demonstrated that all composites exhibited higher degradation performance; among those, $80TiO_2/20SiO_2$

composite calcined at 800 °C exhibited higher degradation values (0.231 min^{-1}) compared with TiO_2-800 (0.00587 min^{-1}) and P25 TiO_2 (0.0671 min^{-1}) photocatalyst in RhB dye degradation. This is attributed to the increase of the size, crystallinity of anatase nanocrystals, the decrease of band gaps and the fixed Ti/Si ratio of 80/20. However, decreasing the Ti/Si ratio to 70/30 and 60/40 resulted in enhancement in the surface area and adsorption percentage of RhB dye, but the photocatalytic activity declines due to the decrease of TiO_2 nanocrystals and increased band gaps. Therefore, the Ti/Si ratio fixed at 80/20 gave better synchronous role of coupled adsorption and photocatalytic oxidation. They also used these same ordered 2-D hexagonal mesoporous anatase TiO_2/SiO_2 nanocomposites for the degradation of cationic dyes (MB, safranin O, crystal violet, brilliant green, basic fuchsin and RhB 6G), anionic dyes (acid fuchsin, orange II, reactive brilliant red X3B and acid red 1) and microcystin-LR [182, 183]. Pal et al., decorated SiO_2 nanosphere with TiO_2 nanostructure having exposed {001} and {101} facets by a simple wet chemical approach for degradation of MB dye under visible light irradiation. The results demonstrated that SiO_2/TiO_2 (1:3)-500 °C composite decomposed ~94% of MB within 40 min whereas in the case of commercial P25 TiO_2 ~70% of dye was degraded. The enhanced photocatalytic activity can be attributed to the three characteristics of the SiO_2 nanospheres: (1) efficient dispersion of TiO_2 and increase in the surface area of TiO_2, (2) facilitate and support the formation of high energy anatase {0 01} surfaces along with the thermodynamically stable {101} surfaces of TiO_2, (3) improve the adsorption and transferring the organic molecules to active sites [184]. Moreover, the photocatalytic activity of TiO_2/SiO_2 composite was further enhanced by suppressing the recombination of photogenerated electron hole pairs by loading of different metal, non-metals and coupling with other semiconductors. Chen et al., synthesized TiO_2/SiO_2 composite (the molar ratio of Si-Ti = 0.20) with different percentage of carbon and silver doping by solvothermal method followed by calcination for RhB dye degradation. The carbon doped TiO_2/SiO_2 with C-Ti molar ratio of 2 showed enhanced RhB dye degradation than the pure TiO_2, this may be due to that the added carbon species distributed in the surface layer of TiO_2/SiO_2 which may promotes the RhB dye degradation efficiency. Moreover, the Ag doping further enhanced the photocatalytic activity of $C_{2.0}$-$TS_{0.20}$, reached maximum activity at Ag-Ti molar ratio of 0.005, and exhibited higher degradation performance than the pure TiO_2 and mono-doped TiO_2. This may be attributed to the formation of Schottky barrier by Ag, thereby photogenerated electron in TiO_2 might transfer into Ag nanoparticles that act as an electron trap, promoting charge separation. Moreover, the $Ag_{0.005}$-$C_{2.0}$-$TS_{0.20}$ possess higher surface area (290 m^2/g) than the TiO_2 (102.8 m^2/g), $C_{2.0}$-TiO_2 (104.6 m^2/g) and $Ag_{0.005}$-TiO_2 (112.3 m^2/g) which enhances the concentration of RhB dye around the TiO_2/SiO_2 composites and promotes the degradation efficiency [185]. Mahesh et al., synthesized

SiO_2 sphere and SiO_2/TiO_2 composite sphere by sol-gel method, deposited the Ag on SiO_2/TiO_2 composite sphere by photodeposition method and studied their activity by photodegradation of acid black 1 (AB 1) dye. The results demonstrated that 100% degradation was observed for 1, 2 and 3wt% Ag-SiO_2/TiO_2 composite within 25, 25 and 20 min whereas it was 60 min for SiO_2/TiO_2 composite. The localized surface plasmon resonance (LSPR) of loaded Ag nanoparticles in the near UV region boosted the electrical field at the near UV region which could increase the UV light absorption by TiO_2 and facilitating the generation of charge carriers in the TiO_2. Moreover, the Ag nanoparticles helped suppress the recombination of the charge carriers and thus improved the photoelectric conversion efficiency [186]. Jiang et al., prepared cerium doped SiO_2/P25 TiO_2 composites (CSP) with different contents of cerium (1, 3, 5 and 7 mM) by hydrothermal method and investigated their visible light photocatalytic catalytic activity by MB and reactive red 4 (RR4) dyes degradation. The results reveal that degradation efficiency was increased with increase in the cerium percentage and reached maximum for CSP-5 photocatalyst which showed best photogradation rate for MB (91.8%) and RR4 (90.2%) dyes compared with the pure P25 TiO_2 and SiO_2/P25. This is attributed to the higher surface area (226.50 m^2/g) and pore diameter (0.412 nm) of CSP-5 compared with P25 TiO_2 (84.67 m^2/g, 0.201 nm) and SiO_2/P25 TiO_2 (147.59 m^2/g, 0.241 nm), which enhances the dispersion of TiO_2 in the composite and adsorption of pollutants. Moreover, the presence of the cerium could capture the charge carriers, while the Ce 4f levels improve the formation of electron–hole pairs, which can facilitate the Ce^{4+}/Ce^{3+} pairs to interact directly with dyes [187]. Likewise, different literature were available on the TiO_2/SiO_2 based composite photocatalytic materials for the degradation of pollutants [Table 5, 188-203] and still the research is continuing on the development of efficient TiO_2/SiO_2 based composite photocatalysts.

Table 5. TiO_2-SiO_2 based composite photocatalyst for degradation of pollutants.

Catalyst	**SA (m^2/g)**	**Method of preparation**	**P**	**C (mg/L or M)**	**D (%)**	**Ref**
TiO_2 /SiO_2	92-264	Sol-gel	MO	0.03×10^{-3} M	74	188
Anatase (TiO_2)/silica	58-124	Hydrothermal; Sol-gel	MB	22 mg/L	NA	189
Nanocrystalline TiO_2 /SiO_2	123-165	Sol-gel	Azo dyes	20 mg/L	NA	190

mesoporous anatase TiO_2-SiO_2	73-236, 50 (P25)	Organic-solvent evaporation induced co-assembly	RhB	1×10^{-5} M	NA	182
TiO_2 /SiO_2	707.59, (uncalcined), 142.38-13.72	Sol-gel	CR	5 mg/L	98	191
TiSBA-15	455-918	Sol-gel	MB	(0.05-0.5) $\times10^{-3}$ M	NA	192
Ag modified hollow SiO_2/TiO_2 hybrid	150.7-267	Hydrothermal	RhB	10 mg/L	NA	193
TiO_2–Pt@SiO_2	NA	Solvothermal and reverse micro-emulsion	RhB	0.13×10^{-3} M	~100	194
Cu-TiO_2-SiO_2	108.9, 46.5 (TiO_2)	Sol-gel	RhB	5 mg/L	96	195
Ag-C/TS	306.0, 102 (TiO_2)	Solvothermal	RhB	30 mg/L	NA	185
mesoporous TiO_2–SiO_2	181.8	Modified solvent EISA method	Organic dyes	1.0-4.0×10^{-5} M	NA	183
Silica/titania	403-429	Template assisted	MO	50×10^{-3} M	NA	196
Ag@SiO_2@TiO_2	NA	Sol-gel	MB	2.0×10^{-5} M	NA	197
Ag-SiO_2@TiO_2	NA	Sol-gel	AB 1	10 mg/L	100	186
SiO_2-TiO_2	482.7-914	Sol-gel	RhB	15×10^{-6} M	95	198
Ni-SiO_2/TiO_2	NA	Hydrolysis and condensation	AB 1 dye	5 mg/L	100	199
TiO_2 /SiO_2	327.9,	Template-	RhB	10 mg/L	95	200

	45.5 (TiO_2)	hydrothermal				
SiO_2–TiO_2	NA	Wet chemical	MB	10×10^{-6} M	94 70 (P25)	201
TiO_2/Brij-35/Si-NPs	95, 62 (TiO_2/Brij-35)	Template assisted sol-gel	MO	20 mg/L	80	202
TiO_2/silica	-	Immobilization	RB19	25 mg/L	97.9	203

*NA-Data not available, P-Pollutants, C- concentration, D-Degradation $: uncalcined, &:calcined

5. Conclusion

The decomposition of pollutants using TiO_2 semiconductor as a photocatalyst under light irradiation has been widely accepted technique and different modification on TiO_2 have been performed over many decades. Specifically, the supporting of TiO_2 on the high surface area materials considerably improved the degradation efficiency via enhancing the adsorption capacity, lifetime of charge carriers and extends the visible light adsorption of TiO_2, respectively. Among those, the literature on TiO_2 supported on AC, zeolite, GO, CNTs and SiO_2 photocatalytic materials and their influence on the degradation efficiency is thoroughly discussed. However, in some cases, the surface area of composite photocatalysts not influencing the degradation efficiency but their charge carrier separation and transport enhanced the degradation efficiency. However, the in-depth study on charge carrier dynamics in the composites and the insignificant of their surface area are not studied. Moreover, the visible light activity of TiO_2-high surface area materials based composites have been investigated mainly under artificial visible light irradiation, thus, the utilization of the direct sunlight is lacking behind the commercial expectation which is one of the challenges standing ahead of the scientific community. In addition, the reusability of composites and the study on degradation of real wastewater are not up to the standards required for commercial application. Therefore, it is understood that future research should be concentrated on the in-depth understanding of photogenerated charge carriers dynamics, positive and negative impact of surface area on the degradation efficiency. Finally, the design of composite photocatalytic materials for real wastewater treatment under direct sunlight irradiation and their reusability in real scale application should be performed.

Abbreviations

2,4-D	2,4-dichlorophenoxyacetic acid
4-CP	4-chlorophenol
AB 1	Acid Black 1
AC or ASAC	Activated carbon
ACP	Acetophenone
Ag-C/TS	Ag-Carbon doped titania-silica
AMX	Amoxicillin
AMP	Ampicillin
AMR	Amaranth
AO7	Acid orange 7
AOP	Advanced oxidation processes
a-TNR	Amorphous TiO_2 layer covered crystalline anatase TiO_2 core
BPA	Bisphenol A
BTEX	Benzene, toluene, ethylbenzene and o-xylene
BZ	Benzene
BZD	Benzene derivatives
CB	Conduction band
CK	activated carbon Kureha
CLI	Clinoptilolite
CNT	Carbon nanotube
CR	Congo red
CVD	Chemical Vapor Deposition
CW	Westvaco activated carbon
DDVP	Dichlorvos
DIC	Diclofenace
DNPC	2,6-dinitro-p-cresol
EISA	Evaporation-induced self-assembly
FAU	Faujasite
GAC	Granular activated carbon
HA	Humic acid
HCHO	Formaldehyde
hGO-TNT	Reduced graphene/TiO_2 nanotube
IPS	Industrial phosphoric acid solution
MB	Methylene blue
MCM-41	Mobil Composition of Matter No. 41
MG	Malachite green

MO	Methyl orange
MOCVD	Metal-Organic Chemical Vapour Deposition
MOFs	Metal-organic frameworks
MWCNTs or MWNT	Multi-walled carbon nanotubes
NO	nitrogen oxide
NST	N, S co-doped TiO_2
PCM	Paracetamol
PAC	Powdered activated carbon
2-POH	2-propanol
PRX	Propoxur
RB19	Reactive Blue 19
RGO	Reduced graphene oxide
RhB	Rhodamine B
RO16	Reactive Orange 16
RR 198	Reactive red 198
SAC	Spherical activated carbon
SBA-15	Santa Barbara Amorphous-15
SY	Sunset yellow
TC	Tetracycline
TC NCs	TiO_2-MWCNTs nanocomposites
TCE	Trichloroethylene
TLE	Toluene
TiSBA-15	Titanium-substituted mesoporous silica
TMA	Tetramethylammonium
TNR	Fully crystalline anatase TiO_2
TNTs	Titania nanotube
TO/GS	TiO_2 nanospheres on graphene sheets
TPA	Terephthalic acid
TPt-rGO	TiO_2-Pt/reduced graphene oxide
TAZ	Tartrazine
VB	Valence band
ZSM5	Zeolite Socony Mobil–5

References

[1] M.A. Montgomery, M. Elimelech, Water and sanitation in developing countries: Including health in the equation, Environ. Sci. Technol. 41 (2007) 17-24. https://doi.org/10.1021/es072435t

[2] http://www.un.org/waterforlifedecade/background.shtml

[3] E. Khan, M. Li, C.P. Huang, Hazardous waste treatment technologies, Water Environ. Res. 79 (2007) 1858-1902. https://doi.org/10.2175/106143007X218601

[4] V.K. Gupta, Suhas, Application of low-cost adsorbents for dye removal – A review, J. Environ. Manage. 90 (2009) 2313-2342. https://doi.org/10.1016/j.jenvman.2008.11.017

[5] I. Ali, New Generation Adsorbents for Water Treatment, Chem. Rev. 112 (2012) 5073-5091. https://doi.org/10.1021/cr300133d

[6] A. Ahmad, S.H. Mohd-Setapar, C.S. Chuong, A. Khatoon, W.A. Wani, R. Kumar, M. Rafatullah, Recent advances in new generation dye removal technologies: novel search for approaches to reprocess wastewater, RSC Adv. 5 (2015) 30801-30818. https://doi.org/10.1039/C4RA16959J

[7] W.H. Glaze, J.W. Kang, D.H. Chapin, The chemistry of water treatment processes involving ozone, hydrogen peroxide and ultraviolet radiation, Ozone Sci. Eng. 9 (1987) 335-352. https://doi.org/10.1080/01919518708552148

[8] R. Andreozzi, V. Caprio, A. Insola, R. Marotta, Advanced oxidation processes (AOP) for water purification and recovery, Catal. Today 53(1999) 51-59. https://doi.org/10.1016/S0920-5861(99)00102-9

[9] A. Fujishima, K. Honda, Electrochemical photolysis of water at a semiconductor electrode, Nature 238(1972) 37-38. https://doi.org/10.1038/238037a0

[10] J.H. Carey, J. Lawrence, H.M. Tosine, Photodechlorination of PCB's in the presence of titanium dioxide in aqueous suspensions. Bull. Environ. Contam. Toxicol. 16 (1976) 697-701. https://doi.org/10.1007/BF01685575

[11] S.N. Frank, A.J. Bard, Heterogeneous photocatalytic oxidation of cyanide ion in aqueous solutions at titanium dioxide powder. J. Am. Chem. Soc. 99 (1977) 303-304. https://doi.org/10.1021/ja00443a081

[12] J.-M. Herrmann, C. Guillard, P. Pichat, Heterogeneous photocatalysis: an emerging technology for water treatment, Catal. Today 17 (1993) 7-20. https://doi.org/10.1016/0920-5861(93)80003-J

[13] A.L. Linsebigler, G.Q. Lu, J.T. Yates, Photocatalysis on TiO_2 surfaces: principles, mechanisms, and selected results, Chem. Rev. 95 (1995) 735-758. https://doi.org/10.1021/cr00035a013

[14] R.J. Tayade, R.G. Kulkarni, R.V. Jasra, Photocatalytic degradation of aqueous nitrobenzene by nanocrystalline TiO_2. Ind. Eng. Chem. Res. 45 (2006) 922-927. https://doi.org/10.1021/ie051060m

[15] X. Chen, S.S. Mao, Titanium dioxide nanomaterials: Synthesis, properties, modification, and applications, Chem. Rev. 107 (2007) 2891-2959. https://doi.org/10.1021/cr0500535

[16] R.J. Tayade, T.S. Natarajan, H.C. Bajaj, Photocatalytic degradation of methylene blue dye using ultraviolet light emitting diodes, Ind. Eng. Chem. Res. 48 (2009) 10262-10267. https://doi.org/10.1021/ie9012437

[17] A.A. Ismail, D.W. Bahnemann, Mesoporous titania photocatalysts: preparation, characterization and reaction mechanisms. J. Mater. Chem. 21 (2011) 11686-11707. https://doi.org/10.1039/c1jm10407a

[18] C. McCullagh, N. Skillen, M. Adams, P.K.J. Robertson, Photocatalytic reactors for environmental remediation: a review. J. Chem. Technol. Biotechnol. 86 (2011) 1002-1017. https://doi.org/10.1002/jctb.2650

[19] T.S. Natarajan, K. Natarajan, H.C. Bajaj, R.J. Tayade, Ultraviolet light emitting diode (UV-LED) source-based photocatalytic reactors for environmental remediation, in: Inamuddin (Eds), Advanced Functional Polymers and Composites. Volume 2, Nova Science Publishers, Inc. 2013, pp. 33-91.

[20] T.S. Natarajan, H.C. Bajaj, R.J. Tayade, Preferential adsorption behavior of methylene blue dye onto surface hydroxyl group enriched TiO_2 nanotube and its photocatalytic regeneration. J. Colloid Interface Sci. 433 (2014) 104-114. https://doi.org/10.1016/j.jcis.2014.07.019

[21] U.G. Akpan, B.H. Hameed, Parameters affecting the photocatalytic degradation of dyes using TiO_2-based photocatalysts: A review, J. Hazard. Mater. 170 (2009) 520-529. https://doi.org/10.1016/j.jhazmat.2009.05.039

[22] S. Ahmed, M.G. Rasul, R. Brown, M.A. Hashib, Influence of parameters on the heterogeneous photocatalytic degradation of pesticides and phenolic contaminants in wastewater: A short review, J. Environ. Manage. 92 (2011) 311-330. https://doi.org/10.1016/j.jenvman.2010.08.028

[23] S. Sarkar, R. Das, H. Choi, C. Bhattacharjee, Involvement of process parameters and various modes of application of TiO_2 nanoparticles in heterogeneous photocatalysis of pharmaceutical wastes- a short review, RSC Adv. 4 (2014) 57250-57266. https://doi.org/10.1039/C4RA09582K

[24] H. Cheng, J. Wang, Y. Zhao, X. Han, Effect of phase composition, morphology, and specific surface area on the photocatalytic activity of TiO_2 nanomaterials, RSC Adv. 4 (2014) 47031-47038. https://doi.org/10.1039/C4RA05509H

[25] K. M. Reza, A. Kurny, F. Gulshan, Parameters affecting the photocatalytic degradation of dyes using TiO_2: a review, Appl. Water Sci. 7 (2017) 1569-1578. https://doi.org/10.1007/s13201-015-0367-y

[26] W. Choi, A. Termin, M.R. Hoffmann, The role of metal ion dopants in quantum-sized TiO_2: correlation between photoreactivity and charge carrier recombination dynamics, J. Phys. Chem. 98 (1994) 13669-13679. https://doi.org/10.1021/j100102a038

[27] R. Asahi, T. Morikawa, T. Ohwaki, K. Aoki, Y. Taga, Visible-light photocatalysis in nitrogen-doped titanium oxides, Science 293 (2001) 269-271. https://doi.org/10.1126/science.1061051

[28] M. Pelaez, N.T. Nolan, S.C. Pillai, M.K. Seery, P. Falaras, A.G. Kontos, P.S.M. Dunlop, J.W.J. Hamilton, J.A. Byrne, K. O'Shea, M.H. Entezarig, D.D. Dionysiou, A review on the visible light active titanium dioxide photocatalysts for environmental applications. Appl. Catal. B. 125 (2012) 331-349. https://doi.org/10.1016/j.apcatb.2012.05.036

[29] L.G. Devi, R. Kavitha, A review on non-metal ion doped titania for the photocatalytic degradation of organic pollutants under UV/solar light: Role of photogenerated charge carrier dynamics in enhancing the activity, Appl. Catal. B. 140-141 (2013) 559-587. https://doi.org/10.1016/j.apcatb.2013.04.035

[30] T.S. Natarajan, K. Natarajan, H.C. Bajaj, R.J. Tayade, Enhanced photocatalytic activity of bismuth-doped TiO_2 nanotubes under direct sunlight irradiation for degradation of Rhodamine B dye. J. Nanopart. Res. 15:1669 (2013) 1-18.

[31] E.P. Reddy, L. Davydov, P. Smirniotis, TiO_2-loaded zeolites and mesoporous materials in the sonophotocatalytic decomposition of aqueous organic pollutants: the role of the support, Appl. Catal. B. 42 (2003) 1-11. https://doi.org/10.1016/S0926-3373(02)00192-3

[32] S. Anandan, M. Yoon, Photocatalytic activities of the nano-sized TiO_2-supported Y-zeolites, J. Photochem. Photobio. C: Photochem. Rev. 4 (2003) 5-18. https://doi.org/10.1016/S1389-5567(03)00002-9

[33] A. Corma, H. Garcia, Zeolite-based photocatalysts, Chem. Commun., (2004) 1443- 1459. https://doi.org/10.1039/b400147h

[34] J.L. Faria, W. Wang, Carbon materials in photocatalysis. In: Serp P, Figueiredo JL, editors. Carbon materials for catalysis. Hoboken, NJ: John Wiley & Sons; 2009. p. 481-506.

[35] K. Woan, G. Pyrgiotakis, W. Sigmund, Photocatalytic Carbon-Nanotube-TiO_2 Composites, Adv. Mater. 21 (2009) 2233-2239. https://doi.org/10.1002/adma.200802738

[36] H. Zhang, X.J. Lv, Y.M. Li, Y. Wang, J.H. Li, P25-Graphene Composite as a High Performance Photocatalyst. ACS Nano 4 (2010) 380-386. https://doi.org/10.1021/nn901221k

[37] C.-C. Wang, J.-R. Li, X.-L. Lv, Y.-Q. Zhang, G. Guo, Photocatalytic organic pollutants degradation in metal-organic frameworks, Energy Environ. Sci. 7 (2014) 2831-2867. https://doi.org/10.1039/C4EE01299B

[38] T.S. Natarajan, H.C. Bajaj, R.J. Tayade, Palmyra tuber peel derived activated carbon and anatase TiO_2 nanotube based nanocomposites with enhanced photocatalytic performance in rhodamine 6G dye degradation. Process Saf. Environ. Prot. 104 (2016) 346-357. https://doi.org/10.1016/j.psep.2016.09.021

[39] T.S. Natarajan, J.Y. Lee, H.C. Bajaj, W.K. Jo, R.J. Tayade, Synthesis of multiwall carbon nanotubes/TiO_2 nanotube composites with enhanced photocatalytic decomposition efficiency. Catal. Today 282 (2017) 13-23. https://doi.org/10.1016/j.cattod.2016.03.018

[40] A.H. Ramırez, I.M. Ramırez (Eds), Photocatalytic Semiconductors: Synthesis, Characterization, and Environmental Applications, Springer International Publishing, Switzerland 2015.

[41] C.J. Brinker, G.W. Scherer, Sol-gel science. Academic, San Diego, (1990).

[42] M. Niederberger, N. Pinna, Metal oxide nanoparticles in organic solvents. Synthesis, formation, assembly and application. Springer, London (2009). https://doi.org/10.1007/978-1-84882-671-7

[43] K. Byrappa, M. Yoshimura, Handbook of hydrothermal technology. A Technology for Crystal Growth and Materials Processing. Noyes, New York, (2001).

[44] S. Guo, Z. Wu, H. Wang, F. Dong, Synthesis of mesoporous TiO_2 nanorods via a mild template-free sonochemical route and their photocatalytic performances. Catal Commun 10 (2009) 1766-1770. https://doi.org/10.1016/j.catcom.2009.05.027

[45] S. Komarneni, R.K. Rajha, H. Katsuki, Microwave-hydrothermal processing of titanium dioxide. Mater. Chem. Phys. 61 (1999) 50-54. https://doi.org/10.1016/S0254-0584(99)00113-3

[46] G.L. Puma, A. Bono, D. Krishnaiah, J.G. Collin, Preparation of titanium dioxide photocatalyst loaded onto activated carbon support using chemical vapor deposition: A review paper, J. Hazard. Mater. 157 (2008) 209-219. https://doi.org/10.1016/j.jhazmat.2008.01.040

[47] R. Leary, A. Westwood, Carbonaceous nanomaterials for the enhancement of TiO_2 photocatalysis, Carbon 49 (2011) 741-772. https://doi.org/10.1016/j.carbon.2010.10.010

[48] J. Matos, J. Laine, J.-M. Herrmanna, Synergy effect in the photocatalytic degradation of phenol on a suspended mixture of titania and activated carbon, Appl. Catal. B. 18 (1998) 281-291. https://doi.org/10.1016/S0926-3373(98)00051-4

[49] J. Matos, J. Laine, J.-M. Herrmann, Effect of the Type of Activated Carbons on the Photocatalytic Degradation of Aqueous Organic Pollutants by UV-Irradiated Titania, J. Catal. 200 (2001) 10-20. https://doi.org/10.1006/jcat.2001.3191

[50] C.H. Ao, S.C. Lee, Enhancement effect of TiO_2 immobilized on activated carbon filter for the photodegradation of pollutants at typical indoor air level, Appl. Catal. B. 44 (2003) 191-205. https://doi.org/10.1016/S0926-3373(03)00054-7

[51] S. X. Liu, X.Y. Chen, X. Chen, A TiO_2/AC composite photocatalyst with high activity and easy separation prepared by a hydrothermal method, J. Hazard. Mater. 143 (2007) 257-263. https://doi.org/10.1016/j.jhazmat.2006.09.026

[52] L.F. Velasco, J.B. Parra, C.O. Ania, Role of activated carbon features on the photocatalytic degradation of phenol, Appl. Surf. Sci. 256 (2010) 5254-5258. https://doi.org/10.1016/j.apsusc.2009.12.113

[53] B. Huang, S. Saka, Photocatalytic activity of TiO_2 crystallite-activated carbon composites prepared in supercritical isopropanol for the decomposition of formaldehyde. J. Wood. Sci. 49 (2003) 79-85. https://doi.org/10.1007/s100860300013

[54] A. Tryba, A.W. Morawski, M. Inagaki, Application of TiO_2-mounted activated carbon to the removal of phenol from water, Appl. Catal. B., 41 (2003) 427-433. https://doi.org/10.1016/S0926-3373(02)00173-X

[55] Y. Li, X. Li, J. Li, J. Yin, Photocatalytic degradation of methyl orange in a sparged tube reactor with TiO_2-coated activated carbon composites. Catal. Commun. 6 (2005) 650-655. https://doi.org/10.1016/j.catcom.2005.06.008

[56] E. Carpio, P. Zúñiga, S. Ponce, J. Solis, J. Rodriguez, W. Estrada, Photocatalytic degradation of phenol using TiO_2 nanocrystals supported on activated carbon, J. Mol. Catal. A: Chem. 228 (2005) 293-298. https://doi.org/10.1016/j.molcata.2004.09.066

[57] Y. Li, X. Li, J. Li, J. Yin, Photocatalytic degradation of methyl orange by TiO_2-coated activated carbon and kinetic study, Water Res. 40 (2006) 1119-1126. https://doi.org/10.1016/j.watres.2005.12.042

[58] Y. Liu, S. Yang, J. Hong, C. Sun, Low-temperature preparation and microwave photocatalytic activity study of TiO_2-mounted activated carbon. J. Hazard. Mater. 142 (2007) 208-215. https://doi.org/10.1016/j.jhazmat.2006.08.020

[59] M.-L. Chen, C.-S. Lim, W.-C. Oh, Preparation with different mixing ratios of anatase to activated carbon and their photocatalytic performance. J. Ceram. Pro. Res. 8 (2007) 119-124.

[60] H.E. Zhong, Y. Shaogui, J. Yongming, S. Cheng, Microwave photocatalytic degradation of Rhodamine B using TiO_2 supported on activated carbon: Mechanism implication. J. Environ. Sci. 21(2009) 268-272. https://doi.org/10.1016/S1001-0742(08)62262-7

[61] X. Wang, Z. Hu, Y. Chen, G. Zhao, Y. Liu, Z. Wen, A novel approach towards high-performance composite photocatalyst of TiO_2 deposited on activated carbon, Appl. Surf. Sci. 255 (2009) 3953-3958. https://doi.org/10.1016/j.apsusc.2008.10.083

[62] M. Baek, W. Jung, J. Yoon, J. Hong, Y. Lee, J. Suh, Preparation, characterization and photocatalytic activity evaluation of micro- and mesoporous TiO_2/spherical activated carbon, J. Ind. Eng. Chem. 19 (2013) 469-477. https://doi.org/10.1016/j.jiec.2012.08.026

[63] M. Baek, J. Yoon, J. Hong, J. Suh, Application of TiO_2-containing mesoporous spherical activated carbon in a fluidized bed photoreactor-Adsorption and photocatalytic activity, Appl. Catal. A: Gen. 450 (2013) 222-229. https://doi.org/10.1016/j.apcata.2012.10.018

[64] A. Omri, S.D. Lambert, J. Geens, F. Bennour, M. Benzina, Synthesis, surface characterization and photocatalytic activity of TiO_2 supported on almond shell activated carbon, J. Mater. Sci. Technol. 30 (2014) 894-902. https://doi.org/10.1016/j.jmst.2014.04.007

[65] M. Ouzzine, A. J. Romero-Anaya, M.A. Lillo-Rodenas, A. L.-Solano, Spherical activated carbon as an enhanced support for TiO_2/AC photocatalysts, Carbon, 67 (2014) 104-118. https://doi.org/10.1016/j.carbon.2013.09.069

[66] R.C. Asha, M. Kumar, Photocatalytic degradation of poultry wastewater using activated carbon-supported titanium dioxide, Desalin. Water Treat. 54 (2015) 3279-3290. https://doi.org/10.1080/19443994.2014.909332

[67] X. Li, H. Lin, X. Chen, H. Niu, T. Zhang, J. Liu, F. Qu, Fabrication of TiO_2/porous carbon nanofibers with superior visible photocatalytic activity. New J. Chem. 39 (2015) 7863-7872. https://doi.org/10.1039/C5NJ01189B

[68] C. Zhang, D. Yang, X. Jiang, W. Jiang, Desulphurization performance of TiO_2-modified activated carbon by a one-step carbonization activation method, Environ. Technol. 37 (2016), 15, 1-4 (doi:10.1080/09593330.2015.1135991).

[69] P. Singha, M.C. Vishnu, K.K. Sharma, A. Borthakur, P. Srivastava, D.B. Pal, D. Tiwary, P.K. Mishra, Photocatalytic degradation of Acid Red dye stuff in the presence of activated carbon-TiO_2 composite and its kinetic enumeration, J. Water Process Eng. 12 (2016) 20-31. https://doi.org/10.1016/j.jwpe.2016.04.007

[70] M.G. Alalm, A. Tawfik, S. Ookawara, Enhancement of photocatalytic activity of TiO_2 by immobilization on activated carbon for degradation of pharmaceuticals, J. Environ. Chem. Eng. 4 (2016) 1929-1937. https://doi.org/10.1016/j.jece.2016.03.023

[71] A. Eshaghi, S. Hayeripour, A. Eshaghi, Photocatalytic decolorization of reactive red 198 dye by a TiO_2–activated carbon nano-composite derived from the sol-gel method, Res. Chem. Intermed. 42 (2016) 2461-2471. https://doi.org/10.1007/s11164-015-2161-8

[72] A.C. Martins, A.L. Cazetta, O. Pezoti, J.R.B. Souza, T. Zhang, E.J. Pilau, T. Asefa, V. C. Almeida, Sol-gel synthesis of new TiO_2/activated carbon photocatalyst and its application for degradation of tetracycline, Ceram. Int. 43 (2017) 4411-4418. https://doi.org/10.1016/j.ceramint.2016.12.088

[73] M. Doi, S. Saka, H. Miyafuji, D.A.I. Gorang, Development of Carbonized TiO_2-woody Composites for Environmental Cleaning, Mater. Sci. Res. Inter. 6 (2000) 15-21. https://doi.org/10.2472/jsms.49.3Appendix_15

[74] C. S. Chuang, M.-K. Wang, C.-H. Ko, C.-C. Ou, C.-H. Wu, Removal of benzene and toluene by carbonized bamboo materials modified with TiO_2, Bioresource Technol. 99 (2008) 954-958. https://doi.org/10.1016/j.biortech.2007.03.003

[75] H.A. Le, L.T. Linh, S. Chin, J. Jurng, Photocatalytic degradation of methylene blue by a combination of TiO_2-anatase and coconut shell activated carbon, Powder Technol. 225 (2012) 167-175. https://doi.org/10.1016/j.powtec.2012.04.004

[76] A.H. El-Sheikh, A.P. Newman, H. Al-Daffaee, S. Phull, N. Cresswell, S. York, Deposition of anatase on the surface of activated carbon, Surf. Coat. Technol. 187 (2004) 284-292. https://doi.org/10.1016/j.surfcoat.2004.03.012

[77] W.C. Oh, Synthesis and Characterization of Fe-containing AC/TiO_2 Composites and Their Photodegradation Effect for the Piggery Waste, Environ. Eng. Res. 13 (2008) 85-92. https://doi.org/10.4491/eer.2008.13.2.085

[78] J. Matos, A. Garcia, T. Cordero, J.M. Chovelon, C. Ferronato, Eco-friendly TiO_2-AC Photocatalyst for the Selective Photooxidation of 4-Chlorophenol, Catal. Lett. 130 (2009) 568-574. https://doi.org/10.1007/s10562-009-9989-8

[79] Z.D. Meng, K. Zhang, W.C. Oh, Preparation and Characterization and Visible Light Photocatalytic Activity of Fe-Treated AC/TiO_2 Composites for Methylene Blue, J. Korean Ceramic Soc. 46 (2009) 621-626. https://doi.org/10.4191/KCERS.2009.46.6.621

[80] J. Kim, B.S. Kwak, M. Kang, TiO_2/Carbon Composites Prepared from Rice Husk and the Removal of Bisphenol A in Photocatalytic Liquid System, Bull. Korean Chem. Soc. 31 (2010) 344-350. https://doi.org/10.5012/bkcs.2010.31.02.344

[81] M. Asilturk, S. Sener, TiO_2-activated carbon photocatalysts: Preparation, characterization and photocatalytic activities, Chem. Eng. J 180 (2012) 354-363. https://doi.org/10.1016/j.cej.2011.11.045

[82] M.X. Zeng, Y.J. Li, M. Y. Ma, W. Chen, L.Y. Li, Photocatalytic activity and kinetics for acid yellow degradation over surface composites of TiO_2-coated activated carbon under different photocatalytic conditions, Trans. Nonferrous Met. Soc. China 23 (2013) 1019-1027. https://doi.org/10.1016/S1003-6326(13)62561-3

[83] L. Youji, Z. Xiaoming, C. Wei, L. Leiyong, Z. Mengxiong, Q. Shidong, S. Shuguo, Photodecolorization of Rhodamine B on tungsten-doped TiO_2/activated carbon under visible-light irradiation, J. Hazard. Mater. 227-228 (2012) 25-33. https://doi.org/10.1016/j.jhazmat.2012.04.071

[84] D. Li, X. Ma, X. Liu, L. Yu, Preparation and characterization of Nano-TiO_2 loaded bamboo-based activated carbon fibers by H_2O activation, BioResources. 9 (2012) 602-612.

[85] B. Wang, R. Karthikeyan, X.Y. Lu, J. Xuan, M.K.H. Leung, High photocatalytic activity of immobilized TiO_2 nanorods on carbonized cotton fibers J. Hazard. Mater. 263 (2013) 659-669. https://doi.org/10.1016/j.jhazmat.2013.10.029

[86] S. Iijima, Helical microtubules of graphitic carbon, Nature 354 (1991) 56-58. https://doi.org/10.1038/354056a0

[87] P. Vincent, A. Brioude, C. Journet, S. Rabaste, S.T. Purcell, J.L. Brusq, J.C. Plenetm Inclusion of carbon nanotubes in a TiO_2 sol-gel matrix, J. Non-Crystalline Solids 311 (2002) 130-137. https://doi.org/10.1016/S0022-3093(02)01371-6

[88] A. Jitianu, T.Cacciaguerra, R.Benoit, S. Delpeux, F. Béguin, S. Bonnamy, Synthesis and characterization of carbon nanotubes-TiO_2 nanocomposites, Carbon, 42, 2004, 1147-1151. https://doi.org/10.1016/j.carbon.2003.12.041

[89] L. Chen, B.-L. Zhang, M.-Z. Qu, Z.-L. Yu, Preparation and characterization of CNTs-TiO_2 composites, Powder Technol. 154 (2005) 70-72. https://doi.org/10.1016/j.powtec.2005.04.028

[90] B. Gao, C. Peng, G.Z. Chen, G.L. Puma, Photo-electro-catalysis enhancement on carbon nanotubes/titanium dioxide (CNTs/TiO_2) composite prepared by a novel surfactant wrapping sol-gel method, Appl. Catal. B. 85 (2008) 17-23. https://doi.org/10.1016/j.apcatb.2008.06.027

[91] C.-Y. Yen, Y.-F. Lin, C.-H. Hung, Y.-H. Tseng, C.-C. Ma, M.-C. Chang, H. Shao, The effects of synthesis procedures on the morphology and photocatalytic activity of multi-walled carbon nanotubes/TiO_2 nanocomposites, Nanotechnology 19 (2008) 045604 (11pp). https://doi.org/10.1088/0957-4484/19/04/045604

[92] T. An, J. Chen, X. Nie, G. Li, H. Zhang, X. Liu, H. Zhao Synthesis of Carbon Nanotube-Anatase TiO_2 Sub-micrometer-sized Sphere Composite Photocatalyst for Synergistic Degradation of Gaseous Styrene, ACS Appl. Mater. Interfaces 4 (2012) 5988-5996. https://doi.org/10.1021/am3016476

[93] J.-Y. Jung, D. Lee, Y.-S. Lee, CNT-embedded hollow TiO_2 nanofibers with high adsorption and photocatalytic activity under UV irradiation, J. Alloys Compds. 622 (2015) 651-656. https://doi.org/10.1016/j.jallcom.2014.09.068

[94] Z. Lu, X. Xiang, L. Zou, J. Xie, Fluffy-ball-shaped carbon nanotube-TiO_2 nanorod nanocomposites for photocatalytic degradation of methylene blue, RSC Adv. 5 (2015) 42580-42586. https://doi.org/10.1039/C5RA05641A

[95] Z.-R. Tang, F. Li, Y. Zhang, X. Fu, Y.-J. Xu, Composites of Titanate Nanotube and Carbon Nanotube as Photocatalyst with High Mineralization Ratio for Gas-

Phase Degradation of Volatile Aromatic Pollutant, J. Phys. Chem. C, 115 (2011) 7880-7886. https://doi.org/10.1021/jp1115838

[96] B.K. Vijayan, N.M. Dimitrijevic, D.F.-Shapiro, J. Wu, K.A. Gray, Coupling Titania Nanotubes and Carbon Nanotubes To Create Photocatalytic Nanocomposites, ACS Catal. 2 (2012) 223-229. https://doi.org/10.1021/cs200541a

[97] T. Jiang, L. Zhang, M. Ji, Q. Wang, Q. Zhao, X. Fu, H. Yin, Carbon nanotubes/TiO_2 nanotubes composite photocatalysts for efficient degradation of methyl orange dye, Particuology 11 (2013) 737-742. https://doi.org/10.1016/j.partic.2012.07.008

[98] W. Wang, P. Serp, P. Kalck, J.L. Faria, Visible light photodegradation of phenol on MWNT-TiO_2 composite catalysts prepared by a modified sol-gel method. J. Mol. Catal. A: Chem. 235 (2005) 194-199. https://doi.org/10.1016/j.molcata.2005.02.027

[99] W. Wang, P. Serp, P. Kalck, J.L. Faria, Photocatalytic degradation of phenol on MWNT and titania composite catalysts prepared by a modified sol-gel method. Appl. Catal. B. 56 (2005) 305-312. https://doi.org/10.1016/j.apcatb.2004.09.018

[100] Y. Yu, J. C. Yu, J.-G. Yu, Y.-C. Kwok, Y.-K. Che, J.-C. Zhao, L. Ding, W.-K. Ge, P.-K. Wong, Enhancement of photocatalytic activity of mesoporous TiO_2 by using carbon nanotubes. App. Catal. A: Gen. 289 (2005) 186-196. https://doi.org/10.1016/j.apcata.2005.04.057

[101] G. An, W. Ma, Z. Sun, Z. Liu, B. Han, S. Miao, Z. Miao, K. Ding, Preparation of titania/carbon nanotube composites using supercritical ethanol and their photocatalytic activity for phenol degradation under visible light irradiation. Carbon 45 (2007) 1795-1801. https://doi.org/10.1016/j.carbon.2007.04.034

[102] B.Liu, H.C. Zeng, Carbon nanotubes supported mesoporous mesocrystals of anatase TiO_2. Chem. Mater. 20 (2008) 2711-2718. https://doi.org/10.1021/cm800040k

[103] H. Wang, H.-L. Wang, W.-F. Jiang, Solar photocatalytic degradation of 2,6-dinitro-p-cresol (DNPC) using multi-walled carbon nanotubes (MWCNTs)-TiO_2 composite photocatalysts. Chemosphere 75 (2009) 1105-1111. https://doi.org/10.1016/j.chemosphere.2009.01.014

[104] C.G. Silva, J.L. Faria, Photocatalytic oxidation of benzene derivatives in aqueous suspensions: Synergic effect induced by the introduction of carbon nanotubes in a

TiO_2 matrix. App. Catal. B. 101 (2010) 81-89. https://doi.org/10.1016/j.apcatb.2010.09.010

[105] W. Zhou, K. Pan, Y.Qu, F. Sun, C. Tian, Z. Ren, G. Tian, H. Fu, Photodegradation of organic contamination in wastewaters by bonding TiO_2/single-walled carbon nanotube composites with enhanced photocatalytic activity. Chemosphere 81 (2010) 555-561. https://doi.org/10.1016/j.chemosphere.2010.08.059

[106] Z. Li, B. Gao, G.Z. Chen, R. Mokaya, S. Sotiropoulos, G.L. Puma, Carbon nanotube/titanium dioxide (CNT/TiO_2) core-shell nanocomposites with tailored shell thickness, CNT content and photocatalytic/photoelectrocatalytic properties. Appl. Catal. B. 110 (2011) 50-57. https://doi.org/10.1016/j.apcatb.2011.08.023

[107] S. Wang, S. Zhou, Photodegradation of methyl orange by photocatalyst of CNTs/P-TiO_2 under UV and visible-light irradiation, J. Hazard. Mater. 185 (2011) 77-85. https://doi.org/10.1016/j.jhazmat.2010.08.125

[108] L. Tian, L. Ye, K. Deng, L. Zan, TiO_2/carbon nanotube hybrid nanostructures: Solvothermal synthesis and their visible light photocatalytic activity. J. Solid State Chem. 184 (2011)1465-1471. https://doi.org/10.1016/j.jssc.2011.04.014

[109] A. Saleh, V.K. Gupta, Photo-catalyzed degradation of hazardous dye methyl orange by use of a composite catalyst consisting of multi-walled carbon nanotubes and titanium dioxide. J. Colloid Interface Sci. 371 (2012) 101-106. https://doi.org/10.1016/j.jcis.2011.12.038

[110] L. Zhu, Z. Meng, K. Cho, W. Oh, Synthesis of CdS/CNT-TiO_2 with a high photocatalytic activity in photodegradation of methylene blue. New Carbon Mater. 27 (2012) 166-174. https://doi.org/10.1016/S1872-5805(12)60011-0

[111] S.D. Dalt, A.K. Alves, C.P. Bergmann, Photocatalytic degradation of methyl orange dye in water solutions in the presence of MWCNT/TiO_2 composites. Mater. Res. Bull. 48 (2013) 1845–1850. https://doi.org/10.1016/j.materresbull.2013.01.022

[112] D. Zhao, X. Yang, C. Chen, X. Wang, Enhanced photocatalytic degradation of methylene blue on multiwalled carbon nanotubes-TiO_2. J. Colloid Interface Sci. 398 (2013) 234-239. https://doi.org/10.1016/j.jcis.2013.02.017

[113] A. Nourbakhsh, S. Abbaspour, M. Masood, S.N. Mirsattari, A. Vahedi, K. J.D. Mackenzie, Photocatalytic properties of mesoporous TiO_2 nanocomposites modified with carbon nanotubes and copper. Ceram. Int. 42 (2016)11901-11906. https://doi.org/10.1016/j.ceramint.2016.04.112

[114] R. Zouzelka, Y. Kusumawati, M. Remzova, J. Rathousky, T. Pauporté, Photocatalytic activity of porous multiwalled carbon nanotube-TiO_2 composite layers for pollutant degradation. J. Hazard. Mater. 317 (2016) 52-59. https://doi.org/10.1016/j.jhazmat.2016.05.056

[115] M. Bozic, V. Vivod, R. Vogrincic, I. Ban, G. Jaksa, S. Hribernik, D. Fakin, V. Kokol, Enhanced catalytic activity of the surface modified TiO_2-MWCNT nanocomposites under visible light. J. Colloid Interface Sci. 465 (2016) 93-105. https://doi.org/10.1016/j.jcis.2015.11.051

[116] H. M. Yang, S.-J. Park, Effect of incorporation of multiwalled carbon nanotubes on photodegradation efficiency of mesoporous anatase TiO_2 spheres. Mater. Chem. Phys. 186 (2017) 261-270. https://doi.org/10.1016/j.matchemphys.2016.10.052

[117] V.B. Koli, A.G. Dhodamani, S.D. Delekar, S.H. Pawar, In situ sol-gel synthesis of anatase TiO_2-MWCNTs nanocomposites and their photocatalytic applications. J. Photochem. Photobiol. A: Chem. 333 (2017) 40-48. https://doi.org/10.1016/j.jphotochem.2016.10.008

[118] N. Shaari, S.H. Tan, A.R. Mohamed, Synthesis and characterization of CNT/Ce-TiO_2 nanocomposite for phenol degradation, J. Rare Earths, 30 (2012) 651-658. https://doi.org/10.1016/S1002-0721(12)60107-0

[119] K. Ouyang, S. Xien, X. Ma, Effect of key operational factors on decolorization of methyl orange by multi-walled carbon nanotubes (MWCNTs)/TiO_2/CdS composite under simulated solar light irradiation, Ceramics Inter. 39 (2013) 8035-8042. https://doi.org/10.1016/j.ceramint.2013.03.073

[120] P. Zhang, Z. Mo, L. Han, Y. Wang, G. Zhao, C. Zhang, Z. Li, Magnetic recyclable TiO_2/multi-walled carbon nanotube nanocomposite: Synthesis, characterization and enhanced photocatalytic activity, J. Mol. Catal. A: Chem. 402 (2015) 17-22. https://doi.org/10.1016/j.molcata.2015.03.005

[121] M.M. Mohamed, G. Osman, K.S. Khairou, Fabrication of Ag nanoparticles modified TiO_2-CNT heterostructures for enhanced visible light photocatalytic degradation of organic pollutants and bacteria, J. Environ. Chem. Eng. 3 (2015) 1847-1859. https://doi.org/10.1016/j.jece.2015.06.018

[122] S. Li, Z. Zhao, Y. Huang, J. Di, Y. Jia, H. Zheng, Hierarchically structured WO_3-CNT@TiO_2NS composites with enhanced photocatalytic activity, J. Mater. Chem. A 3 (2015) 5467-5473. https://doi.org/10.1039/C4TA06883A

[123] S. Stankovich, D. A. Dikin, G.H.B. Dommett, K.M. Kohlhaas, E.J. Zimney, E.A. Stach, R.D. Piner, S.T. Nguyen, R.S. Ruoff, Graphene-based composite materials, Nature 442 (2006) 282-286. https://doi.org/10.1038/nature04969

[124] X. Huang, X. Qi, F. Boey, H. Zhang, Graphene-based composites, Chem. Soc. Rev. 41 (2012) 666-686. https://doi.org/10.1039/C1CS15078B

[125] Q. Xiang, J. Yu, Graphene-Based Photocatalysts for Hydrogen Generation, J. Phys. Chem. Lett. 4 (2013) 753-759. https://doi.org/10.1021/jz302048d

[126] Y. Zhang, Z.-R. Tang, X. Fu, Y.-J. Xu, TiO_2-Graphene Nanocomposites for Gas-Phase Photocatalytic Degradation of Volatile Aromatic Pollutant: Is TiO_2-Graphene Truly Different from other TiO_2-Carbon Composite Materials? ACS Nano 4 (2010) 7303-7314. https://doi.org/10.1021/nn1024219

[127] G. Jiang, Z. Lin, C. Chen, L. Zhu, Q. Chang, N. Wang, W. Wei, H. Tang, TiO_2 nanoparticles assembled on graphene oxide nanosheets with high photocatalytic activity for removal of pollutants, Carbon 49 (2011) 2693-2701. https://doi.org/10.1016/j.carbon.2011.02.059

[128] S.D. Perera, R.G. Mariano, K. Vu, N. Nour, O. Seitz, Y. Chabal, K.J. Balkus, Jr., Hydrothermal Synthesis of Graphene-TiO_2 Nanotube Composites with Enhanced Photocatalytic Activity, ACS Catal. 2 (2012) 949-956. https://doi.org/10.1021/cs200621c

[129] Y. Gu, M. Xing, J. Zhang, Synthesis and photocatalytic activity of graphene based doped TiO_2 nanocomposites Appl. Surf. Sci. 319 (2014) 8-15. https://doi.org/10.1016/j.apsusc.2014.04.182

[130] M. Sun, Y. Fang, Y. Wang, S. Sun, J. He, Z. Yan, Synthesis of Cu_2O/graphene/rutile TiO_2 nanorod ternary composites with enhanced photocatalytic activity, J. Alloys Compds. 650 (2015) 520-527. https://doi.org/10.1016/j.jallcom.2015.08.002

[131] B. Appavu, S. Thiripuranthagan, Visible active N, S co-doped TiO_2/graphene photocatalysts for the degradation of hazardous dyes, J. Photochem. Photobio. A: Chem. 340 (2017) 146-156. https://doi.org/10.1016/j.jphotochem.2017.03.010

[132] Y. Liang, H. Wang, H. S. Casalongue, Z. Chen, H. Dai, TiO_2 nanocrystals grown on graphene as advanced photocatalytic hybrid materials. Nano Res. 3 (2010) 701-705. https://doi.org/10.1007/s12274-010-0033-5

[133] H. Zhang, P. Xu, G. Du, Z. Chen, K. Oh, D. Pan, Z. Jiao, A facile one-step synthesis of TiO_2/graphene composites for photodegradation of methyl orange, Nano Res. 4 (2011) 274-283. https://doi.org/10.1007/s12274-010-0079-4

[134] V. Stengl, D. Popelková, P. Vlacil, TiO_2-Graphene Nanocomposite as High Performace Photocatalysts. J. Phys. Chem. C 115 (2011) 25209-25218. https://doi.org/10.1021/jp207515z

[135] N. Li, G. Liu, C. Zhen, F. Li, L. Zhang, H.-M. Cheng, Battery performance and photocatalytic activity of mesoporous anatase TiO_2 nanospheres/graphene composites by template-free self-assembly. Adv. Funct. Mater. 21 (2011) 1717-1722.

[136] M.S.A.S. Shah, A.R. Park, K. Zhang, J.H. Park, P.J. Yoo, Green synthesis of biphasic TiO_2-reduced graphene oxide nanocomposites with highly enhanced photocatalytic activity. ACS Appl. Mater. Interfaces 4 (2012) 3893-3901. https://doi.org/10.1021/am301287m

[137] X. Pan, Y. Zhao, S. Liu, Carol L. Korzeniewski, S.Wang, Z. Fan, Comparing graphene-TiO_2 nanowire and graphene-TiO_2 nanoparticle composite photocatalysts. ACS Appl. Mater. Interfaces 4 (2012) 3944–3950. https://doi.org/10.1021/am300772t

[138] Q. Huang, S. Tian, D. Zeng, X. Wang, W. Song, Y. Li, W. Xiao, C. Xie, Enhanced photocatalytic activity of chemically bonded TiO_2/graphene composites based on the effective interfacial charge transfer through the C−Ti bond. ACS Catal. 3 (2013) 1477-1485. https://doi.org/10.1021/cs400080w

[139] W.-K. Jo, Y. Won, I. Hwang, R.J. Tayade, Enhanced photocatalytic degradation of aqueous nitrobenzene using graphitic carbon-TiO_2 composites. Ind. Eng. Chem. Res. 53 (2014) 3455-3461. https://doi.org/10.1021/ie500245d

[140] B. Appavu, K. Kannan, S. Thiripuranthagan, Enhanced visible light photocatalytic activities of template free mesoporous nitrogen doped reduced graphene oxide/titania composite catalysts. J. Ind. Eng. Chem. 36 (2016) 184-193. https://doi.org/10.1016/j.jiec.2016.01.042

[141] R. Wang, R. Yang, B. Wang, W. Gao, Efficient degradation of methylene blue by the nano TiO_2-functionalized graphene oxide nanocomposite photocatalyst for wastewater treatment. Water Air Soil Pollut. 227:2 (2016) 1-9. https://doi.org/10.1007/s11270-015-2720-z

[142] Y. Yang, L. Xu, H. Wang, W. Wang, L. Zhang, TiO_2/graphene porous composite and its photocatalytic degradation of methylene blue. Mater. Des. 108 (2016) 632-639. https://doi.org/10.1016/j.matdes.2016.06.104

[143] J. Hu, H. Li, S. Muhammad, Q. Wu, Y. Zhao, Q. Jiao, Surfactant-assisted hydrothermal synthesis of TiO_2/reduced graphene oxide nanocomposites and their photocatalytic performances. J. Solid State Chem. 253 (2017) 113-120. https://doi.org/10.1016/j.jssc.2017.05.034

[144] G. Zerjav, M. S. Arshad, P. Djinovic, I. Junkar, J. Kovac, J. Zavasnik, A. Pintar, Improved electron-hole separation and migration in anatase TiO_2 nanorod/reduced graphene oxide composites and their influence on photocatalytic performance. Nanoscale 9 (2017) 4578-4592. https://doi.org/10.1039/C7NR00704C

[145] M.-C. Rosu, M. Coros, F. Pogacean, L. Magerusan, C. Socaci, A. Turza, S. Pruneanu, Azo dyes degradation using TiO_2-Pt/graphene oxide and TiO_2-Pt/reduced graphene oxide photocatalysts under UV and natural sunlight irradiation. Solid State Sci. 70 (2017) 13-20. https://doi.org/10.1016/j.solidstatesciences.2017.05.013

[146] X. Liu, K.-K. Iu, J.K. Thomas, Preparation, characterization and photoreactivity of titanium(IV) oxide encapsulated in zeolites, J. Chem. Soc. Faraday Trans. 89 (1993) 1861-1865. https://doi.org/10.1039/ft9938901861

[147] K.J. Green, R. Rudham, Photocatalytic oxidation of propan-2-ol by semiconductor-zeolite composites, J. Chem. Soc. Faraday Trans. 89 (1993) 1867-1870. https://doi.org/10.1039/FT9938901867

[148] A. Corma, H. Garcia, Zeolite-based photocatalysts, Chem. Commun., (2004) 1443-1459. https://doi.org/10.1039/b400147h

[149] X. Yu, C.H. Langford, Enhanced Photoactivity of a Titanium(IV) Oxide Supported on ZSMS and Zeolite A at Low Coverage, J. Phys. Chem. 99 (1995) 11501-11507. https://doi.org/10.1021/j100029a031

[150] C. Zhu, L. Wang, L. Kong, X. Yang, L.Wang, S. Zheng, F. Chen, F. MaiZhi, H. Zong, Photocatalytic degradation of AZO dyes by supported TiO_2 + UV in aqueous solution, Chemosphere 41 (2000) 303-309. https://doi.org/10.1016/S0045-6535(99)00487-7

[151] J. Chen, L. Eberlein, C.H. Langford, Pathways of phenol and benzene photooxidation using TiO_2 supported on a zeolite, J. Photochem. Photobiol. A: Chem. 148 (2002) 183 189. https://doi.org/10.1016/S1010 6030(02)00041 2

[152] R.J. Tayade, R.G. Kulkarni, R.V. Jasra, Enhanced Photocatalytic Activity of TiO_2-Coated NaY and HY Zeolites for the Degradation of Methylene Blue in Water, Ind. Eng. Chem. Res. 46 (2007) 369-376. https://doi.org/10.1021/ie060641o

[153] R.J. Tayade, P.K. Surolia, M.A. Lazar, R.V. Jasra, Enhanced Photocatalytic Activity by Silver Metal Ion Exchanged NaY Zeolite Photocatalysts for the Degradation of Organic Contaminants and Dyes in Aqueous Medium, Ind. Eng. Chem. Res. 47 (2008) 7545-7551. https://doi.org/10.1021/ie800441c

[154] C.-C. Wang, C.-K. Lee, M.-D. Lyu, L.-C. Juang, Photocatalytic degradation of C.I. Basic Violet 10 using TiO_2 catalysts supported by Y zeolite: An investigation of the effects of operational parameters, Dyes Pigm. 76 (2008) 817-824. https://doi.org/10.1016/j.dyepig.2007.02.004

[155] G. Zhang, W. Choi, S.H. Kim, S.B. Hong, Selective photocatalytic degradation of aquatic pollutants by titania encapsulated into FAU-type zeolites, J. Hazard. Mater. 188 (2011) 198-205. https://doi.org/10.1016/j.jhazmat.2011.01.105

[156] W. Zhang, X. Xiao, L. Zheng, C. Wan, Fabrication of TiO_2/MoS_2@zeolite photocatalyst and its photocatalytic activity for degradation of methyl orange under visible light, Appl. Surf. Sci. 358 (2015) 468-478. https://doi.org/10.1016/j.apsusc.2015.08.054

[157] N. Setthaya, P. Chindaprasirt, S. Yin, K. Pimraksa, TiO_2-zeolite photocatalysts made of metakaolin and rice husk ash for removal of methylene blue dye, Powder Technol., 313 (2017) 417-426. https://doi.org/10.1016/j.powtec.2017.01.014

[158] M.V. Shankar, S. Anandan, N. Venkatachalam, B. Arabindoo, V. Murugesan, Fine route for an efficient removal of 2,4-dichlorophenoxyacetic acid (2,4-D) by zeolite-supported TiO_2. Chemosphere 63 (2006) 1014-1021. https://doi.org/10.1016/j.chemosphere.2005.08.041

[159] M. Huang, C. Xu, Z. Wu, Y. Huang, J. Lin, J. Wu, Photocatalytic discolorization of methyl orange solution by Pt modified TiO_2 loaded on natural zeolite, Dyes Pigm. 77 (2008) 327-334. https://doi.org/10.1016/j.dyepig.2007.01.026

[160] S. Koa, P. D. Fleming, M. Joyce, P. Ari-Gur, High performance nano-titania photocatalytic paper composite. Part II: Preparation and characterization of natural zeolite-based nano-titania composite sheets and study of their photocatalytic activity. Mater. Sci. Eng. B 164 (2009) 135–139. https://doi.org/10.1016/j.mseb.2009.08.010

[161] S. Artkla, W. Kim, W. Choi, J. Wittayakun, Highly enhanced photocatalytic degradation of tetramethylammonium on the hybrid catalyst of titania and MCM-41 obtained from rice husk silica. Appl. Catal. B. 91 (2009) 157-164. https://doi.org/10.1016/j.apcatb.2009.05.019

[162] M. Mahalakshmi, S.V. Priya, B. Arabindoo, M. Palanichamy, V. Murugesan, Photocatalytic degradation of aqueous propoxur solution using TiO_2 and Hβ-zeolite-supported TiO_2. J. Hazard. Mater. 161 (2009) 336-343. https://doi.org/10.1016/j.jhazmat.2008.03.098

[163] S.B. Rasmussen, R. Portela, S. Suarez, J.M. Coronado, M.-L. Rojas-Cervantes, P. Avila, B. Sánchez, Hybrid TiO_2-$SiMgO_X$ composite for combined chemisorption and photocatalytic elimination of gaseous H_2S. Ind. Eng. Chem. Res. 49 (2010) 6685-6690. https://doi.org/10.1021/ie901733p

[164] W. Zhang, K. Wang, Y. Yu, H. He, TiO_2/HZSM-5 nano-composite photocatalyst: HCl treatment of NaZSM-5 promotes photocatalytic degradation of methyl orange. Chem. Eng. J. 163 (2010) 62-67. https://doi.org/10.1016/j.cej.2010.07.042

[165] C. Lazau, C. Ratiu, C. Orha, R. Pode, F. Manea, Photocatalytic activity of undoped and Ag-doped TiO_2-supported zeolite for humic acid degradation and mineralization. Mater. Res. Bull.46 (2011) 1916-1921. https://doi.org/10.1016/j.materresbull.2011.07.026

[166] L. You-ji, C. Wei, Photocatalytic degradation of Rhodamine B using nanocrystalline TiO_2-zeolite surface composite catalysts: effects of photocatalytic condition on degradation efficiency. Catal. Sci. Technol. 1 (2011) 802-809. https://doi.org/10.1039/c1cy00012h

[167] C. Wang, H. Shi, Y. Li, Synthesis and characterization of natural zeolite supported Cr-doped TiO_2 photocatalysts. Appl. Surf. Sci. 258(2012) 4328–4333. https://doi.org/10.1016/j.apsusc.2011.12.108

[168] Y. Kuwahara, J. Aoyama, K. Miyakubo, T. Eguchi, T. Kamegawa, K. Mori, H. Yamashita, TiO_2 photocatalyst for degradation of organic compounds in water and air supported on highly hydrophobic FAU zeolite: Structural, sorptive, and photocatalytic studies. J. Catal. 285 (2012) 223-234. https://doi.org/10.1016/j.jcat.2011.09.031

[169] S. Gomez, C. L. Marchena, L. Pizzio, L. Pierella Preparation and characterization of TiO_2/HZSM-11 zeolite for photodegradation of dichlorvos in aqueous solution. J. Hazard. Mater. 258-259 (2013) 19-26. https://doi.org/10.1016/j.jhazmat.2013.04.030

[170] S. Liu, M. Lim, R. Amal, TiO_2-coated natural zeolite: Rapid humic acid adsorption and effective photocatalytic regeneration. Chem. Eng. Sci. 105 (2014) 46–52. https://doi.org/10.1016/j.ces.2013.10.041

[171] D. Kanakaraju, J. Kockler, C.A. Motti, B.D. Glass, M. Oelgemöller, Titanium dioxide/zeolite integrated photocatalytic adsorbents for the degradation of amoxicillin. Appl. Catal. B: Environ. 166-167 (2015) 45-55. https://doi.org/10.1016/j.apcatb.2014.11.001

[172] M.N. Chong, Z.Y. Tneu, P.E. Poh, B. Jin, R. Aryald, Synthesis, characterization and application of TiO_2 zeolite nanocomposites for the advanced treatment of industrial dye wastewater. J. Taiwan Inst. Chem. Eng. 50 (2015) 288-296. https://doi.org/10.1016/j.jtice.2014.12.013

[173] S. Wardhani, M.F. Rahman, D. Purwonugroho, R.T. Tjahjanto, C.A. Damayanti, I. O. Wulandari, Photocatalytic Degradation of Methylene Blue Using TiO_2-Natural Zeolite as A Photocatalyst. J. Pure App. Chem. Res. 2016, 5 (1), 19-27. https://doi.org/10.21776/ub.jpacr.2016.005.01.232

[174] H.B. Yener, M. Yılmaz, Ö. Deliismail, S.F. Özkan, Ş.Ş. Helvacl, Clinoptilolite supported rutile TiO_2 composites: Synthesis, characterization, and photocatalytic activity on the degradation of terephthalic acid. Sep. Purif. Technol. 173 (2017) 17–26. https://doi.org/10.1016/j.seppur.2016.09.010

[175] I. Jansson, S. Suárez, F.J. García-García, B. Sánchez, ZSM-5/TiO_2 Hybrid Photocatalysts: Influence of the Preparation Method and Synergistic Effect. Top. Catal. (2017), DOI:10.1007/s11244-017-0805-1. https://doi.org/10.1016/j.seppur.2016.09.010

[176] C.T. Kresge, M.E. Leonowicz, W.J. Roth, J.C. Vartuli, J.S. Beck, Ordered mesoporous molecular sieves synthesized by a liquid-crystal template mechanism, Nature 359 (1992) 710-712. https://doi.org/10.1038/359710a0

[177] J.S. Beck, J.C. Vartuli, W.J. Roth, M.E. Leonowicz, C.T. Kresge, K.D. Schmitt, C.T.W. Chu, D.H. Olson, E.W. Sheppard, S.B. McCullen, J.B. Higgins, J.L. Schlenker, A new family of mesoporous molecular sieves prepared with liquid crystal templates, J. Am. Chem. Soc. 114 (1992) 10834-10843. https://doi.org/10.1021/ja00053a020

[178] S.A. Bagshaw, E. Prouzet, T.J. Pinnavaia, Templating of mesoporous molecular sieves by nonionic polyethylene oxide surfactants, Science 269 (1995) 1242-1244. https://doi.org/10.1126/science.269.5228.1242

[179] E. Prouzet, T.J. Pinnavaia, Assembly of mesoporous molecular sieves containing wormhole Motifs by a nonionic surfactant pathway: control of pore size by synthesis temperature, Angew. Chem. Int. Ed. Engl., 36 (1997) 516-518. https://doi.org/10.1002/anie.199705161

[180] D. Zhao, J. Feng, Q. Huo, N. Melosh, G.H. Fredrickson, B.F. Chmelka, G.D. Stucky, Triblock copolymer syntheses of mesoporous silica with periodic 50 to 300 angstrom pores, Science 279 (1998) 548-552. https://doi.org/10.1126/science.279.5350.548

[181] T. Torimoto, S. Ito, S. Kuwabata, H. Yoneyama, Effects of adsorbents used as supports for titanium dioxide loading on photocatalytic degradation of propyzamide, Environ. Sci. Technol. 30 (1996) 1275-1281. https://doi.org/10.1021/es950483k

[182] W. Dong, C. W. Lee, X. Lu, Y. Sun, W. Hua, G. Zhuang, S. Zhang, J. Chen, H. Hou, D. Zhao, Synchronous role of coupled adsorption and photocatalytic oxidation on ordered mesoporous anatase TiO_2-SiO_2 nanocomposites generating excellent degradation activity of RhB dye, Appl. Catal. B. 95 (2010) 197-207. https://doi.org/10.1021/es950483k

[183] W. Dong, Y. Sun, Q. Ma, L. Zhu, W. Hua, X. Lu, G. Zhuang, S. Zhang, Z. Guo, D. Zhao, Excellent photocatalytic degradation activities of ordered mesoporous anatase TiO_2-SiO_2 nanocomposites to various organic contaminants, J. Hazard. Mater. 229-230 (2012) 307-320. https://doi.org/10.1016/j.jhazmat.2012.06.002

[184] A. Pal, T.K. Janan, K. Chatterjee, Silica supported TiO_2 nanostructures for highly efficient photocatalytic application under visible light irradiation, Mater. Res. Bulletin 76 (2016) 353-357. https://doi.org/10.1016/j.materresbull.2015.12.040

[185] Q. Chen, H. Shi, W. Shi, Y. Xu, D. Wu, Enhanced visible photocatalytic activity of titania-silica photocatalysts: effect of carbon and silver doping, Catal. Sci. Technol. 2 (2012) 1213-1220. https://doi.org/10.1039/c2cy00545j

[186] K.P.O. Mahesh, D.-H. Kuo, B.-R. Huang, Facile synthesis of heterostructured Ag-deposited SiO_2@TiO_2 composite spheres with enhanced catalytic activity towards the photodegradation of AB 1 dye, J. Mol. Catal. A: Chem. 396 (2015) 290-296. https://doi.org/10.1016/j.molcata.2014.10.017

[187] Y. Jiang, Z. Jin, C. Chen, W. Duan, B. Liu, X. Chen, F. Yang, J. Guo, Cerium-doped mesoporous-assembled SiO_2/P25 nanocomposites with innovative visible-light sensitivity for the photocatalytic degradation of organic dyes, RSC Adv. 7 (2017) 12856-12870. https://doi.org/10.1039/C7RA00191F

[188] J. Yu, J. C. Yu, X. Zhao, The effect of SiO_2 addition on the grain size and photocatalytic activity of TiO_2 thin films. J. Sol-Gel Sci. Technol. 24 (2002) 95-103. https://doi.org/10.1023/A:1015258105966

[189] M. Hirano, K, Ota, Direct formation and photocatalytic performance of anatase (TiO_2)/Silica (SiO_2) composite nanoparticles, J. Am. Ceram. Soc., 87 (2004) 1567-1570. https://doi.org/10.1111/j.1551-2916.2004.01567.x

[190] J. Marugan, M.-J. Lopez-Munoz, R.V. Grieken, J. Aguado, Photocatalytic decolorization and mineralization of dyes with nanocrystalline TiO_2/SiO_2 materials. Ind. Eng. Chem. Res. 46 (2007) 7605-7610. https://doi.org/10.1021/ie070093u

[191] A. Nilchi, S. Janitabar-Darzi, S. Rasouli-Garmarodi, Sol-gel preparation of nanoscale TiO_2 /SiO_2 composite for eliminating of congo red azo dye, Mater. Sci. Appl. 2 (2011) 476-480.

[192] C.-H. Huang, K.-P. Chang, H.-D. Ou, Y.-C. Chiang, E.-E. Chang, C.-F. Wang, Characterization and application of Ti-containing mesoporous silica for dye removal with synergistic effect of coupled adsorption and photocatalytic oxidation. J. Hazard. Mater. 186 (2011) 1174-1182. https://doi.org/10.1016/j.jhazmat.2010.11.125

[193] W. Zhao, L. Feng, R. Yang, J. Zheng, X. Li, Synthesis, characterization, and photocatalytic properties of Ag modified hollow SiO_2/TiO_2 hybrid microspheres. Appl. Catal. B. 103 (2011) 181-189. https://doi.org/10.1016/j.apcatb.2011.01.025

[194] H.-S. Wu, L.-D. Sun, H.-P. Zhou, C.-H. Yan, Novel TiO_2-Pt@SiO_2 nanocomposites with high photocatalytic activity. Nanoscale 4 (2012) 3242-3247. https://doi.org/10.1039/c2nr30523b

[195] J. Li, D. Zhen, G. Sui, C. Zhang, Q. Deng, L. Jia, Nanocomposite of Cu-TiO_2-SiO_2 with high photoactive performance for degradation of rhodamine B dye in aqueous wastewater. J. Nanosci. Nanotechnol. 12 (2012) 6265-6270. https://doi.org/10.1166/jnn.2012.6438

[196] X. Zhang, Y. Zhu, X. Yang, S. Wang, J. Shen, B. Lin, C. Li, Enhanced visible light photocatalytic activity of interlayer-isolated triplex Ag@SiO_2@TiO_2 core-shell nanoparticles. Nanoscale 5 (2013) 3359-3366. https://doi.org/10.1039/c3nr00044c

[197] B. Yu, X. Jiang, J. Yin, Silica/titania sandwich-like mesoporous nanosheets embedded with metal nanoparticles templated by hyperbranched poly(ether amine) (hPEA). Nanoscale 5 (2013) 5489-5498. https://doi.org/10.1039/c3nr01336g

[198] Y. Yu, M. Zhu, W. Liang, S. Rhodes, J. Fang, Synthesis of silica-titania composite aerogel beads for the removal of Rhodamine B in water. RSC Adv. 5 (2015) 72437-72443. https://doi.org/10.1039/C5RA13625C

[199] K.P.O. Mahesh, Dong-Hau Kuo, Synthesis of Ni nanoparticles decorated SiO_2/TiO_2 magnetic spheres for enhanced photocatalytic activity towards the degradation of azo dye. Appl. Surf. Sci. 357 (2015) 433-438. https://doi.org/10.1016/j.apsusc.2015.08.264

[200] Z.-Y. Yang, G.-Y. Shen, Y.-P. He, X.-X. Liu, S.-J. Yang, Preparation of TiO_2/SiO_2 composite oxide and its photocatalytic degradation of rhodamine B, J. Porous Mater. 23 (2016) 589-599. https://doi.org/10.1007/s10934-015-0114-7

[201] A. Pal, T. K. Jana, K. Chatterjee, Silica supported TiO_2 nanostructures for highly efficient photocatalytic application under visible light irradiation. Mater. Res. Bulletin 76 (2016) 353-357. https://doi.org/10.1016/j.materresbull.2015.12.040

[202] M.U.D. Sheikh, G.A. Naikoo, M. Thomas, M. Bano, F. Khan, Solar-assisted photocatalytic reduction of methyl orange azo dye over porous TiO_2 nanostructures. New J. Chem. 40 (2016) 5483-5494. https://doi.org/10.1039/C5NJ03513A

[203] D. Maucec, A. Suligoj, A. Ristic, G. Drazic, A. Pintar, N.N. Tusar, Titania versus zinc oxide nanoparticles on mesoporous silica supports as photocatalysts for removal of dyes from wastewater at neutral pH. Catal. Today 2017. https://doi.org/10.1016/j.cattod.2017.05.061

Chapter 3

Preparation, Characterization and Applications of Visible Light Responsive Photocatalytic Materials

Arpita Pandey[1], Sangeeta Kalal[1], Nutan Salvi[1], Chetna Ameta[1], Rakshit Ameta[2] and Pinki B. Punjabi[1*]

[1]Photochemistry Laboratory, Department of Chemistry, University College of Science, M. L. Sukhadia University, Udaipur - 313002, Rajasthan, India

[2]Department of Chemistry, J.R.N. Rajasthan Vidyapeeth Deemed to be University, Udaipur - 313001, Rajasthan, India

E-mail: pb_punjabi@yahoo.com

Abstract

Today, scientists all over the world are looking for eco-friendly methods to treat polluted water for its reuse. Photocatalytic activity (PCA) depends on the ability of the catalyst to create electron–hole pairs, which generate free radicals (e.g. hydroxyl radicals: $^{\bullet}OH$), which undergo in secondary reactions as efficient oxidant under irradiation of light. Various applications of visible light active (VLA) photocatalytic materials, in terms of environmental remediation are- Elimination of several pollutants (e.g. alkanes, alkenes, phenols, pesticides, etc.) particularly in water treatment, disinfection and air purification, self-cleaning glass, photocatalytic concrete and paints; photoreduction of carbon dioxide; outdoor and indoor coatings of photocatalysts for roads and buildings. Therefore, the main objective of this proposal is to review the eco-friendly heterogeneous catalytic systems with low cost starting compounds with the benefit that no sludge formation is there and catalysts are reusable too.

Keywords

Photocatalyst, Semiconductor, Band Gap, Photocatalytic Degradation, Water Pollution, Advanced Oxidative Processes, Binary, Ternary and Quaternary Photocatalyst

Contents

1. Introduction

Environment is the surrounding which includes living things and natural forces. The environment of living things provides the conditions for development and growth, as well as it also causes danger and damage, if not cared for. Pollution is the introduction of contaminants into the environment that causes adverse changes. Water pollution is due to the presence of non-biodegradable dyes along with some other toxic pollutants like metals, acid, alkali, carcinogenic aromatic amines, etc. in the industrial effluents. Water pollution due to effluents coming from textile dyeing industry is a cause of serious concern. The techniques for detection of dyes are cost intensive and futile because the dyes undergo chemical changes under environmental conditions and the transformation products may be more toxic and carcinogenic than the parent molecule.

Dye wastes represent one of the most problematic groups of pollutants because they can be easily identified by the human eye and are not easily biodegradable. Dyes may be of a number of structural varieties like acidic, basic, disperse, azo, anthraquinone based and metal complex dyes. Weber and Adams [1] studied that during the coloration process; a large percentage of the synthetic dye does not bind and is lost to the waste stream. Approximately 10-15% dyes are released into the environment during dyeing process making the effluent, highly colored and aesthetically unpleasant. Wang et al. [2] reported that the effluent coming from textile industries carries a large number of dyes and other additives, which are added during the coloring processes. These are difficult to remove in conventional water treatment procedures and can be transported easily through sewers and rivers especially because they are designed to have high water solubility.

The detection of dyes is a difficult process because of the large variety of functional groups in different dyes and their diverse properties. Analytical procedures, which are used for determination of dyes at parts per million (ppm) levels, are very limited. Most of the dyes are non-volatile; and hence, gas chromatography cannot be used. Tincher and Robertson [3] studied the use of high pressure liquid chromatography (HPLC) and mass spectrometry for analysis of some of the dyes. Various processes, which are used for the treatment of dye waste, include biological treatment, catalytic oxidation, filtration, and sorption process and combination treatments. The textile industry is one of the largest polluters in the world. The World Bank estimates that almost 20% of global industrial water pollution comes from the dyeing of textiles.

A large amount of fresh water is being used for dyeing, rinsing, and treatment of textiles. A single T-shirt made from conventional cotton requires 2700 liters of water, and a third of a pound of chemicals. Millions of gallons of wastewater are discharged each year by mills and industries, which contain chemicals such as formaldehyde (HCHO), chlorine and heavy metals such as lead and mercury. These chemicals cause both environmental damage and cause human disease.

1.1 Textile manufacturing dyes release

- Aromatic amines (benzidine and toluidine), heavy metals, ammonia, alkali salts, toxic solids and large amounts of pigments
- Chlorine, a known carcinogen

Untreated dyes cause chemical and biological changes in our aquatic system, which threaten life of many fishes and aquatic plants. The enormous amount of water which is required by textile production competes with the growing daily water requirements of the half billion people that live in drought-prone regions of the world. By 2025, the number of inhabitants of drought-prone areas is projected to increase to almost one-third of the world's population. If global consumption of fresh water continues to double every 20 years, the polluted waters resulting from textile production will pose a great threat to human lives.

Phenols, also known as total phenols or phenolics, are important due to their widespread use in many manufacturing processes. Phenols pose a serious threat to many ecosystems, water supplies and human health because of their inertness, toxicity, endocrine disrupting abilities and carcinogenic behavior [4, 5]. The United States, Canada and the European Union have included some phenols in their list of priority pollutants [6–8]. Phenol is commonly employed in the manufacturing of phenolic resins, bisphenol A, caprolactam and chlorophenols such as pentachlorophenol [4]. Cresols are isomeric mono-substituted phenols. Commercially, cresol is produced as a by-product from the fractional distillation of crude oil and coal tars and the gasification of coal. Phenol and its derivatives have been identified as effluents coming from petroleum refining [9], pulp and paper manufacturing [10], coal processing [11] and chemical production facilities [12]. Oil-shale processing is another industry that produces effluents containing phenol and cresols [13]. Removing phenolic compounds from wastewaters and drinking water supplies has received widespread attention recently because of their toxic and endocrine disrupting properties [14]. Phenols can be removed by physical processes such as flocculation, precipitation, granular activated carbon (GAC) or reverse osmosis (RO) [15].

The presence of dyes or their degradation products in water even at very low concentration can cause human health disorders like nausea, haemorrhage, ulceration of skin and mucous membranes and can cause severe damage to kidneys, reproductive system, liver, brain and central nervous system. Hence, it is essential to devise some methods to successfully remove them.

Photocatalysis is a surface phenomenon initiated by the irradiation of UV/ visible light of higher energy than the band gap of the photocatalyst used. The surface properties involving the nature of active sites and its number play a key role in determining the photocatalytic reactivity of the catalysts. Photocatalysis has been regarded as a green, simple, and low-cost method to degrade dyes.

Gogate and Pandit [16] reviewed advanced oxidative processes (AOP), these are successful in removing complex organic contaminants because they can achieve complete oxidation. AOP offer a distinct advantage over many conventional treatment methods, such as biological processes, because faster degradation rates are accomplished and contaminants are degraded completely rather than being transferred from one phase to another. Wang et al. [17] observed that in advanced oxidative processes (AOP), there is no requirement for by-product disposal. Matilainena and Sillanpaa [18] studied that AOP processes can be configured using a combination of chemical and physical agents such as a combination of oxidizing agents, an oxidizing agent plus ultraviolet, catalyst or ultrasound and a catalyst plus ultraviolet. Glaze et al. [19] reported that the degradation of organics is mediated by the generation of •OH radicals in all AOP processes.

1.2 Visible light photoactive materials

Light is the electromagnetic radiation within a certain portion of the electromagnetic spectrum. The word usually refers to "visible light", which is visible to the human eye and is responsible for the sense of sight. Visible light is usually defined as having a wavelength in the range of 400 to 700 nm; between the infrared (with longer wavelengths) and the ultraviolet (with shorter wavelengths). The main source of light on Earth is the Sun. Heterogeneous photocatalytic technologies have been applied to control the organic pollutants and microorganisms in water. Development of narrow band-gap photocatalysts which function in the visible light remains a challenge in the wastewater treatment processes.

1.3 Binary photocatalysts

Binary photocatalysts are those catalysts, which are composed of two catalysts, which are active in the presence of light. Binary mixed TiO_2:SiO_2 was reported to have higher photocatalytic activity than a titania only. Jung and Park [20] reported that addition of silica into titania enhances the thermal stability as well as increases the specific surface area and acidity. Guillard et al. [21] have observed that the process of preparation for TiO_2:SiO_2 films can avoid the filtration problems of the slurry. Oh et al. [22] studied the photo-oxidative transparent TiO_2:SiO_2 films on glass which form the basis for self-cleaning indoor windows, lamps, and automotive windshields.

Do et al. [23] and Papa et al. [24] found that the degradation rates of 1, 4-dichlorobenzene was increased by a factor of 3 by calibrating 3 mole % addition of WO_3 and MoO_3 in titania. They found a strong correlation between surface acidity and reactivity. Surface acidity is thought to take the form of stronger surface hydroxyl groups. These hydroxyl groups accept the holes generated by illumination and as a result, oxidize adsorbed molecules. Hole traps such as the hydroxyl groups prevent the electron-hole recombination and therefore, increases quantum yield. Thus, the greater number of surface hydroxyl groups yields the higher reaction rate [25].

TiO_2–Al_2O_3 binary oxide surfaces were utilized in order to develop an alternative photocatalytic NO_x abatement approach, where TiO_2 sites were used for ambient photocatalytic oxidation of NO with O_2 and alumina sites were exploited for NO_x storage. To perform complete photocatalytic reduction of NO_x, an alternative NO_x abatement strategy has been demonstrated, which includes photocatalytic oxidation of NO_x on a TiO_2/Al_2O_3 binary oxide photocatalyst surface and its storage in the solid state in the form of nitrates and nitrites. This alternative strategy was inspired by some studies on NSR technology (NO_x storage and reduction), which is used for the thermal catalytic after treatment of automotive NO_x emissions [26-28].

Individual CdS photocatalyst has very low separation efficiency of photogenerated electron-hole pairs and undergoes photocorrosion, which limits its practical application [29]. In order to improve its photoactivity and to inhibit the photocorrosion, cadmium sulfide is usually coupled with other semiconductors, including CdS/TiO_2 [30–32], CdS/ZnO [33, 34], CdS/ZnS [35], CdS/WO_3 [36] as well as CdS/MoS_2 [37]. The rate of hydrogen evolution on MoS_2/CdS was higher than on CdS particles loaded with other metal catalysts such as Pt, Ru, Rh, Pd and Au. This result was explained by the better electron transfer between MoS_2 and CdS [37].

A novel Ti(IV) / Fe(III) mixed oxide catalyst has shown an increased photocatalytic activity for destruction of dichloroacetic acid at 450 nm [38]. The photocatalytic activity of iron doped TiO_2 or mixed oxide of different proportions of Fe and Ti, prepared by impregnation and coprecipitation methods has been shown to be active in the visible light for the photoreduction of N_2 [39], degradation of oligocarboxylic acid [40], 4-nitrophenol [41] and $CHCl_3$ oxidation [42] etc. Activity of these catalysts was higher than that of pure hydrous oxide prepared under similar conditions. Navio et al. [43] have prepared the specimen with higher Fe (III) content and showed its photoactivity in the visible light. It has also been found that the catalytic activity of Fe/Ti mixed oxides largely depends on preparation methods, iron content, sintering temperature and phase composition.

The measurement of photoluminescence of the photocatalysts is one of the most important and useful ways to elucidate the surface properties related to adsorption, catalysis, and photocatalysis and reactivity of photocatalyst [44]. The photoluminescence spectrum of TiO_2 nanoparticles is efficiently quenched by the addition of oxygen onto the surface through an increase in the extent of the band bending of TiO_2 photocatalyst due to the adsorption of O_2^- species [45, 46]. The relative reactivity of photocatalysts toward oxygen molecule was estimated by the quenching degree of the photoluminescence by the addition of O_2. The intensity of the photoluminescence was decreased to 89% of its original intensity for SiO_2/TiO_2, 37% for TiO_2 and 26% for boron-SiO_2/TiO_2, respectively in the presence of 20 Torr of O_2 [47].

Yang and Swisher [48] reported that the photocatalytic efficiency of the photocatalysts can be enhanced by coupling ZnO with TiO_2. Wang et al. [49] reported that binary oxides can provide a more efficient charge separation, increased lifetime of charge carriers and enhanced interfacial charge transfer to absorbed substrates. Liao et al. [50] studied the preparation and photocatalytic activity of the binary oxide photocatalyst ZnO/TiO_2 in the degradation of methyl orange dye. It was shown that the addition of ZnO enhanced the photocatalytic activity of TiO_2 significantly.

1.4 Ternary photocatalysts

Graphene is proposed to serve as an efficient acceptor for the photogenerated electrons in graphene–TiO_2 nanocomposites and thus, significantly suppresses the charge recombination and enhancing the photocatalytic rate of the nanocomposite as compared to that of pure TiO_2 nanoparticles (NPs) [51,52]. Lin et al. [53] reported that a ternary nanocomposite of graphene–TiO_2– Fe_3O_4 (GTF) possesses the integrated functions as well as a low-cost, recollectable and stable photocatalyst for the degradation of organic dyes namely rhodamine B (Rh B), methyl orange and acid blue 92.

Bi_2WO_6 possess interesting physical properties such as ferroelectric piezoelectricity, pyroelectricity, catalytic behavior, and a non-linear dielectric susceptibility [54,55]. Bi_2WO_6 show a great potential as a visible-light-active photocatalyst for organic compound degradation and O_2 evolution under visible light irradiation [56, 57]. Tang et al. [58] reported that Bi_2WO_6 showed the activity of mineralizing both; $CHCl_3$ and CH_3CHO contaminants under visible light irradiation.

The sphere-like bismuth tungstate (Bi_2WO_6) was synthesized with silica protection calcination by using a facile hydrothermal method. The flower-like bismuth tungstate (Bi_2WO_6) was also synthesized without silica protection. Natrajan et al. [59] reported the use of sphere and flower-like Bi_2WO_6 for the degradation of reactive black 5 (RB 5) in the presence of visible light. The bandgap of mesoporous nanocrystalline TiO_2 was modified with Li, Mg, Pd and Sr, metal ion salt solution using wet impregnation. Ion impregnated samples were used as photocatalysts for the degradation of organic contaminants [60]. Pathak et al. [61] reported that the nanocrystals of Mg-Mn ferrites were used for the degradation of nitrobenzene.

Ternary chalcogenide nanocrystals such as $Cd_{1-x}Zn_xS$ [62], $Cd_{1-x}Zn_xSe$ [63], and $CdS_{1-x}Se_x$ [64] have received considerable attention because their composition-tunable band edge offers new opportunities to harvest light energy in the entire visible region of solar spectrum. A ternary chalcogenide nanocrystal shows various potential applications in a wide array of fields, which include photovoltaics [65], photocatalysis [66], and photoelectrochemical cells [67]. When photons are completely absorbed, photon energy that exceeds semiconductor band gap is dissipated as heat due to the vibrational relaxation of excitons [68]. Because of the heat loss, a substantial amount of solar energy has already been consumed, before it can be converted into other accessible energies. This obstacle needs to be circumvented to promote the advancement of photocatalysis technology for solar fuel production.

Aguiar et al. [69] reported that the rare earth oxides are broadly used in luminous materials, polishing powder, electronic ceramics, and efficient catalyst in photocatalytic field. Heteropoly compounds like $H_3PW_{12}O_{40}/TiO_2$ [70], $NdPW_{12}O_{40}/TiO_2$ [71], $BiPW_{12}O_{40}$ [72], $Cu_3(PW_{12}O_{40})_2$ [73] had high photocatalytic activities in photocatalytic elimination of formaldehyde, acetone, and methanol.

Wide band gap p-block metal semiconductors like Ga_2O_3, In $(OH)_3$, InOOH and $Sr_2Sb_2O_7$ [74] show high activity and stability in the photocatalytic degradation of benzene. The high photocatalytic performance observed over these wide band gap p-block metal semiconductors is related to their peculiar electronic structure.

Wide band gap endows the photogenerated charge carriers in these semiconductors with redox ability strong enough to react with surface adsorbed H_2O to produce $^{\bullet}OH$ radicals. In this way, the degradation of benzene can proceed via the $^{\bullet}OH$ radical pathway, which makes these wide band gap p-block semiconductors capable of maintaining a clean catalyst surface and a long stability during the photocatalytic reaction studies. The wide band gap p-block metal semiconductors are a new generation of photocatalysts for benzene degradation. Besides this, except the common characteristics like wide band gap and the dispersive conduction band, other factors influencing their photocatalytic activity for benzene degradation remain largely unclear. For this purpose, more wide-band gap p-block metal semiconductors should be investigated especially ternary ones due to their diversified crystallographic and electronic structure and their photocatalytic activity.

Ag_3PO_4 also has a quantum efficiency of 90% at wavelengths longer than 420 nm [75] and shows high photocatalytic activity owing to its absorption in the visible portion of the solar spectrum and high charge carrier mobility because of the delocalized charge distribution of the conduction-band minimum, which results in a small electron effective mass, beneficial for the surface carrier mobility [76]. Graphitic carbon nitride (g-C_3N_4) is based on the stacked two-dimensional structure analogous of graphite with N replacing non-adjacent carbon atoms. Wang et al. [77] used graphitic carbon nitride (g-C_3N_4) in water splitting application as a metal-free photocatalyst operating under visible light irradiation. Various g-C_3N_4 based hybrid photocatalysts including g-C_3N_4/$BiPO_4$ [78], graphene/g-C_3N_4 [79], g-C_3N_4/ Bi_2WO_6 [80], Fe-g-C_3N_4-LUS-1 [81], g-C_3N_4/SiO_2-HNb_3O_8 [82], and g-C_3N_4/TaON [83] have been developed to further extend the visible light absorption range and the photogenerated carriers separation efficiency. Shen et al. [84] reported that Ag/Ag_3PO_4/g-C_3N_4 hybrid system exhibits superior photocatalytic performance for rhodamine B degradation compared with a binary Ag_3PO_4/g-C_3N_4 photocatalyst, or Ag_3PO_4 and g-C_3N_4 as individual components. It was found that this photodegradation strongly depends on the proportion of silver nanoparticles on the Ag_3PO_4 surface and the ratio of the components in the Ag_3PO_4/g-C_3N_4 hybrid. The enhanced photoactivity can be attributed to the surface plasmon resonance (SPR) originating from silver nanoparticles and the heterojunction-like interface between Ag_3PO_4 and g-C_3N_4.

Kim et al. [85] reported the synthesis of CdS/TiO_2/WO_3 ternary hybrid systems as new photoactive composites and found that the ternary hybrid exhibited much higher photocatalytic activity than that of CdS alone or binary hybrids. Lin et al. [86] prepared CdS nanoparticles/ZnO shell/TiO_2 nanotube (NT) arrays for water splitting and observed that the conversion efficiency increased from 0.39% to 1.30%. Chen et al. [87] described a new method to form a ZnO energy barrier layer between TiO_2 NTs and CdS quantum dots, which exhibited improved efficiency of the quantum dots-sensitized solar cells.

1.5 Quaternary photocatalysts

Quaternary thin films were reported for use as the photoabsorber in solar cells because of direct band gap and high optical absorption coefficient. Cu_2ZnSnS_4 (CZTS) thin films are used as the photoabsorber in photovoltaic cells because of its low cost, non-toxic constituents, direct band gap and high optical absorption coefficient as high as 104 cm^{-1}.

Bwamba et al. [88] reported the optical absorption studies of the annealed CZTS films. The absorption spectra indicated that annealed films have a high absorbance of light in the visible region. Yeh and Cheng [89] reported the preparation of the Ag-Zn-Sn-S quaternary thin films using chemical bath deposition technique. The X-ray diffraction patterns of the samples reveal that tetragonal Ag_2ZnSnS_4 phase, with small amount of impurities such as Ag_8SnS_6 and SnS can be obtained using TEA as the chelating agent and deposition temperature 70 °C.

Quaternary metal oxide photocatalysts such as $Ni_xIn_{1-x}TaO_4$ [90], $K_3Ta_3Si_2O_{13}$ [91], and $Sm_2Ti_2O_5S_2$ [92] display enhanced photocatalytic activity in contrast to the conventional photocatalysts viz. $TiO_{2-x}N_x$ [93], TaON [94] and $(Ga_{1-x}Zn_x)$ $(N_{1-x}O_x)$ [95] because such crystal lattice possess large population of catalytically active sites [92]. Because of its low band gap, Pb_2FeNbO_6 (PFNO) is used in ferroelectrics, sensors, photocatalysis, photoelectrochemical hydrogen production [96].

$PbBiO_2Br$ (anisotropic 2D structure) has been used as a photocatalyst due to a band gap of 2.47 eV and its unique layered structure [97]. In anisotropic 2D structure, photoinitiated charge carriers experience two kinds of the confinement. The strong confinement (thickness) is necessary to increase sufficiently the free energy of conduction band electrons for photocatalytic reaction; meanwhile, the weak confinement (length and width) [98] is needed to facilitate effective delocalization of longer-living excitons and separated charges [99]. Therefore, the probability of photoinduced electron-hole recombination is effectively minimized.

Nb-containing oxides such as $CsBiNb_2O_7$, $CsBi_2Nb_5O_{16}$, and $PbBi_2Nb_2O_9$ were used for the decomposition of gaseous 2-propanol in the presence of visible light. The photocatalytic activity of $CsBi_2Nb_5O_{16}$ was higher than that of $CsBiNb_2O_7$ [100] and $PbBi_2Nb_2O_9$ (Eg = 2.88 eV) was much more active than $TiO_{2-x}N_x$ (Eg = 2.73 eV) [101]. $RbBi_2Nb_5O_{16}$ and $RbBiNb_2O_7$ decomposed gaseous acetaldehyde and $RbBi_2Nb_5O_{16}$ showed a higher photoactivity than $RbBiNb_2O_7$ in accordance with their different absorption properties [102]. $LiBi_4M_3O_{14}$ (M = Nb, Ta) was used for the degradation of various dyes and phenolic compounds under UV irradiation [103]. $LiBi_4Nb_3O_{14}$ was more efficient for the degradation of the phenolic compounds, due to its lower band gap (3.0 eV) and preferential affinity of niobium for the phenolic functional groups. Bi_2InTaO_7 obtained by the sol–gel method showed a high efficiency for the UV-induced degradation of alizarin red S [104]. Li et al. [105] synthesized polycrystalline Ag_2ZnGeO_4 at 220 °C by the cation exchange method with Na_2ZnGeO_4 as the parent compound. The Ag_2ZnGeO_4 sample showed good activity for the photodegradation of rhodamine B and orange II. $PbBiO_2Br$, tested for the degradation of methylene blue and methyl orange, and it was found more photocatalytically more active than $PbBi_2Nb_2O_9$, $TiO_{2-x}N_x$ and BiOBr under visible light [106].

2. Preparation and characterization of various photocatalysts:

2.1 Binary photocatalysts

2.1.1 Preparation of TiO_2:SiO_2 thin films

Solutions of $(NH_4)_2TiF_6$ (0.1 mol·L^{-1}) and $(NH_4)_2SiF_6$ with a Si/Ti atomic ratio of 15% in treated solutions were mixed as a parent solutions for the liquid phase deposition (LPD) process [107]. Very small amounts of nanocrystalline TiO_2 and amorphous SiO_2 seeds were dissolved in the mixed solutions. The solutions were stirred for 24 h and then filtered. After the introduction of H_3BO_3 having a concentration of 0.3 mol·L^{-1} into the precursor solutions, the chemical reactions cause the formation of TiO_2 and SiO_2 precipitates. The slide-glass substrates were ultrasonically cleaned with diluted nitric acid, ethanol, and distilled water. These were then immersed vertically into solutions of the reactive precursor for 9 h. During the deposition, the precursor solutions were kept at a constant temperature of 35 °C. The TiO_2 and the SiO_2 precipitates in the solutions were deposited on glass substrates. The deposited films were washed in distilled water, dried naturally at ambient temperature, and annealed at temperatures ranging from 100 to 500 °C in intervals of 100 °C for 1 h in air.

From SEM of annealed TiO_2:SiO_2 thin films, it was concluded that many interspaces among the particles and nanopores on the top of TiO_2 particles are present at low temperatures due to the stress at the interface between TiO_2 and SiO_2. Higher was the annealing temperature, the larger is the particle size, and the fewer will be the interspaces and nanopores. The specific surface areas of the samples gradually decrease with increasing annealing temperature due to the formation of larger clusters and to the disappearance of nanopores as a result of stress release after annealing. A decrease in the surface area means a declined number of active sites, where the holes are bound with hydroxyls and participate in the redox reaction, which implies a decrease in the photocatalytic activity [108].

In FTIR spectra of binary TiO_2:SiO_2 thin films, a band was observed at 3700 cm^{-1} due to the stretching vibration of hydroxyl groups, but the Ti-OH stretching vibration bands become much weaker when the annealing temperature was higher than 100 °C, which means that a certain number of OH groups is progressively removed during the annealing. The disappearance of the Ti-OH vibration modes after annealing at temperature above 300 °C indicates directly an elimination of hydroxyl groups.

2.1.2 Preparation of TiO_2/Al_2O_3 binary oxides

Titanium (IV) isopropoxide (TIP) and aluminum-tri-sec-butoxide (ASB) were used as the main ingredients in the preparation of the TiO_2/Al_2O_3 binary oxides via sol–gel method [109-111]. The samples are labeled as "xTi/Al-y", where x represents the TiO_2 to Al_2O_3 mole ratio (i.e. 0.25, 0.5 and 1.0) and y represents the calcination temperature (150–1000 °C) of the sample. Depending on the corresponding TiO_2–Al_2O_3 mole ratio, an appropriate amount of ASB was mixed with propan-2-ol and acetylacetone for 30 min during the synthesis of TiO_2/Al_2O_3 binary oxide. TIP was added in a drop wise fashion to the mixture over the course of another 30 min. All of the synthesis steps were carried out at room temperature under vigorous stirring. The co-precipitation of the obtained hydroxides was accomplished after the gradual addition of 0.5 M HNO_3 (aq) to the solution, which leads to the formation of a gel. The resulting yellow gel was aged under ambient conditions for 2 days and the dried sample was ground to form a fine powder. Now, synthesized TiO_2/Al_2O_3 binary oxides were calcined in air for 2 h at various temperatures ranging from 150 to 1000 °C.

2.1.3 Characterization of TiO_2/Al_2O_3 binary oxides

Diffuse reflectance UV–VIS (DR-UV–VIS) spectra were utilized in order to obtain electronic band gap values (Fig. 1). Fig. 1 indicates that for the 1.0 Ti/Al photocatalyst family, no significant photocatalytic activity was detected up to 800 °C, while at this calcination temperature, a remarkable increase in the activity was observed, though this catalyst is not as effective as the 05 Ti/Al-900 catalyst in total NO_x abatement, due to the significant the significant NO_2 (g) generation of the former. It was observed that for calcination temperatures above 800 °C, NO_x abatement starts to fall, as evident by the increased NO_2 (g) slip into the atmosphere as well as decreasing NO_x storage in the solid state. 0.5 Ti/Al-900 binary oxide catalyst shows the highest NO_x abatement performance among all of the analyzed photocatalysts, where it performs 160% higher NO_x storage and 55% lower NO_2 (g) release to the atmosphere compared to the Degussa P25.

All of the NO_x adsorption/storage (i.e. Al_2O_3) sites cannot be utilized at low TiO_2 loadings, due to limited photo-oxidation capability of the inadequate number of TiO_2 oxidation sites on the surface. At very high TiO_2 loadings, TiO_2 covers most of the Al_2O_3 surface and upon calcination above 800 °C, SSA of the catalyst sample falls drastically together with the formation of crystalline anatase and rutile mixture; limiting the available number of NO_x storage sites that are available after photooxidation. Onset of photocatalytic activity was observed in a rather sharp manner at 950, 900 and 800 °C for the 0.25 Ti/Al, 0.5 Ti/Al and 1.0 Ti/Al samples, respectively. As the relative TiO_2 loading in the TiO_2/Al_2O_3 binary oxide samples increases, onset temperature for the photocatalytic activity shifts to lower temperatures.

Electronic band gap values for the relatively inactive amorphous Ti/Al samples reveal a characteristically high value within 3.4–3.6 eV, which are calcined at lower temperatures. With the onset of the photocatalytic activity, a very sharp fall in the electronic band gap values were observed, where the band gap decreases to a typical value of 3.05–3.10 eV, in line with the formation of ordered anatase and rutile phases. Typical band gap values for bulk anatase and rutile phases are ca 3.2 and 3.0 eV, respectively [112]. Thus, for the active photocatalyst samples, the band gap value is in between that of anatase and rutile, being closer to the latter, in accordance with the fact that in the most active photocatalyst, rutile exists as the predominant phase together with anatase as the minority phase. Electronic band gap cannot be used as a sole indicator for the estimation of the photocatalytic activity trends due to the fact that once the photocatalytically active structure is obtained leading to a drastic decrease in the electronic band gap. Band gap values cease to change at higher calcination temperatures although photocatalytic activity starts to decline.

2.1.4 Preparation of binary CdS-MoS_2 composites

A typical hydro/solvothermal mixed solution process was used for the preparation of CdS-MoS_2 composites with various molar ratios of semiconductors [113]. The prepared mixture was placed in the autoclave and heated at 200°C for 24 h. The resulting precipitate was washed with distilled water and ethanol, respectively and dried in the oven at 70°C for 8 h.

2.1.5 Characterization of binary CdS-MoS_2 composites

SEM images of binary CdS-MoS_2 composites with different CdS and MoS_2 molar ratios are shown in Fig. 1. In the case of CdS-MoS_2 5-1 sample containing the highest molar ratio of CdS, nanoleaf structure indicating the presence of CdS was observed. The increase of MoS_2 ratio resulted in the formation of hexagonal shaped nanostructures with average edge size of about 100–125 nm (Fig. 1b). Further increase in molar ratio of MoS_2 to CdS causes a large change in microstructures (Fig. 1c and 1d). Bonded structures of microspheres with diameters ranging from 0.08 to 1 μm were observed.

Fig. 2 depicts the spectra for pure CdS and MoS_2 semiconductors and their binary composites with varying molar ratio between CdS and MoS_2. The absorption edge of single CdS is about 510 nm, which coincides with the literature. It was previously reported that the absorption properties of CdS are strongly shape-dependent [114]. Liu et al. [115] reported enhanced absorption properties for CdS-MoS_2 composites for photocatalytic H_2 production. As compared with single CdS, composites with excess of CdS exhibited a red shift and a less steep absorption edge as well as absorption was more intense for composites containing excess MoS_2.

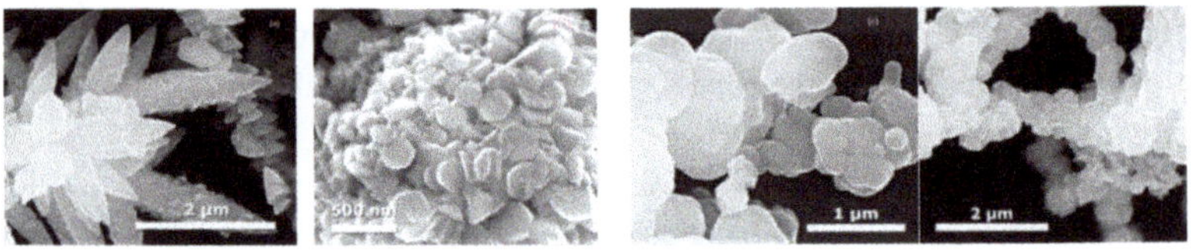

(a) (b) (c) (d)

Figure 1. SEM images of binary CdS-MoS$_2$ composites obtained by solvothermal mixed solution methods with different molar ratio of CdS: (a) CdS:MoS$_2$ = 5:1; (sample CdS-MoS$_2$ 5-1); (b) CdS:MoS$_2$ = 4:1 (sample CdS-MoS$_2$ 4-1); (c) CdS:MoS$_2$ = 1:1 (sample CdS-MoS$_2$ 1-1); and (d) CdS:MoS$_2$ = 1:5 (sample CdS-MoS$_2$ 1-5). (Reproduced with permission from Beata Bajorowicz, Anna Cybula, Michał J. Winiarski, Tomasz Klimczuk and Adriana Zaleska, Surface properties and photocatalytic activity of KTaO$_3$, CdS, MoS$_2$ semiconductors and their binary and ternary semiconductor composites, Molecules 2014, 19, 15339-15360).

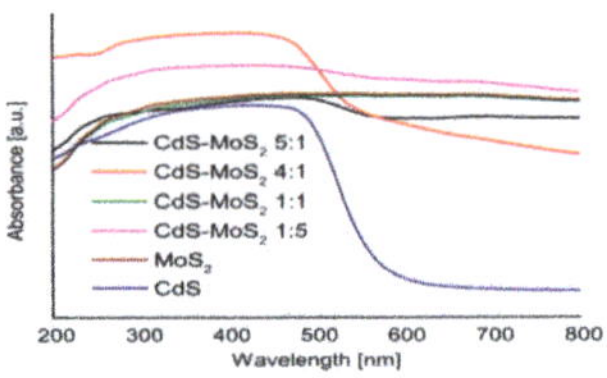

Figure 2. The UV-Vis diffuse reflectance spectra of single CdS, MoS$_2$, and binary CdS-MoS$_2$ nanocomposites. (Reproduced with permission from Beata Bajorowicz, Anna Cybula, Michał J. Winiarski, Tomasz Klimczuk and Adriana Zaleska, Surface properties and photocatalytic activity of KTaO$_3$, CdS, MoS$_2$ semiconductors and their binary and ternary semiconductor composites, Molecules 2014, 19, 15339-15360).

2.2 Ternary photocatalysts

2.2.1 Preparation of $KTaO_3$-CdS-MoS_2 composites

A typical hydro/solvothermal mixed solutions process was also used for the preparation of $KTaO_3$-CdS-MoS_2 composites with various molar ratios of semiconductors [113]. The prepared mixture was placed in the autoclave and heated at 200°C for 24 h. The resulting precipitate was washed with distilled water and ethanol, respectively and dried in the oven at 70°C for 8 h.

2.2.2 Characterization of $KTaO_3$-CdS-MoS_2 composites

SEM of $KTaO_3$-CdS-MoS_2 composites are given in Fig. 3. The transition of the $KTaO_3$ structure from cubic to octahedral was observed for the samples $KTaO_3$-CdS-MoS_2 10-1-1 obtained by solvothermal mixed solutions. Solvothermal mix solutions method did not cause dramatic changes in the structure cubes nor was nano leaf observed (Fig. 3c). In calcinated $KTaO_3$-CdS-MoS_2 10-5-1 composite, nano leaves of CdS was deposited on the surface of large cubes of $KTaO_3$ (Fig. 3d).

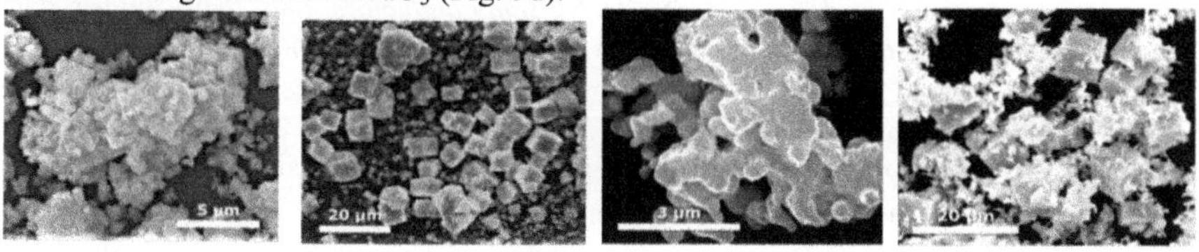

(a) (b) (c) (d)

Figure 3. SEM images of ternary $KTaO_3$-CdS-MoS_2 composites obtained with different molar ratio and using two different preparation route: (a) $KTaO_3$-CdS-MoS_2 (10:1:1) obtained by solvo thermal mixed solutions (sample $KTaO_3$-CdS-MoS_2 10-1-1_MS); (b) $KTaO_3$-CdS-MoS_2 (10:1:1) obtained by calcination of single previously synthesized semiconductors (sample $KTaO_3$-CdS-MoS_2 10-1-1_C); (c) $KTaO_3$-CdS-MoS_2 (10:5:1) obtained by solvo thermal mixed solutions (sample $KTaO_3$-CdS-MoS_2 10-5-1_MS); and (d) $KTaO_3$-CdS-MoS_2 (10:5:1) obtained by calcination of single previously synthesized semiconductors (sample $KTaO_3$-CdS-MoS_2 10-5-1_C). (Reproduced with permission from Beata Bajorowicz, Anna Cybula, Michał J. Winiarski, Tomasz Klimczuk and Adriana Zaleska, Surface properties and photocatalytic activity of $KTaO_3$, CdS, MoS_2 semiconductors and their binary and ternary semiconductor composites, Molecules 2014, 19, 15339-15360).

Fig. 4 shows the spectra for ternary $KTaO_3$-CdS-MoS_2 composites containing different amount of CdS prepared by various methods. A steep absorption edge (at about 510 nm) was observed only for a calcined composite. The best adsorption properties could probably be achieved for the ternary $KTaO_3$-CdS-MoS_2 composites containing appropriate molar ratio between semiconductors.

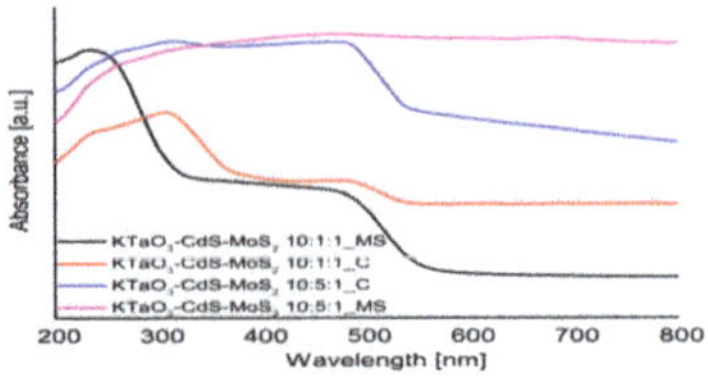

Figure 4. The UV-VIS diffuse reflectance spectra of ternary $KTaO_3$-based nanocomposites. (Reproduced with permission from Beata Bajorowicz, Anna Cybula, Michał J. Winiarski, Tomasz Klimczuk and Adriana Zaleska, Surface properties and photocatalytic activity of $KTaO_3$, CdS, MoS_2 semiconductors and their binary and ternary semiconductor composites, Molecules 2014, 19, 15339-15360).

2.2.3 Graphene–TiO_2–Fe_3O_4 (GTF)

A ternary, hybrid graphene–semiconductor–magnetic nanocomposite (graphene–TiO_2–Fe_3O_4; GTF), possesses the integrated functions [116] as:

(i) TiO_2 nanoparticles (NPs) act as a semiconductor photocatalyst to degrade the dye,

(ii) Graphene provides an effective electron pathway to suppress the charge recombination in TiO_2 and enhance its photocatalytic activity, and

(iii) Fe_3O_4 nanoparticles (NPs) act as a magnetic material for magnetic separation.

2.2.4 Preparation of graphene–TiO_2–Fe_3O_4 (GTF)

For the synthesis of GTF, graphene–TiO_2 (43 mg) was suspended in deionized water (30 mL) in a round-bottomed flask under N_2. An aqueous solution (1.56 mL) containing $FeCl_3{\cdot}6H_2O$ (3.95 mg) and $FeCl_2{\cdot}4H_2O$ (60 mg) was then injected into the flask using a pipette [117]. After stirring for 5 h under N_2, the flask was sealed after adding NH_4OH aqueous solution (4.5 mL, 1.5 m). The reaction was allowed to continue at 65 °C for 2.5 h. The product was collected by centrifugation and washed three times with water to remove excess ions before characterization. The amount of graphene–TiO_2 precursor employed was kept the same and the amounts of $FeCl_3{\cdot}6H_2O$ and $FeCl_2{\cdot}4H_2O$ were varied to obtain GTF with different ratios of Fe_3O_4 to TiO_2.

2.2.5 Characterization of graphene–TiO_2–Fe_3O_4 (GTF)

In as-obtained GTF nanocomposite, Fe_3O_4 and TiO_2 NPs can be easily distinguished by TEM image based on the differences in their size and contrast. The average size of TiO_2 and Fe_3O_4 NPs was estimated to be ca. 20 and ca. 80 nm, respectively. The powder XRD pattern of the GTF confirmed the cubic magnetite crystal structure for Fe_3O_4 and the anatase phase with tetragonal crystal structure for TiO_2.

2.2.6 Preparation of Bi_2WO_6 nanoplates

Bi_2WO_6 nanoplates were synthesized through a hydrothermal process. The starting mixture was allowed to react in a teflon-lined autoclave at different temperatures to obtain well-crystallized nanoplates. Crystal diffraction peaks were found when the temperature was not less than 120 °C. This photocatalyst belongs to the orthorhombic system, space group Pca21. The Bi_2WO_6 crystal with a layered structure includes the corner-shared WO_6. Bi atom layers are sandwiched between WO_6 octahedral layers [118,119].

2.2.7 Characterization of Bi_2WO_6 nanoplates

Zhang et al. [120] prepared flower-like Bi_2WO_6 spherical superstructures by employing the similar hydrothermal method, but with an acidic (pH = 1) precursor. The flower-like microstructures were found to be constructed from nanoplates with single crystal structure. The formation of this flower-like structure was proposed to follow a three-step process: self-aggregation, Ostwald ripening and self-organization (Fig. 5).

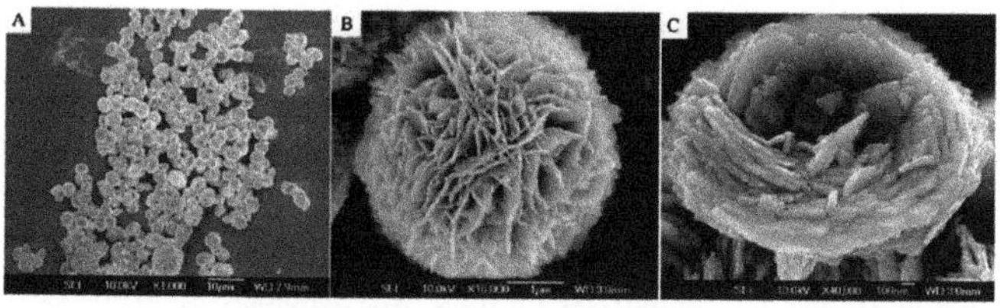

Figure 5. SEM images of flower-like Bi_2WO_6 superstructures [120]. (Reproduced with permission from Liwu Zhang and Yongfa Zhu, A review of controllable synthesis and enhancement of performances of bismuth tungstate visible-light-driven photocatalysts, Catal. Sci. Technol., 2012, 2, 694–706).

Zhang et al. have studied the photocatalytic properties of Bi_2WO_6 micro/nano-structures, including nanoplates, tyre/helix-like structure, disintegrated-flower-like and flower-like superstructures [121,122] These Bi_2WO_6 micro/nanostructures were found to exhibit different photocatalytic activities under visible light irradiation (Fig. 6). Among these photocatalysts, the uncalcined flower-like Bi_2WO_6 superstructure possesses the highest photocatalytic performance, while the nanoplates structure shows the lowest activity in the photocatalytic degradation of rhodamine B under visible light irradiation. The calcination process can be further improved the photocatalytic performance of the flower-like Bi_2WO_6 superstructure.

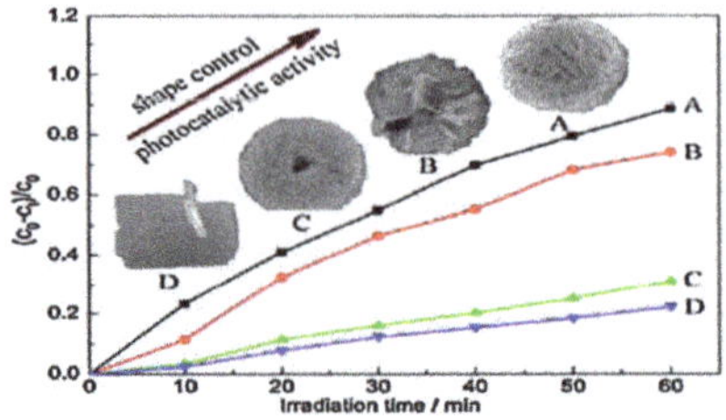

Figure 6. The photocatalytic activities of different Bi_2WO_6 micro/nanostructures: (A) uncalcined flower-like structure, (B) uncalcined disintegrated-flower-like structure, (C) calcined tyre/helix-like structure, (D) uncalcined nanoplates structure [116]. (Reproduced with permission from Liwu Zhang and Yongfa Zhu, A review of controllable synthesis and enhancement of performances of bismuth tungstate visible-light-driven photocatalysts, Catal. Sci. Technol., 2012, 2, 694–706).

2.3 Quaternary photocatalysts

2.3.1 Synthesis of $FeNbO_4$ and Pb_2FeNbO_6 photocatalysts

The $FeNbO_4$ (FNO), Pb_2FeNbO_6 (PFNO) and PFNO:FNO composite photo catalysts were prepared by the conventional solid state reaction method [123-125]. Thus, respective precursors viz. PbO, Fe_2O_3, Bi_2O_3 and Nb_2O_5 were mixed in desired stoichiometric ratio and ground in presence of isopropyl alcohol, and then calcined at required temperature. (i) The pelletized mixture was calcined at 1150 °C for 4 hrs to obtain a pure phase monoclinic phase FNO. (ii) The FNO powder obtained in previous step was used. A stoichiometric number of moles of FNO and PbO were mixed and calcined at 800 °C for 3 hrs. Further this mixture was pelletized and calcined again at 1125 °C for 2 hrs to get PFNO. (iii) A stoichiometric ratio of Pb: Fe: Nb: : 1 : 2 : 2 was mixed. The pelletized mixture was heated at 800 °C followed by sintering in the temperature range of 1000-1215 °C for more than 24 hrs under static air atmosphere to obtain PFNO: FNO composite system.

2.3.2 Characterization of $FeNbO_4$, and Pb_2FeNbO_6 photocatalysts

The crystal structure of photocatalyst was characterized by X-ray diffractometer. The X-ray diffraction spectra were compared with the standard data in the JCPDS (Joint Committee Powder Diffraction Standards) for identification of the phase. The diffuse reflectance spectrum (DRS) of the photocatalyst powder was recorded in the wavelength range of 300-800 nm. The X-ray diffraction (XRD) spectra of the photocatalysts viz. FNO, PFNO, and PFNO:FNO indicate the formation of respective phases of monoclinic, cubic and composite systems. XRD spectra for the PFNO:FNO precursors calcined at various temperatures between 1000-1200 °C exhibited a mixture of respective PFNO and FNO. However, some impurity peaks were also observed due to Fe_2O_3.

UV-Vis diffuse reflectance spectra of the PFNO:FNO samples calcined at various temperatures exhibit a sharp absorption edge. This absorption edge at still larger wavelength indicates the larger absorption capability of the photocatalyst. The effect of calcination temperature was also observed. The band gap lies in the range of 1.92 eV (PFNO) to 2.04 eV (FNO), the constituent counterparts for PFNO:FNO.

2.3.3 About HfTiErO

High-k gate dielectrics were developed to replace conventional SiO_2 as alternative gate dielectric materials to resolve the problem of device compatibility, reliability and high gate leakage currents [126]. HfO_2 is a suitable high-k gate dielectrics material for integration in complementary metal oxide semiconductor (CMOS) devices because of its good thermal stability and wide band gap on Si [127]. RF sputtering is PVD based technique combined with plasma, it offers a low temperature processing and blocks the oxygen from the ambience, which prevents the formation of interfacial layer (IL) [128]. Therefore, RF sputtering technique explores the possibility of obtaining the good quality high-k gate dielectric thin films, which is the most important requirement of advanced CMOS technology.

2.3.4 Preparation of HfTiErO

Hafnium disk of 99.99% purity with the diameter of 60 mm was used as the sputtering main target, while other targets titanium and erbium were fixed on the main target disk [129]. The sputtering chamber was evacuated with lowest pressure about 3×10^{-4} Pa before Ar and O_2 gases were used. The target was pre-sputtered in argon (Ar) ambient for 10 min in order to remove surface oxide on the target, prior to HfTiErO deposition. During the deposition process, the RF power, substrate temperature, working pressure, substrate-to-target distance, and total gas-flow rate were kept at 100 W, 250 °C, 0.7 Pa, 5.5 cm, and 0.4 ratio, respectively in order to obtain HfTiErO films with different Ti and Er contents. Different Ti and Er materials were introduced on hafnium oxides, after deposition of the ultra-thin films annealed in N_2 atmosphere for 60 s at 500 °C and 700 °C temperatures.

2.3.5 Characterization of HfTiErO

The microstructure and chemical composition of the deposited material were characterized by using atomic force microscopy (AFM) and X-ray photoelectron spectroscopy (XPS).

From AFM, it was concluded that the surface of both as-deposited and annealed films (high-k materials), was very significant because this affects the electrical properties of dielectric thin films and induces shifts in electronic energy levels [130]. It was observed that the surface roughness of films was improved slightly by increasing the erbium concentration. The films display a homogeneous and smooth surface structure, by means of a homogeneous material distribution with a low surface roughness after annealing temperatures at 500 °C and 700 °C. From AFM analysis of both; as deposited and annealed films, it was observed that the root mean square (RMS) surface roughness for S1, S2, S3, S4 and S5 samples are approximately 5.0 nm, 4.0 nm, 3.8 nm, 3.0 nm and 3.2 nm, respectively. The surface of S1 and S2 samples consist of dense narrow spikes in shape, whereas the samples S4 and S5 are more mountain like, flat and void free. Cho *et al.* [131] reported that the amorphous structure of the thin film was changed into the polycrystalline structure after annealing; due to the increase of the grain size, which resulted in an increase of the surface roughness. This shows that the surface of films becomes smoother after annealing process. The diffusion and mobility of the surface atoms can be increased by increasing the annealing temperature, which provides energy to surface atoms. Due to transfer of such atoms to existing voids and defects, the surface gets smoother and as a consequence, the surface roughness is reduced.

From XPS the samples S1, S2, S3, S4 and S5 (S1- $HfTi_6O$, S2- $HfTi_6Er_2O$, S3- $HfTi_6Er_4O$, S4- $HfTi_6Er_4O$ at 500 °C and S5- $HfTi_6Er_4O$ at 700 °C) were analysed. It was observed that a series of peaks from Hf 4f, Er 4d, Hf 4d, C 1s, Ti 2p and O1s arises from the surface contamination of adventitious carbon. Annealing leads to the removal of C 1s peaks for the samples S2 and S4. Effects of experimental charging were corrected by setting the C 1s peak for adventitious carbon at 284.6 eV.

2.3.6 Synthesis of $PbBiO_2Br$ nanosheets samples

Assembly ultrathin $PbBiO_2Br$ nanosheets samples were synthesized using a solvothermal method [132]. In a typical procedure, 0.5 mmol $Bi(NO_3)_3{\cdot}5H_2O$ was added into 20 mL of ethanol containing stoichiometric amounts of hexadecyltrimethylammonium bromide (CTAB) and $Pb(NO_3)_2$ with continuous stirring, and then 5 mL ammonia water was added into this solution. The mixture was stirred for at least 30 min and then poured into a 50 mL teflon-lined stainless autoclave. The autoclave was heated at 180°C for 12 h under autogenously pressure and then cooled to room temperature. The resulting precipitates were collected and washed with ethanol and deionized water thoroughly and dried at 70 °C in air.

2.3.7 Characterization of $PbBiO_2Br$ nanosheets samples

A novel hierarchical $PbBiO_2Br$ assembled with ultra-thin nanosheets was characterized by scanning electron microscopy (SEM) and transmission electron microscopy (TEM). $PbBiO_2Br$ nanosheets are well-crystallized and have strong absorption in the visible range up to 500 nm. The assembly $PbBiO_2Br$ nanosheets show excellent photocatalytic activity for degradation of organic contaminants under visible light irradiation, which is primarily attributed to the synergistic effect of the high surface area, specially assembled structure and the low recombination rate of charges in the ultrathin nanosheets.

From SEM, it was concluded that the sample consists of massive uniform flower like spheres with an average diameter of 2 μm [133, 134]. The spheres are constructed of a large number of nanosheets, which stand almost vertically and connect with each other to form a hierarchical structure. $PbBiO_2Br$ nanosheets are 8~9 nm in thickness, which is consistent with the concept of ultra-thin defined by Cademartiri and Ozin [135].

3. Applications of visible light photoactive catalyst

3.1 Photocatalytic reduction of carbon dioxide

The combustion of fossil fuels leads to the emission of CO_2, which is considered as a main source for global warming caused by the greenhouse effect [136]. Mainly rhenium- and ruthenium-based systems have been reported for their ability to electrochemically or photochemically accelerate the reduction of CO_2 to CO. Carbon monoxide itself can be used as a precursor compound for fuel synthesis processes, where CO and H_2 are mixed as syngas to form hydrocarbons such as methane or methanol [137]. The synthesis and characterisation of the rhenium complex 2 (5, 50-bisphenylethynyl-2, 20-bipyridyl) Re $(CO)_3Cl$ were carried out by Oppelt et al. [138]. Portenkirchner et al. [139] reported that rhenium complex fac-(5, 50-bisphenylethynyl-2, 20-bipyridyl) Re $(CO)_3Cl$ was used as a novel catalyst for the electro- and photochemical reduction of CO_2 to CO in homogeneous solution. (5, 50-Bisphenylethynyl-2, 20-bipyridyl)Re$(CO)_3Cl$ showed a 6.5-fold increase in current density under CO_2 at 1750 mV versus normal hydrogen electrode (NHE) as compared to the operation without CO_2. 2 (2,20-Bipyridyl) Re $(CO)_3Cl$ showed high photocatalytic activity with a quantum yield of 8.7% [140, 141]. The new catalyst 2 (5, 50-bisphenylethynyl-2, 20-bipyridyl) Re $(CO)_3$ Cl showed a 22.8 fold lesser generation of CO under the same conditions as 1 (2,20-bipyridyl) Re $(CO)_3Cl$.

The best estimate for the quantum yield of catalyst 2 (5, 50-bisphenylethynyl-2, 20-bipyridyl) Re $(CO)_3$ Cl (above 360 nm irradiation) is in the order of 0.4%.

3.2 Photocatalytic water splitting

Photocatalytic water splitting is an artificial photosynthesis process. In this process, a photoelectrochemical cell is used for the dissociation of water into its constituents, hydrogen and oxygen, using either artificial or natural light. Photocatalytic water splitting has the simplicity of using a powder in solution and sunlight to produce H_2 and O_2 from water and it can provide a clean, renewable energy, without producing greenhouse gases or having many adverse effects on the atmosphere.

Photocatalytic water splitting has been investigated to produce hydrogen fuel, which burns cleanly and can be used in a hydrogen fuel cell. Water splitting is beneficial since it utilizes water, an inexpensive renewable resource.

When H_2O is split into O_2 and H_2, the stoichiometric ratio of its products is 2:1:

$$2\ H_2O \overset{\text{photon energy} > 1.23eV}{\rightleftharpoons} 2\ H_2 + \ O_2$$

The process of water splitting is a highly endothermic process ($\Delta H > 0$). Water splitting occurs naturally in photosynthesis when photon energy is absorbed and converted into the chemical energy through a complex biological pathway. Production of hydrogen from water requires large amounts of input energy, making it incompatible with existing energy generation. For this reason, most of commercially produced hydrogen gas is from natural gas.

Three significant properties of water splitting photocatalysts are (i) Crystal structure and its thermodynamic phase stability (versus competing solids and gases); (ii) Band gap; and (iii) Conduction band (CB) and valence band (VB) edge positions relative to the H_2/H_2O and O_2/H_2O levels in water.

$NaTaO_3$:La yields the highest water splitting rate by photocatalysts without using sacrificial reagents [142]. This UV-based photocatalyst was shown to be highly effective with water splitting rates of 9.7 mmol h^{-1} and a quantum yield of 56%. The nanostep structure of the material promotes water splitting as edges functioning as H_2 production sites and the grooves functioning as O_2 production sites. Addition of NiO particles as co-catalysts assisted in H_2 production; this step was done by using an impregnation method with an aqueous solution of $Ni(NO_3)_2{\bullet}6H_2O$ and evaporating the solution in the presence of the photocatalyst. $NaTaO_3$ has a conduction band higher than that of NiO, so photogenerated electrons are more easily transferred to the conduction band of NiO for H_2 evolution [143].

$K_3Ta_3B_2O_{12}$ has the ability to split water without the assistance of co-catalysts and gives a quantum yield of 6.5% along with a water splitting rate of 1.21 mmol h^{-1}. This ability is due to the pillared structure of the photocatalyst, which involves TaO_6 pillars connected by BO_3 triangle units. Loading with NiO did not assist the photocatalyst due to the highly active H_2 evolution sites [144]. GaN, Ge_3N_4 and Ta_3N_5 are known water splitting photocatalysts [145, 146]. The band edge positions have been experimentally measured only for Ta_3N_5 [147]. Zr_2ON_2 has been reported as a promising material for photoelectrochemical water splitting [148]. Cu_3N, AgN_3, and Zr_3N_4 are known compounds but these have not been reported as photocatalysts yet. $Ti_3O_3N_2$ has the potential to be better water splitting photocatalysts than TaON. $Na_4WO_2N_2$ and $Ca_5WO_2N_4$ have a band gap too large for visible light absorption and can work under UV light illumination.

3.3 Photocatalytic pervious concrete with TiO_2 and paints

Direct interaction of TiO_2 with UV light is very critical. Mixing TiO_2 into traditional concrete can only have limited NO_x reduction effectiveness at the air/solid interface. The process was observed to improve after the concrete material was abraded (some cement paste was peeled off and more TiO_2 was exposed at the surface) [149]. The durability of the photocatalytic effect becomes another challenge if TiO_2 is applied to highly trafficked highways through surface material adhesion. The dynamic tire-pavement interaction under shear and abrasion impact can dislodge coated TiO_2 particles at the surface, leaving untreated pavements. Therefore, coating TiO_2 on the substrate of pervious concrete could have a number of benefits, to maximize the effect of air purification in pavements through the TiO_2 photocatalytic reaction. As compared to traditional concrete pavements which have low porosities and relatively smooth surface textures, pervious concrete pavements have much higher porosities and rougher surface features. The higher void ratio and the increased concave surface texture (due to surface voids) with more surface area could enhance the bonding and durability of the applied TiO_2 at the surface, reduce impacts due to traffic abrasion and climate (snow, ice, water, heat, etc.), and increase the direct contact between TiO_2 and natural light. At the same time, pervious concrete pavement allows water to infiltrate completely through it so that rainwater can filter into the ground and replenish groundwater resources [150]. Installing pervious concrete may reduce costs in installing drainage and storm water systems, reduce the urban heat island effect and noise, improve roadway skid resistance, and prevent hydroplaning. TiO_2 treated pervious concrete pavement can be widely used for pedestrian sidewalks, bike lanes, parking lots, roadway shoulders, and urban low traffic streets for its storm water benefits and air quality purification, resulting in a greener urban living environment.

Two most effective coatings, the commercial water-based TiO_2 (CWB) and the driveway protector mix (DPM), were painted on sections of an actual newly paved pervious concrete sidewalk located between two sports playfields in Pullman, Washington for scaling observation. The sidewalk endured the same type of weathering at the same time of year as the weathering samples. Both coatings were painted on two different 2 x 2 ft. areas each for observation. The DPM coating adds an obvious white color to the pavement, while the CWB coating turns the pavement to slightly lighter gray.

Both coatings had faded in colour with time. After 3 months, the appearance of the CWB coating was not as distinguishable from the regular uncoated pervious concrete as it was originally. The DPM coating, though slightly more faded than it was originally, still had a white colour. The white colour of the DPM coating indicates that the coating is still there and still working to effectively remove pollutants, whereas with the CWB coating, it is unknown whether the coating has come off due to abrasive forces or if it is still working. If these coatings were applied in the field, the DPM coating could be reapplied as a maintenance fulfillment, whenever it was obvious that the colour had worn away.

3.4 Photocatalytic materials for environmental remediation

Due to the position of the valence band of ZnO, the photogenerated holes have strong enough oxidizing power to decompose most of the organic compounds [143]. ZnO has been used to decompose aqueous solutions of several dyes [152], and many other environmental pollutants [153]. Chen et al. [154] reported that ZnO is more efficient than TiO_2 because of its good optoelectronic, catalytic and photochemical properties along with its low cost.

Cu_2O octahedra show better photocatalytic activity than cubes, because the {1 1 1} facets are more active than {1 0 0} facets due to the dangling bonds of {1 1 1} surfaces, whilst {1 0 0} facets have saturated chemical bonds and no dangling bands exist [155, 156]. Huang et al. [157] prepared Cu_2O nanoparticles and microparticles for the photodegradation of methyl orange. Nanoporous WO_3 films anodically grown on tungsten foil substrates were photoactive for the oxidation of methylene blue and the reductive conversion of Cr(VI) under visible light illumination [158]. WO_3 films deposited on a Pt substrate showed a higher photoelectrocatalytic activity for the photodegradation of naphthol blue black than TiO_2 nanoparticulate film electrodes [159].

Bessekhouad et al. [160] found that Bi_2O_3 was able to degrade orange II but the efficiency of the photocatalytic reaction was rather low. Monoclinic α-Bi_2O_3, synthesized via calcination of hydrothermally prepared $(BiO)_2CO_3$ was much more active than commercial Bi_2O_3 for the degradation of NO and formaldehyde at typical indoor air concentration [161].

The research on ternary and quaternary oxides is an efficient strategy to overcome the intrinsic limitations of the binary metal oxides and new materials have been obtained, which are suitable to exploit the visible component of sunlight. Kohtani et al. [162] reported that $BiVO_4$ can efficiently decompose long-chain alkyl phenols and polycyclic aromatic hydrocarbons under visible light irradiation. Ag_3VO_4 powders synthesized by precipitation [163] were evaluated for the decolorization of acid red B under visible light irradiation. $CdWO_4$ is investigated for the photodegradation of organic compounds under UV light irradiation [164,165]. Monoclinic $CdWO_4$ short rods synthesized via a hydrothermal process exhibited a high photocatalytic activity for the degradation of methyl orange and rhodamine B [164].

Single-phase quaternary oxide materials have been recently studied with the aim to develop novel photocatalysts showing high activity in the visible light region. Shan et al. [106] tested layered Bi-based oxyhalides for the degradation of methyl orange under UV and visible illumination. Bi_4NbO_8Cl (Eg = 2.38 eV) showed an excellent visible light efficiency and was more active than the ternary oxychloride Bi_3O_4Cl (Eg = 2.80 eV) [166].

4. Conclusion

TiO_2 nanocrystallites are uniformly dispersed throughout the amorphous SiO_2 matrix, and nanopores form on top of TiO_2 particles, resulting in an improved photoactivity through increasing the effective absorption sites. The specific surface areas and the surface states of the binary films decrease gradually with increasing annealing temperature; thus, deteriorating the photocatalytic activity after annealing. With the onset of the photocatalytic activity for NO_x abatement is concomitant to the switch between amorphous to a crystalline phase with an electronic band gap within 3.05–3.10 eV, where the most active photocatalyst revealed predominantly rutile phase together with anatase as the minority phase.

The ternary semiconductor hybrid prepared by calcination of $KTaO_3$, CdS and MoS_2 powders at the 10:5:1 molar ratio exhibited very good stability in toluene degradation and excellent photocatalytic performance in phenol degradation among all obtained photocatalysts. The activity reached 80% under UV-Vis and 42% under Vis light 60 min irradiation. This relatively high photoactivity under visible light is due to the presence of heterojunction between semiconductors.

GTF exhibited a higher photocatalytic activity during the degradation of rhodamine B as compared to the unsupported TiO_2–Fe_3O_4 composite and pure TiO_2 NPs. The electron transfer between the TiO_2 NPs and RGO help in achieving the high durability of GTF. GTF can be easily recollected from water using a magnet. Due to its attractive features such as easy recollection and reusability, the GTF hybrid nanocomposite is used in wastewater treatment.

The Bi_2WO_6 nanoplates showed high activity as photocatalysts under visible light irradiation. Quaternary photocatalyst (Pb_2FeNbO_6) show significant photocatalytic activity for methylene blue degradation due to its suitable band energetics. The composite photocatalyst (Pb_2FeNbO_6) is found to be twice efficient than $TiO_{2-x}N_x$. XPS Hf 4f and O1 s results shows that ultra-thin HfTiErO films microstructure is formed, while the annealed temperature also affected the microstructures that caused the increase in intensity and binding energy. These improved material and electrical results indicate that HfTiErO may be used for the generation of higher-k gate dielectrics. The electrocatalytic activity of the new compound 2 (5, 50-bisphenylethynyl-2, 20-bipyridyl) Re $(CO)_3$ Cl is in the range of 45% for Faradaic efficiency for CO formation and ranks equally with the benchmark molecule 1 (2,20-bipyridyl) Re $(CO)_3Cl$.

References

[1] E.J. Weber, R.L. Adams, Chemical and sediment mediated reduction of the azo dye: Disperse blue 79, Environ. Sci. Technol. 29 (1995) 1163-1170. https://doi.org/10.1021/es00005a005

[2] C. Wang, A. Yediler, D. Linert, Z. Wang, A. Kettrup, Toxicity evaluation of reactive dye stuff, auxiliaries and selected effluents in textile finishing industry to luminescent bacteria vibrio fisheri, Chemosphere 46 (2002) 339-344. https://doi.org/10.1016/S0045-6535(01)00086-8

[3] W.C. Tincher, J.R. Robertson, Analysis of dyes in textile dyeing waste water, Text. Chem. Color. 43 (1982) 269-275.

[4] World Health Organization, Phenol: Environmental Health Criteria 161, World Health Organization: Geneva, Switzerland (1994).

[5] World Health Organization, Cresols: Environmental Health Criteria 168, World Health Organization: Geneva, Switzerland (1995).

[6] European Union. The list of priority substances in the field of water policy and amending directive, Council directive 2455/2001/ECC. Official Journal of the European Communities L331, 20 November 2001, 1–5.

[7] Environment Canada, The Second Priority Substances List (PSL2) of the Canadian Environmental Protection Act (CEPA), Environment Canada: Gatineau, Canada (1995).

[8] United States Environmental Protection Agency (USEPA), Sampling and Analysis Procedure for Screening of Industrial Effluents for Priority Pollutants, Environment Monitoring and Support Laboratory: Cincinnati, OH, USA (1977).

[9] F.M. Pfeffer, The 1977 screening survey for measurement of organic priority pollutants in petroleum refinery wastewaters, ASTM Specif. Tech. Publ. 686 (1979) 181–190. https://doi.org/10.1520/STP35014S

[10] L.H. Keith, Identification of organic compounds in unbleached treated kraft paper mill wastewaters, Environ. Sci. Technol. 10 (1976) 555–564. https://doi.org/10.1021/es60117a009

[11] B.R. Parkhurst, A.S. Bradshaw, J.L. Forte, An evaluation of the acute toxicity to aquatic biota of a coal conversion effluent and it major components, Bull. Environ. Contam. Toxicol. 23 (1979) 349–356. https://doi.org/10.1007/BF01769968.

[12] G. Jungclaus, V. Avila, R. Hites, Organic compounds in an industrial wastewater: A case study of their environmental impact, Environ. Sci. Technol. 12 (1978) 88–96. https://doi.org/10.1021/es60137a015

[13] I. Kamenev, R. Munter, L. Pikkov, L. Kekisheva, Wastewater treatment in oil shale chemical industry, Oil Shale 20 (2003) 443–457.

[14] Z. Guo, R. Ma, G. Li, Degradation of phenol by nanomaterial TiO_2 in wastewater, Chem. Eng. J. 119 (2006) 55–59. https://doi.org/10.1016/j.cej.2006.01.017

[15] M. Auriol, Y. Filali-Meknassi, R.D. Tyagi, C.D. Adams, R.Y. Surampalli, Endocrine disrupting compounds removal from wastewater, a new challenge, Process Biochem. 41 (2006) 525–539. https://doi.org/10.1016/j.procbio.2005.09.017

[16] P.R. Gogate, A.B. Pandit, A review of imperative technologies for wastewater treatment I: oxidation technologies at ambient conditions, Adv. Environ. Res. 8 (2004) 501–551. https://doi.org/10.1016/S1093-0191(03)00032-7

[17] K.H. Wang, Y.H. Hsieh, L.J. Chen, The heterogeneous photocatalytic degradation, intermediates and mineralization for the aqueous solution of cresols and nitrophenols, J. Hazard. Mater. 59 (1998) 251–260. https://doi.org/10.1016/S0304-3894(97)00151-9

[18] A. Matilainena, M. Sillanpaa, Removal of natural organic matter from drinking water by advanced oxidation processes, Chemosphere 80 (2010) 351–365. https://doi.org/10.1016/j.chemosphere.2010.04.067

[19] W.H. Glaze, J.W. Kang, D.H. Chapin, Chemistry of water treatment processes involving ozone, hydrogen peroxide and ultraviolet radiation, Ozone Sci. Eng. 9 (1987) 335–352. https://doi.org/10.1080/01919518708552148

[20] K.Y. Jung, S.B. Park, Photoactivity of SiO_2\$TiO_2$ and ZrO_2\$TiO_2$ mixed oxide prepared by sol-gel method, Mater. Lett. 58 (2004) 2897-2900. https://doi.org/10.1016/j.matlet.2004.05.015

[21] Ch. Guillard, B. Beaugiraud, C. Dutriez, Physicochemical properties and photocatalytic activities of TiO_2-films prepared by sol-gel methods, Appl. Catal. B: Environ. 39(4) (2002) 331-342. https://doi.org/10.1016/S0926-3373(02)00120-0

[22] S.H. Oh, D. Jin Kim and S.H. Hahn, Comparison of optical and photocatalytic properties of TiO_2 thin films prepared by electron-beam evaporation and sol–gel dip-coating, Mater. Lett. 57, 4151-4155 (2003). https://doi.org/10.1016/S0167-577X(03)00281-7

[23] Y.R. Do, W. Lee, K. Dwight, A. Wold, The effect of WO_3 on the photocatalytic activity of TiO_2, J. Solid State Chem. 108 (1994) 198-201. https://doi.org/10.1006/jssc.1994.1031

[24] J. Papa, S. Soled, K. Dwight, A. Wold, Surface acidity and photocatalytic activity of TiO_2, WO_3/TiO_2, and MoO_3/TiO_2 photocatalysts, Chem. Mater 1994, 6, 496-500. https://doi.org/10.1021/cm00040a026

[25] Y. Oasawa, A. Gratzel, Effect of surface hydroxyl density on photocatalytic oxygen generation in aqueous TiO_2 suspensions, J. Chem. Soc., Faraday Trans.1 84(1) (1998) 197-205. https://doi.org/10.1039/f19888400197

[26] E. Ozensoy, J. Szanyi, C.H.F. Peden, Model NO_x storage systems: storage capacity and thermal aging of BaO/ θ-Al_2O_3/NiAl (1 0 0), J. Catal. 243 (2006) 149–157. https://doi.org/10.1016/j.jcat.2006.06.028

[27] E. Kayhan, S. M. Andonova, G.S. Senturk, C.C. Chusuei, E. Ozensoy, Iron promoted NO_x storage materials: Structural properties and NO_x uptake, J. Phys. Chem. C 114 (2010) 357–369. https://doi.org/10.1021/jp907982q

[28] E. Emmez, E.I. Vovk, V.I. Bukhtiyarov, E. Ozensoy, Direct evidence for the instability and deactivation of mixed-oxide systems: Influence of surface

segregation and subsurface diffusion, J. Phys. Chem. C 115 (2011) 22438–22443. https://doi.org/10.1021/jp206969h

[29] H. Zhang, Y. Zhu, Significant visible photoactivity and antiphotocorrosion performance of CdS photocatalysts after monolayer polyaniline hybridization, J. Phys. Chem. C 114 (2010) 5822–5826. https://doi.org/10.1021/jp910930t

[30] M. Zhong, J. Shi, F. Xiong, W. Zhang, C. Li, Enhancement of photoelectrochemical activity of nanocrystalline CdS photoanode by surface modification with TiO_2 for hydrogen production and electricity generation, Sol. Energy 86 (2012) 756–763. https://doi.org/10.1016/j.solener.2011.12.006

[31] X. Li, T. Xia, C. Xu, J. Murowchick, X. Chen, Synthesis and photoactivity of nanostructured CdS–TiO_2 composite catalysts, Catal. Today 225 (2014) 64–73. https://doi.org/10.1016/j.cattod.2013.10.086

[32] D. He, M. Chen, F. Teng, G. Li, H. Shi, J. Wang, M. Xu, T. Lu, X. Ji, Y. Lv, Enhanced cyclability of CdS/TiO_2 photocatalyst by stable interface structure, Superlattices Microstruct. 51 (2012) 799–808. https://doi.org/10.1016/j.spmi.2012.03.026

[33] S. Panigrahi, D. Basak, Morphology driven ultraviolet photosensitivity in ZnO–CdS composite, J. Colloid Interface Sci. 364 (2011) 10–17. https://doi.org/10.1016/j.jcis.2011.08.001

[34] T.K. Jana, A. Pal, K. Chatterjee, Self-assembled flower like CdS–ZnO nanocomposite and its photo catalytic activity, J. Alloys Compd. 583 (2014) 510–515. https://doi.org/10.1016/j.jallcom.2013.08.184

[35] S. Liu, H. Li, L. Yan, Synthesis and photocatalytic activity of three-dimensional ZnS/CdS composites, Mater. Res. Bull. 48 (2013) 3328–3334. https://doi.org/10.1016/j.materresbull.2013.05.055

[36] X. Liu, Y. Yan, Z. Da, W. Shi, C. Ma, P. Lv, Y. Tang, G. Yao, Y. Wu, P. Huo, Significantly enhanced photocatalytic performance of CdS coupled WO_3 nanosheets and the mechanism study, Chem. Eng. J. 214 (2014) 243–250. https://doi.org/10.1016/j.cej.2013.12.058

[37] X. Zong, H. Yan, G. Wu, G. Ma, F. Wen, L. Wang, C. Li, Enhancement photocatalytic H_2 evolution on CdS by loading MoS_2 as co-catalyst under visible light irradiation, J. Am. Chem. Soc. 130 (2008) 7176–7177. https://doi.org/10.1021/ja8007825

[38] D.W. Bahnemann, D. Bockelmann, R. Goslich, M. Hilgendorff, D. Weichgrebe, in: D.F. Ollis et al. (Eds.), Photocatalytic purification and treatment of water and air, Elsevier, Amsterdam 22 (1993) 301-320.

[39] G.N. Schrauzer, T.D. Guth, Photocatalytic reactions: Photolysis of water and photoreduction of nitrogen on titanium dioxide, J. Am. Chem. Soc. 99(22) (1977) 7189-7193. https://doi.org/10.1021/ja00464a015

[40] M.I. Litter, J.A. Navio, Comparison of the photocatalytic efficiency of TiO_2, iron oxides and mixed Ti (IV) and Fe (III) oxides: Photodegradation of oligocarboxylic acids, J. Photochem. Photobiol. A: Chem. 84(2) (1994) 183-193. https://doi.org/10.1016/1010-6030(94)03858-9

[41] L. Palmisano, M. Schiavello, A. Sclafani, C. Martin, I. Martin, V. Rives, Surface properties of iron-titania photocatalysts employed for 4-nitrophenol photodegradation in aqueous TiO_2 dispersion, Catal. Lett. 24 (3-4) (1994) 303-315. https://doi.org/10.1007/BF00811803

[42] W. Choi, A. Termin, M.R Hoffmann, Role of metal-ion dopants in quantum-sized TiO_2 co-relation between photoreactivity and charge-carrier recombination dynamics, J. Phys. Chem. 98 (1994) 13669-13679. https://doi.org/10.1021/j100102a038

[43] J.A. Navio, G. Colon, M.I. Litter, G.N. Bianco, Synthesis, characterization and photocatalytic properties of iron-doped titania semiconductors prepared from TiO_2 and iron (III) acetylacetonate, J. Mol. Catal. A. Chem. 106(3) (1996) 267-276. https://doi.org/10.1016/1381-1169(95)00264-2

[44] M. Anpo, M. Che, Applications of photoluminescence techniques to the characterization of solid surfaces in relation to adsorption, catalysis, and photocatalysis, Adv. Catal. 44 (2000) 119-257. https://doi.org/10.1016/S0360-0564(08)60513-1

[45] M. Anpo, M. Tomonari, M.A. Fox, In situ photoluminescence of titania as a probe of photocatalytic reactions, J. Phys. Chem. 93(21) (1989) 7300-7302. https://doi.org/10.1021/j100358a008

[46] M.A. Fox, M.T. Dulay, Heterogeneous photocatalysis, Chem. Rev. 93(1) (1993) 341-357. https://doi.org/10.1021/cr00017a016

[47] S.-I. Nishimoto, B. Ohtani, H. Kajiwara, T. Kagiya, Correlation of the crystal structure of titanium dioxide prepared from titanium tetra-2-propoxide with the photocatalytic activity for redox reactions in aqueous propan 2 ol and silver salt

solutions, J. Chem. Soc. Faraday Trans. I, 81 (1985) 61-68. https://doi.org/10.1039/f19858100061

[48] J. Yang, J. H. Swisher, The phase stabilization of $Zn_2Ti_3O_8$, Mater. Charact. 37 (1996) 153-159. https://doi.org/10.1016/S1044-5803(96)00098-8

[49] C. Wang, J. Zhao, X. Wang, B. Mai, G. Sheng P. Peng J. Fu, Preparation, characterization and photocatalytic activity of nano-sized ZnO/SnO_2 coupled photocatalysts, Appl. Catal., B: Environ. 39 (2002) 269–279. https://doi.org/10.1016/S0926-3373(02)00115-7

[50] S. Liao, D. Huang, D. Yu, Y. Su, G. Yuan, Preparation and characterization of ZnO/TiO_2, $SO_4^{2-}/ZnO/TiO_2$ photocatalyst and their photocatalysis, J. Photochem Photobiol., A: Chem. 168 (2004) 7-13. https://doi.org/10.1016/j.jphotochem.2004.05.010

[51] J.T. Zhang, Z.G. Xiong, X. S. Zhao, Graphene–metal–oxide composites for the degradation of dyes under visible light irradiation, J. Mater. Chem. 21 (2011) 3634–3640. https://doi.org/10.1039/c0jm03827j

[52] C. Chen, W.M. Cai, M.C. Long, B.X. Zhou, Y.H. Wu, D.Y. Wu, Y.J. Feng, Synthesis of visible-light responsive graphene oxide/TiO_2 composites with p/n heterojunction, ACS Nano 4(11) (2010) 6425–6432. https://doi.org/10.1021/nn102130m

[53] Y. Lin, Z. Geng, H. Cai, L. Ma, J. Chen, J. Zeng, N. Pan, X. Wang, Ternary graphene–TiO_2–Fe_3O_4 nanocomposite as a recollectable photocatalyst with enhanced durability, Eur. J. Inorg. Chem. 28 (2012) 4439–4444. https://doi.org/10.1002/ejic.201200454

[54] Castro, P. Begue, B. Jimenez, J. Ricote, R. Jimenez, J. Galy, New $Bi_2Mo_{1-x}W_xO_6$ solid solution: Mechanosynthesis, structural study, and ferroelectric properties of the $x = 0.75$ member, Chem. Mater. 15(17) (2003) 3395-3401. https://doi.org/10.1021/cm030224r

[55] S. Luo, Y. Noguchi, M. Miyayama, T. Kudo, Rietveld analysis and dielectric properties of Bi_2WO_6–$Bi_4Ti_3O_{12}$ ferroelectric system, Mater. Res. Bull. 36 (2001) 531-540. https://doi.org/10.1016/S0025-5408(01)00516-5

[56] H.B. Fu, C.S. Pan, W.Q. Yao, Y.F. Zhu, Visible-light-induced degradation of rhodamine B by nanosized Bi_2WO_6, J. Phys. Chem. B 109(47) (2005) 22432-22439. https://doi.org/10.1021/jp052995j

[57] C. Zhang, Y.F. Zhu, Synthesis of square Bi_2WO_6 nanoplates as high-activity visible-light-driven photocatalysts, Chem. Mater. 17(13) (2005) 3537-3545. https://doi.org/10.1021/cm0501517

[58] J.W. Tang, Z.G. Zou, J.H. Ye, Photocatalytic decomposition of organic contaminants by Bi_2WO_6 under visible light irradiation, Catal. Lett. 92(1-2) (2004) 53-56. https://doi.org/10.1023/B:CATL.0000011086.20412.aa

[59] T.S. Natarajan, H.C. Bajaj, R.J. Tayade, Synthesis of homogeneous sphere-like Bi_2WO_6 nanostructure by silica protected calcination with high visible-light-driven photocatalytic activity under direct sunlight, Cryst .Eng. Comm. 17(5) (2015) 1037-1049. https://doi.org/10.1039/C4CE01839G

[60] R.J. Tayade, H.C. Bajaj, R.V. Jasra, Photocatalytic removal of organic contaminants from water exploiting tuned bandgap photocatalysts, Desalination 275(1) (2011) 160-165. https://doi.org/10.1016/j.desal.2011.02.047

[61] T.K. Pathak, N.H. Vasoya, T.S. Natarajan, K.B. Modi, R.J. Tayade, Photocatalytic degradation of aqueous nitrobenzene solution using nanocrystalline Mg-Mn ferrites, Materials Sci. Forum 764 (2013) 116-129. https://doi.org/10.4028/www.scientific.net/MSF.764.116

[62] S. Sadhu, A. Patra, Lattice strain controls the carrier relaxation dynamics in Cd_xZn_{1-x} S alloy quantum dots, J. Phys. Chem. C 116(28) (2012) 15167–15173. https://doi.org/10.1021/jp304901w

[63] X. Zhong, Y. Feng, Y. Zhang, Z. Gu, L. Zou, A facile route to violet- to orange-emitting $Cd_xZn_{1-x}Se$ alloy nanocrystals via cation exchange reaction, Nanotechnol. 18(38) (2007) 385606-385612. https://doi.org/10.1088/0957-4484/18/38/385606

[64] L.Y. Chen, P.A. Yang, C.H. Tseng, B.J. Hwang, C.H. Chen, Internal structure of tunable ternary $CdSe_xS_{1-x}$ quantum dots unraveled by X-ray absorption spectroscopy, App. Phys. Lett. 100(16) (2012) 163113-16116. https://doi.org/10.1063/1.3703123

[65] G. Kartopu, A.J. Clayton, W.S.M. Brooks, S.D. Hodgson, V. Barrioz, A. Maertens, D.A. Lamb, S.J.C. Irvine, Effect of window layer composition in Cd_{1-x} Zn_xS/CdTe solar cells, Prog. Photovolt. Res. 22(1) (2012) 18-23. https://doi.org/10.1002/pip.2272

[66] X. Wu, Y. Yu, Y. Liu, Y. Xu, C. Liu, B. Zhang, Synthesis of hollow $Cd_xZn_{1-x}Se$ nanoframes through the selective cation exchange of inorganic–organic hybrid

ZnSe–amine nanoflakes with cadmium ions, Angew. Chem. Int. Ed. 51(13) (2012) 3211–3215. https://doi.org/10.1002/anie.201108098

[67] C.I. Wang, Z. Yang, A.P. Periasamy, H.T. Chang, Photoluminescent C-dots RGO probe for sensitive and selective detection of acetylinecholine, Anal. Chem. 85(6) (2013) 3263-3270. https://doi.org/10.1021/ac303613d

[68] W.A. Tisdale, K.J. Williams, B.A. Timp, D.J. Norris, E.S. Aydil, X.Y. Zhu, Hot-electron transfer from semiconductor nanocrystals, Science 328 (2010) 1543–1547. https://doi.org/10.1126/science.1185509

[69] P. Aguiar, D. Chadwick, L. Kershenbaum, Modelling of an indirect internal reforming solid oxide fuel cell, Chem. Eng. Sci. 57 (2002) 1665–1677. https://doi.org/10.1016/S0009-2509(02)00058-1

[70] Q. Deng, S.L. Jiang, T.J. Cai, Z.S. Peng, Z.J. Fang, Selective oxidation of isobutane over $H_xFe_{0.12}Mo_{11}VPAs_{0.3}O_y$ heteropoly compound catalyst, J. Mol. Catal. A: Chem. 229 (2005) 165–170. https://doi.org/10.1016/j.molcata.2004.11.013

[71] T.J. Cai, M. Yue, X.W. Wang, Q. Deng, Z.S. Peng, W.H. Zhou, Preparation, characterization and photocatalytic performance of $NdPW_{12}O_{40}/TiO_2$ composite catalyst, Chin. J. Catal. 28(1) (2007) 10–16. https://doi.org/10.1016/S1872-2067(07)60007-2

[72] X.J. Wu, Y.C. Liao, X.W. Wang, Z.S. Peng, Y.H. Jiang, X.Y. Liu, T.J. Cai, Microwave solid-phase synthesis of bismuth phosphotungstate nanoparticle catalysts and photocatalytic elimation of methanol, Acta Sci. Circumst. 28(5) (2008) 965–970.

[73] Q. Deng, W.H. Zhou, M. X. Li, Z.S. Peng, S.L. Jiang, M. Yue, T.J. Cai, Microwave radiation solid-phase synthesis of phosphotungstate nanoparticle catalysts and photocatalytic degradation of formaldehyde, J. Mol. Catal. A: Chem. 262 (2007) 149–155. https://doi.org/10.1016/j.molcata.2006.08.043

[74] H. Xue, Z.H. Li, L. Wu, Z.X. Ding, X.X. Wang, X.Z. Fu, Nanocrystalline ternary wide band gap p-block metal semiconductor $Sr_2Sb_2O_7$: Hydrothermal synthesis and photocatalytic benzene degradation, J. Phys. Chem. C 112(15) (2008) 5850-5855. https://doi.org/10.1021/jp712186r

[75] Z. G. Yi, J.H. Ye, N. Kikugawa, T. Kako, S.X. Ouyang, H. Stuart-Williams, H. Yang, J.Y. Cao, W.J. Luo, Z.S. Li, Y. Liu, R. Withers, An orthophosphate

semiconductor with photooxidation properties under visible-light irradiation, Nat. Mater. 9 (2010) 559-564. https://doi.org/10.1038/nmat2780

[76] N. Umezawa, S.X. Ouyang, J.H. Ye, Theoretical study of high photocatalytic performance of Ag_3PO_4, Phys. Rev. B 83(3-1) (2011) 035202.

[77] X.C. Wang, K. Maeda, A. Thomas, K. Takanabe, G. Xin, J.M. Carlsson, K. Domen, M. Antonietti, A metal-free polymeric photocatalyst for hydrogen production from water under visible light, Nat. Mater. 8 (2009) 76-80. https://doi.org/10.1038/nmat2317

[78] C.S. Pan, J. Xu, Y.J. Wang, D. Li, Y.F. Zhu, Dramatic activity of $C_3N_4/BiPO_4$ photocatalyst with core/shell structure formed by self-assembly, Adv. Funct. Mater. 22(7) (2012) 1518-1524. https://doi.org/10.1002/adfm.201102306

[79] Q.J. Xiang, J.G. Yu, M. Jaroniec, Preparation and enhanced visible-light photocatalytic H_2-production activity of graphene/C_3N_4 composites, J. Phys. Chem. C 115 (2011) 7355-7363. https://doi.org/10.1021/jp200953k

[80] L. Ge, C.C. Han, J. Liu, Novel visible light-induced g-C_3N_4/Bi_2WO_6 composite photocatalysts for efficient degradation of methyl orange, Appl. Catal. B 108-109 (2011) 100-107. https://doi.org/10.1016/j.apcatb.2011.08.014

[81] G. Shiravand, A. Badiei, G.M. Ziarani, M. Jafarabadi, M. Hamzehloo, Photocatalytic synthesis of phenol by direct hydroxylation of benzene by modified nanoporous silica (LUS-1) under sunlight, Chin. J. Catal. 33 (2012) 1347-1353. https://doi.org/10.1016/S1872-2067(11)60422-1

[82] H. Pan, X. K. Li, Z. J. Zhuang, C. Zhang, g-C_3N_4/SiO_2–HNb_3O_8 composites with enhanced photocatalytic activities for rhodamine B degradation under visible light, J. Mol. Catal. A: Chem 345(1-2) (2011) 90-95. https://doi.org/10.1016/j.molcata.2011.05.024

[83] S.C. Yan, S.B. Lv, Z.S. Li, Z.G. Zou, Organic–inorganic composite photocatalyst of g-C_3N_4 and TaON with improved visible light photocatalytic activities, Dalton Trans. J. Inorg. Chem. 39 (2010) 1488-1491. https://doi.org/10.1039/B914110C

[84] K. Shen, M.A. Gondal, R.G. Siddique, S. Shi, S. Wang, J. Sun, Q. Xu, Preparation of ternary Ag/Ag_3PO_4/g-C_3N_4 hybrid photocatalysts and their enhanced photocatalytic activity driven by visible light, Chin. J. Catal. 35 (2014) 78–84. https://doi.org/10.1016/S1872-2067(12)60712-8

[85] H.I. Kim, J. Kim, W. Kim, W. Choi, Enhanced photocatalytic and photoelectrochemical activity in the ternary hybrid of CdS/TiO_2/WO_3 through the

cascadal electron transfer, J. Phys. Chem. C. 115 (2011) 9797–9805. https://doi.org/10.1021/jp1122823

[86] C.J. Lin, Y.T. Lu, C.H. Hsieh, S.H. Chien, Surface modification of highly ordered TiO_2 nanotube arrays for efficient photoelectrocatalytic water splitting, Appl. Phys. Lett. 94 (2009) 113102-113110. https://doi.org/10.1063/1.3099338

[87] C. Chen, Y. Xie, G. Ali, S.H. Yoo, S.O. Cho, Improved conversion efficiency of CdS quantum dots-sensitized TiO_2 nanotube array using ZnO energy barrier layer, Nanotechnol. 22 (2011) 015202-015210. https://doi.org/10.1088/0957-4484/22/1/015202

[88] A.J. Bwamba, N. Alu, A.K. Kenneth, Z. Abdullahi, I.U. Unwana, C.E. Augustine, O.A. Anthony, Characterization of CZTS absorbent material prepared by field assisted spray pyrolysis, Am. J. Mater. Sci. 4 (2014) 127-132.

[89] L.Y. Yeh, K.W. Cheng, Preparation of the Ag-Zn-Sn-S quaternary photoelectrodes using chemical bath deposition for photoelectrochemical applications, Thin Solid Films 558 (2014) 289-293. https://doi.org/10.1016/j.tsf.2014.02.046

[90] Z. Zou, J. Ye, K. Sayama, H. Arakawa, Direct splitting of water under visible light irradiation with an oxide semiconductor photocatalyst, Nature 414 (2001) 625-627. https://doi.org/10.1038/414625a

[91] A. Kudo, H. Kato, Photocatalytic decomposition of water into H_2 and O_2 over novel photocatalyst $K_3Ta_3Si_2O_{13}$ with pillared structure consisting of three TaO_6 chains, Chem. Lett. 20 (1997) 867-868. https://doi.org/10.1246/cl.1997.867

[92] A. Ishikawa, T. Takata, J.N. Kondo, M. Hara, H. Kobayashi, K. Domen, K., Oxysulfide $Sm_2Ti_2S_2O_5$ as a stable photocatalyst for water oxidation and reduction under visible light irradiation ($\lambda \leq 650$ nm), J. Am. Chem. Soc. 124(45) (2002) 13547-13553. https://doi.org/10.1021/ja0269643

[93] R. Asahi, T. Morikawa, T. Ohwaki, K. Aoki, Y. Taqo, Visible-light photocatalysis in nitrogen-doped titanium oxides, Sci. 293 (2001) 269-271. https://doi.org/10.1126/science.1061051

[94] G. Hitoki, T. Takata, J. Kondo, M. Hara, H. Kobayashi, K. Domen, An oxynitride, TaON, as an efficient water oxidation photocatalyst under visible light, Chem. Commun. 21(16) (2002) 1698-1699. https://doi.org/10.1039/B202393H

[95] K. Maeda, K. Teramura D. Lu, T. Takata, N. Saito, Y. Inoue, K. Domen, Photocatalyst releasing hydrogen from water, Nature 440 (2006) 295-305. https://doi.org/10.1038/440295a

[96] K. Vijayasankara, N.Y. Hebalkara, H.G. Kimb, P.H. Borse, Controlled band energetics in Pb-Fe-Nb-O metal oxide composite system to fabricate efficient visible light photocatalyst, J. Ceram. Process. Res. 14(4) (2013) 557-562.

[97] M. Cherevatskaya, M. Neumann, S. Füldner, C. Harlander, S. Kümmel, S. Dankesreiter, A. Pfitzner, K. Zeitler, B. König, Visible-light-promoted stereoselective alkylation by combining heterogeneous photocatalysis with organocatalysis, Angew. Chem. Int. Ed. 51(17) (2012) 4062-4066. https://doi.org/10.1002/anie.201108721

[98] A. Thibert, F. A. Frame, E. Busby, D.S. Larsen, Primary photodynamics of water-solubilized two-dimensional CdSe nanoribbons, J. Phys. Chem. C 115(40) (2011) 19647-19658. https://doi.org/10.1021/jp206828y

[99] Y.-H. Liu, V.L. Wayman, P.C. Gibbons, R.A. Loomis, W.E. Buhro, Origin of high photoluminescence efficiencies in CdSe quantum belts, Nano Lett. 10(1) (2009) 352-357.

[100] T. Kako, Z. Zou, J. Ye, Photocatalytic oxidation of 2-propanol in the gas phase over cesium bismuth niobates under visible light irradiation, Res. Chem. Intermed. 31 (2005) 359–364. https://doi.org/10.1163/1568567053956563

[101] H.G. Kim, D.W. Hwang, J.S. Lee, An undoped, single-phase oxide photocatalyst working under visible light, J. Am. Chem. Soc. 126 (2004) 8912–8913. https://doi.org/10.1021/ja049676a

[102] T. Kako, J. Ye, Photocatalytic decomposition of acetaldehyde over rubidium bismuth niobates under visible light irradiation, Mater. Trans. 46 (2005) 2699–2703. https://doi.org/10.2320/matertrans.46.2694

[103] B. Muktha, M.H. Priya, G. Madras, T.N. Guru Row, Synthesis, structure, and photocatalysis in a new structural variant of the aurivillius phase: $LiBi_4M_3O_{14}$ (M = Nb, Ta), J. Phys. Chem. B 109 (2005) 11442–11449. https://doi.org/10.1021/jp051228h

[104] L.M. Torres-Martínez, I. Juárez-Ramírez, J.S. Ramos-Garza, F. Vázquez-Acosta, S.W. Lee, Sol-gel preparation of Bi_2InTaO_7 and its photocatalytic behavior for organic compounds degradation, Mater. Sci. Forum 658 (2010) 491–494. https://doi.org/10.4028/www.scientific.net/MSF.658.491

[105] X. Li, S. Ouyang, N. Kikugawa, J. Ye, Novel Ag_2ZnGeO_4 photocatalyst for dye degradation under visible light irradiation, Appl. Catal. A: Gen. 334 (2008) 51–58. https://doi.org/10.1016/j.apcata.2007.09.033

[106] Z. Shan, W. Wang, X. Lin, H. Ding, F. Huang, Photocatalytic degradation of organic dyes on visible-light responsive photocatalyst $PbBiO_2Br$, J. Solid State Chem. 181 (2008) 1361–1366. https://doi.org/10.1016/j.jssc.2008.03.001

[107] F. Mei, C. Liu, Effect of annealing temperature on binary TiO_2:SiO_2 nanocrystalline thin films, J. Korean Phys. Soc. 48(6) (2006) 1509-1513.

[108] K.Y. Jung, S.B. Park, Photoactivity of SiO_2/TiO_2 and ZrO_2/TiO_2 mixed oxides prepared by sol-gel method, Mater. Lett. 58(22-23) (2004) 2897-2900. https://doi.org/10.1016/j.matlet.2004.05.015

[109] A.M. Soylu, M. Polat, D.A. Erdogan, Z. Say, C. Yıldırım, Ö. Birer, E. Ozensoy, TiO_2–Al_2O_3 binary mixed oxide surfaces for photocatalytic NO_x abatement, Appl. Surf. Sci. 318 (2014) 142–149. https://doi.org/10.1016/j.apsusc.2014.02.065

[110] S.M. Andonova, G.S. Senturk, E. Kayhan, E. Ozensoy, Nature of the Ti–Ba interactions on the $BaO/TiO_2/Al_2O_3$ NO_x storage system, J. Phys. Chem. C 113 (2009) 11014–11026. https://doi.org/10.1021/jp9005026

[111] S.M. Andonova, G.S. Senturk, E. Ozensoy, Fine-tuning the dispersion and the mobility of BaO domains on NO_x storage materials via TiO_2 anchoring sites, J. Phys. Chem. C 114 (2010) 17003–17016. https://doi.org/10.1021/jp102555c

[112] D.A.H. Hanaor, C.C. Sorrell, Review of the anatase to rutile phase transformation, J. Mater. Sci. 46 (2011) 855–874. https://doi.org/10.1007/s10853-010-5113-0

[113] B. Bajorowicz, A. Cybula, M. J. Winiarski, T. Klimczuk, A. Zaleska, Surface properties and Photocatalytic activity of $KTaO_3$, CdS, MoS_2 semiconductors and their binary and ternary semiconductor composites, Molecules 19 (2014)15339-15360. https://doi.org/10.3390/molecules190915339

[114] S. Zhong, L. Zhang, Z. Huang, S. Wang, Mixed-solvothermal synthesis of CdS micro/nanostructures and their optical properties, Appl. Surf. Sci. 257 (2011) 2599–2603. https://doi.org/10.1016/j.apsusc.2010.10.029

[115] Y. Liu, H. Yu, X. Quan, S. Chen, Green synthesis of feather-shaped MoS_2/CdS photocatalyst for effective hydrogen production, Int. J. Photoenergy (2013) 247516–24752.

[116] Y. Lin, Z. Geng, H. Cai, L. Ma, J. Chen, J. Zeng, N. Pan, X. Wang, Ternary graphene–TiO_2–Fe_3O_4 nanocomposite as a recollectable photocatalyst with enhanced durability, Eur. J. Inorg. Chem. (2012) 4439–4444. https://doi.org/10.1002/ejic.201200454

[117] L. Zhang, Y. Zhu, A review of controllable synthesis and enhancement of performances of bismuth tungstate visible-light-driven photocatalysts, Catal. Sci. Technol. 2 (2012) 694–706. https://doi.org/10.1039/c2cy00411a

[118] C. Zhang,Y. F. Zhu, Synthesis of square Bi_2WO_6 nanoplates as high-activity visible-light-driven photocatalysts, Chem. Mater. 17(13) (2005) 3537-3545. https://doi.org/10.1021/cm0501517

[119] H.B. Fu, L.W. Zhang, W.Q. Yao, Y.F. Zhu, Photocatalytic properties of nanosized Bi_2WO_6 catalysts synthesized via a hydrothermal process, Appl. Catal. B 66 (2006) 100-110. https://doi.org/10.1016/j.apcatb.2006.02.022

[120] L. Zhang, W. Wang, Z. Chen, L. Zhou, H. Xu, W. Zhu, Fabrication of flower-like Bi_2WO_6 superstructures as high performance visible-light driven photocatalysts, J. Mater. Chem. 17 (2007) 2526-2532. https://doi.org/10.1039/b616460a

[121] L. Zhang, H. Wang, Z. Chen, P.K. Wong, J. Liu, Bi_2WO_6 micro/nano-structures: Synthesis, modifications and visible-light-driven photocatalytic applications, Appl. Catal., B 106(1-2) (2011) 1-13. https://doi.org/10.1016/j.apcatb.2011.05.008

[122] L. Zhang, W.Z. Wang, L. Zhou, H.L. Xu, Bi_2WO_6 nano-and microstructures: shape control and associated visible-light-driven photocatalytic activities, Small 23(9) (2007) 1618-1624. https://doi.org/10.1002/smll.200700043

[123] K. Vijayasankar, N.Y. Hebalkar, H.G. Kim, P.H. Borse, Controlled band energetics in Pb-Fe-Nb-O metal oxide composite system to fabricate efficient visible light photocatalyst, J. Ceram. Process. Res. 14(4) (2013) 557-562.

[124] S.P. Singh, A.K. Singh, D. Pandeya, H. Sharma, and Om Parkash, Crystallographic phases, phase transitions and barrier layer formation in (1-x)[$Pb(Fe_{1/2}Nb_{1/2})O_3$]-x$PbTiO_3$, J. Mater. Res. 18 (2003) 2677-2687. https://doi.org/10.1557/JMR.2003.0374

[125] R. Sun, W. Tan, B. Fang, Perovskite phase formation and electrical properties of $Pb(Fe_{1/2}Nb_{1/2})O_3$ ferroelectric ceramicse, Phys. Status Solidi A. 206(2) (2009) 326-331. https://doi.org/10.1002/pssa.200824432

[126] R.R. Schaller, Technological innovation in the semiconductor industry: a case study of the international technology roadmap for semiconductors (ITRS), George Mason University (2004).

[127] G.D. Wilk, R.M. Wallace, J.M. Anthony, High-κ gate dielectrics: current status and materials properties considerations, J. Appl. Phys. 89 (2001) 5243-5249. https://doi.org/10.1063/1.1361065

[128] F.B. Ergin, R. Turan, S.T. Shishiyanu, Effect of γ-radiation on HfO_2 based MOS capacitor, Nucl. Inst. Meth. Phys. Res. B: Beam Interactions with Materials and Atoms 268 (2010) 1482-1485.

[129] M.A. Khaskheli, P. Wu, A.M. Soomro, M. Khan, M.E. Asadullah, M.S. Kalhoro, Fabrication and electrical characteristic of quaternary ultrathin HfTiErO thin films for MOS devices grown by RF sputtering, Chem. Mater. Res. 3(4) (2013) 41-47.

[130] Q. Tao, A. Kueltzo, M. Singh, Atomic layer deposition of HfO_2, TiO_2, and $Hf_xTi_{1-x}O_2$ using metal (diethylamino) precursors and HfO_2, J. Electrochem. Soc. 158 (2008) G27-G33. https://doi.org/10.1149/1.3522758

[131] D.Y. Cho, S.J. Oh, Y. Chang Y, Role of oxygen vacancy in $HfO_2/SiO_2/Si$ (100) interfaces. Appl. Phys. Lett. 88 (2006) 193502-193503. https://doi.org/10.1063/1.2201050

[132] F.Y. Xiao, J. Xing, L. Wu, Z.P. Chen, X.L. Wang, H.G. Yang, Assembly ultrathin $PbBiO_2Br$ nanosheets with enhanced visible light Photocatalytic property, R. Soc. Chem. (2013).

[133] Z. Deng, D. Chen, B. Peng, F. Tang, From bulk metal Bi to two-dimensional well-crystallized BiOX (X = Cl, Br) micro-and nanostructures: Synthesis and characterization, Cryst. Growth Des. 8(8) (2008) 2995-3003. https://doi.org/10.1021/cg800116m

[134] F.Y. Xiao, J. Xing, L. Wu, Z.P. Chen, X.L. Wanga, H. G. Yang, Assembly ultrathin $PbBiO_2Br$ nanosheets with enhanced visible light photocatalytic property, RSC Adv. 3(27) (2013) 10687-10690. https://doi.org/10.1039/c3ra41324a

[135] L. Cademartiri, G.A. Ozin, Ultrathin nanowires—A materials chemistry perspective, Adv. Mater. 21(9) (2009) 1013-1020. https://doi.org/10.1002/adma.200801836

[136] K. Richardson, W. Steffen, H.J. Schellnhuber, J. Alcamo, T. Barker, D.M. Kammen, R. Leemans, D. Liverman, M. Munasinghe, B. Osman-Elsha, N. Stern, O. Waever, Synthesis report from climate change: global risks, Challenges & Decisions, University of Copenhagen (2009).

[137] G.A. Olah, A. Goeppert, G.K.S. Prakash, Beyond oil and gas: The methanol economy, Wiley-VCH, Weinheim (2006) 212.

[138] K. Oppelt, D.A.M. Egbe, U. Monkowius, M. List, M. Zabel, N.S. Sariciftci, G. Knör, Luminescence and spectroscopic studies of organometallic rhodium and rhenium multichromophore systems carrying polypyridyl acceptor sites and

phenylethynyl antenna subunits, J. Organomet. Chem. 696 (2011) 2252-2258. https://doi.org/10.1016/j.jorganchem.2010.11.008

[139] E. Portenkirchner, K. Oppelt, C. Ulbricht, A.M. Egbe , H. Neugebauer, G. Knör, N. S. Sariciftci, Electrocatalytic and photocatalytic reduction of carbon dioxide to carbon monoxide using the alkynyl-substituted rhenium(I) complex (5,50-bisphenylethynyl-2,20-bipyridyl)Re$(CO)_3$Cl, J. Organomet. Chem. 716 (2012) 19-25.

[140] J. Hawecker, J.M. Lehn, R. Ziessel, H. Chim, Photochemical and electrochemical reduction of carbon dioxide to carbon monoxide mediated by (2, 2′-Bipyridine) tricarbonylchlororhenium(I) and related complexes as homogeneous catalysts, Acta 69(8) (1986) 1990-2012. https://doi.org/10.1002/hlca.19860690824

[141] K. Koike, H. Hori, M. Ishizka, J.R. Westwell, K. Takeuchi, T. Ibusuki, K. Enjouji, H. Konno, K. Sakamoto, O. Ishitani, Key process of the photocatalytic reduction of CO_2 using $[Re(4,4'\text{-}X_2\text{-bipyridine})(CO)_3PR_3]^+$ (X = CH_3, H, CF_3; PR_3 = phosphorus ligands): dark reaction of the one-electron-reduced complexes with CO_2, Organomet. 16(26) (1997) 5724-5729. https://doi.org/10.1021/om970608p

[142] A. Fujishima, Electrochemical photolysis of water at a semiconductor electrode, Nature 238 (1971) 37–38. https://doi.org/10.1038/238037a0

[143] F. Del Valle, F. Del Valle, J.A. Villoria De La Mano, M. C. Álvarez-Galván, J.L. G. Fierro, Photocatalytic water splitting under visible light: concept and materials requirements, Adv. Chem. Eng. 36 (2009) 111–143.

[144] H. Kato, K. Asakura, A. Kudo, Highly efficient water splitting into H_2 and O_2 over lanthanum-doped NaTaO photocatalysts with high crystallinity and surface nanostructure, J. Am. Chem. Soc. 125 (2003) 3082-3092. https://doi.org/10.1021/ja027751g

[145] D. Yokoyama, H. Hashiguchi, K. Maeda, T. Minegishi, T. Takata, R. Abe, J. Kubota, K. Domen, Ta_3N_5 photoanodes for water splitting prepared by sputtering, Thin Solid Films 519(7) (2011) 2087–2092. https://doi.org/10.1016/j.tsf.2010.10.055

[146] K. Maeda, T. Takata, M. Hara, N. Saito, Y. Inoue, H. Kobayashi, K. Domen, GaN: ZnO solid solution as a photocatalyst for visible-light-driven overall water splitting, J. Am. Chem. Soc. 127 (2005) 8286-8287. https://doi.org/10.1021/ja0518777

[147] K. Maeda, K. Domen, New non-oxide photocatalysts designed for overall water splitting under visible light, J. Phys. Chem. C 111(22) (2007) 7851-7861.

[148] T. Mishima, M. Matsuda, M. Miyake, Visible-light photocatalytic properties and electronic structure of Zr-based oxynitride, Zr_2ON_2, derived from nitridation of ZrO_2, Appl. Catal. A: Gen. 324(1-2) (2007) 77–82. https://doi.org/10.1016/j.apcata.2007.03.017

[149] M.M. Hassan, H. Dylla, L.N. Mohammad, T. Rupnow, Effect of application methods on the effectiveness of titanium dioxide as a photocatalyst compound to concrete pavement, Presented at 89th Annual Meeting of the Transportation Research Board, Washington, D.C. (2010).

[150] J. Yang, G. Jiang, G, Experimental study on properties or pervious concrete pavement materials. Cem. Concr. Res. 33 (2003) 381-386. https://doi.org/10.1016/S0008-8846(02)00966-3

[151] M. Miyauchi, A. Nakajima, T. Watanabe, K. Hashimoto, Photocatalysis and photoinduced hydrophilicity of various metal oxide thin films, Chem. Mater. 14 (2002) 2812–2816. https://doi.org/10.1021/cm020076p

[152] S. Navarro, J. Fenoll, N. Vela, E. Ruiz, G. Navarro, Photocatalytic degradation of eight pesticides in leaching water by use of ZnO under natural sunlight, J. Hazard. Mater. 172 (2009) 1303–1310. https://doi.org/10.1016/j.jhazmat.2009.07.137

[153] A.A. Abdel, S.A. Mahmoud, A.K. Aboul-Gheit, Sol-gel and thermally evaporated nanostructured thin ZnO films for photocatalytic degradation of trichlorophenol, Nanoscale Res. Lett. 4 (2009) 627–634. https://doi.org/10.1007/s11671-009-9290-1

[154] C.C. Chen, Degradation pathways of ethyl violet by photocatalytic reaction with ZnO dispersions, J. Mol. Catal. A: Chem. 264 (2006) 82–92. https://doi.org/10.1016/j.molcata.2006.09.013

[155] H. Xu, W. Wang, W. Zhu, Shape evolution and size-controllable synthesis of Cu_2O octahedra and their morphology-dependent photocatalytic properties, J. Phys. Chem. B 110 (2006) 13829–13834. https://doi.org/10.1021/jp061934y

[156] J.-Y Ho, M.H. Huang, Synthesis of submicrometer-sized Cu_2O crystals with morphological evolution from cubic to hexapod structures and their comparative photocatalytic activity, J. Phys. Chem. C 113 (2009) 14159–14164. https://doi.org/10.1021/jp903928p

[157] L. Huang, F. Peng, H. Yu, H. Wang, Preparation of cuprous oxides with different sizes and their behaviours of adsorption, visible-light driven photocatalysis and photocorrosion, Solid State Sci. 11 (2009) 129–138. https://doi.org/10.1016/j.solidstatesciences.2008.04.013

[158] A. Watcharenwong, W. Chanmanee, N.R. de Tacconi, C.R. Chenthamarakshan, P. Kajitvichyanukul, K. Rajeshwar, Anodic growth of nanoporous WO_3 films: morphology, photoelectrochemical response and photocatalytic activity for methylene blue and hexavalent chrome conversion, J. Electroanal. Chem. 612 (2008) 112–120. https://doi.org/10.1016/j.jelechem.2007.09.030

[159] J. Luo, M. Hepel, Photoelectrochemical degradation of naphthol blue black diazo dye on WO_3 film electrode, Electrochim. Acta 46 (2001) 2913–2922. https://doi.org/10.1016/S0013-4686(01)00503-5

[160] Y. Bessekhouad, D. Robert, J.-V. Weber, Photocatalytic activity of Cu_2O/TiO_2, Bi_2O_3/TiO_2 and $ZnMn_2O_4/TiO_2$ heterojunctions, Catal. Today 101 (2005) 315–321. https://doi.org/10.1016/j.cattod.2005.03.038

[161] Z. Ai, Y. Huang, S. Lee, L. Zhang, Monoclinic α-Bi_2O_3 photocatalyst for efficient removal of gaseous NO and HCHO under visible light irradiation, J. Alloys Compd. 509 (2011) 2044–2049. https://doi.org/10.1016/j.jallcom.2010.10.132

[162] S. Kohtani, J. Hiro, N. Yamamoto, A. Kudo, K. Tokumura, R. Nakagaki, Adsorptive and photocatalytic properties of Ag-loaded $BiVO_4$ on the degradation of 4-n-alkylphenols under visible light irradiation, Catal. Commun. 6 (2005) 185-195. https://doi.org/10.1016/j.catcom.2004.12.006

[163] X. Hu, C. Hu, Preparation and visible-light photocatalytic activity of Ag_3VO_4 powders, J. Solid State Chem. 180 (2007) 725–732. https://doi.org/10.1016/j.jssc.2006.11.032

[164] D. Ye, D. Li, W. Zhang, M. Sun, Y. Hu, Y. Zhang, X. Fu, A new photocatalyst $CdWO_4$ prepared with a hydrothermal method, J. Phys. Chem. C 112 (2008) 17351–17356. https://doi.org/10.1021/jp8059213

[165] T. Yan, L. Li, W. Tong, J. Zheng, Y. Wang, G. Li, $CdWO_4$ polymorphs: Selective preparation, electronic structures, and photocatalytic activities, J. Solid State Chem. 184 (2011) 357–364. https://doi.org/10.1016/j.jssc.2010.12.013

[166] X. Lin, T. Huang, F. Huang, W. Wang, J. Shi, Photocatalytic activity of a Bi-based oxychloride Bi_4NbO_8Cl, J. Mater. Chem. 17 (2007) 2145–2150. https://doi.org/10.1039/b615903f

Chapter 4

Enhanced Photocatalytic Activity of TiO_2 Supported on Different Carbon Allotropes for Degradation of Pharmaceutical Organic Compounds

Rajesh J. Tayade [1,2*] and Wan-Kuen Jo [1]

[1]Department of Environmental Engineering, Kyungpook National University, 80 Universtiy Road, Bukgu, Daegu 702-701, Republic of Korea

[2]Inorganic Materials and Catalysis Division, Central Salt and Marine Chemicals Research Institute, Council of Scientific and Industrial Research (CSIR-CSMCRI), G. B. Marg, Bhavnagar 364002, Gujarat, India

Email: rtayade@gmail.com

Abstract

TiO_2-carbon composites with different amount of TiO_2 (5%, 10%, 20%, 30% and 50%) supported on different carbon allotropes such as activated charcoal, graphite, and graphene were synthesized by hydrothermal method. The synthesized catalysts were characterized by X-ray diffraction (XRD), ultraviolet-visible diffuse reflectance spectroscopy (UV-vis DRS), N2 adsorption and scanning electron microscopy (SEM) techniques. XRD result demonstrated that, the loading of TiO_2 has not significantly altered the structure of the support carbon allotropes. However, the bandgap and surface area of the composite was varied with respect to the amount of TiO_2 loading in the composites. The photocatalytic activity of the synthesized TiO_2-carbon composites was evaluated by photocatalytic degradation of isoniazide in aqueous medium. All the synthesized catalyst were found easy to separate from the reaction mixture. The result demonstrated that the composites synthesized using activated charcoal showed enhanced photocatalytic activity as compared to the other allotropes of carbon. The highest photocatalytic activity was obtained using a composite having 30% TiO_2 supported on activated charcoal.

Keywords

Activated Charcoal, Graphite, Graphene, TiO_2, Isoniazide, Photocatalysis

Contents

1. Introduction

In recent year, there has been great interest in the use of heterogeneous photocatalysis for the degradation of hazardous and pharmaceutical organic compounds [1-2]. Nano-TiO_2 semiconductor photocatalysts received the main focuse due to its unique properties like non-toxicity, low-cost, high reactivity, high photochemical stability and bio-compatibility for the degradation of various organic compounds [3-4]. However, the disadvantages of conventional nano powder photocatalyst are related to low efficiency and recovery or separation of catalyst after the photocatalytic treatment of the water [5-6]. These disadvantages of nano-TiO_2 result in low photocatalytic degradation of organic compounds and difficulties in separation of catalyst in practical applications. To achieve faster and efficient degradation of organic pollutant, high internal efficiency associated to the separation, it may be useful to load photocatalytic material on suitably stable adsorbent to set in contact with external pollutant molecules and its easy separation.

In order to obtain easy separation of the photocatalysts from the treated water various type photocatalyst were developed including floating catalysts such as TiO_2 coated cenospheres [7], TiO_2 grafted on polystyrene beads [8], cenosphere supported AgCl/TiO_2 [9], TiO_2 coated on mosquito net [10] and also some effort were done towards the synthesis of magnetically separable titania composites [11]. In addition, composites using different inorganic materials such as zeolite [12-13], layered double hydroxides (LDH) [14], clay [15] and carbon [16] were used as support materials

Among these, carbon is one of the most common element in our environment and is more environmentally and biologically friendly. There are several allotropes of carbon such as graphite, activated carbon, and amorphous carbon. The physical properties of carbon vary with the allotropic form. The carbon based nanomaterials recently used in various applications, such as high-strength composite materials, biomaterials, and electronics due their excellent mechanical strength, electrical and thermal conductivity, and optical properties [16-17]. From the literature it is evident that carbon materials can be used for synthesis of photocatalytic materials [19-23]. However major work has been carried out on the doping of carbon on TiO_2 for enhancement in the absorption of light toward visible range and to increase the rate of electron hole recombination in order to enhance the photocatalytic activity [24]. It is well known that the electrical conductivity in carbon varies with respect allotropes of the carbon and may be useful to develop eco-friendly photocatalytic materials. However till today, the photocatalytic activity of the various allotropes of carbon as support materials has not been investigated.

In the present study, the easy separable TiO_2-carbon composites were synthesized by loading of TiO_2 (5%, 10%, 20%, 30%, and 50%) on three carbon allotropes such as activated charcoal, graphite, and graphene to obtain TiO_2-carbon composites. The TiO_2-carbon composite materials were synthesized by dispersing carbon materials in dilute titanium isopropoxide solution. The synthesized TiO_2-carbon materials were applied for the photocatalytic degradation of isonizide, which is an organic compounds used for the medication of tuberculosis and found in pharmaceutical waste water. To the best of our knowledge till today there is no report on degradation of this compound. The prepared TiO_2-carbon catalysts were characterized by various sophisticated techniques to determine the structural, textural, and electronic properties. The photocatalytic performance of the synthesized catalysts was studied by photocatalytic degradation of isoniazide in aqueous medium under irradiation of ultraviolet light.

2. Experimental

2.1 Chemicals and materials

Titanium isopropoxide (TIP), graphite, isoniazide, activated charcoal (Darko, 4-12 mesh) were procured from Aldrich. The properties of isoniazide are given in Table- 1. Standards for chemical oxygen demand (COD) determination were purchased from Humas Co. Ltd, Daegon, South Korea. During the preparation of the catalyst and isoniazide solution distilled water was used.

Table 1. *Properties of Isoniazide.*

Molecular structure	Properties	
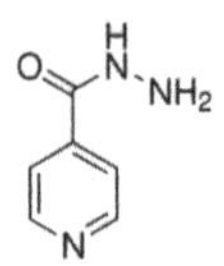	Molecular formula	$C_6H_7N_3O$
	Formula weight	137.14 g/mol
	Appearance	White
	Water solubility	14 g/100 mL (25 ºC)

2.2 Catalysts preparation

2.2.1 TiO_2 coated activated charcoal composites

The TiO_2-activated charcoal (AC) composites were prepared by dispersing activated charcoal powder in dilute titanium isopropoxide solution. Prior to the loading of TiO_2 on activated charcoal, the activated charcoal was ground to obtain fine power and pass 70 mesh sieve and to remove the moisture during the grinding the charcoal was activated by calcinations at 450°C for 4 h under nitrogen flow in a tubular furnace. To coat TiO_2 on the activated charcoal, appropriate amount of titanium isopropoxide to obtain 5, 10, 20, 30 and 50 % of loading of TiO_2 on activated charcoal was added to 50 ml absolute ethanol in a round bottom flask (250 mL) and stirred for 30 min to obtain a transparent solution. After complete dissolution of titanium isopropoxide, one gram of activated charcoal was added to the solution and the mixture was stirred for 12 h to obtain uniform distributed TiO_2 particles on the activated charcoal. Further appropriate amount of water was added to the mixture for the hydrolysis of TiO_2 and stirred for 30 min. The solvent from the mixture was slowly removed using rotavapour at 343 K and obtained TiO_2-activated charcoal composites were dried in an oven at 373 K for 12 h. Thus obtained TiO_2-activated charcoal were further calcined at 273 K for 4 h in a tubular furnace in air atmosphere to convert the amorphous TiO_2 into anatase crystalline TiO_2 in the composites. The catalysts obtained were designated as ACT-1, ACT-2, ACT-3, ACT-4, and ACT-5 for 5, 10, 20, 30, and 50% loading of TiO_2 respectively, on the activated charcoal.

2.2.2 TiO_2-graphite composites

As mentioned above the process for the TiO_2-activated charcoal composite was adapted for the synthesis of TiO_2- graphite composites. The catalysts obtained were designated as GCT-1, GCT-2, GCT-3, GCT-4, and GCT-5 for 5, 10, 20, 30, and 50% loading of TiO_2 on graphite respectively.

2.2.3 TiO_2-graphene composites

For the synthesis of TiO_2-graphene composites, first graphene oxide was prepared by the modified Hummer-Offeman method [25]. In brief, graphite powder (10 g) was added in mixture of cold concentrated sulphuric acid (230 mL, 98 wt.%) and then potassium permanganate (KMnO4, 30 g) gradually added with continuous stirring and keeping the mixture in a dry ice bath to maintain a temperature of below 293K. Further, the dry ice bath was replaced by a water bath and the mixture was heated to 308 K for 0.5 h with gas release under continuous stirring, which was followed by slow addition of de-ionized water (460 mL), which produced a rapid increase in solution temperature up to a maximum of 371 K. The reaction was carried out for 40 min in order to increase the oxidation degree of the graphite oxide product and then the resultant bright-yellow suspension was finished by addition of more distilled water (140 mL) followed by hydrogen peroxide solution (H_2O_2, 30 %, 30 mL). Then by application of centrifugation at 4000 rpm the solid product was separated and washed initially with 5% HCl until sulphate ions were no longer detectable with barium chloride. The solid product was then washed several times with acetone and air dried overnight at 338 K. The dried product was ground using mortal and pistol. To load the TiO_2 on graphene appropriate amount of synthesized anatase TiO_2 was added to 50 ml absolute ethanol and the synthesized graphene oxide was added to it. To convert graphene oxide to graphene utrasonication was applied and mixture was ultrasonicated using a horn type ultrasonicator (Power:750 W, Frequency:20KW) keeping an amplitude of 30% without pulse for 1 h. Then the solvent was removed using rotavapour at 343 K and obtained TiO_2 loaded graphene composites were dried in an oven at 373 K for 12 h. The catalysts obtained were designated as GOT-1, GOT-2, GOT-3, GOT-4, and GOT-5 for 5, 10, 20, 30, and 50% laoding of TiO_2 on graphene respectively. TiO_2 powder was also prepared from titanium isopropoxide using a similar synthetic procedure without carbon material addition and was used as a reference sample referred to as pristine TiO_2.

2.3 Catalyst characterization:

Powder X-ray diffraction patterns of the synthesized TiO_2 coated composites were recorded with a Regaku diffractometer (D/Max-2500) system using CuKα radiation (λ = 0.154056nm) in the 2θ range from 10°-80° at a scan rate of 0.1°sec^{-1}. The standard anatase and rutile diffractograms were used for comparison with X-ray diffraction pattern of synthesized composites. (*ICDD Reference Pattern Database*; File No. 01-075-2547 & 01-087-0920, 2014) The anatase and rutile phase percentage formed were calculated from the integrated intensity peak at 2θ = 25.3 (1 0 1) for anatase and peak at 2θ = 27.4 (1 1 0) for rutile. The percentage of anatase, A (%) was determined using Equation 1 given below [26]

$$A\ (\%) = 100/\ (1+1.265 I_R/I_A) \qquad (1)$$

I_R is the intensity of rutile peak at 2θ = 27.4

I_A is the intensity of anatase peak at 2θ = 25.3

BET surface area, pore size and volume distributions of synthesized TiO_2-carbon composites were determined from N_2 adsorption-desorption isotherms at 77K (ASAP 2010, Micromeritics, USA). Surface area and pore size distribution were determined using the BET equation and BJH method respectively [27]. The TiO_2 supported activated charcoal and graphite catalysts were degassed under vacuum (10^{-3} Torr) at 623 K for 4 h, and TiO_2 supported catalysts were degassed under vacuum at 373 K prior to measurement.

The band-gap energy and absorption edge of the TiO_2-carbon composite was determined using the UV-Visible Diffuse reflectance spectroscopy (UV-DRS) (Scinco Co. Ltd, S-3100) equipped with an integrating sphere and $BaSO_4$ was used as a reference [12]. The spectra were recorded at room temperature in the wavelength range of 250-600 nm. The bandgap energy of synthesized TiO_2-carbon composites were determined using UV-Vis-DRS method and calculated according to Equation 2 given below

$$\text{Bandgap (EG)} = hc/\lambda \qquad (2)$$

where EG is the band gap energies (eV), h the Planck's constant, c the light velocity (m/s) and λ is the wavelength (nm).

The electron microscopic study has been done with scanning electron microscope (Hitachi S-4300/EDX-350) to determine the morphology. The sample powder was supported on aluminium stubs using silver paint and then coated with gold by plasma prior to measurement.

2.4 Adsorption study (in dark)

To evaluate the adsorption capacities of the synthesized TiO_2-carbon composites a 50 ppm isoniazide solution (250 mL) was combined with the synthesized TiO_2-carbon composites (50 mg) under continuous stirring. To avoid light shining onto the mixture, the 500 ml conical flask was covered using aluminum foil for 4 h at 25°C. Aliquots (7.5 mL) were withdrawn hourly and centrifuged to separate the catalyst from the supernatant. The concentration of isoniazide in the solution was then determined using a UV-vis spectrophotometer.

2.5 Photocatalytic irradiation system

The quartz immersion well photocatalytic reactor fabricated locally was used to determine the photocatalytic efficiency of synthesized TiO_2 coated composites [18]. The photocatalytic reactor was composed of inner quartz double wall jacket with an empty chamber at the center to hold the 250 W mercury vapour lamp (Woosung electric Co. Ltd., South Korea) having inlet and outlet provision for the circulation of water to maintain the temperature of the reaction mixture and outer pyrex glass container (volume 500 ml after insertion of inner part) in which the reaction mixture is taken and irradiated. The temperature of the reaction mixture was maintained by circulating cold water through the inner quartz double wall jacket. The concentration of isoniazide in bulk solution prior to irradiation was used as the initial value for the measurement of degradation of isoniazide. At each interval of 30 min, 7.5 ml of the sample was withdrawn with a syringe from the irradiated suspension. Before commencing irradiation, a suspension 100 mg of synthesized TiO_2 –carbon composite and 500 mL aqueous solution of about 50 ppm isoniazide was ultrasonicated for 2 minutes and then stirred for 30 minutes in the dark to get the adsorption of the isoniazide on the surface of the composite.

2.6 Chemical analysis

The concentration of isoniazide in aqueous solution during the photocatalytic reaction was determined by a UV-visible spectrophotometer and the degradation was studied by the analysis of the chemical oxygen demand. The UV-visible absorbance of aqueous isoniazide solution was measured at λ_{max}=262 nm with a UV-Visible spectrophotometer (Agilent 8453) equipped with a quartz cell having a path length of 1 cm. The spectral absorbance was measured with a baseline correction at scan rate of 400 nm min^{-1} and a data-point interval of 1 nm. The concentration of nitrobenzene in solution was determined using a calibration curve of nitrobenzene (concentration vs absorbance) prepared with known concentrations.

The oxygen equivalent of the organic matter of each sample, i.e. chemical oxygen demand (COD), was measured using a Hach COD analyser (DRB-200-PD). 2 ml of degraded solution was added to the reagent and digested at 421 K for 2h and after cooling the mixture was determined using a COD analyser.

3. Result and discussions

3.1 X-ray diffraction

To evaluate the presence of TiO_2 in the composites, TiO_2-carbon composites were analyzed by X-ray diffraction analyses and compared to the virgin carbon materials used in the synthesis. The crystal structures of TiO_2-carbon composites prepared using activated charcoal, graphite, and graphene oxide are shown in Fig. 1. The XRD pattern (Fig. 1A) for activated carbon has main peak at 26.69°, which are assigned to the 002 plane (ICDD Reference Pattern Database; File No.01-071-3739, 2014). The sharp peak indicates the crystalline nature of activated charcoal. For the TiO_2 the major peak at 2θ = 25.3°, 27.4° and at 30.7° corresponds to the diffraction from the (101) crystal plane of anatase phase were observed. In case of the TiO_2-activated charcoal composite the major peak at 25.3° was observed after addition of 30 % TiO_2 which clearly indicates the presence of TiO_2 in the composite and the peak intensity was found to increase with increase in the amount of TiO_2.

The XRD pattern of the TiO_2-graphite composite can be seen in Fig. 1B. The peak at ca. 26.6°, which is assigned for the 003 plane of graphite (ICDD Reference Pattern Database; File No. 01-075-2078, 2014). From the XRD diffraction pattern it is observed that the peak correspond to the anatase phase of TiO_2 and was observed in GCT-5 catalyst only without any structural change in the graphite. However the peak intensity of the peak corresponding to the 003 crystal plane was found reduced with increase in amount of TiO_2.

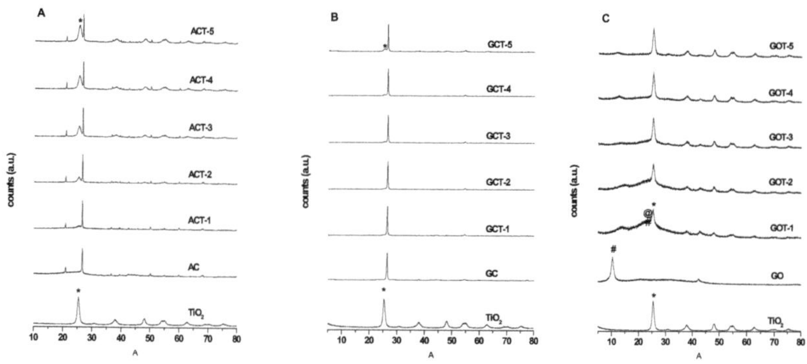

Figure 1: *XRD pattern of TiO_2-carbon composites (*:Anatase, #:graphene oxide, @:graphene)*

For the synthesis of the TiO_2-graphene composite materials, we have used the anatase TiO_2 to avoid the loss of graphene due to post synthesis calcinations. Fig. 1C shows that the graphene oxide used for the synthesis of TiO_2 graphene composite. From the XRD pattern it is clear that the graphene oxide was converted to graphene after ultrasonifcation during the synthesis of the TiO_2-graphene composites. The XRD pattern of GO shows a strong peak located in the low-angle region at about 10°. This peak completely disappears and after ultrasonication treatment a new peak emerged at about 25 which correspond to graphene in the GOT-1 catalyst. Similar disappearance of the GO peak and emergence of graphene peak after thermal treatment was reported by Ding et al. [28]. Also a peak corresponding to the anatase phase of TiO_2 was also observed at about 25.3°. However with increase in the TiO_2 amount the anatase peak become dominant and the graphene peak could not be seen in the GOT-3, GOT-4, and GOT-5 catalysts.

3.2 UV-visible diffuse reflectance spectroscopy

The diffuse reflectance spectra of the TiO_2-carbon composites are shown in Fig. 2. The reflectance of all the carbon materials without addition of TiO_2 clearly indicated that there was no reflectance in the range of 200-600 nm. Further, it is found that after loading of different percentage of TiO_2 the reflectance of the composites was enhanced in the range of 350-800 nm which clearly indicate the presence of the TiO_2 in the composites. Similar results were also obtained for the different amount of TiO_2 coated zeolite was reported [12-13]. The increase in the reflectance in the different form of carbon composite was observed in the order of activated carbon>graphite >graphene oxide. The synthesized TiO_2-carbon composites are given in Table-2.

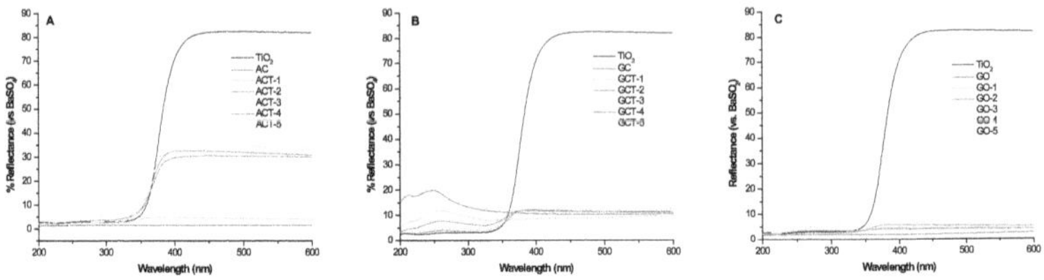

Figure 2: *UV-vis diffuse reflectance spectra of TiO_2-carbon composites using A) Activated charcoal B) Graphite C) Graphene.*

Table 2: *Band edge and bandgap of TiO_2-carbon composite materials.*

Sr. No	Catalyst Name	TiO_2 in composite (%)	Band edge (nm)	Bandgap (eV)
1	ACT-1	5	344	3.56
2	ACT-2	10	361	3.43
3	ACT-3	20	367	3.37
4	ACT-4	30	367	3.37
5	ACT-5	50	367	3.37
6	GCT-1	5	348	3.56
7	GCT-2	10	356	3.48
8	GCT-3	20	358	3.46
9	GCT-4	30	358	3.46
10	GCT-5	50	359	3.45
11	GOT-1	5	358	3.46
12	GOT-2	10	359	3.45
13	GOT-3	20	359	3.45
14	GOT-4	30	359	3.45
15	GOT 5	50	362	3.42

The band edge and bandgap of parallel synthesized TiO_2 was 374 nm and 3.31 eV respectively. It is clear that the change in the bandgap of the carbon composite materials is due to the presence of TiO_2 in the composite materials. The bandgap of the TiO_2-carbon composite prepared using activated charcoal, graphite, and graphene oxide was in the range of 3.56-3.37 eV, 3.56-3.45 eV, and 3.46-3.42 eV respectively.

3.3 Surface area analysis

It is observed that the composites were synthesized by incorporating the titanium source in the higher surface area and porous support materials such as zeolite, clay, layered double hydroxide (LDH) and carbon, results new composite with less surface area as compared to the pristine porous materials [12-13]. This may be due to the crystallization of TiO_2 in/on the surface of the support materials taking place which result in the reduction of surface area of the support material. However the reduction of the surface area of the composite depends on the percentage of TiO_2 loading and the position of TiO_2 in the pore or the outer side of the support materials. When the titanium source enters in the porous material cages or layers, it has a small crystallite size and amorphous nature. After hydrolysis and calcinations, the crystallite size increases and the amorphous phase changes to anatase phase below 450 °C [12].

Fig. 3 shows the adsorption-desorption isotherm of the TiO_2-carbon composites. There was no change in the shape of the isotherm plot after coating of TiO_2 has been observed in case of activated charcoal and graphite composites. In this study we have found that the reduction in the surface area of activated charcoal was 2.5 and 6.2 times at the loading of 5 % and 10 % of TiO_2 loading. This may be due to the lower amount of TiO_2 blocking the pores of the activated charcoal. Further addition of TiO_2 showed increase in the surface area gradually. This may be because some of the TiO_2 particles crystallize on the activated carbon **(Table 3).**

The graphite has very small surface area prior to TiO_2 loading, which indicate the fewer pores on the surface of graphite so in case of GCT-1, after getting all pores filled with TiO_2, additional TiO_2 was formed in the composite and with increase in the amount of TiO_2 surface area of the TiO_2-graphite composite was increased. The result demonstrated that the GCT-5 catalyst has about 4 times higher surface area than that of the pristine graphite. For synthesis of TiO_2-graphene composite, graphene oxide is used, it is found that the composite have less surface area than that of the graphene oxide. The BET surface area, pore volume, and pores size of the parallel synthesized TiO_2 was 23.6 m^2g^{-1}, 0.063 cm^3g^{-1} and 3.77 nm.

Table 3: Surface area of TiO_2 coated carbon composites.

Property	Support	Percentage of TiO_2					
		0%	5%	10%	20%	30%	50%
Surface area (m^2g^{-1})	Activated Charcoal	452.1	174.9	72.6	83.7	101.3	170.8
	Graphite	12.4	21.6	37.6	52.5	72.75	47.3
	Graphene	41.0*	7.5	37.3	51.32	90.55	103.4
Total pore volume (cm^3g^{-1})	Activated Charcoal	0.5732	0.5241	0.3661	0.3336	0.3228	0.4191
	Graphite	0.0856	0.0786	0.0962	0.1233	0.1566	0.1888
	Graphene	0.3081 *	0.0252	0.0500	0.0745	0.132	0.1793
Mean pore diameter (nm)	Activated Charcoal	5.07	11.98	20.396	15.936	12.882	9.8113
	Graphite	27.58	14.58	10.22	9.38	8.62	10.16
	Graphene	30.04*	13.37	5.354	5.807	5.4663	6.9347

*:graphene oxide

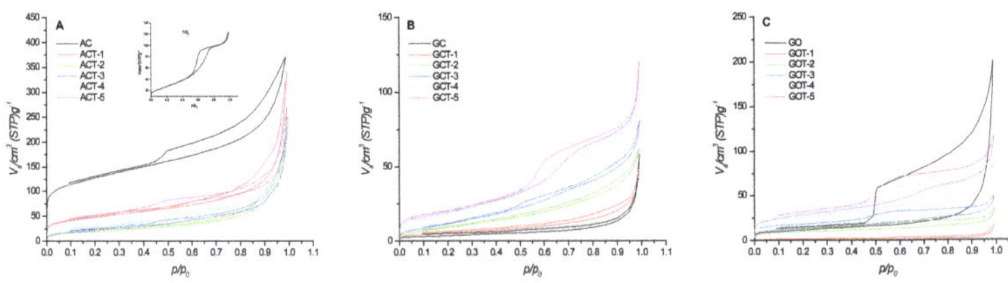

***Figure 3:** Isotherm plot of TiO_2 TiO_2-carbon composites using A) Activated charcoal B) Graphite C) Graphene.*

Fig 4. shows the pore size distribution of the TiO_2-carbon composites synthesized using different carbon allotropes. It is seen in Fig. 4A, there were no additional pores being developed in composites synthesized using activated charcoal. This may be due to the pore filling of activated carbon by TiO_2. In case of composite synthesized using graphite support it can be seen that additional pores in the range of 2.5 -10 nm is due to the TiO_2 (Fig. 4B). Similar it is observed in case of TiO_2-graphene composites and it is found the pore volume is increase with increased in the amount of TiO_2 which clearly indicate that the addition pores are due to the TiO_2 (Fig 4C).

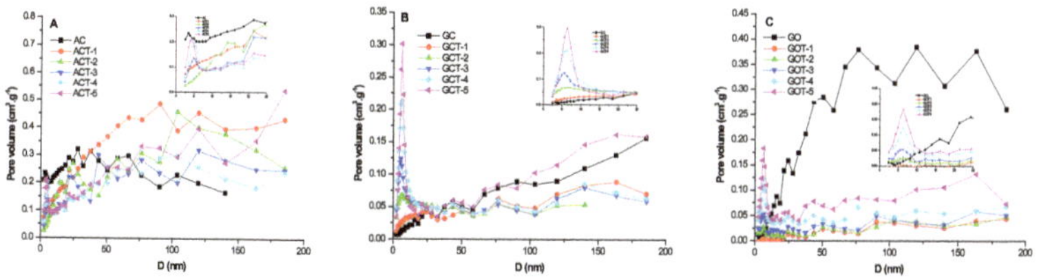

Figure 4: *Pore size distribution of TiO_2-carbon composites using A) Activated charcoal B) Graphite C) Graphene.*

3.4 Scanning electron microscopy

The images of TiO_2-carbon composites are shown in Fig. 5. The increase in the amount of TiO_2 in the carbon materials showed the increase in the spherical TiO_2 in the carbon materials.

The morphology of the activated charcoal was irregular, whereas the graphite and graphene oxides were in sheet form. From the images it can be seen that TiO_2 was anchored on the surface of the activated charcoal and graphite carbon materials. However in case of graphene, the GO is clearly seen in layered sheet structure and on application of ultrasonication these layered got wrapped around the TiO_2 particle which can be seen clearly in the GOT-3 composite image.

3.5 Isoniazide adsorption study

The heterogeneous adsorption of different organic molecule on the surface of the carbon material such as activated charcoal, graphite, carbon nanotube is one of the important parameter useful for the photocatalytic degradation of organic compounds. The adsorption of organic compound on the surface of carbon materials is due to the presence of high energy adsorption sites, electrostatic and non-electrostatic interaction, functional groups, pore size and surface area of carbon materials [19, 29-30].

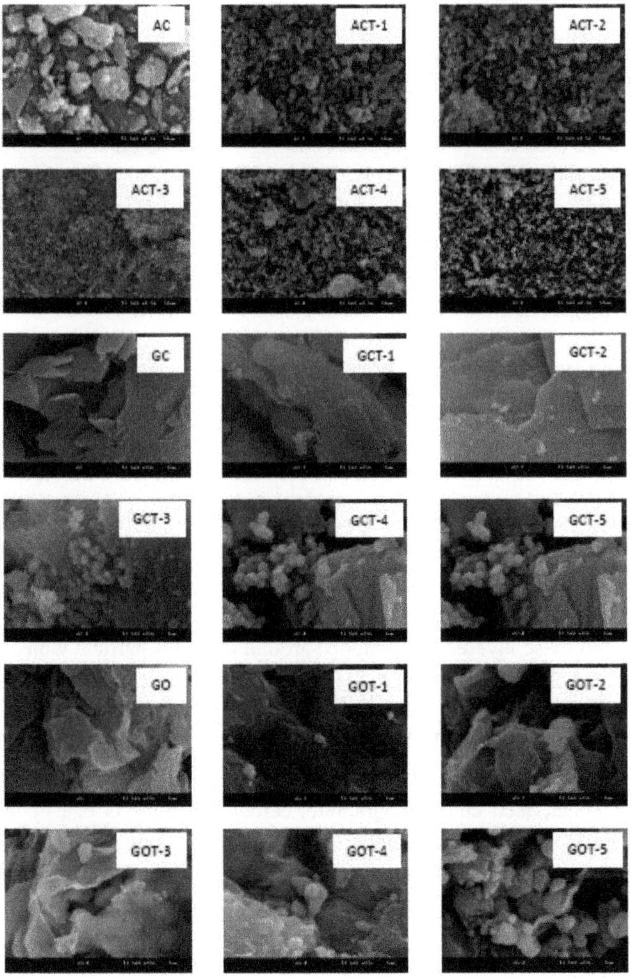

Figure 5: *SEM images of pristine carbon materials and composites.*

The adsorption of isoniazide on TiO_2-carbon composites was determined spectrophotometrically. Prior to the adsorption studies, the composites were dried for 12 h at 398 K to remove adsorbed water. The dried catalyst was added to 250 mL of isoniazide (50 ppm) solution and kept in the dark for 4 h. Samples were removed intialy after every 10 min up to 1 hr and then after every 30 min, and analyzed by UV-vis spectrophotometry after separation of the composites to determine the isoniazide concentration. It is found that in case of TiO_2-activated charcoal composite, the maximum adorption was reached within half an hour and it was in the range of 2-5%, whereas in case of TiO_2-graphite composite the lowest adsorption of isoniazide in the range of 1-2% observed. This may be due to the smaller surface area of the composites as compared to the activated charcoal composite and the surface functional groups. However, the adsorption on the surface of TiO_2-graphene composites was higher as compared to both the composites and it was in the range of 20-29 % and the highest adsorption (29%) was obtained for the lowest TiO_2 loaded composite. This indicates that the surface properties of all the carbon materials are different, and depends on the interaction of organic compound and the surface of carbon materials. However, the final degradation of organic compound can get enhanced if the organic compound comes in contact with hydroxyl radicated generated by TiO_2.

3.6 Photocatalytic activity

The photocatalytic activity of the TiO_2- carbon composites materials was studied by degradation of isoniazide under irradiation of ultraviolet light. During the photocatalytic reaction temperature was maintained at 303K by circulating cold water. After separation of the composite from the reaction mixture at different interval of time, the UV-visible absorbance spectrum of the reaction mixture containing only isoniazide was recorded to determine the concentration of isoniazide. It is observed that the trends of UV-visible spectra of the reaction mixture taken at different time interval during the photocatalytic reaction were the same for all the TiO_2-carbon composites catalysts and parallel synthesized TiO_2. The decrease in the peak at 262 nm was observed with respect to irradiation time and no additional peak was observed during the photocatalytic degradation of isoniazide. The results of photocatalytic degradation of isoniazide using 5 mg, 10 mg, 20 mg, 30 mg and 50 mg catalyst and without catalyst is shown in Fig. 6.

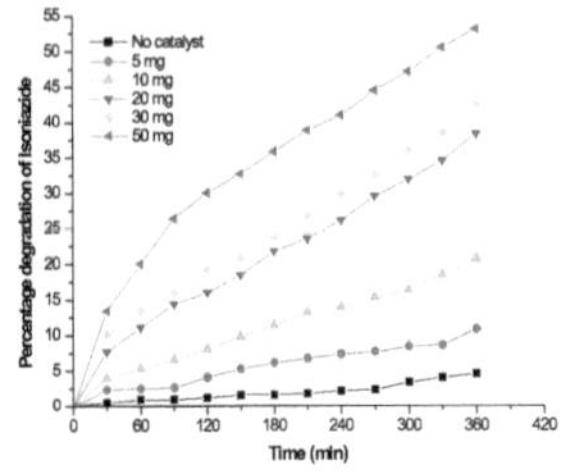

Figure.6: *Photocatalytic degradation of isoniazide using different amount of pristine TiO_2 under irradiation of ultraviolet light.*

From the results, it is clear that there was nearly 4 % degradation obtained without catalyst and with increase in the amount of TiO_2 photocatalyst the degradation percentage was also increased. The degradation using 5 mg, 10 mg, 20 mg, 30 mg, and 50 mg weight of TiO_2 in reaction mixture resulted in the 10.9%, 20.8%, 38.4%, 42.61%, and 53.2 % respectively. To confirm the degradation of isoniazide the chemical oxygen demand of the samples were carryout. It is found there was 80% of reduction of chemical oxygen demand. This clearly indicates that the degradation of isoniazide has been taken place due to presence of TiO_2 photocatalysts.

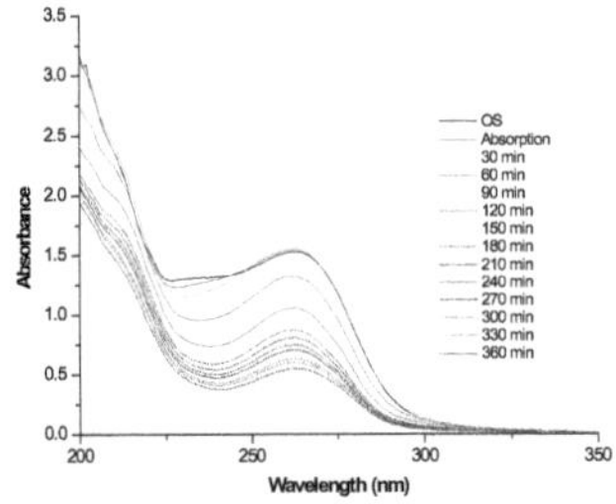

Figure 7: *UV-visible spectra of photocatalytic degradation of isoniazide using ACT-4 catalyst.*

By comparing the UV-visible spectra of degradaded samples using pristine TiO_2 and TiO_2-carbon composite, it is observed that the trend of the degradation was the same. Fig. 7 shows the UV-Visible spectra of the reaction mixture during the photocatalytic reaction taken at different time interval of degradation of isoniazide using ACT-4 catalyst. Fig. 8 shows the percentage of photocatalytic degradation of isoniazide using TiO_2-carbon composite materials. It is found that the highest percentage of degradation of isoniazide was achieved using TiO_2-activated charcoal composites. The highest photocatalytic activity was attributed using ACT-4 catalyst and the percentage degradation of isoniazide was it was 76.98 %. The adsorption of the isoniazide on the surface of TiO_2-activated charcoal composites was in the range of 1-6 %. The order of photocatalytic degradation of TiO_2-activated charcoal catalysts was ACT-4>ACT-5>ACT-3>ACT-2>ACT-1. This indicates that the optimum value of TiO_2 loading on activated charcoal to achieve highest photocatalytic activity is 30%. Further it is found that the photocatalytic activity of the TiO_2 coated photocatalyst was higher than that of the parallel systemized TiO_2.

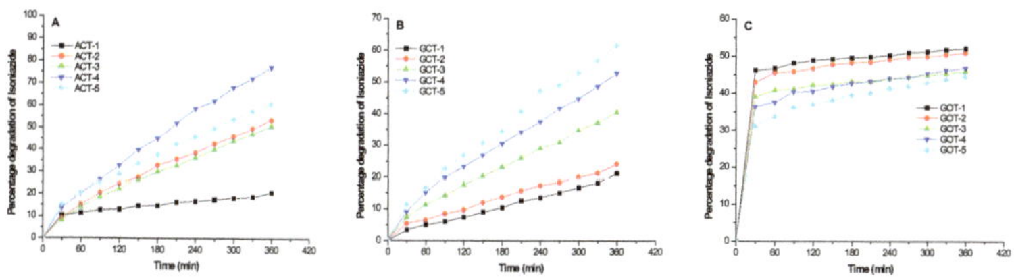

Figure 8. *Photocatalytic degradation using TiO_2-carbon composites materials :A) TiO_2-Activated charcoal B) TiO_2-graphite C) TiO_2-Graphene.*

In case of TiO_2 coated on graphite, the highest photocatalytic activity was attributed to the GCT-5 catalyst which was having maximum amount of TiO_2 loading. The adsorption of isoniazide on TiO_2-graphite catalyst was less as compared to the TiO_2- activated charcoal composites and it was in the range of 0.5-2.5 %. The order of degradation of isoniazide using TiO_2-graphite composites was GCT-5>GCT-4>GCT-3>GCT-2>GCT-1 and the highest percentage of degradation was reached at 61.6 % using the GCT-5 catalyst. The TiO_2 coated graphite composite showed better photocatalytic activity as compared to pristine TiO_2 but it was lower than that of the TiO_2 coated activated charcoal composite.

The photocatalytic activity of the TiO_2-graphene composites was lower than that of other synthesized composites, although the adsorption of isoniazide on the surface of TiO_2 coated graphene catalyst was higher. The adsorption of isoniazide on the surface of GOT-1, GOT-2, GOT-3, GOT-4, and GOT-5 was 41.9, 36.9, 30.5, 28.4 and 26.5 %. This indicates that the decrease in adsorption of isoniazide was found with increase in the amount of TiO_2. However the highest photocatalytic degradation was obtained 44.6 % using GOT-1 catalyst. From the results it is clear that the adsorption of the isoniazide on the surface of the catalyst was higher as compared to the other catalysts but it did not support to enhance the photocatalytic activity as it was found in case of other forms of carbon composite catalysts. This may be due to the graphene layer surrounded TiO_2 has inhibited the contact between isoniazide and hydroxyl radical generated by TiO_2.

4. Conclusion

The TiO_2-carbon composites were synthesized using different carbon allotropes as support by loading different amount of TiO_2 using the hydrothermal method. The synthesized catalysts were separated easily after photocatalytic reaction. The TiO_2 supported on activated charcoal and graphite showed better photocatalytic activity as compared to the pristine TiO_2 for degradation of isoniazide in aqueous medium under irradiation of ultraviolet light. It is observed that the optimum percentage of TiO_2 in activated charcoal and graphite was 30 and 50 % respectively. It is observed that the adsorption of isoniazide on the surface of TiO_2-graphene catalysts was higher as compared to other synthesized composites but could not show efficient photocatalytic activity. The photocatalytic activity was slow as compared to the other TiO_2 supported composites reported in this study. This result demonstrates that the higher adsorption may not favor the photocatalytic degradation if organic compound does not come in contact with the hydroxyl radicals generated by TiO_2.

Acknowledgements

This study was carried out with the support of the Ministry of Science, ICT and Planning (MSIP) (Project No. 132S-5-3-0610) and the National Research Foundation of Korea (NRF), by the Korean Government (MEST) (No. 2011-0027916). The authors thank Mr. Ajit Dale for his help in carring out photocatalytic experiments. Mr. Joon Yoob Lee, Mr. Go Tae Park, and Mr. Ingyu Hwang for their support to carry out the sample analysis.

References

[1] U.I. Gaya, A.H. Abdullah, Heterogeneous photocatalytic degradation of organic contaminants over titanium dioxide: A review of fundamentals, progress and problems. J Photochem Photobio C., 9 (2008) 1-12. https://doi.org/10.1016/j.jphotochemrev.2007.12.003

[2] H. Yang, G. Li, T. An, Y. Gao, J. Fu, Photocatalytic degradation kinetics and mechanism of environmental pharmaceuticals in aqueous suspension of TiO_2: A case of sulfa drugs. Catal .Today, 153 (2010) 200-207. https://doi.org/10.1016/j.cattod.2010.02.068

[3] M.R. Hoffmann, S.T. Martin, W. Choi, D.W. Bahnemann, Environmental applications of semiconductor photocatalysis, Chem. Rev., 95 (1995) 69-96. https://doi.org/10.1021/cr00033a004

[4] P.K. Surolia, M.A. Lazar, R.J. Tayade, R.V. Jasra, Photocatalytic degradation of 3, 3′-Dimethylbiphenyl-4, 4′-diamine (o-Tolidine) over nanocrystalline TiO_2 synthesized by sol– gel, solution combustion, and hydrothermal methods, Ind. Eng. Chem. Res., 47 (2008) 5847-5855. https://doi.org/10.1021/ie800073j

[5] Z. Ding, X. Hu, G.Q. Lu, P.-L. Yue, P.F. Greenfield, Novel silica gel supported TiO_2 photocatalyst synthesized by CVD method, Langmuir, 16 (2000) 6216-6222. https://doi.org/10.1021/la000119l

[6] Z. Liu, D.D. Sun, P. Guo, J.O. Leckie, An efficient bicomponent TiO_2/SnO_2 nanofiber photocatalyst fabricated by electrospinning with a side-by-side dual spinneret method, Nano Lett.,7 (2007) 1081-1085. https://doi.org/10.1021/nl061898e

[7] P.K. Surolia, R.J. Tayade, R.V. Jasra, TiO_2-coated cenospheres as catalysts for photocatalytic degradation of methylene blue, p-nitroaniline, n-decane, and n-tridecane under solar irradiation, Ind. Eng. Chem. Res., 49 (2010) 8908-8919. https://doi.org/10.1021/ie100388m

[8] P. Huo, Y. Yan, S. Li, H. Li, W. Huang, Floating photocatalysts of fly-ash cenospheres supported $AgCl/TiO_2$ films with enhanced Rhodamine B photodecomposition activity, Desalination, 256 (2010) 196-200. https://doi.org/10.1016/j.desal.2010.01.012

[9] M. Holgado, A. Cintas, M. Ibisate, C. J. Serna, C. Lopez, F. Meseguer, Three-dimensional arrays formed by monodisperse TiO_2 coated on SiO_2 spheres. J. Colloid. Interface Sci., 229 (2000) 6-11. https://doi.org/10.1006/jcis.2000.6973

[10] W.K Jo, R.J. Tayade, Facile photocatalytic reactor development using nano-TiO_2 immobilized mosquito net and energy efficient UVLED for industrial dyes effluent treatment, Journal of Environmental Chemical Engineering, 4 (2016) 319-327. https://doi.org/10.1016/j.jece.2015.11.024

[11] T.K. Pathak, N.H. Vasoya, T.S. Natarajan, K.B. Modi, R.J. Tayade, Photocatalytic degradation of aqueous nitrobenzene solution using nanocrystalline Mg-Mn ferrites, Materials Science Forum 764 (2013) 116-129. https://doi.org/10.4028/www.scientific.net/MSF.764.116

[12] R.J. Tayade, R.G. Kulkarni, R.V. Jasra, Enhanced photocatalytic activity of TiO_2-coated NaY and HY zeolites for the degradation of methylene blue in water. Ind. Eng. Chem. Res., 46 (2007) 369-376. https://doi.org/10.1021/ie060641o

[13] R.J. Tayade, P.K. Surolia, M.A. Lazar, R.V. Jasra, Enhanced photocatalytic activity by silver metal ion exchanged NaY zeolite photocatalysts for the degradation of organic contaminants and dyes in aqueous medium, Ind. Eng. Chem. Res., 47 (2008) 7545-7551. https://doi.org/10.1021/ie800441c

[14] L. Shao, Y. Yao, S. Quan, H. Wei, R. Wang, Z. Guo, One-pot in situ synthesized TiO_2/layered double hydroxides (LDHs) composites toward environmental remediation, Mater. Lett. 114 (2014) 111-114. https://doi.org/10.1016/j.matlet.2013.09.121

[15] C. Ooka, H. Yoshida, K. Suzuki, T. Hattori, Highly hydrophobic TiO_2 pillared clay for photocatalytic degradation of organic compounds in water, Micropor. Mesopor. Mat., 67 (2004) 143-150. https://doi.org/10.1016/j.micromeso.2003.10.011

[16] T.S. Natarajan, H.C. Bajaj, R.J. Tayade, Palmyra tuber peel derived activated carbon and anatase TiO 2 nanotube based nanocomposites with enhanced photocatalytic performance in rhodamine 6G dye degradation, Process Safety and Environmental Protection 104 (2016) 346-357. https://doi.org/10.1016/j.psep.2016.09.021

[17] Pierson, H. 1993, Handbook of carbon graphite, diamond, and fullerenes: properties, processing, and applications. Park Ridge, NJ, Noyes Publications.

[18] Kruger A. 2010 Carbon materials and nanotechnology. Weinheim, Germany, Wieley-VCH.

[19] W.K. Jo, Y. Won, I. Hwang, R.J. Tayade, Enhanced photocatalytic degradation of aqueous nitrobenzene using graphitic carbon-TiO_2 composites, Ind. Eng. Chem. Res., 53 (2014) 3455-3461. https://doi.org/10.1021/ie500245d

[20] F.A. Filho, A.J.G. Zarbin, Hollow porous carbon microspheres obtained by the pyrolysis of TiO_2/poly(furfuryl alcohol) composite precursors, Carbon. 44 (2006) 2869–2876. https://doi.org/10.1016/j.carbon.2006.06.002

[21] M. Ouzzine, A.J. Romero-Anaya, M.A. Lillo-Rodenas, A. Linares-Solano, Spherical activated carbon as an enhanced support for TiO_2/AC photocatalysts, Carbon, 67 (2014) 104-118. https://doi.org/10.1016/j.carbon.2013.09.069

[22] B. Yin, J.T. Wang, W. Xu, D.–H. Long, W.M. Qiao, L.-C. Ling, Preparation of TiO_2/mesoporous carbon composites and their photocatalytic performance for methyl orange degradation, Carbon, 56 (2013) 393–394. https://doi.org/10.1016/j.carbon.2013.01.050

[23] R. Leary, A. Westwood, Carbonaceous nanomaterials for the enhancement of TiO_2 photocatalysis, Carbon, 49 (2011) 741-772. https://doi.org/10.1016/j.carbon.2010.10.010

[24] Z. He, W. Que, J. Chen, Y. He, G. Wang, Surface chemical analysis on the carbon-doped mesoporous TiO_2 photocatalysts after post-thermal treatment: XPS and FTIR characterization, J. Phys. Chem. Solids, 74 (2013) 924-928. https://doi.org/10.1016/j.jpcs.2013.02.001

[25] W.C. Oh, and F.J. Zhang, Preparation and characterization of graphene oxide reduced from a mild chemical method, Asian J. Chem., 23 (2011) 875-79

[26] R.A. Spurr, H. Myers, Quantitative analysis of anatase-rutile mixture with an X-ray diffractometer, Anal. Chem. 29 (1957) 760-762. https://doi.org/10.1021/ac60125a006

[27] Gregg S.J., Sing K.S.W. 1982, *Adsorption, Surface Area and Porosity*, 2nd ed., New York, Academic Press.

[28] S. Ding, J.S. Chen, D. Luan, F.Y.C. Boey, S. Madhavi, and X.W. Lou, Graphene-supported anatase TiO_2 nanosheets for fast lithium storage, Chem. Commun. 47(2011) 5780–5782. https://doi.org/10.1039/c1cc10687b

[29] B. Pan, and B. Xing, Adsorption mechanisms of organic chemicals on carbon nanotubes, Environ. Sci. Technol., 42 (2008) 9005-9013. https://doi.org/10.1021/es801777n

[30] C. Moreno-Castilla, Adsorption of organic molecules from aqueous solutions on carbon materials, Carbon, 42 (2004) 83–94. https://doi.org/10.1016/j.carbon.2003.09.022

Chapter 5

Effect of TiO_2 Nanotube Calcination Temperature and Oxygen Pressure to Photocatalytic Oxidation of Phenol

Farid F. Orudzhev[1], Abdulgalim B. Isaev[1]*, Nabi N. Shabanov[1,2]

[1]Department of Environmental Chemistry and Technology, Dagestan State University, M. Gadjieva, 43a, 367001, Makhachkala, Russian Federation

[2]Analytical Center of Common Access DSC RAS, 94 M. Yaragsky St., Makhachkala 367003, Russia

Email: abdul-77@yandex.ru

Abstract

The influences of oxygen pressure and TiO_2 nanotubes calcination temperature on the photocatalytic degradation phenol were investigated. According to experimental results the dissolved oxygen at different pressure and TiO_2 calcination temperature was a determining parameter for the photocatalytic degradation of phenol. The calcination temperature of TiO_2 nanotubes affects the anatase phase, crystallite size, surface area and pore volume of TiO_2 powder and respectively to rate of photodegradation of phenol. The kinetics of photocatalytic degradation of phenol in presence of TiO_2 nanotubes at high pressure of oxygen is investigated. The initial rate of photodegradation phenol were increased from 0.21 to 0.52 $mg \cdot l^{-1} \cdot min^{-1}$ when the initial oxygen pressure was increased from 0.1 to 0.6 MPa and have linear relationship between phenol oxidation rate and the oxygen pressure. The dissolved oxygen acted as an electron scavenger with formation reactive oxygen species such as the superoxide ion and the hydroxyl radical.

Keywords

Calcinations Temperature, Oxygen Pressure, Phenol, Photocatalytic Degradation, TiO_2 Nanotubes

Contents

1. Introduction

Phenol and phenolic compounds are major pollutants of industrial waste water. Phenol can enter the environment due to its widespread use in various industries such as paper mills, textile and petrochemical industries, paints, pesticide plants, etc. and is a toxic and slowly degradable pollutant [1, 2]. The traditional methods for phenol and phenolic compounds degradation such as flocculation [3], electrocoagulation [4], adsorption [5], osmosis and chemical oxidation has gained wide application [6]. These physical-chemical technologies have some limitations for application due the transportation of reagents and high cost. The biological treatment is not valid on account of biorefractory characteristics of pollutants [7]. Recently, the heterogeneous photocatalytic oxidation processes have attracted considerable research interests for their high efficiency, environmentally friendly property and cost effectiveness [8–12].

Heterogeneous photocatalysis is widely used for removal of organics and shows a high potential for purification of air, water and on solid surfaces [13]. Over the past few years, there has been an enormous amount of work that has been done regarding applications of heterogeneous photocatalysis for the removal of pollutants [8–10,12,14–20]. The heterogeneous photocatalysis is based on a semiconductor used as catalyst which is activated by UV light. The titania (TiO_2) is mainly used as the catalyst, because of its high activity at UV light irradiation, nontoxity, photostability and low cost [19, 20]. Nanocrystalline TiO_2 has been widely employed in photocatalytic or photoelectrochemical systems because of its semiconductor characteristics capable of generating charge carriers [21, 22]. TiO_2 nanotubes are an intensively studied nanosized materials [23-25].

The photocatalytic oxidation begins when catalyst absorbs the photon energy equal or higher than its band gap energy and electron can inject from valence band to conduction band with photogeneration holes [8, 26, 27]. Most of the electrons and holes are recombining within nanoseconds and the energy is dissipated as heat. Only a few photo-generated electrons and holes have been migrate to the catalyst surface, where they initiate the redox reaction. In an aqueous TiO_2 suspension the valence band holes can react with water to generate hydroxyl radicals which are strong oxidizing agents. Furthermore, the photogenerated electrons react with adsorbed molecular oxygen and reducing it to superoxide radical anion which, in turn, reacts with protons to form peroxide radicals [28].

In this study, we prepared the TiO_2 nanotubes by using a low-temperature hydrothermal method without templates and investigated the effect of annealing temperature, dissolved oxygen to photocatalytic activity of prepared nanotubes.

2. Experimental

2.1 Preparation TiO_2 nanotubes

All the reagents used in the experiments were of analytical grade, procured from Ecros Analit, (St-Peterburg, Russia) and used without further purification. In a typical synthesis, 50 g P25 TiO_2 was added into the 80 ml 10 M NaOH and the mixture were kept in an autoclave at 1300 °C for 24 h without stirring. After hydrothermal reaction the products were washed with hot distilled water until neutral pH. TiO2 nanotubes are obtained after the precipitates are washed with a dilute HCl aqueous solution and distilled water.

2.2 Characterization TiO_2 nanotubes

Characteristic obtained TiO_2 nanotubes samples were performed using electron microscopy, X-ray analysis. The morphology of the TiO_2 was characterized by scanning electron microscopy microscope LEO 1450 with analyzer ISYS and EDX system (Leica Micro-systems Wetzlar GmbH, Germany). X-ray diffraction (XRD) studies of synthesized catalysts were carried out at 25 °C using diffractometer Empyrean series 2 firms PANalytical with Cu $K\alpha_1$ radiation (k = 0,15406 nm). The operating voltage and current were 40 kV and 30 mA, respectively. The diffraction patterns were measured in the 2θ range of 5-90° with step size and step time of 0.026 and 197 s was used for data collection. The data processing was performed using the HighScore Plus software and diffraction database PDF-2. Using Debye–Scherer's equation, the crystallite size was calculated [29]. The BET surface area of TiO_2 catalyst was measured according to the N_2 adsorption isotherm at 77 K using a Sorbi-MS, ZAO "Meta" sorption analyzer.

2.3 Photocatalytic activity

The photocatalytic activities of prepared TiO_2 nanotubes were determined by studying the degradation of an aqueous solution of phenol in a teflon cylindrical vessel photocatalytic reactor with volume 200 ml described in more detail in our previous work [30, 31]. The oxygen pressure created by feeding from a high pressure cylinder to the cell. The cell was depressurized before sampling; then, oxygen was pumped into the cell once again to a specified pressure and the subsequent treatment was performed (2–3 h). The suspension was intensely stirred with a magnetic stirrer in the course of photocatalysis to prevent the sedimentation of titanium dioxide. For the convenience of stirring, the cell was turned and irradiation was performed from the top (Fig. 1). Teflon vessel *6* with a volume of 200 ml served as a cell, a side of which was made of quartz glass 10 mm thick (*1*). The quartz glass was fixed with bronze rings *2* and screws *7*. The solution was poured through orifices *3*, which were than stoppered. Oxygen was pumped through valve *4*, and the pressure was measured with manometer *5*.

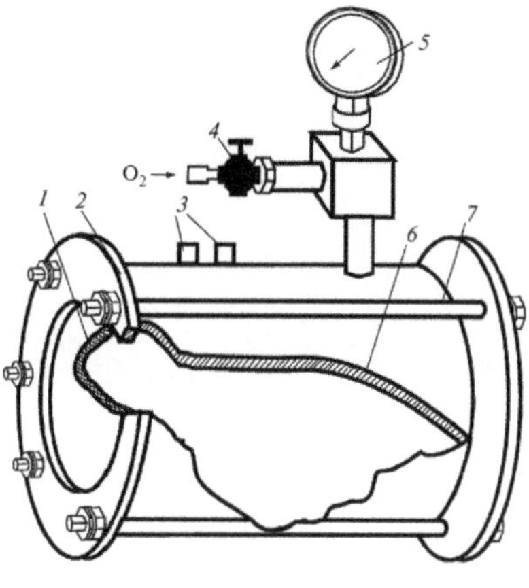

Figure 1. Schematic representation of the cell for photocatalytic oxidation of phenol: (1) quartz window 10 mm thick, (2) bronze rings, (3) nipples for sampling and loading a suspension of titanium dioxide and a phenol solution, (4) oxygen supply valve, (5) manometer, (6) teflon case 20 mm thick, and (7) screw.

The phenol concentration was 10 mg·l^{-1} and the amount of catalyst added into reactor was 0.1 g·l^{-1}. The liquid samples were withdrawn at preset time intervals and measured the phenol concentration by fluorimetric method in spectrofluorimeter Model "Fluorat 02 Panorama" (Russia). All the experiments were performed under irradiation 400-W high-pressure mercury UV-lamp. The UV-lamp as light source vertically irradiates the reaction mixture.

3. Results and discussion

3.1 TiO_2 nanotubes characterization

Structure and morphology of TiO_2 nanotubes was investigated using scanning electron microscopy. As shown in Fig. 2, the powder of TiO_2 are of tube morphology with length about 1.0–2.0 μm and represent nanotubes agglomeration. The sample has a porous structure which can provide a larger specific surface area. It can be seen that as-formed nanotubes with inner and outer diameters of about 20 and 30 nm respectively are open at the top end and at the bottom end, just as shown in Fig. 2(b).

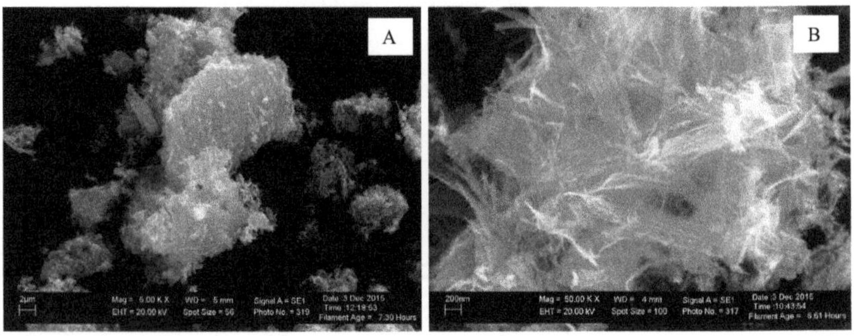

Figure 2. Typical SEM image of samples TiO_2 nanotubes.

Fig. 3 shows typical XRD patterns of the TiO_2 nanotubes samples prepared at calcinations at different temperature. XRD measurements show that the as-prepared and calcining at 100-200°C TiO_2 nanotubes powder are amorphous. Annealing at 300°C during 1 hour was carried out to crystallize the TiO_2 nanotubes to anatase. The XRD results confirm successful conversion, and the evaluation of the main anatase (101) peak of 25°. From the X-ray results, the average crystallite size TiO_2 nanotubes was quantitatively calculated. These calculated results are presented in Fig. 4, where lists the weight percents of anatase (A) phases in these TiO_2 nanotubes, which were determined according to the integrated intensities of anatase (101) peaks in the XRD patterns.

The nanotubes, which calcinations beginning at 300°C, contain only anatase crystalline structure and at 800°C the amorphous phase has been completely transformed into an anatase structure. An increase in calcination time decreases the weight percent of anatase TiO_2. This indicates that the formation of the anatase phase of TiO_2 is affected by not only calcination temperature but also calcination time [32]. The average size of the anatase crystallites in the TiO_2 nanotubes was estimated using the Scherrer equation and the XRD anatase (101) peaks. The calculated crystallite size was increasing when increasing the calcinations temperature. The crystallite size for samples calcined at 300°C was 13 nm and increases to 93 nm for samples calcined at 800°C (Fig. 3), but the surface area decreases when the calcined temperature increases (Table 1). This phenomena have indicated that the brookite and anatase TiO_2 are metastable, and will transform exothermally and irreversibly to the rutile TiO_2 over a range of temperature at room temperature to 750°C [32].

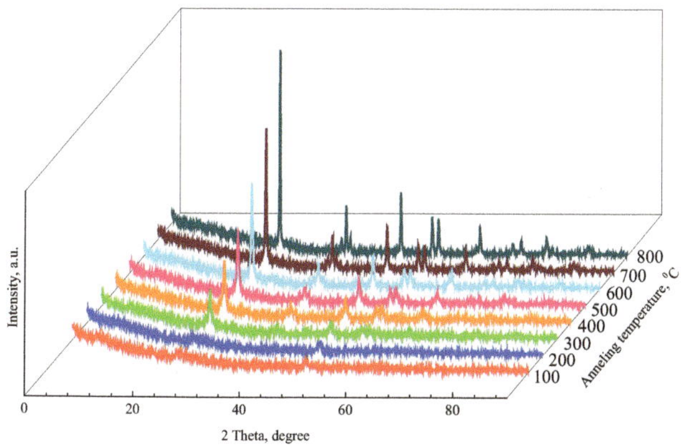

Figure 3. XRD patterns of TiO_2 nanotubes prepared by hydrothermal method at different calcination temperatures.

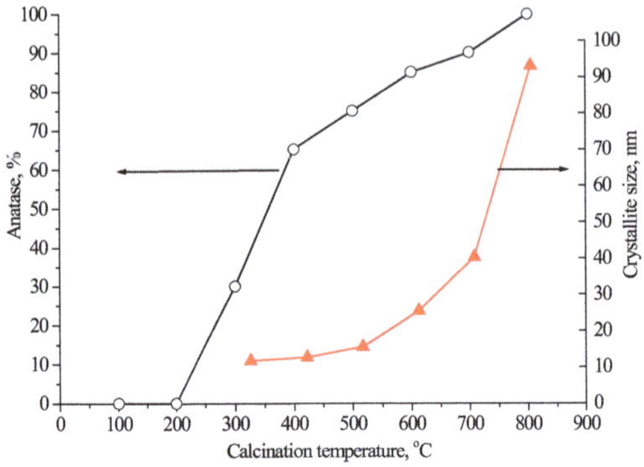

Figure 4. Effect of calcination temperature to percents of anatase phase and crystallite size.

The measure of the specific area of the nanotubes was obtained using the BET method with the nitrogen adsorption-desorption process. Fig. 5 shows an isotherm of TiO_2 nanotubes previously heat treated at 100°C with a surface area of 265 m^2g^{-1}. According to the IUPAC classification, the hysteresis loops in Fig. 5 correspond to Type H2, which represents poorly defined pore size and shape distributions [33, 34]. The pore size distribution measurement indicated that the TiO_2 nanotubes had a pronounced mesoporosity. As can be seen, from Table 1 the surface area and pore volume in TiO_2 nanotubes powder are decreases when increase the calcinations temperature.

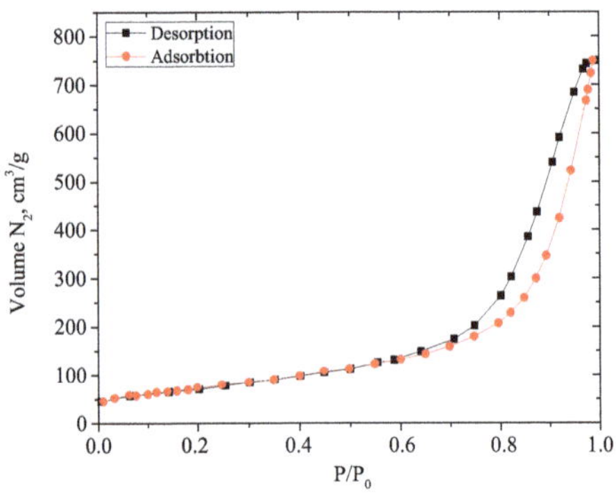

Figure 5. N_2 adsorption–desorption isotherms of the TiO_2 nanotubes calcined at 1000°C

Table 1. Effect of calcined temperature to TiO_2 nanotube powder surface area and pore volume

Calcined temperature, [°C]	Surface area, [$m^2 \cdot g^{-1}$]	Pore volume, [$cm^3 \cdot g^{-1}$]
100	265	0.89
200	230	0.71
300	185	0.41
400	133	0.34
500	90	0.28
600	82	0.25
700	74	0.22
800	56	0.18

The high specific surface area and pore volume for samples TiO_2 nanotubes calcined at 100°C may be associated with the formation of larger mesoporous structure for amorphous phase.

3.2 Effect of calcinations temperature to photocatalytic activity

The phenol photocatalytic degradation under UV light irradiation using the prapared TiO_2 nanotubes sample as catalyst is shown in Fig. 6. The phenol photodegradation followed by the fluorimetric method.

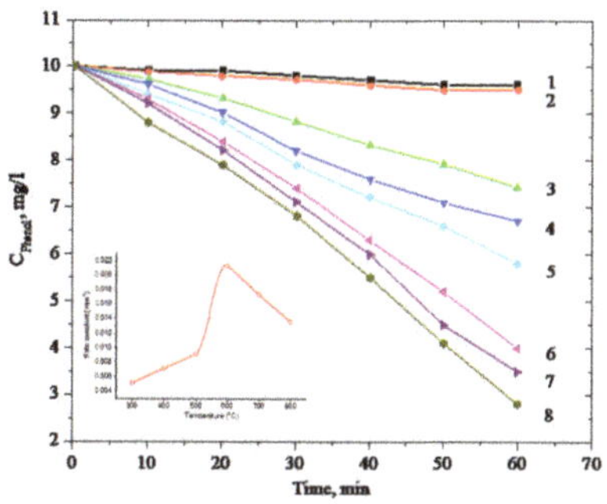

Figure 6. Effect of calcinations temperature of TiO_2 nanotubes to photocatalytic oxidation of phenol (°C): 1 – 100° C, 2 – 200° C, 3 – 300° C, 4 – 400° C, 5 – 500° C, 6 – 800° C, 7 – 700° C, 8 – 600° C. (Insert: temperature dependence of constant rate) ($[Ph]_0$ – 10 $mg \cdot l^{-1}$, $[TiO_2]_0$ – 0.1 $g \cdot l^{-1}$).

The photodegradation of phenol at presence TiO_2 nanotube calcined at 100 °C evidences only 6% degradation due to UV light alone. What is more, photoactivity of TiO_2 is increased due to the calcination temperature increase. Table 2 shows the effect of TiO_2 calcination on the phenol degradation rate, after 60 min of reaction. The constant rate and phenol photodegradation rate were calculated from kinetic curves presented in insert Fig. 6. As observed, the TiO_2 photoactivity is increased by increasing calcination temperature up to 600 °C, after which, there is a decrease in the photodegradation of phenol. Higher crystallinity of TiO_2 result in the promotion of recombination centers for the photogenerated charges [35]. Thus, TiO_2 calcination temperature is a determinant parameter in the design of TiO_2 nanotube photocatalysts. The increase in photoactivity due to increase of the crystallite size and existence of only anatase phase which is highly photoactivity.

Table 2. The constants rate and initial rates of photocatalytic oxidation of phenol under UV light irradiation.

Calcined temperature, [°C]	Constant rate [min^{-1}]	Initial rate, [$mg \cdot l^{-1} \cdot min^{-1}$]	Degree, [%]	Linear correlation, R^2
600	0.0213±0.0006	0.213±0.006	72.4±1.1	0.99
700	0.0173±0.0005	0.173±0.005	65.5±0.9	0.99
800	0.0136±0.0003	0.136±0.003	60.6±0.9	0.99
500	0.0091±0.0004	0.091±0.004	42.8±0.8	0.98
400	0.0071±0.0003	0.071±0.003	33.7±0.9	0.99
300	0.0051±0.0002	0.051±0.002	26.6±0.9	0.99

3.3 Effect of oxygen pressure to photocatalytic activity

The availability of oxidating agent, as oxygen, in the water phenol solution play an important role in heterogeneous photocatalysis and is one of the determinant factors of the rate of photodegradation of phenol. The dependency of the photocatalytic degradation of phenol on the initial oxygen pressure was studied in the range from 0.1 to 0.6 MPa. Typical time-dependent phenol concentration during photocatalytic degradation is illustrated in Fig. 7. As can be seen, when phenol solution with TiO2 nanotube calcined at 600 °C was exposed to UV light at oxygen pressure 0,6 MPa, the phenol concentration decreased drastically with illumination time in comparison with the same experiment performed in the oxygen pressure of 0.1 MPa. The difference between direct photolysis and photocatalysis at oxygen pressure revealed that UV light and TiO2 photocatalyst together had a significant effect on the degradation of phenol. Since OH radicals was the key feature of TiO_2 nanotube assisted photocatalytic process, the degradation of phenol was primarily related to the generated OH radicals [36]. As can be seen from Fig. 7, the oxygen pressure was contributed to the photocatalytic degradation. The depletion of the phenol concentration during UV light irradiation in presence of TiO_2 nanotubes, as recorded in Fig. 7, confirmed that oxygen was involved in the photocatalytic degradation.

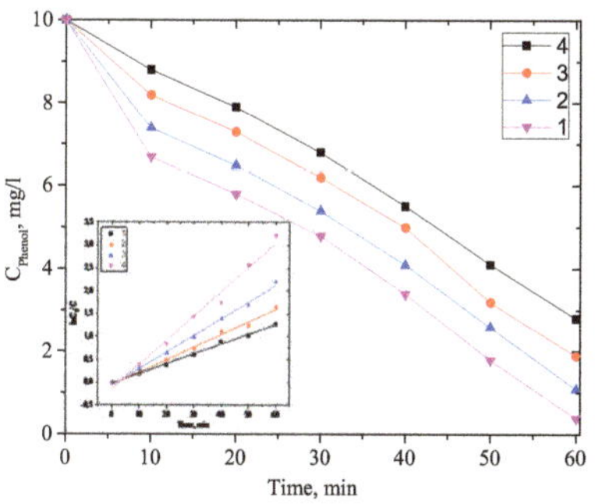

Figure 7. Effect of oxygen pressure to photocatalytic oxidation of phenol (°C): 1 – 0.1 MPa, 2 – 0.2 MPa, 3 – 0.4 MPa, 4 – 0.6 MPa. (Insert: kinetic curves) ($[Ph]_0$ – 10 mg·l^{-1}, $[TiO_2]_0$ – 0.1 g·l^{-1}).

Accordingly, the improvement of the degradation of phenol, which relate to the increasing initial oxygen pressure, can be attributed to the fact that oxygen acted as electron acceptor to trap the photo induced electron [37]

$$O_2 + e \rightarrow O_2^{\cdot -} \quad (1)$$

$$O_2^{\cdot -} + H^+ \rightarrow HO_2^{\cdot} \quad (2)$$

$$HO_2^{\cdot} + HO_2^{\cdot} \rightarrow H_2O_2 + O_2 \quad (3)$$

$$O_2^{\cdot -} + HO_2^{\cdot} + H^+ \rightarrow H_2O_2 + O_2 \quad (4)$$

Through the reduction of oxygen, reactive superoxide radical anions (O2.-) was produced (Eq. 1). Fig. 8 shows the relation of the initial rate of phenol photocatalytic oxidation by oxygen pressure. From Fig. 8 one can see, that the increase of oxygen pressure intensifies the phenol oxidation process. In this pressure range there is a linear relationship between phenol oxidation rate and the oxygen pressure. This relationship is described by a linear equation (Eq. 5) with a linear correlation coefficient of 0.98.

$$R_0 = 0,58 \cdot PO_2 + 0.15 \quad (5)$$

The increasing oxygen pressure in the photocatalytic reactor from 0.2 MPa to 0.6 MPa the initial rate of phenol oxidation increases 1.9 times. These confirm the mechanism of preventing the electron-hole recombination due to the reduction of adsorbed oxygen on the surface TiO^2 nanotube to form highly reactive oxygen species such as the superoxide ion and the hydroxyl radical. Due to its high reactivity and oxidizing power these reactive oxygen species instantly react with the dissolved phenol and to each other, whereby, we observe intensification photodegradation of phenol.

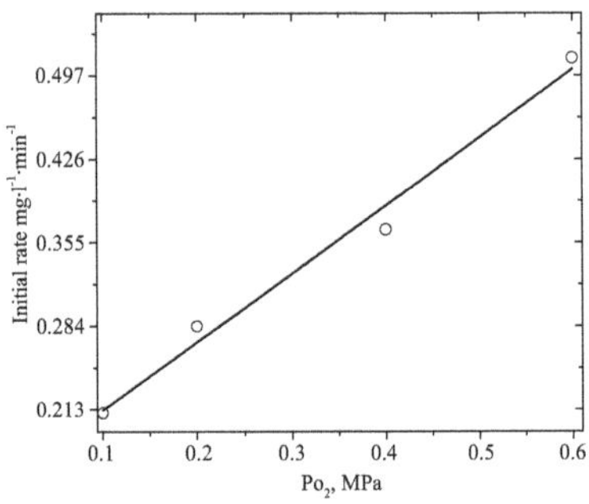

Figure 8. Dependence of initial rate of photocatalytic oxidation of phenol with presence of TiO_2 nanotubes calcined at 600 °C under UV light irradiation on the oxygen pressure ($[Ph]_0 – 10\ mg \cdot l^{-1}$, $[TiO_2]_0 – 0.1\ g \cdot l^{-1}$).

4. Conclusions

The TiO_2 nanotubes calcination temperature and the oxygen pressure were proven to have significant effect on the photodegradation of phenol. Notably, oxygen pressure and TiO_2 nanotubes together showed a marked effect in photocatalytic oxidation of phenol. Increasing the oxygen pressure was beneficial for the photocatalytic degradation of phenol. Correspondingly, the photocatalytic oxidation rate and rate constants increased with an increase in oxygen pressure and TiO_2 nanotubes calcination temperature up to 600 °C. For the photocatalytic oxidation of phenol the dissolved oxygen at different pressure acted as electron acceptor to enhance the degradation efficiency.

References

[1] S. Martha, D.P. Das, N. Biswal, and K.M. Parida, Facile synthesis of visible light responsive V_2O_5/N,S–TiO_2 composite photocatalyst: enhanced hydrogen production and phenol degradation, J. Mater. Chem. 22 (2012) 10695-10703. https://doi.org/10.1039/c2jm30462g

[2] S. Ahmed, M.G. Rasul, W.N. Martens, R. Brown, and M. A. Hashib, Advances in heterogeneous photocatalytic degradation of phenols and dyes in wastewater: A review. Water. Air. Soil Pollut. 215 (2011) 3–29. https://doi.org/10.1007/s11270-010-0456-3

[3] Y.-H. Shen, Removal of phenol from water by adsorption-flocculation using organobentonite. Water Res. 36 (2002) 1107–1114. https://doi.org/10.1016/S0043-1354(01)00324-4

[4] A.S. Fajardo, R.F. Rodrigues, R.C. Martins, L.M. Castro, and R.M. Quinta-Ferreira, Phenolic wastewaters treatment by electrocoagulation process using Zn anode. Chem. Eng. J. 275 (2015) 331–341. https://doi.org/10.1016/j.cej.2015.03.116

[5] A. Dabrowski, P. Podkościelny, Z. Hubicki, and M. Barczak, Adsorption of phenolic compounds by activated carbon-a critical review. Chemosphere. 58 (2005) 1049–1070. https://doi.org/10.1016/j.chemosphere.2004.09.067

[6] A. Mnif, D. Tabassi, M. Ben Sik Ali, and B. Hamrouni, Phenol removal from water by AG reverse osmosis membrane. Environ. Prog. Sustain. Energy. 34 (2015) 982–989.

[7] N.V. Pradeep, S. Anupama, K. Navya, H.N. Shalini, M. Idris, and U.S. Hampannavar, Biological removal of phenol from wastewaters: a mini review. Appl. Water Sci. 5 (2014) 105–112. https://doi.org/10.1007/s13201-014-0176-8

[8] M.A. Fox, and M.T. Dulay, Heterogeneous Photocatalysis. Chem. Rev. 93 (1993) 341–357. https://doi.org/10.1021/cr00017a016

[9] J. Herrmann, Heterogeneous photocatalysis: fundamentals and applications to the removal of various types of aqueous pollutants. Catal. Today. 53 (1999) 115–129. https://doi.org/10.1016/S0920-5861(99)00107-8

[10] U.I. Gaya, and A.H. Abdullah, Heterogeneous photocatalytic degradation of organic contaminants over titanium dioxide: A review of fundamentals, progress and problems. J. Photochem. Photobiol. C Photochem. Rev. 9 (2008) 1–12. https://doi.org/10.1016/j.jphotochemrev.2007.12.003

[11] K. Pirkanniemi, and M. Sillanpää, Heterogeneous water phase catalysis as an environmental application: a review. Chemosphere. 48 (2002) 1047–1060. https://doi.org/10.1016/S0045-6535(02)00168-6

[12] U. Gaya, Heterogeneous photocatalysis using inorganic semiconductor solids (Springer Science & Business Media, 2013).

[13] M. Pelaez, N.T. Nolan, S.C. Pillai, M.K. Seery, P. Falaras, A.G. Kontos, P.S.M. Dunlop, J.W.J. Hamilton, J.A. Byrne, K. O'shea, M.H. Entezari, and D.D. Dionysiou, A review on the visible light active titanium dioxide photocatalysts for environmental applications. Appl. Catal. B Environ. 125 (2012) 331–349. https://doi.org/10.1016/j.apcatb.2012.05.036

[14] M.R. Hoffmann, S.T. Martin, W. Choi, and D.W. Bahnemann, Environmental Applications of Semiconductor Photocatalysis. Chem. Rev. 95 (1995) 69–96. https://doi.org/10.1021/cr00033a004

[15] K. Nakata, and A. Fujishima, TiO_2 photocatalysis: Design and applications. J. Photochem. Photobiol. C Photochem. Rev. 13 (2012) 169–189. https://doi.org/10.1016/j.jphotochemrev.2012.06.001

[16] A.L. Linsebigler, G. Lu, and J.T. Yates, Photocatalysis on TiO_2 Surfaces: Principles, Mechanisms, and Selected Results, Chem. Rev. 95 (1995) 735–758. https://doi.org/10.1021/cr00035a013

[17] A. Fujishima, T.N. Rao, and D.A. Tryk, Titanium dioxide photocatalysis. J. Photochem. Photobiol. C Photochem. Rev. 1 (2000) 1–21. https://doi.org/10.1016/S1389-5567(00)00002-2

[18] J.-M. Herrmann, C. Duchamp, M. Karkmaz, B.T. Hoai, H. Lachheb, E. Puzenat, and C. Guillard, Environmental green chemistry as defined by photocatalysis. J. Hazard. Mater. 146 (2007) 624–629. https://doi.org/10.1016/j.jhazmat.2007.04.095

[19] A. Fujishima, X. Zhang, and D. Tryk, Heterogeneous photocatalysis: From water photolysis to applications in environmental cleanup. Int. J. Hydrogen Energy. 32 (2007) 2664–2672. https://doi.org/10.1016/j.ijhydene.2006.09.009

[20] D. Bahnemann, Photocatalytic water treatment: solar energy applications. Sol. Energy. 77 (2004) 445–459. https://doi.org/10.1016/j.solener.2004.03.031

[21] J. Qiu, S. Zhang, and H. Zhao, Recent applications of TiO_2 nanomaterials in chemical sensing in aqueous media. Sensors Actuators, B Chem. 160 (2011) 875–890. https://doi.org/10.1016/j.snb.2011.08.077

[22] K. Rajeshwar, Photoelectrochemistry and the environment. J. Appl. Electrochem. 25 (1995) 1067–1082. https://doi.org/10.1007/BF00242533

[23] D.V. Bavykin, V.N. Parmon, A.A. Lapkin, and F.C. Walsh, The effect of hydrothermal conditions on the mesoporous structure of TiO_2 nanotubes. J. Mat. Chem. 14 (2004) 3370-3377. https://doi.org/10.1039/b406378c

[24] R.J. Tayade, and D.L. Key, Synthesis and Characterization of Titanium Dioxide Nanotubes for Photocatalytic Degradation of Aqueous Nitrobenzene in the Presence of Sunlight. Materials Science Forum. 657 (2010) 62-74. https://doi.org/10.4028/www.scientific.net/MSF.657.62

[25] T.S. Natarajan, K. Natarajan, H.C. Bajaj, and R.J. Tayade, Energy Efficient UV-LED Source and TiO_2 Nanotube Array-Based Reactor for Photocatalytic Application. Ind. Eng. Chem. Res. 50 (2011) 7753-7762. https://doi.org/10.1021/ie200493k

[26] K. Shankar, J.I. Basham, N.K. Allam, O.K. Varghese, G.K. Mor, X. Feng, M. Paulose, J.A. Seabold, K.Sh. Choi, and C.A. Grimes, Recent Advances in the Use of TiO_2 Nanotube and Nanowire Arrays for Oxidative photoelectrochemistry. J. Phys. Chem. C. 113 (2009) 6327–6359. https://doi.org/10.1021/jp809385x

[27] M.A. Rauf, and S.S. Ashraf, Fundamental principles and application of heterogeneous photocatalytic degradation of dyes in solution. Chem. Eng. J. 151 (2009) 10–18. https://doi.org/10.1016/j.cej.2009.02.026

[28] T. Velegraki, I. Poulios, M. Charalabaki, N. Kalogerakis, P. Samaras, and D. Mantzavinos, Photocatalytic and sonolytic oxidation of acid orange 7 in aqueous solution. Appl. Catal. B Environ. 62 (2006) 159–168. https://doi.org/10.1016/j.apcatb.2005.07.007

[29] A.M. Abdullah, N.J. Al-Thani, K. Tawbi, and H. Al-Kandari, Carbon/nitrogen-doped TiO_2: New synthesis route, characterization and application for phenol degradation. Arab. J. Chem. 9 (2016) 229–237. https://doi.org/10.1016/j.arabjc.2015.04.027

[30] F.F. Orudzhev, F.G. Gasanova, Z.M. Aliev, and A.B. Isaev, Photoelectrocatalytic oxidation of phenol on platinum-modified TiO_2 nanotubes. Nanotechnologies in Russia. 7 (2012) 482-485. https://doi.org/10.1134/S1995078012050102

[31] F.F. Orudzhev, F.G. Gasanova, Z.M. Aliev, A.B. Isaev, and N.S. Shabanov, Photoelectrocatalytic oxidation of phenol on TiO_2 nanotubes under oxygen

pressure. Russ. J. Electrochem. 51 (2015) 1108-1114. https://doi.org/10.1134/S1023193515110130

[32] S.-H. Lin, C.-H. Chiou, C.-K. Chang, and R.-S. Juang, Photocatalytic degradation of phenol on different phases of TiO_2 particles in aqueous suspensions under UV irradiation. J. Environ. Manage. 92 (2011) 3098–3104. https://doi.org/10.1016/j.jenvman.2011.07.024

[33] W.Y. Gan, M.W. Lee, R. Amal, H. Zhao, and K. Chiang, Photoelectrocatalytic activity of mesoporous TiO_2 films prepared using the sol-gel method with tri-block copolymer as structure directing agent. J. Appl. Electrochem. 38 (2008) 703–712. https://doi.org/10.1007/s10800-008-9495-5

[34] K. Ángel-Sánchez, O. Vázquez-Cuchillo, M. Salazar-Villanueva, J.F. Sánchez-Ramírez, A. Cruz-López, and A. Aguilar-Elguezabal, Preparation, characterization and photocatalytic properties of TiO_2 nanostructured spheres synthesized by the Sol–Gel method modified with ethylene glycol. J. Sol-Gel Sci. Technol. 58 (2011) 360–365. https://doi.org/10.1007/s10971-011-2401-3

[35] C.A. Castro, A. Centeno, and S.A. Giraldo, Iron promotion of the TiO_2 photosensitization process towards the photocatalytic oxidation of azo dyes under solar-simulated light irradiation. Mater. Chem. Phys. 129 (2011) 1176–1183. https://doi.org/10.1016/j.matchemphys.2011.05.082

[36] S.L. Orozco, C. A. Arancibia-Bulnes, and R. Suárez-Parra, Radiation absorption and degradation of an azo dye in a hybrid photocatalytic reactor. Chem. Eng. Sci. 64 (2009) 2173–2185. https://doi.org/10.1016/j.ces.2009.01.038

[37] D.-H. Tseng, L.-C. Juang, and H.-H. Huang, Effect of Oxygen and Hydrogen Peroxide on the Photocatalytic Degradation of Monochlorobenzene in Aqueous Suspension. Int. J. Photoenergy. (2012) 1–9. https://doi.org/10.1155/2012/328526

Chapter 6

Understanding Reaction Mechanism in Photon-Assisted Reduction of Carbon Dioxide

N. Hariprasad[1,2], B. Viswanathan[1], K. R.Krishnamurthy[*1], M. V. Harindranathan Nair[2]

[1]National Centre for Catalysis Research, Indian Institute of Technology Madras, Chennai- 600 036, Tamil Nadu, India.
[2]School of Environmental Studies, Cochin University of Science and Technology, Cochin-682 022, Kerala, India.
Email: krkm@iitm.ac.in; krkonda1949@gmail.com

Abstract

Photon-assisted reduction of carbon dioxide has become an emerging field of scientific interest for utilisation of CO_2, which may be helpful in overcoming energy and environment-related problems [1-6]. However, the knowledge assimilation in this area does not consider all the fundamental aspects appropriately, possibly because of the interdisciplinary nature of the research field. The nature and extent of electron transfer are crucial aspects of the reduction process. Most of the studies excluded the interfacial phenomenon across the semiconductor/ electrolyte junctions while interpreting the reaction mechanism. Through this chapter, we are trying to revisit the photophysics and photochemistry of semiconductors fundamentally, to enhance the understanding of electron transfer mechanism of the photon-assisted reduction of carbon dioxide.

Keywords

Carbon Dioxide, Photo-reduction, Solar Energy Conversion, Reaction Mechanism, Electron Transfer, Semiconductor-electrolyte Interface

Contents

1. Introduction

Photon-assisted reduction of carbon dioxide has become an emerging field of scientific interest to utilize carbon dioxide for the sustainable production of chemicals and fuels [1-6]. Even though the research efforts in this regard are overwhelming, the exact electron transfer mechanism is not entirely clear. Various efforts have been expanded to find an appropriate material for the reduction process, but all these efforts have not been successful. Though we have enough materials that satisfy the band edge threshold for the proton-coupled electron transfer reduction of CO_2, till date, we are unable to execute the reaction with better efficiency [7]. This indicates that the problem is not with the material, but with the charge carrier dynamics within the semiconductor. The materials are unable to execute the multiple proton-coupled electron transfers due to the difficulties in successive charge carrier transfer within the semiconductor. The literature on these possibilities is not uniform throughout because of the interdisciplinary nature of the research field. Most of the reports considered only the product yield on a particular material rather than the difficulties in electron transfer. Therefore, a basic understanding of the semiconductor properties has to be enhanced or revisited both regarding physics and chemistry, to resolve this non-uniformity in the interpretation of results. There are several excellent books and reviews available on each of these aspects separately, and we refer the reader attention to that literature [8-15]. In this chapter, we are following the charge carriers and their journey towards the interfacial region of the semiconductor, starting with an introduction to the core concepts of semiconductor physics and the semiconductor/electrolyte interface in presence and absence of irradiation. The concluding remarks involve possible suggestions for the design of new materials based on the emerging concepts.

2. Semiconductor in equilibrium

The knowledge of the system under equilibrium is necessary for the better understanding of any of the energy conversion processes. In this section, we are considering a semiconductor in the absence of any electric field, magnetic field, voltage, and irradiation. Under the equilibrium conditions, the semiconductor properties would be independent of time. Initially, we are considering the intrinsic semiconductor and followed by the extrinsic semiconductor, which is usually employed as a material in all semiconductor-based photon-assisted CO_2 reduction systems.

2.1 Charge carriers in semiconductors

The properties of the semiconductor rely heavily on the distribution of charge carriers such as electrons in the conduction band and holes in the valence band. The overall density of electrons and holes in a semiconductor is closely related to the density of states and Fermi distribution functions. The product of allowed quantum states and probability of that state occupied by an electron gives rise to the distribution of electrons in the conduction band as in the following equation.

$$n(E) = g_c(E) f_F(E) \quad (1)$$

Here g_C is the density of quantum states in the conduction band and f_F is the Fermi-Dirac probability function [8,10]. The integration of the Eq. 1 gives the total concentration of conduction band electrons. In a similar way, the hole distribution in the valence band of the semiconductor can be expressed as,

$$p\,(E) = g_V(E)\,[1 - f_F(E)] \tag{2}$$

In this, the products of allowed quantum states in the valence band and probability of that state not occupied by an electron gives rise to the distribution of holes situated in the valence band. The integration of Eq. 2 gives the total concentration of holes in the valence band [8,10].

To figure out electron and hole concentrations, one needs to determine the exact position of the Fermi energy E_F with reference to the conduction band minimum (E_C) and valence band maximum(E_V). Fermi levels refer to the electrochemical potential of the electron and in equilibrium conditions, its occupancy is assumed or the probability of finding the electrons in that state is 50%. For intrinsic semiconductors, the valence band is completely filled with electrons at absolute temperature $(T = 0\,K)$ and therefore, the Fermi level situated in the mid-point of the band gap if the effective masses of the electron and holes are equal. As the temperature rises, more electrons occupy in the conduction band, leaving an electron vacancy in the valence band, usually coined as holes. The occupation of the hole formed in a bond with the electrons from the adjacent covalent bond create a hole in that bond and this continuous process enables the locomotion of the holes throughout the semiconductor. The concentration of electrons and holes are the same in the case of intrinsic semiconductors with high purity and null lattice defects.

As stated, the thermal equilibrium concentration of electrons (n_0) and holes (p_0) can be calculated from the integral of the Eq. 1 and Eq. 2 within the limit of E_C to ∞ and $-\infty\ to\ E_v$respectively [8-11].

$$n_0 = \int_{E_C}^{\infty} g_C(E) f_F(E)\, dE \tag{3}$$

Similarly, $$p_0 = \int_{-\infty}^{E_V} g_V(E)[1 - f_F(E)]\, dE \tag{4}$$

The solution of Eq.3 and Eq.4 gives the thermal equilibrium concentration of electrons and holes respectively.

$$n_0 = N_C \exp\left[\frac{-(E_C - E_F)}{kT}\right] \tag{5}$$

$$p_0 = N_V \exp\left[\frac{-(E_F - E_V)}{kT}\right] \tag{6}$$

The terms N_C *and* N_V represent the effective density of states of the conduction band and valence band and are constant in any given temperature for a given semiconductor material. k is the Boltzmann constant at a temperature T and it signifies that a reasonable change in the temperature increases the charge carrier concentration by several orders.

$$N_C = 2\left(\frac{2\pi m_n^* kT}{h^2}\right)^{3/2} \tag{7}$$

$$N_V = 2\left(\frac{2\pi m_p^* kT}{h^2}\right)^{3/2} \tag{8}$$

Where m_n^* and m_p^* are the effective masses of electrons and holes in the semiconductor lattice. For an intrinsic semiconductor, the carrier concentrations are equal which means that $n_i = n_0 = p_i = p_0$. Therefore, the product of Eq. 5 and Eq. 6 gives the carrier concentration of an intrinsic semiconductor.

$$n_i^2 = N_C N_V \exp\left[\frac{-(E_C - E_V)}{kT}\right] = N_C N_V \exp\left[\frac{-E_g}{kT}\right] \tag{9}$$

Here, the E_g represents the energy of the bandgap. This equation indicates that the carrier concentration of an intrinsic semiconductor is a constant at a fixed temperature and independent of the energy of the Fermi level. The Fermi level position of an intrinsic semiconductor can easily be derived from the Eq.5 and Eq. 6.

$$N_C \exp\left[\frac{-(E_C - E_{Fi})}{kT}\right] = N_V \exp\left[\frac{-(E_{Fi} - E_V)}{kT}\right] \tag{10}$$

One can solve the above equation for $\boldsymbol{E_{Fi}}$, the Fermi level position of an intrinsic semiconductor, by taking logarithm on both sides of the equations.

$$E_{Fi} = \frac{1}{2}(E_C + E_V) + \frac{1}{2}kT\ln\left(\frac{N_V}{N_C}\right) \tag{11}$$

Applying the definitions of N_V *and* $N_C$in the Eq. 9 would redefine the Fermi level equation in terms of effective masses of electrons and holes.

$$E_{Fi} = E_{midgap} + \frac{3}{4}kT\ln\left(\frac{m_p^*}{m_n^*}\right) \tag{12}$$

$$E_{Fi} - E_{midgap} = \frac{3}{4}kT\ln\left(\frac{m_p^*}{m_n^*}\right) \tag{13}$$

The formulated Eq. 13 indicates that the relative position of Fermi level in an intrinsic semiconductor depends on the effective masses of the charge carriers which in turn is directly related to the density of states. If the effective masses were the same, then the

Fermi level should be at the centre of the band gap. In the case of an n-type semiconductor, the majority charge carriers are electrons and effective masses of electrons greater than that of the holes. Therefore, the Fermi level position lies above the intrinsic Fermi level, $E_F > E_{Fi}$. When the majority charge carrier is hole, the so-called semiconductor termed as p-type and the Fermi level would situate below the intrinsic Fermi level, $E_F < E_{Fi}$(Fig.1).

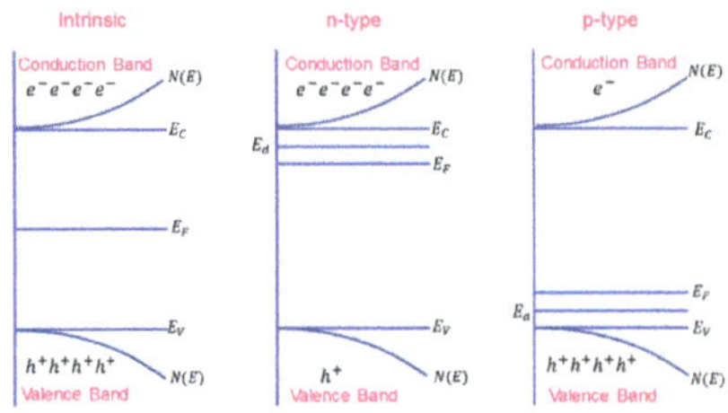

Figure 1. Representation of semiconductors at thermal equilibrium. (a) intrinsic (b) n-type and (c) p-type semiconductor.

In an n-type semiconductor, impurities create an additional donor energy level, E_d, which is present near to the conduction band minimum. The energy required to excite an electron from the donor level to the conduction band is minimal as compared to the band gap excitation. Excitation of the electrons from the donor level creates a positively charged species in the donor sites. This kind of impurity atoms usually is termed as donor impurity atoms and it produces electrons in the conduction band without creating a hole in the valence band. Similarly, in p-type semiconductor, the additional impurity atoms create an energy level very near to the valence band. A small amount of thermal energy is required to excite the valence electrons into the acceptor energy level, E_a. The electron present in the acceptor energy level don't have sufficient energy as conduction band electrons, and the electrons fill the unoccupied orbitals of acceptor impurity atoms, therefore, the impurity atom turned to negatively charged species fixed in the crystal and produces holes in the valence band without producing electrons in the conduction band. One can make n-type or p-type semiconductor from intrinsic semiconductors by adding a small amount of donor or acceptor impurity atoms, which are collectively called extrinsic semiconductors.

For n-type semiconductor, one rewrites the Eq.5 as [8-11],

$$n_0 = N_C \exp\left[\frac{-(E_C - E_{Fi}) + (E_F - E_{Fi})}{kT}\right] \tag{14}$$

$$n_0 = N_C \exp\left[\frac{-(E_C - E_{Fi})}{kT}\right] \exp\left[\frac{(E_F - E_{Fi})}{kT}\right] \quad (15)$$

It is evident from the Eq. 5 and Eq. 8 that,

$$n_i = N_C \exp\left[\frac{-(E_C - E_{Fi})}{kT}\right] \quad (16)$$

Substituting the value of $n_i$in Eq. 15,

$$n_0 = n_i \exp\left[\frac{(E_F - E_{Fi})}{kT}\right] \quad (17)$$

Similarly, we can write,

$$p_0 = p_i \exp\left[\frac{(E_F - E_{Fi})}{kT}\right] \quad (18)$$

The n_0p_0 product of extrinsic semiconductor is similar to the expression of intrinsic semiconductor as described earlier in the Eq. 7. In thermal equilibrium, as like previous cases, the product of concentration of charge carriers is always a constant at a particular temperature.

$$n_0p_0 = n_i^2 \quad (19)$$

Even though this equation seems to be simple, it is one of the prominent principles in semiconductor physics at thermal equilibrium, and this equation is using the Boltzmann approximation. If Boltzmann approximation is not valid, then the equilibrium concentration of electrons and holes could represent regarding Fermi-Dirac integral [8-11].

$$n_o = \frac{4\pi}{h^2}(2m_n^* kT)^{3/2} \int_0^{\infty} \frac{\eta^{1/2}\, d\eta}{1 + \exp(\eta - \eta_F)} \quad (20)$$

$$p_o = \frac{4\pi}{h^2}(2m_p^* kT)^{3/2} \int_0^{\infty} \frac{\eta'^{1/2}\, d\eta'}{1 + \exp(\eta' - \eta_F')} \quad (21)$$

The integral function in the above equation is called the Fermi-Dirac integral, where $\eta_F = \left[\frac{(E_F-E_C)}{kT}\right]$,$\eta_F' = \left[\frac{(E_V-E_F)}{kT}\right]$, $\eta = \left[\frac{(E-E_C)}{kT}\right]$and $\eta = \left[\frac{(E_V-E)}{kT}\right]$. Here, it is important to mention that if the $\eta_F > 0$ or $\eta_F' > 0$, then the Fermi level would be situated in the conduction band or valence band respectively. More precisely, such kind of semiconductor is called degenerate semiconductor.

$$n_0 = \frac{(N_d - N_a)}{2} + \sqrt{\left(\frac{(N_d - N_a)}{2}\right)^2 + n_i^2} \tag{24}$$

The probability density of electrons and holes in the donor and acceptor level is related to the degeneracy factor described in the Pauli Exclusion Principle as [8-11],

$$n_d = \frac{N_d}{1 + \frac{1}{g}\exp\left(\frac{E_d - E_F}{kT}\right)} \tag{22}$$

$$p_a = \frac{N_a}{1 + \frac{1}{g}\exp\left(\frac{E_F - E_a}{kT}\right)} \tag{23}$$

Where g is the degeneracy factor and by definition these equations can be represented in a simpler form as $n_d = N_d - N_d^+$ and $p_a = N_a - N_a^+$.

In n-type or p-type materials having donor or acceptor level, electrons and holes would be present in the conduction and valence bands at room temperature. This phenomenon is called complete ionisation of the charge carriers at room temperature. These ionization events would not happen at $0K$; the electrons and holes completely freeze in the respective energy level and is known as complete freeze out.

The semiconductor is electrically neutral under thermal equilibrium conditions. We have this fundamental principle to arrive at the thermal equilibrium concentration of charge carriers as a function of impurity doping concentration. Compensated semiconductors are another class of semiconductors that contains both donor and acceptor impurities. The spillover of acceptor impurities or donor impurities into n-type or p-type material causes the formation of n-type ($N_d > N_a$) or p-type ($N_d > N_a$) compensated semiconductors. If both carrier concentrations are the same, the compensated semiconductor shows the characteristics of intrinsic semiconductors.

The equilibrium concentration of charge carriers in a compensated semiconductor can be express as a quadratic equation.

$$p_0 = \frac{(N_a - N_d)}{2} + \sqrt{\left(\frac{(N_a - N_d)}{2}\right)^2 + n_i^2} \tag{25}$$

3. Importance of the Fermi level

The Eq. 5 and Eq.6 give the equilibrium concentration of charge carriers of a semiconductor. These equations can be rewritten in the form of logarithmic function as,

$$E_c - E_F = kT \ln\left(\frac{N_c}{n_0}\right) \tag{26}$$

$$E_F - E_V = kT\ln\left(\frac{N_V}{p_0}\right) \tag{27}$$

For n-type and p-type semiconductors, $\boldsymbol{n}_0 = N_d$ and $p_0 = N_a$ respectively. So that,

$$E_C - E_F = kT \ln\left(\frac{N_C}{N_d}\right) \tag{28}$$

$$E_F - E_V = kT\ln\left(\frac{N_V}{N_a}\right) \tag{29}$$

The above equations reveal the logarithmic relationship of the distance between band edge positions and Fermi level and charge carrier concentrations. The increase in the donor or acceptor concentration moves the Fermi level towards the band edges and thereby increases the carrier concentrations in the respective bands. For a compensated semiconductor, N_d and N_a from the Eq. 27 can be replaced by $N_d - N_a$and $N_a - N_d$. For intrinsic semiconductors, the slight modifications required for the Eq. 28 and Eq. 29 and can be represented as,

$$E_F - E_{Fi} = kT \ln\left(\frac{n_0}{n_i}\right) \tag{30}$$

$$E_{Fi} - E_F = kT\ln\left(\frac{p_0}{n_i}\right) \tag{31}$$

It is important to remember that the Fermi level (E_F) is located above the intrinsic Fermi level (E_{Fi}) for n-type and below for p-type semiconductors. The charge carrier concentration in any semiconductor materials depends on the temperature. Therefore, the temperature changes affect the Fermi level position in a semiconductor. The increase in the temperature increases both carrier concentrations and Fermi level moves towards the intrinsic Fermi level. In other words, at higher temperature extrinsic semiconductors behave as intrinsic semiconductors.

When two electronic systems come in contact, the Fermi levels of both systems get equilibrated. Let us consider the case of a $p-n$ junction as given in the Figure 3. In the first semiconductor, the energy states are filled up to E_{F1} and in the second case, up to E_{F2}. If these two materials get in contact with each other, the electrons present in the system occupy the lowest possible energy level and therefore, the electrons are transferred from the system with higher electrochemical potential to the lower potential (or one with lower work function to higher work function). This leads to the creation of a space charge layer and this phenomenon is termed as Fermi level equilibration. Not all electron transfers across the semiconductor hetero-junction are smooth as in the Fig. 2.

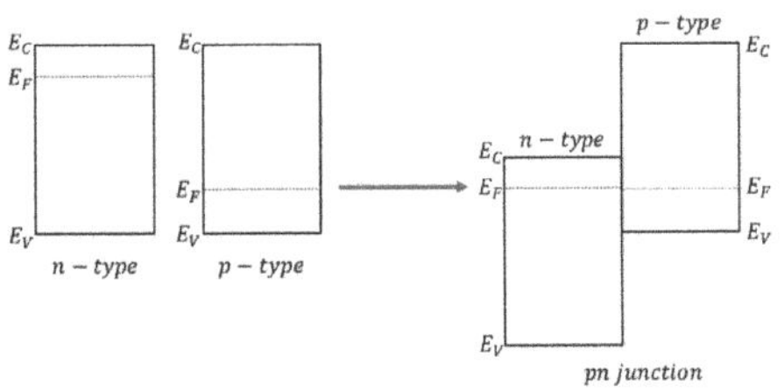

Figure 2. Energy diagram of a p − n junction before and after contact.

There exist different kinds of hetero-junction formulations and the nature of electron transfer across each such junction was different. If we irradiated the hetero junction shown in Fig. 2 with light greater than the band gap, the electron would be transferred from the second one to the first (one with higher work function to lower) and holes move in the opposite direction [16]. Note here that the charge carrier transfer across the hetero junction is not smooth as mentioned in the above situation. There exist more than twenty hetero junction formulations and nature of electron transfer is different in each of these formulations [17]. Also, the Fermi level reaches an equilibrium state when a semiconductor is in direct contact with any metal [18]. Similarly, when a semiconductor material gets in contact with an electrolyte, the Fermi levels of both get equilibrated. The equilibration alters the energy levels in a semiconductor to some extent and leads to the formation space charge layer at the interfaces (Fig. 4). After Fermi level equilibration, there exist an energy difference between the surface interface and that of the bulk of the semiconductor. This energy difference usually termed as band bending [19-20]. The nature of band bending depends on the nature of semiconductor and electrolyte employed. The nature of band bending depends on the kind of semiconductor material whether it belongs to n-type or p-type or any other. It is important to understand here that the term 'band bending' means the interface is at higher (depletion and inversion layers) or lower energy (accumulation layer) level than that of the bulk, and there is no 'bending' of the energy level existing between the surface and bulk. More details are discussed in the next session.

4. Electrical double layer at semiconductor/electrolyte interfaces

The topic semiconductor/electrolyte interface is described extensively in the literature and books as well [21-25]. The concepts are laid out in simple terms in a book by Norio Sato [26] have been followed in the discussions below.

The semiconductor/electrolyte interface composed of three layers namely (a) space charge layer in the semiconductor side with thickness, d_{SC}, of 10-100 nm (b) compact layer at the interface with a thickness, d_H, of 0.4-0.6 nm (c) diffuse layer on the solution side with a thickness, d_d, of 1-10 nm as shown in Fig. 3.

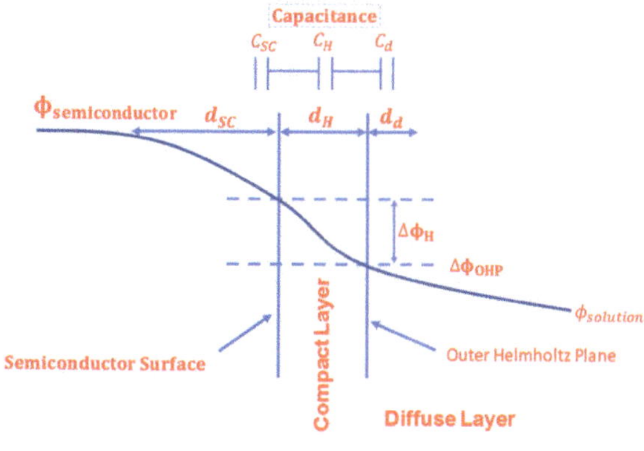

Figure 3. The electrical double layer at the semiconductor-electrolyte interface.

Let us consider the charge of the space charge layer as σ_{SC}, charge of surface states as σ_{SS}, charge of adsorbed layers present in the compact layer as σ_{ad}, and ionic excess charge of hydrated ions in the diffused layer as σ_S. Therefore, electroneutrality across the interface could be represented as,

$$\sigma_{SC} + \sigma_{SS} + \sigma_{ad} + \sigma_S = 0 \tag{32}$$

Further, the total potential across the semiconductor interface, $\Delta\phi$, is given by the sum of potentials that exist in the individual layers at the interface namely; $\Delta\phi_d$, the potential of diffuse layer present in the solution interior; $\Delta\phi_H$, potential across the compact layer in the outer Helmholtz plane; $\Delta\phi_{SC}$, potential across the interior space charge layer of the semiconductor.

$$\Delta\phi = \Delta\phi_{SC} + \Delta\phi_{H} + \Delta\phi_{d} \tag{33}$$

The thickness of the space charge layer and diffuse layer directly related to the total concentration of mobile charge carriers present in the semiconductor, n_i, as well as the ions present in the aqueous solution. The Debye length (L_D) is used for the calculation of the thickness of these respective layers.

$$L_D = \sqrt{\frac{\varepsilon kT}{e^2 \sum n_i z_i^2}} \tag{34}$$

The Debye length is usually corresponding to 100nm for semiconductors with an impurity concentration of the order of $10^{-15} cm^{-3}$ and 10nm in a 0.1M ionic solution. In 1983, Rudiger Memming [22] arrived at a quantitative relationship, connecting the thickness of the space charge layer and Debye length as shown in the Eq. 35.

$$d_{SC} = 2L_d \times \sqrt{\left(\frac{e\Delta\phi_{SC}}{kT}\right) - 1} \tag{35}$$

At the semiconductor-electrolyte interface, the total differential capacitance C can be derived using the equation of a series plate capacitor consisting of C_{SC}, C_H, and C_d; the capacitance of space charge layer, compact layer and diffuse layer respectively. The capacitance of the semiconductor interface is approximately equals to the capacitance of the space charge layer because the total capacitance of any system connected in series is determined by the smallest capacitance value.

$$\frac{1}{C} = \frac{1}{C_{SC}} + \frac{1}{C_H} + \frac{1}{C_d} \tag{36}$$

The total potential across the semiconductor interface about Debye length could be expressed as in the Eq. 37.

$$\Delta\phi = E_{SC}L_{D,SC} + E_H d_H + E_S L_{D,S} \tag{37}$$

Here, E represents the field strength in each layer. If σ_{SS} *and* σ_{ad} were absent, the electrostatic equation can be rewritten as

$$E_{SC}\varepsilon_{SC} = E_H\varepsilon_H = E_S\varepsilon_S \tag{38}$$

So, the overall potential distribution would be represented as;

$$\frac{\delta\Delta\phi_H}{\delta\Delta\phi_{SC}} = \frac{E_H d_H}{E_{SC} L_{D,SC}} \tag{39}$$

Substituting $E_H = E_{SC}\varepsilon_{SC}/\varepsilon_H$. Therefore, Eq. 39 can be modified as;

$$\frac{\delta\Delta\phi_H}{\delta\Delta\phi_{SC}} = \frac{\varepsilon_{SC} d_H}{L_{D,SC}\,\varepsilon_H} \tag{40}$$

Similarly, one can arrive for overall potential across space charge layer and diffuse layer.

$$\frac{\delta\Delta\phi_d}{\delta\Delta\phi_{SC}} = \frac{L_{D,S}\varepsilon_{SC}}{L_{D,SC}\varepsilon_S} \tag{41}$$

These equations reveal that most of the potential changes that occur in the space charge layer and the potentials of other two layers remain unchanged. This characteristic of the semiconductor called Band Edge Pinning. In this state, the band edge potentials of the semiconductor at the interface remains as fixed and therefore the potential across the compact layer. We have assumed that the surface state charge is absent in the above cases. If surface state charge is present, then one can express the electrostatic equation as shown in Eq. 42.

$$E_H\varepsilon_H = E_{SC}\varepsilon_{SC} + \sigma_{SS} \cong \sigma_{SS} \tag{42}$$

If the surface state value σ_{SS} is relatively high, then $\Delta\phi_H$ can be represented as;

$$\Delta\phi_H = E_H d_H = \frac{\sigma_{SS} d_H}{\varepsilon_H} \tag{43}$$

4.1 Band edge pinning and Fermi level pinning

Band edge pinning at the interface of a semiconductor occurs when the Fermi level lies in between the bandgap away from the surface states and band edge positions of higher density of states. During the band edge pinning, the potential across the space charge layer and compact layer remain constant. Whereas, Fermi level pinning occurs when the Fermi level of a semiconductor at the interface approaches to the surface states or band edge positions with a higher density of states. At this condition, the potential across the space charge layer remains unchanged, and a significant change in the potential occurred in the compact layer [27,28]. At Fermi level pinning conditions, the semiconductor interface could behave as a metal, and the phenomenon referred as quasi-metalization. Schematic representation of both phenomena is shown in Fig.4 and Fig.5.

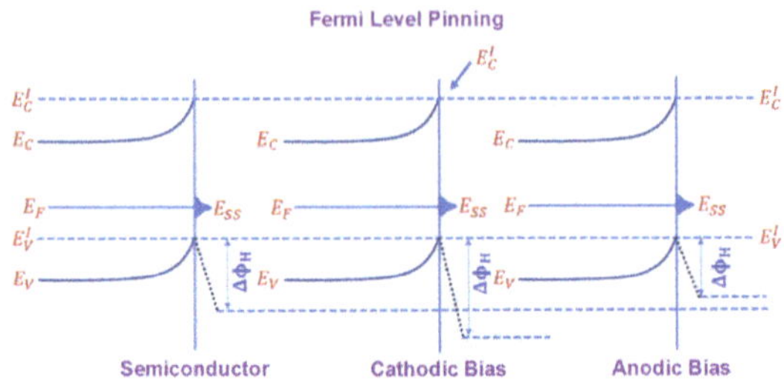

Figure 4. Fermi level pinning at the semiconductor interface under different conditions.

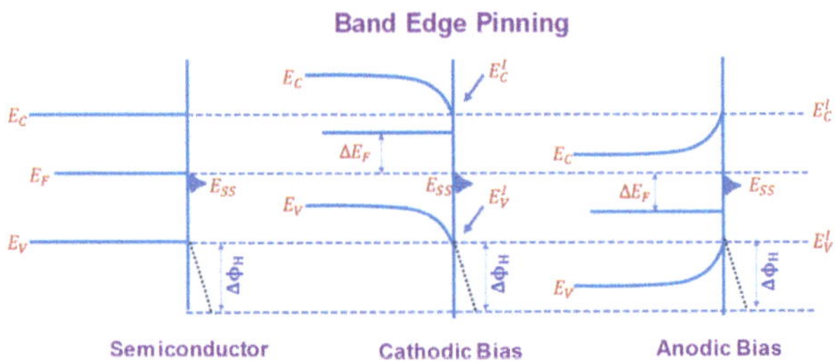

Figure 5. Band edge pinning of the semiconductor under different conditions.

4.2 Space charge layers

When a semiconductor gets in contact with an electrolyte solution, a surface charge is generated because of the interfacial acid-base reactions. The generated surface charge has a profound influence on the energy levels present near to the semiconductor-electrolyte interface. This phenomenon is called band bending. The pH in which the charge of acidic and basic sites present at the interface was equal referred as iso-electric point or point of zero charges. Therefore, at the point of zero charges, the semiconductor exhibits flat band conditions. The interface gets a negative charge when the pH lies above the isoelectric point. In this situation, the excess negative charge present at the interface repels the electrons from the bulk of the semiconductor, and the interface is situated at a higher energy level compared to that of the bulk. This would lead to an upward band bending at the interface. Conversely, a positive surface charge creates a downward band bending. The maximum energy difference between interface and bulk of the semiconductor is about $0.5V$ and this energy difference transports electron and hole in opposite direction within the interfacial zone, and reduces the recombination. The region in which band bending occurs is collectively referred as *space charge layer*.The magnitude of the interfacial charges, dielectric constant of the materials and charge carrier concentration affect the thickness of the space charge layer. The thickness usually of the order of $100\ nm\ to\ 1\ \mu m$. Four different types of space charge layer exist (a) Accumulation layer (b) Depletion layer (c) Inversion layer (d) Deep depletion layer (Fig. 6).

The total differential capacitance, C_{SC}, of space charge region of an intrinsic semiconductor can be represented as,

$$C_{SC} = \frac{\varepsilon}{L_D} \cosh\left(\frac{e\Delta\phi_{SC}}{2kT}\right) \tag{44}$$

Here, L_D is the Debye length (Eq. 34), n_i is the concentration of charge carrier and $\Delta\phi_{SC}$ is the potential across the space charge layer. The capacitance of space charge layer of the n-type or p-type semiconductor can be expressed in general form as,

$$C_{SC} = \frac{\varepsilon}{L_{D_{eff}}} \left(\frac{1 - \exp\left(\frac{\pm e\,\Delta\phi_{SC}}{kT}\right)}{\sqrt{\left|1 + \frac{\pm e\,\Delta\phi_{SC}}{kT} - \exp\left(\frac{\pm e\,\Delta\phi_{SC}}{kT}\right)\right|}} \right) \tag{45}$$

In this equation, $L_{D_{eff}}$ refers to the effective Debye length and equals to $\sqrt{\frac{\varepsilon kT}{2\,N\,e^2}}$,

N denotes the concentration of donor or acceptor. In the $\pm$ sign, $+$ is for p-type semiconductor and $-$ is for n-type semiconductor. This is the general capacitance equation for space charge layer in a semiconductor from which one can derive the subsequent capacitance for all the four types of space charge layer classification.

4.2.1 Accumulation layer

The accumulation layer transports the majority charge carriers to the interface. The capacitance of the accumulation layer can derive from the preceded equation as a function of Boltzmann distribution. In this, one assumes that the Fermi level is situated away from the band edge positions. If the Fermi level is located near to the band edge position, Fermi level pinning occurs. Under this condition, the capacitance,C_{SC}, does not depend on the potential of the space charge layer.

$$C_{SC} = \frac{\varepsilon}{L_{D_{eff}}} \sqrt{\exp\left(\frac{e\,\Delta\phi_{SC}}{kT}\right) - 1} \cong \frac{\varepsilon}{2\,L_{D_{eff}}} \exp\left(\frac{|e\,\Delta\phi_{SC}|}{kT}\right) \tag{46}$$

Here, $|e\,\Delta\phi_{SC} > 3kT|$ and $\Delta\phi_{SC} < 0$ for n-type semiconductor and $\Delta\phi_{SC} > 0$ for p-type semiconductor.

4.2.2 Depletion layer

The formation of a depletion region depletes the majority charge carriers from the semiconductor/electrolyte interface and transports the minority charge carrier to the interface [29]. In the capacitance equation, $|e\,\Delta\phi_{SC} > 3kT|$ and $\Delta\phi_{SC} > 0$ for n-type semiconductor and $\Delta\phi_{SC} < 0$ for p-type semiconductor.

$$C_{SC} = \frac{\varepsilon}{2L_{D_{eff}}}\frac{1}{\sqrt{\left(\frac{|e\,\Delta\phi_{SC}|}{kT}\right) - 1}} \tag{47}$$

Rearranging the Eq. 47,

$$\frac{1}{C_{SC}^2} = \left(\frac{2L_{D,eff}}{\varepsilon}\right)^2\left(\frac{\Delta\phi_{SC}}{kT} - 1\right) = \left(\frac{2L_{D,eff}}{\varepsilon}\right)^2\frac{e}{kT}\left(E - E_{fb} - \frac{kT}{e}\right) \tag{48}$$

This equation is called Mott-Schottky equation. Here, E is the potential of semiconductor at the interface and E_{fb} is the flat band potential. The plot of $1/C_{SC}^2$ versus E is called the Mott-Schottky plot and it gives the flat band potential of a semiconductor and effective Debye length. The Mott-Schottky plot of n-type semiconductor gives the valence band edge position and for p-type semiconductor gives the conduction band edge position. The energy barrier associated with depletion region usually referred to as Schottky Barrier [30].

4.2.3 Inversion layer

An inversion layer is formed when the Fermi level at the interface approaches to the band edge position corresponds to the minority charge carriers, and this would increase the change in the potential of the depletion region. The accumulation of minority charge carriers at the interface inverses the interfacial property from n-type to p-type or p-type to n-type. Therefore, such kind of space charge layer is called inversion layer. The capacitance of the inversion layer is similar to the accumulation layer capacitance equation.

$$C_{SC} = \frac{\varepsilon}{2\,L_{D_{min}}}\exp\left(\frac{|e\,\Delta\phi_{SC}|}{2kT}\right) \tag{49}$$

The Debye length of minority charge carriers, $L_{D_{min}} = \sqrt{\frac{\varepsilon kT}{2\,n_{min}\,e^2}}$ and n_{min} represents the concentration of minority charge carriers. Here, $\Delta\phi_{SC} > 0$ and $\Delta\phi_{SC} < 0$for n-type and p-type semiconductors respectively. The Fermi level pinning occurs in inversion layer as the potential $\Delta\phi_{SC}$ increases. In this particular situation, the band edge position and capacitance remain as constants.

4.2.4 Deep depletion layer

It is an extended form of the depletion layer and commonly found in large band gap semiconductors. In the deep depletion layer, the minority charge carriers get consumed as a result of interface reactions being faster than the carrier generation. Thus, producing an insulating layer at the interface. The Mott-Schottky relation gives the capacitance of the deep depletion layer.

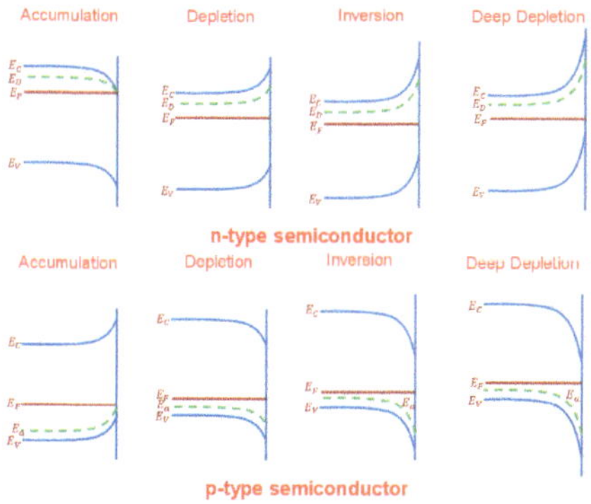

Figure 6. Different types of Space Charge Layer in an n-type and p-type Semiconductor.

4.3 Compact layer

The hydroxylation of semiconductor produces acid and base site on the surface of the semiconductor, and these sites attract other ions to the surface and form a compact layer at the interface of the semiconductor. This hydroxylated semiconductor surface shows a positive charge in acidic solution and negative charge in basic solutions. The potential across the compact layer is represented as $\Delta\phi_H$.

$$\Delta\phi_H = \text{Constant} + \frac{kT}{e}\ln[H^+_{aq}] = \text{Constant} - 2.3\frac{kT}{e}(pH) \tag{50}$$

The equation gives a clear-cut indication of the linear dependence of solution pH on the potential across the compact layer. The compact layer potential ($\Delta\phi_H$)is the adduct of both interfacial potentials ($\Delta\phi_\sigma$)formed because of the interfacial charge (σ_H) and dipole potential ($\Delta\phi_{dipole}$) produced because of the interfacial dipole.

$$\Delta\phi_H = \Delta\phi_\sigma + \Delta\phi_{dipole} \tag{51}$$

The pH in which the interfacial charge (σ_H) becomes zero is called point of zero charge (PZC) or iso-electric point (IEP) of a semiconductor. We can represent the Eq. 50 in terms of iso-electric point as,

$$\Delta\phi_{H} = \Delta\phi_{dipole} - 2.3\frac{kT}{e}(pH - pH_{IEP}) \quad (52)$$

$$\frac{\Delta\phi_{H}}{dpH} = 59mV/pH \quad (53)$$

Therefore, the potential across the compact layer changes 59mV/pH. Rearranging the electro neutrality equation at the interface with interfacial hydroxyl charge(σ_H) yields,

$$\sigma_{SC} + \sigma_{SS} + \sigma_{H} + \sigma_{ad} + \sigma_{S} = 0 \quad (54)$$

In flat band conditions, we know that $\sigma_{SC} = 0$. Therefore, the preceded equation becomes,

$$\sigma_{SS} + \sigma_{H} + \sigma_{ad} + \sigma_{S} = 0 \quad (55)$$

In the case of typical semiconductors, the charge of surface states (σ_{SS}) and the charge of adsorption(σ_{ad}) were zero or remains as constant. In such semiconductors, the value of flat band can be can be shown as follows,

$$E_{FB} = E_{FB,IEP} - 2.3\frac{kT}{e}(pH - pH_{IEP}) \quad (56)$$

Here, $E_{FB,IEP}$ is the flat band potential at the point of zero charge (PZC) or iso electric point (IEP). From this equation, it is evident that an increase in the pH of the solution shifts the conduction band position to more negative values at rate of 59mV/pH. In the case of simple and compound semiconductors, the flat band potential changes linearly with pH of the solution except in the case of WeS_2. The flat band potential of WeS_2 does not change with pH and depends on the concentration of hydrated selenide ions. Such kind of ions that controls the potential across the compact layer is called potential-determining ion. In most of the cases, the potential determining ion is the adsorbed hydroxylated group (hydrated protons and hydrated hydroxyl groups). Therefore, the flat band potential has a linear dependence on pH. However, in the case of metal chalcogenides such as metal sulphides, selenides and tellurides, the flat band potentials do not vary with pH of the solution because the concentration of hydrated chalcogenides is the potential determining ion.

5. Semiconductor in non equilibrium

According to G.N. Lewis and M. Randall [31], it is not the coal but the combustion of coal which causes a steam engine to operate. As a rule, one system affects another system, in consequence of changes which are going on within it. A system far removed from its condition of equilibrium is the one chosen if we wish to harness its processes for doing useful work. Consequently, we are disturbing the semiconductor under equilibrium in order to make useful energy, executing redox reactions or electricity generation in solar cells.

5.1 Charge carrier recombination mechanisms

The illumination of semiconductors with threshold light disturbs the thermal equilibrium and produces electrons and holes in the conduction and valence bands. During non equilibrium situations, excess charge carriers were present at the respective bands. Therefore, Eq. 19 is not fulfilled in these cases.

$$np > n_i^2 \qquad (57)$$

Here, n and p are the non equilibrium excess concentrations of electrons and holes. After illumination, the electron and hole recombine and regain the thermal equilibrium state. Up to some extent recombination of charge carriers is beneficial to the photocatalysts because it prevents the charge accumulation in that particular band. In direct band gap materials, the valence band and conduction bands are vertically aligned to each other. The absorption coefficient, therefore, is very high for direct band gap materials since it requires only the threshold photon energy to excite an electron from the valence band to conduction band. However, in indirect bandgap materials, bands are not vertically aligned, and therefore extra energy from phonons also required for the excitation in addition to the photon energy. The same also applies to recombination. In direct band gap materials, the direct recombination process is spontaneous while in indirect bandgap material the direct recombination is not spontaneous. Crystalline silicon is an example for indirect bandgap materials. In such kind of materials, augur recombination mechanism is more predominant than radiative recombination mechanisms. Depending upon the semiconductor properties various types of recombination processes exist. We have devoted this section to such recombination mechanisms [10, 32].

5.1.1 Direct recombination

Radiative generation and recombination mechanism predominantly exist in direct band gap materials. In such mechanism, the generation of excitons requires band gap energy and emits the same amount of energy when the excitons recombine. At thermal equilibrium conditions, $np = n_i^2$. At temperatures above 0K, some of the bonds break and electron-hole pair generates at the rate of $G_{thermal}$ or G_{th}. In other words, the rate of recombination ($R_{thermal}$ or R_{th}) and generation is the same at thermal equilibrium. So, in thermal equilibrium situations, the rate of recombination is directly proportional to the concentration of charge carriers.

$$R_{th} = G_{th} \qquad (58)$$

$$R_{th} = \beta n_0 p_0 \quad (59)$$

Where β is the proportionality constant. The similar equation can be rewritten at the non equilibrium conditions as,

$$R^* = \beta np \quad (60)$$

We can express the total recombination rate regarding equilibrium concentration (n_0, p_0) and excess carrier concentrations $(\Delta n = n - n_0; \Delta p = p - p_0)$.

$$R^* = \beta np = \beta(n_0 + \Delta n)(p_0 + \Delta p) \quad (61)$$

The total generation rate of the excitons can represent as the adduct of generation rate at thermal equilibrium(G_{th}) and at non-equilibrium conditions(G_L).

$$G = G_{th} + G_L \quad (62)$$

At steady state condition where the rate of generation and recombination is similar.

$$G_L = R^* - G_{th} = \beta(np - n_0 p_0) = R_d \quad (63)$$

Here, R_d corresponds to net radiative recombination rate. For n-type semiconductors,$\Delta n \ll n$ and $p \ll n$. Therefore, R_d can be represented as,

$$R_d = \beta n_0(p - p_0) = \frac{p - p_0}{\tau_p} \quad (64)$$

Here τ_p represents the life time of minority charge carrier (holes). The excess charge carrier concentration can be expressed as product of generation rate and the life time of minority charge carriers.

$$p - p_0 = G_L \tau_p \quad (65)$$

If no excess charge carriers were present in the semiconductor, then the net recombination rate, $R_d = 0$.Once the irradiation is stopped, the generation rate of charge carriers becomes zero. Then, the excess charge carrier concentration can be expressed as,

$$\frac{dp}{dt} = -\frac{p(t) - p_0}{\tau_p} \quad (66)$$

The solution of the above equation yields,

$$p(t) = p_0 + G_L \tau_p \exp\left(-\frac{1}{\tau_p}\right) \quad (67)$$

Accordingly, the exact meaning of lifetime of minority charge carrier can be defined as the time constant at which the excess carrier concentrations decays exponentially if the generation rate of charge carriers were zero.

Similarly, for p-type semiconductors ($\Delta p \ll p\ and\ n \ll p$), the recombination rate can be expressed as,

$$R_d = \beta p_0(n - n_0) = \frac{n - n_0}{\tau_n} \tag{68}$$

Here τ_n represents the life time of electrons.

The diffusion of length of any minority carrier can be expressed regarding lifetime as,

$$L_n = \sqrt{D_n \tau_n} \text{ for electrons in p-type semiconductors} \tag{69}$$

$$L_p = \sqrt{D_p \tau_n} \text{ for holes in n-type semiconductors} \tag{70}$$

Here, $D_n, D_p \& L_n, L_p$ indicates the diffusion coefficients and diffusion length of minority charge carriers.

5.1.2 Shockley-Read-Hall (SRH) recombination

In this type of process, the recombination takes place through trap states (E_T) created as a result of lattice defects or impurity atoms. The electron trapped in the trap state attracts the hole and dissipates heat to the lattice as a result of recombination. There are two types of trap states according to the nature of impurities or defects, donor-type traps and acceptor-type traps.

For an intrinsic semiconductor, the SRH recombination rate can be stated as,

$$R_{SRH} = \upsilon_{th} \sigma N_T \frac{np - n_i^2}{n + p + 2n_i \cosh\left(\frac{E_T - E_{Fi}}{k_B T}\right)} \tag{71}$$

For an n-type semiconductor, the excess electron concentration is almost similar to the thermal equilibrium electron concentrations, $n \approx n_0$ and $n \gg p$.

$$R_{SRH} = \upsilon_{th} \sigma N_T \frac{p - p_0}{1 + 2\frac{n_i}{n_0} \cosh\left(\frac{E_T - E_{Fi}}{k_B T}\right)} = c_p N_T (p - p_0) = \frac{p - p_0}{\tau_{p,SRH}} \tag{72}$$

c_P and $\tau_{p,SRH}$ represents the hole capture coefficient and hole life time in an n-type semiconductor and $\upsilon_{th}$is the thermal velocity.

Similarly, the rate of recombination in the p-type semiconductor can be written as regarding electron capture coefficient (c_n) and electron life time ($\tau_{p,SRH}$).

$$R_{SRH} = \upsilon_{th}\sigma N_T \frac{p - p_0}{1 + 2\frac{n_i}{p_0}\cosh\left(\frac{E_T - E_{Fi}}{k_B T}\right)} = c_n N_T(n - n_0) = \frac{n - n_0}{\tau_{p,SRH}} \quad (73)$$

In all the above cases, the life of the minority charge carrier is inversely proportional to the density of trapped state.

$$\tau_{p,SRH} = \frac{1}{c_p N_T}\ ;\ \tau_{n,SRH} = \frac{1}{c_n N_T} \quad (74)$$

For more detailed information on charge carrier recombination statistics of Shockley-Read-Hall Recombination can be found in the references [32-34].

5.1.3 Auger recombination

Auger recombination is a three particles process involving electrons, holes, and a neighbour electron or hole. Thus, carrier concentrations have a strong dependence on the Auger recombination process. In this process, the energy and momentum during the recombination process are transfers to the next electron or hole and excites to higher energy levels. The relaxation of this third carrier dissipates energy in the form of vibrational energy of the lattice.

The auger recombination rate can be express as,

$$R_{eeh} = C_n n^2 p = C_n N_D^2 p \quad (75)$$

$$R_{ehh} = C_p n p^2 = C_p N_A^2 n \quad (76)$$

Where C_n and C_p are the temperature dependent proportionality constants and R_{eeh}&R_{ehh} are the Auger recombination rate of an n-type semiconductor and a p-type semiconductor respectively.

The Auger lifetime in n-type and p-type semiconductors can be expressed as the following.

$$R_{eeh} = \frac{1}{C_n N_d^2}\ ;\ R_{ehh} = \frac{1}{C_p N_A^2} \quad (77)$$

The Auger recombination rate is depended on the square of donor or acceptor density.

Therefore, at higher doping levels Auger recombination is predominant than others.

5.1.4 Surface recombination

All recombination mechanism discussed above happens in the bulk of a semiconductor. In the case of highly pure semiconductors, surface recombination predominates over bulk recombination processes. The surface trap state generated within the band gap because of the dangling bond present at the surfaces. The surface recombination rate of n-type and p-type semiconductor can be stated as,

$$R_{sn} = \upsilon_{th}\sigma_p N_{ST}(p_s - p_0) \tag{78}$$

$$R_{sp} = \upsilon_{th}\sigma_n N_{ST}(n_s - n_0) \tag{79}$$

Where σ_p *and* σ_n represents the capture cross section of holes and electrons correspondingly; N_{ST} indicates the surface trap density; $p_s$and n_s are the hole and electron concentration at the surface; p_0 and n_0 are the thermal equilibrium concentration of holes and electrons.

5.2 Concept of quasi-Fermi level

When a semiconductor under thermal equilibrium irradiated with a light of energy larger than its bandgap, electron excites from the valence band to the conduction band and produces excitons in the respective bands. In the case of an n-type semiconductor, the irradiation hardly influences the concentration of the conduction band electrons (majority charge carrier) and increases the hole (minority charge carrier) concentration in the valence band. Similarly, in p-type semiconductors, the irradiation enhances the concentration of electrons (minority charge carrier) and hardly influences the concentration of holes (majority charge carrier). So that, generally we can say that the irradiation of a semiconductor enhances the concentration of minority charge carriers.

$$n^* = n + \Delta n \tag{80}$$

$$\mathbf{p}^* = \mathbf{p} + \Delta\mathbf{p} \tag{81}$$

n, p and n^*, p^*represents the overall concentration of electrons and holes in the dark and in presence of irradiation respectively. Δn^* and Δp^* is the increase in the concentration of charge carriers after irradiation. During photoexcitation, the charge carriers get equilibrated with phonons in the respective bands. The irradiation splits the Fermi level into two, the quasi-Fermi level of electrons and quasi-Fermi level of holes. The electrochemical potential of electrons and holes in the photo-stationary state is therefore referred as the quasi-Fermi level of electrons $\left(E_F^{n^*}\right)$ and holes $(E_F^{p^*})$ correspondingly. In the absence of any irradiation (under thermal equilibrium), the quasi-Fermi level and the Fermi level of the semiconductor was the same. Under photoexcitation conditions, the quasi Fermi level of electrons situates above the Fermi level and that of, holes situate below the Fermi level (Fig. 7).

$$E_F^{n^*} = E_C - kT \ln\left(\frac{N_C}{n^*}\right) = E_F + kT \ln\left(\frac{n + \Delta n^*}{n}\right) \tag{82}$$

$$E_F^{p^*} = E_V + kT \ln\left(\frac{N_V}{p^*}\right) = E_F - kT \ln\left(\frac{p + \Delta p^*}{p}\right) \tag{83}$$

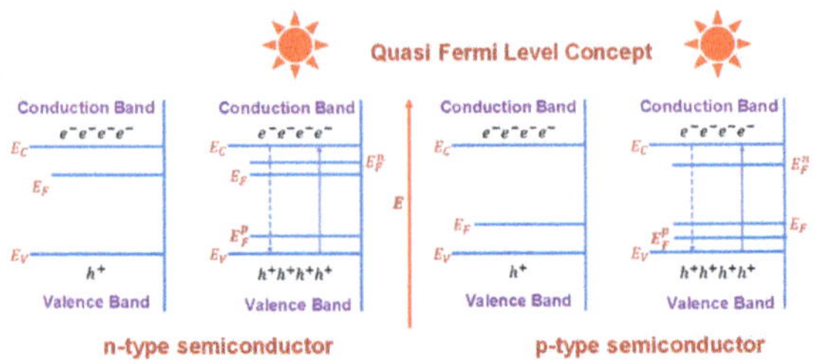

Figure 7. Splitting of Fermi level to quasi-Fermi levels in the presence of irradiation and quasi-Fermi level formation in n- and p-type semiconductors.

In an n-type semiconductor, $n \gg p$ and $n \gg \Delta n^*$. Therefore, the quasi-Fermi level of electrons is approximately equal to the Fermi level and the quasi-Fermi level of holes $(p < \Delta p^*)$ situates far away and lower than the Fermi level. Generally, we can say that under photo excitation conditions the quasi-Fermi level of the majority charge carriers is closer to or approximately the same as that of the Fermi level.

Under equilibrium conditions, the Fermi levels of redox species and the semiconductor were in the balance. So, there is no electron transfer taking place between them and therefore no reaction occurs in the absence of any irradiation. If electron transfer could occur between semiconductor and redox species, the Fermi level of the semiconductor should be above the Fermi level of the redox species. This criterion attains when the semiconductor irradiated with threshold light. The necessary pre-require for a reduction reaction is that the quasi-Fermi level of the electron should be higher than that of the redox species. In such conditions, the electrons from the conduction band flow to the oxidant species. Likewise, the quasi-Fermi level position of the hole must be lower than that of the reductant species.

$$E_F^{n^*} > E_{F_{redox}} \tag{84}$$

$$E_F^{p^*} < E_{F_{redox}} \tag{85}$$

We can consider the water-splitting reaction as an example to explain what we mentioned previously. When an n-type semiconductor is immersed in water, the Fermi level of both species gets equilibrated. Under this situation, the semiconductor does not have sufficient potential to reduce or oxidize the redox species, and therefore the reaction is thermodynamically not feasible under dark conditions. The presence of irradiation split the Fermi level into two quasi-Fermi levels and aligned above and below of the redox potential of water. Now, the necessary prerequisites for the reduction and oxidation get satisfied. Consequently, water undergoes redox reaction under these circumstances (Fig. 8).

$$E_F^{n^*} > E_{F(H_2O/H_2)} \tag{86}$$

$$E_F^{p^*} < E_{F(O_2/H_2O)} \tag{87}$$

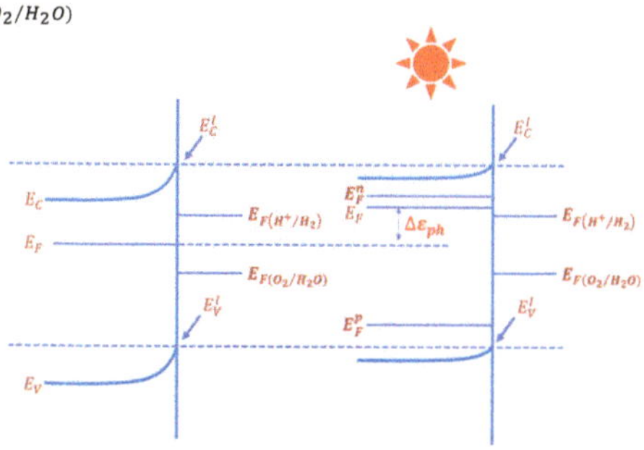

Figure 8. Band characteristics of the semiconductor in presence and absence of illumination concerning water redox potential.

This is the reason why the water-splitting reaction or any other photocatalytically driven reaction is thermodynamically impossible in the absence of any irradiation field.

5.3 Photo-potential

We have already seen that the flat band condition of the semiconductor promotes the charge carrier recombination and the space charge layer formed at the interface separates the charge carriers via transporting them in opposite directions. Such kind of migration of charge carriers induce an inverse potential in the semiconductor and reduces the potential across the space charge layer. The decrease in the potential across the space charge layer retards migration of charges in the opposite direction. The inverse potential generated in the space charge layer as result of photo-excitation is referred as the photopotential.

$$\Delta E_{ph} = \frac{-\Delta \varepsilon_{ph}}{e} \tag{88}$$

When the band bends in an upward direction, at the time of photo-excitation, the Fermi level moves upwards by an energy of $\Delta\varepsilon_{ph}$ and vice-versa when the band bends in a downward direction. Therefore, the maximum photo-potential (ΔE^0_{ph}) of a semiconductor is the difference between the potential of the semiconductor in the dark (E) and the flat band potential (E_{fb}) of the semiconductor during irradiation in which the band bending disappears.

$$|\Delta E_{ph}| \leq |\Delta E^0_{ph}| = |E - E_{fb}| \tag{89}$$

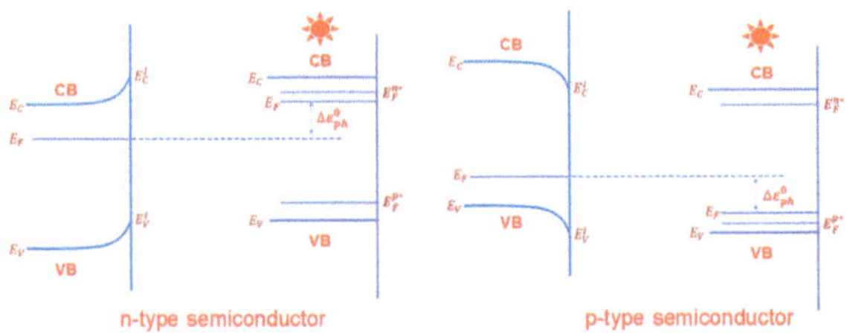

Figure 9. Generation of photopotential under illumination in n-type & p-type semiconductors.

The photopotential value of a semiconductor with high impurity concentration can approximately arrive at via the charge carrier concentration if there is no interfacial chemical reaction present (Fig. 9).

$$\Delta E^{n^*}_{ph} = -\frac{kT}{e}\ln\left(\frac{\Delta p^*}{p}\right) \tag{90}$$

$$\Delta E^{p^*}_{ph} = \frac{kT}{e}\ln\left(\frac{\Delta n^*}{n}\right) \tag{91}$$

6. Understanding of band-edge positions and its calculations

The position of Fermi level at flat band conditions gives the flat band potential of a semiconductor. In this context, it is important to mention that flat band potential of a semiconductor is not the same as that of the band edge potentials [35,36]. As we have already seen that the flat band potential of materials depends on the impurity concentration, but the band edge potentials is an intrinsic property of a semiconductor and do not rely on any impurity concentration. The relation that connects band edge potential and flat band potential are given below. In ordinary semiconductors, the values $E_C - E_F^n = 0.1$ eV and $E_V - E_F^p = -0.1$ eV.

$$E_C = E_{FB}^n - \frac{E_C - E_F^n}{e} \text{ for n} - \text{type} \tag{92}$$

$$E_V = E_{FB}^p - \frac{E_V - E_F^p}{e} \text{ for p} - \text{type} \tag{93}$$

In any of the photocatalytic processes, the position of band edges determines the redox behaviour of the charge carriers. Calculation of band edge position has a significant role in determining the extent and feasibility of a particular photo-catalytic reaction. In this section, we are describing the conventional ways of determining the band positions of a semiconductor material.

Manually one can arrive at the flat band potential of the semiconductors from the equation connecting electronegativity of the semiconductor (χ), energy of free electron on hydrogen scale ($E^e = 4.5$eV), conduction band position (E_{CB}) and bandgap(E_g)[37].

$$E_{CB} = \chi - E^e - 0.5E_g \tag{94}$$

The geometric mean of the individual electronegativities of constituent atoms in a semiconductor would give the electronegativity of a particular semiconductor. Considering TiO_2 as an example, the band edge calculations were derived and given below (Table 1).

Experimental calculation of band edge positions involves a combination of instrumental methods. The valence band XPS and UPS directly give the valence band and conduction band positions of material. If we have any of the VB XPS or CB UPS values, the addition of band gap of the semiconductor determined from Diffuse Reflectance Spectrometric (DRS) measurements or Tauc's extrapolation gives the valence or conduction band position [38]. These band position values can be used to interpret the reaction mechanism involving gaseous phase photo-catalytic reactions. In liquid phase reactions, the band edge position changes with the solvent due to the Fermi level equilibration. So that one can utilize electrochemical techniques such as impedance or cyclic voltammetry for determining the band positions [39-44]. The use impedance measurement for this purpose is called Mott-Schottky plot [45-49]. It gives the 'flat band potential' of a semiconductor. For n-type semiconductors, the flat band potential is closer to the conduction band edge and would give the approximate position of the conduction band, and in p-type semiconductors, the flat band potential is just above the valence band and gives a rough position of the valence band. The addition of bandgap value to the flat band potential gives the corresponding approximate valence or conduction band position.

Table 1. Band edge calculations of titanium dioxide.

<table>
<tr><td>Material</td><td>Bandgap (eV)</td><td>Constituents Elements</td><td>Electron Affinity (eV)</td><td>Ionisation Energy (eV)</td><td>Elements Electro negativity
Arithmetic Mean</td></tr>
<tr><td>TiO_2</td><td>3.2</td><td>Ti
O</td><td>0.08
1.46</td><td>6.81
13.6</td><td>$\frac{0.08+6.81}{2}=3.44$
$\frac{1.46+13.6}{2}=7.53$</td></tr>
<tr><td colspan="2">Semiconductor Electronegativity (Geometric Mean)</td><td colspan="2">Conduction Band Position
E_{CB}</td><td colspan="2">Valence Band Position
E_{VB}</td></tr>
<tr><td colspan="2">No. of Elements =3
$=\sqrt[3]{3.44\times(7.53)^2}$
$=5.79$</td><td colspan="2">$E_{CB}=\chi-E^e-0.5E_g$
$E_{CB}=5.79-4.5-0.5\times 3.2$
$=-0.31V$</td><td colspan="2">$E_{VB}=E_{CB}+E_g$
$E_{VB}=-0.31V+3.2V$
$E_{VB}=2.89V$</td></tr>
</table>

7. Mechanism of semiconductor mediated photon-assisted carbon dioxide reduction

Mechanism of carbon dioxide reduction involves the multiple proton-coupled electron transfer reactions. The nature of product formed during the process depends on the nature of the material used for photo-reduction because each material has different capabilities for executing the proton-coupled reduction. We have already discussed in the previous section, the density and transport properties of charge carriers and thereby semiconductors vary according to the nature of the materials. Consequently, different photo-catalysts offer the same or different reduction products during the reaction. It is evident from the above sections that the electron transport from the generation site to surface sites is not an easy process, and number hurdles are to be overcome in order for one single electron to reach the surface site and to carry out the reaction. If single electron assists such problems, one can imagine how difficult it is to execute multiple electron transfer reactions. More detailed information on the mechanistic aspects of photo-reduction of carbon dioxide is discussed in detail elsewhere [50-51].

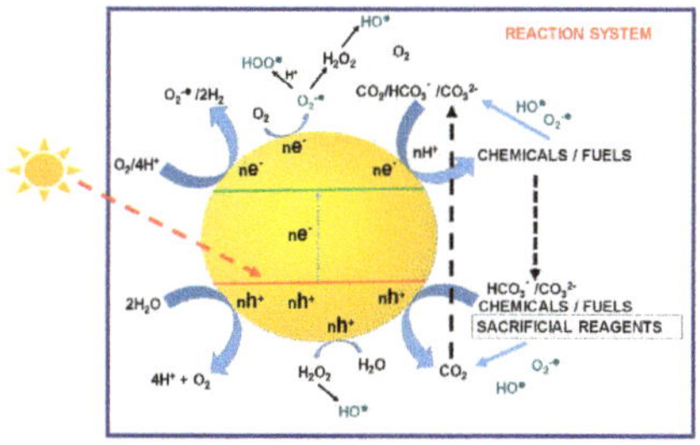

Figure 10. Various processes involved in the carbon dioxide reduction system.

Let us consider the whole photo-reduction system as represented in Fig. 10. The system involves both reduction and oxidation reactions that occur on the same surface in adjacent surface sites, and one knows that the reduced product gets oxidized during the reduction process. Therefore, the rate of the reaction continuously fluctuates while the reaction is progressing for hours. This particular aspect has been discussed in many articles. However, none mentioned why this is happening. The rate of formation of products is high in the initial hours and the rate declines while the reaction proceeds. During the first half of the reaction, more reduction processes are happening because of that reduction products are formed at a significant rate. During the second half, oxidation processes predominates over the reduction and decline the rate of the formation of reduction products. As we know, the light irradiation recreates almost flat band conditions in the semiconductor. As the irradiation progresses, we have seen in many results that the oxidation of reduced products increased after continuous irradiation, indicating that more holes reached the surface interface than that of the electrons. This situation happens only when the band bend in an upward direction and transports the holes to the interface. The light-induced changes in the semiconductor is a least explored research area, but the literature on this aspect observed a significant shift in the band edge characteristics and termed it as light soaking effect [52-54].

In Fig. 11, one can see that a significant reduction of products after 20 hours of irradiation [55,56]. After 20 hours, the oxidation process predominates over the reduction process thereby the product formed as result of reduction gets oxidized. This kind of detrimental effect can be reduced or ignored by stopping the illumination at constant time intervals for a few minutes. It enables the semiconductor to get back to the thermal equilibrium state via recombination the excess charge carriers present in the space charge layer. The illumination of the semiconductor after a small break favours the reduction process and optimization of the time interval also needed in this case. This helps to maintain the reaction rate and increases the product yield after prolonged irradiation for producing more reduction products of carbon dioxide. Such an effect has never been considered in the case of carbon dioxide photo-reduction. Maybe champion photocatalysts yield more products in the initial hours but that does not mean that it would yield products at a constant rate throughout the prolonged irradiation. So, the light-induced changes in the semiconductor are an exciting research area and must be studied in detail especially in the case of carbon dioxide photo-reduction.

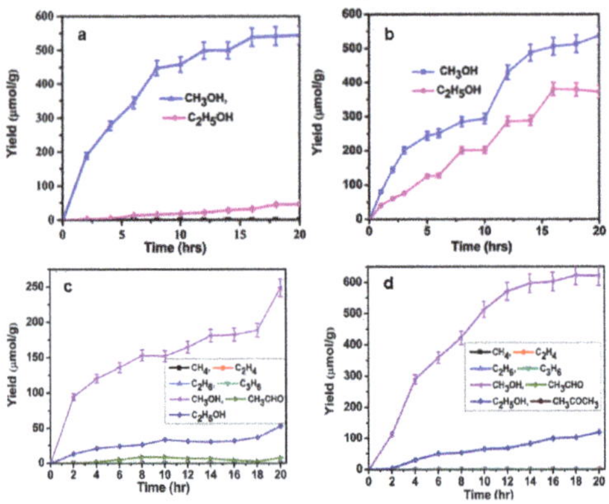

Figure 11. Generation of CO_2 reduction products after 20-hours of irradiation. In each of the figure, one can see that generation of fuels reached a plateau after 20-hour irradiation which indicates the domination of oxidation over reduction under prolonged irradiation. a,b-Reproduced with permission from Ref. 56, ©Elsevier 2016. c,d-Reproduced with permission from Ref. 55, ©Royal Society of Chemistry 2015.

8. Design of materials and futuristic perceptions

The very first criterion of designing material for any photocatalytic reactions is that it should have sufficient band gap for light absorption with appropriate band edge potentials to execute the redox reactions. The use of co-catalyst must be highly appreciated because it reduces the energy barrier in transferring electrons or holes from the surface to the adsorbed species. Usually, metals are used as co-catalysts for reduction reactions and metal oxides such as RuO_2 for oxidation. [57]. For the selection of appropriate eco-catalyst, optimization should be performed. For that, one needs to know the basic semiconductor and metal parameters such as work function, ionization energy, chemical potential and electron affinity. The evaluation of these parameters gives an idea of the nature of the interface between the metal/semiconductor and the semiconductor/semiconductor and this allows the prediction of the nature of the electron transfer between them. These materials should be stable under any photo-catalytic circumstances. Elimination of electron-hole recombination is the one main challenge in the field of photocatalysis. The use of material with the intrinsic electric field within the system should be utilized for this purpose which automatically separates the charge carries. While using that system, the separated charge carriers should be utilized as fast as possible which means the interfacial reaction should be faster. Otherwise, the accumulation of same charge carriers in either part of the semiconductor increases the repulsion among them. This leads to increase in the band bending across the interface which restricts the flow of one of the charge carrier and enhances the surface recombination.

The next aspect is related to the particle size of the material. As we have seen that the thickness of space charge region is about 100 nm to 1 μm. The photon impinging on the semiconductor absorbed in the thin surface layer of thickness ranging from 100 nm to few microns. The design of materials with particle size double the width of space charge layer offers the production of charge carries within the space charge layer and reduces the carrier transport issues existing in the bulk of the semiconductor. Be beware of the facts that reducing the particle size increase the band edge positions due to quantum confinement. And, it increases the electron-hole surface recombination. The ionicity of the bond also has a pronounced effect on the electron transfer. Most of the currently designed photocatalysts are either too ionic or too covalent. The designed material for CO_2 should have the just right iconicity which means neither too ionic nor too covalent. Preferably, ionicity of the bond must be in the range of 30-40%; this should automatically take care of the materials to be active in the visible region of the solar spectrum. The medium ionicity of the bond also enhances the carrier transport properties of the semiconductor. Since any improvement in the photocatalyst also enhances the competing chemical reactions; a round table compromise meeting is required within the photocatalysts with different parameters that affect the CO_2 reduction process!

Even though we have enough materials and designs that satisfy the above materials and design perspectives, we are unable to achieve a product yield at a significant level. The main reason for this is the difficulties in the electron transfer at the interface. The properties of band edge potential and electron transfer across the interface are also dependent on the nature of electrolyte used. So, the design of materials alone could not offer a better solution. We need to design appropriate material-electrolyte combos that offer less resistance to the interfacial electron transfer. Finally, it must be remembered that the fundamental thermodynamics of the carbon dioxide conversion to more reduced products will not be changed and considerable energy is needed to take an additional step in this regard.

References

[1] P.V. Kamat, Semiconductor surface chemistry as holy grail in photocatalysis and photovoltaics,Acc. Chem. Res.50 (2017), 527–531. https://doi.org/10.1021/acs.accounts.6b00528

[2] T. Hisatomi, K. Domen, Introductory lecture: sunlight-driven water splitting and carbon dioxide reduction by heterogeneous semiconductor systems as key processes in artificial photosynthesis,Faraday Discuss. 198 (2017)11-35. https://doi.org/10.1039/C6FD00221H

[3] M.C. Beard, J. L. Blackburn, J. C. Johnson, G. Rumbles, Status and prognosis of future-generation photoconversion to photovoltaics and solar fuels, ACS Energy Lett. 1 (2016)344–347. https://doi.org/10.1021/acsenergylett.6b00204

[4] S. Chu, Y. Cui, N. Liu, The path towards sustainable energy, Nat. Mater. 16 (2016) 16–22. https://doi.org/10.1038/nmat4834

[5] A. Dibenedetto, Across the board: angeladibenedetto, ChemSusChem 9 (2016) 3124–3127. https://doi.org/10.1002/cssc.201601431

[6] W. Tu, Y. Zhou, Z. Zou, Photocatalytic conversion of CO_2 into renewable hydrocarbon fuels: State-of-the-art accomplishment, challenges, and prospects,Adv. Mater. 26 (2014) 4607–4626. https://doi.org/10.1002/adma.201400087

[7] X. Chang, T. Wang, J. Gong, Effective Charge Carrier Utilizationin Visible-Light-Driven CO_2Conversion, in: Z. Mi, L. Wang, C. Jagadish (Eds.), Semiconductors and Semimetals, Academic Press, Burlington, 97 (2017) pp. 429-467.

[8] D. A. Neamen, Semiconductor Physics and Devices: Basic Principles, Fourth Edition, McGraw-Hill Education, New York, 2011.

[9] Y. Y. Peter, M. Cardona, Fundamentals of Semiconductors: Physics and Materials Properties, Fourth Edition, Springer-Verlag, Berlin-Heidelberg, 2010.

[10] S. M. Sze and Kwok K. Ng, Physics of Semiconductor Devices, Third Edition, John Wiley & Sons, New Jersey, 2007.

[11] R. F. Pierret, Advanced Semiconductor Fundamentals (Modular Series on Solid State Devices, Vol. 6), Pearson Prentice Hall, New Jersey, 2002.

[12] K. W. Böer, Introduction to Space Charge Effects in Semiconductors, Springer-Verlag Berlin Heidelberg, 2010. https://doi.org/10.1007/978-3-642-02236-4

[13] A. I. Kokorin, D. W. Bahnemann, Chemical Physics of Nanostructured Semiconductors, CRC Press, The Netherlands, 2003.

[14] W. Shockley, Electrons and Holes in Semiconductors with Application to Transistor Electronics, Van Nostrand Reinhold Inc., U.S, 1950.
[15] X. Yang, D. Wang, Photophysics and photochemistry at the semiconductor/electrolyte interface for solar water splitting, in: Z. Mi, L. Wang, C. Jagadish (Eds.), Semiconductors and Semimetals, Academic Press, Burlington, 97 (2017) pp. 47-80.
[16] J. Bisquert, Nanostructured Energy Devices: Equilibrium Concepts and Kinetics, CRC Press, Boca Raton, 2015.
[17] A. G. Milnes, D. L. Feucht, Heterojunction and Metal Semiconductor Junctions, Academic Press, New York, 1972.
[18] B. L. Sharma, Metal-Semiconductor Schottky Barrier Junctions and Their Applications, Plenum Press, New York, 1984. https://doi.org/10.1007/978-1-4684-4655-5
[19] M. X. Tan, P. E. Laibinis, S.T. Nguyen, J. M. Kesselman, C. E. Stanton, N. E. Lewis, Principles and Applications of Semiconductor Photoelectrochemistry, in:K. D. Karlin (Eds.), Progress in Inorganic Chemistry, John Wiley & Sons, Inc., Hoboken, NJ, USA, 41 (1994) pp. 21-144. https://doi.org/10.1002/9780470166420.ch2
[20] M. Gratzel, Photoelectrochemical cells, Nature 414(2001) 338–344. https://doi.org/10.1038/35104607
[21] J. O'M. Bockris, A. K.N. Reddy, M. E. Gamboa-Aldeco, Modern Electrochemistry 2A: Fundamentals of Electrodics,Springer-Verlag US, 2000.
[22] R. Memming, Semiconductor Photoelectrochemistry, Wiley VCH, Weinheim, 2015.
[23] A. J. Bard, M. Stratmann, S. Licht, Encyclopedia of Electrochemistry, Semiconductor Electrodes and Photoelectrochemistry, Volume 6, Wiley-VCH, 2002.
[24] S. Srinivasan, Electrode/Electrolyte Interfaces: Structure and Kinetics of Charge Transfer, in: Fuel Cells from Fundamental to Application, Springer US, Boston, 2006, pp. 27–92.
[25] W. Plieth, Electrochemistry of Material Science, Elsevier B.V., UK, 2008.
[26] N. Sato, Electrochemistry at Metal and Semiconductor Electrodes, Elsevier B.V. The Netherlands, 1998.
[27] A. J. Bard, A. B. Bocarsly, F. R. F. Fan, E. G. Walton, M. S. Wrighton, The concept of Fermi level pinning at semiconductor/liquid junctions. Consequences for energy conversion efficiency and selection of useful solution redox couples in solar devices, J. Am. Chem. Soc. 102 (1980) 3671-3677. https://doi.org/10.1021/ja00531a001

[28] J.E. Thorne, S. Li, C. Du, G. Qin, D. Wang, Energetics at the surface of photoelectrodes and its influence on the photoelectrochemical properties, J. Phys. Chem. Lett. 6 (2015) 4083–4088. https://doi.org/10.1021/acs.jpclett.5b01372
[29] W. W. Gartner, Depletion layer photoeffects in semiconductor, 116 (1959) 84-87.
[30] Z. Zhang, J. T. Yates, Band bending in semiconductors: chemical and physical consequences at surfaces and interfaces, Chem. Rev. 112 (2012) 5520-5551. https://doi.org/10.1021/cr3000626
[31] G. N. Lewis, Merle Randall, Thermodynamics and the Free Energy of Chemical Substances, McGraw-Hill, United States, 1923.
[32] A. Smets, K. Jager, O. Isabella, R. V. Swaaij, M. Zeman, Solar Energy: The Physics and Engineering of Photovoltaic Conversion, Technologies and Systems, UTI Cambridge, 2016.
[33] W. Shockley, W. T. Read, Statistics of the recombinations of holes and electrons, Phys. Rev. 87 (1952) 835. https://doi.org/10.1103/PhysRev.87.835
[34] R. N. Hall, Electron-hole recombination in germanium, Phys. Rev. 87 (1952) 387. https://doi.org/10.1103/PhysRev.87.387
[35] J. Bisquert, P. Cendula, L. Bertoluzzi, S. Gimenez, Energy diagram of semiconductor/electrolyte junctions, J. Phys. Chem. Lett. 5 (2014) 205–207. https://doi.org/10.1021/jz402703d
[36] W.A. Smith, I.D. Sharp, N.C. Strandwitz, J. Bisquert, Interfacial band-edge energetics for solar fuels production, Energy Environ. Sci. 8 (2015) 2851–2862. https://doi.org/10.1039/C5EE01822F
[37] B. Chai, T. Peng, P. Zeng and J. Mao, Synthesis of floriated In_2S_3 decorated with TiO_2 nanoparticles for efficientphotocatalytic hydrogen production under visible light, J. Mater. Chem. 21 (2011)14587–14593. https://doi.org/10.1039/c1jm11566a
[38] Z. He, D. Wang, J. Tang, S. Song, J. Chen, X. Tao, A quasi-hexagonal prism-shaped carbon nitride for photoreduction of carbon dioxide under visible light, Environ. Sci. Pollut. Res. 24 (2017) 8219-8229. https://doi.org/10.1007/s11356-017-8497-4
[39] W. J. Chun, A. Ishikawa, H. Fujisawa, T. Takata, J. N. Kondo, M. Hara, M. Kawai, Y. Matsumoto, K. Domen, Conduction and valence band positions of Ta_2O_5, TaON, and Ta_3N_5 by UPS and electrochemical methods, J. Phys. Chem. B 107 (2003) 1798-1803. https://doi.org/10.1021/jp027593f
[40] S. Gimenez, H.K. Dunn, P. Rodenas, F. Fabregat-Santiago, S.G. Miralles, E.M. Barea, R. Trevisan, A. Guerrero, J. Bisquert, Carrier density and interfacial kinetics of mesoporous TiO_2 in aqueous electrolyte determined by impedance

spectroscopy, J. Electroanal. Chem. 668 (2012) 119–125. https://doi.org/10.1016/j.jelechem.2011.12.019

[41] L. Bertoluzzi, P. Lopez-Varo, J.A. Jimenez Tejada, J. Bisquert, Charge transfer processes at the semiconductor/electrolyte interface for solar fuel production: insight from impedance spectroscopy, J. Mater. Chem. A. 4 (2016) 2873–2879. https://doi.org/10.1039/C5TA03210E

[42] D. B. Bonham, M. E. Orazem, A Mathematical model for the influence of deep-level electronic states on photoelectrochemical impedance spectroscopy I. Theoretical development, J. Electrochem. Soc. 139 (1992)118-126. https://doi.org/10.1149/1.2069155

[43] S.K. Haram, A. Kshirsagar, Y.D. Gujarathi, P.P. Ingole, O.A. Nene, G.B. Markad, S.P. Nanavati, Quantum confinement in CdTe quantum dots: Investigation through cyclic voltammetry supported by density functional theory (DFT), J. Phys. Chem. C. 115 (2011) 6243–6249. https://doi.org/10.1021/jp111463f

[44] L. Bertoluzzi, L. Badia-Bou, F. Fabregat-Santiago, S. Gimenez, J. Bisquert, Interpretation of cyclic voltammetry measurements of thin semiconductor films for solar fuel applications, J. Phys. Chem. Lett. 4 (2013) 1334–1339. https://doi.org/10.1021/jz400573t

[45] K. Gelderman, L. Lee, S.W. Donne, Flat-band potential of a semiconductor: Using the Mott–Schottky equation, J. Chem. Educ. 84 (2007) 685. https://doi.org/10.1021/ed084p685

[46] W.J. Albery, G.J. O'Shea, A.L. Smith, Interpretation and use of Mott-Schottky plots at the semiconductor/electrolyte interface, J. Chem. Soc. Faraday Trans. 92 (1996) 4083–4085. https://doi.org/10.1039/FT9969204083

[47] A. Mani, C. Huisman, A. Goossens, J. Schoonman, Mott−Schottky analysis and impedance spectroscopy of TiO_2/6T and ZnO/6T devices, J. Phys. Chem. B. 112 (2008) 10086–10091. https://doi.org/10.1021/jp8013964

[48] O. Almora, C. Aranda, E. Mas-Marzá, G. Garcia-Belmonte, On Mott-Schottky analysis interpretation of capacitance measurements in organometal perovskite solar cells, Appl. Phys. Lett. 109 (2016) 173903. https://doi.org/10.1063/1.4966127

[49] A. Lasia, Semiconductors and Mott-Schottky Plots, in: Electrochemical Impedance Spectroscopy and its Applications,Springer, New York, 2013, pp.251-255.

[50] H. Narayanan, B. Viswanathan, S. Yesodharan, Photocatalytic reduction carbon dioxide: Issues and prospects, Curr. Catal. 5 (2016) 79-107. https://doi.org/10.2174/2211544705666160427113828

[51] H. Narayanan, H. Nair,B. Viswanathan, On the current status of the mechanistic aspects of photocatalyticreduction of carbon dioxide, Indian J. Chem., Sect. A. 56A (2017) 251-269.

[52] E.Y. Liu, J.E. Thorne, Y. He, D. Wang, Understanding photocharging effects on bismuth vanadate, ACS Appl. Mater. Interfaces. 9 (2017) 22083–22087. https://doi.org/10.1021/acsami.7b06528

[53] J. Hu, R. Gottesman, L. Gouda, A. Kama, M. Priel, S. Tirosh, J. Bisquert, A. Zaban, Photovoltage behavior in perovskite solar cells under light-soaking showing photoinduced interfacial changes, ACS Energy Lett. 2 (2017) 950–956. https://doi.org/10.1021/acsenergylett.7b00212

[54] R. Gottesman, P. Lopez-Varo, L. Gouda, J.A. Jimenez-Tejada, J. Hu, S. Tirosh, A. Zaban, J. Bisquert, Dynamic phenomena at perovskite/electron-selective contact interface as interpreted from photovoltage decays, Chem. 1 (2016) 776–789. https://doi.org/10.1016/j.chempr.2016.10.002

[55] V. Jeyalakshmi, R. Mahalakshmy, K. Ramesh, P.V.C. Rao, N. V Choudary, G. Sri Ganesh, K. Thirunavukkarasu, K.R. Krishnamurthy, B. Viswanathan, Visible light-drivenreduction of carbon dioxide with water on modified $Sr_3Ti_2O_7$ catalysts, RSC Adv. 5 (2015) 5958–5966. https://doi.org/10.1039/C4RA11985A

[56] V. Jeyalakshmi, S. Tamilmani, R. Mahalakshmy, P. Bhyrappa, K.R. Krishnamurthy, B. Viswanathan, Sensitization of La-modified $NaTaO_3$ with cobalt tetra phenyl porphyrin for photocatalytic reduction of CO_2 by water with UV–visible light, J. Mol. Catal. A Chem. 420 (2016) 200–207. https://doi.org/10.1016/j.molcata.2016.04.027

[57] L. Li, P.A. Salvador, G.S. Rohrer, Photocatalysts with internal electric fields, Nanoscale. 6 (2014) 24–42. https://doi.org/10.1039/C3NR03998F

Chapter 7

Photo-Electrochemical Reduction of CO_2 to Solar Fuel: A Review

Kinjal J. Shah [a, b,*], Shu-Yuan Pan [a, b], Vimal Gandhi [c], Pen-Chi Chiang [a, b,*]

[a] Graduate Institute of Environmental Engineering, National Taiwan University, Taiwan

[b] Carbon Cycle Research Center, National Taiwan University, Taipei City 10673, Taiwan

[c] Department of Chemical Engineering & Shah-Schulman Center for Surface Science and Nanotechnology, Dharmsinh Desai University, Nadiad 387001, India

Email: kinjalshah8@gmail.com; pcchiang@ntu.edu.tw.

Abstract

Green chemistry and sustainable technology for CO_2 capture, sequestration and utilization are urgently required to control CO_2 growth. Amongst the available major methods (including chemical, photochemical, electrochemical and enzymatic methods), the photo-electrochemical method, especially development for solar to chemical or fuel technology, offers a green and potential alternative for efficient CO_2 capture and conversion. This is a very active research area covering wide concepts and ideas under investigation with many barriers considering system engineering. This review will discuss the recent progress in this field as well as give a brief background on solar fuel conversion from captured CO_2, suitable procedure for the solar fuel production and some of their critical components with case studies of methanol production.

Keywords

Green Chemistry, Solar Fuels, Photo-Electrochemical Solar Devices, Methanol, Carbon Utilization

Contents

1. Introduction:

Green Chemistry (GC) is a relatively new emerging field with an objective to achieve environmental sustainability through design of a chemical product (chemistry) and processes (engineering) so as to reduce (a) generation of pollution at the source and (b) risk to human health and the environment [1-3]. GC basically works on 12 principles namely, prevention, atom economy, less hazardous chemical synthesis, designing safer chemicals, safer solvent, energy efficient, renewable feedstock, reduce derivatives, catalysis, design for degradation, pollution prevention and accident prevention [1,4]. If we correlate all these principles, the major focus of GC is on waste prevention and pollution prevention. In waste and pollution, emission of greenhouse gases is mostly reported, in which CO_2 is a major concern (60% contributed in greenhouse gases), as it is responsible for increasing earth temperature [5-7]. To mitigate global warming, Kyoto Protocol [8], Copenhagen accord [9, 10] and Paris agreements [11] are the major action taken by the world. Apart from those agreements, the International Energy Agency pointed out that in order to achieve the ± 2°C goal, CO_2 capture and storage (CCS) technologies are neede [9, 12]. It is therefore essential to develop a CCS technology and adding U (utilization) so that CCSU technologies can cope with the global demand of CO_2 reduction [13]. Table 1 represents the advantages and disadvantages of the different CO_2 capture and storage technologies.

Table 1 Advantages and disadvantages of the different CO_2 capture and storage technologies.

Capture process	Advantages	Disadvantages	References
Post-combustion	Easy to retrofit into existing plants.	Capture efficiency depends on CO_2 concentration.	[14,15]
Pre-combustion	Fully developed technology and commercially accepted in many industries.	High capital and operating costs for the system.	[14, 16]
Oxyfuel-combustion	Very high CO_2 concentration that enhances absorption efficiency.	Cryogenic O_2 production is costly and corrosion problem arises.	[17]
Chemical looping combustion	CO_2 is the main combustion without mixing of N_2 and thus intensive air separation is not require.	Large scale operation is expensive and under control conditions.	[18]
Absorption	Most natural and highly efficient technology.	Absorption efficiency depends on CO_2 concentration.	[13,19]
Adsorption	Process is reversible and adsorbent can be recycled.	Higher energy cost for CO_2 desorption and reutilization of sorbent.	[19, 20,21]
Membrane separation	Separation of gases could be possible.	Operational problems such as low fluxes and fouling.	[22,23]
Hydrate based separation	Low energy requirement.	Cost and storage is an issue.	[24]
Cryogenic distillation	Adopted for many years in industry for CO_2 recovery.	High energy consumption.	[14 ,25]
Unmineable coal bed storage	CO_2 can be injected into deep coal beds to recover methane.	Many of coal have low permeability that make process not applicable.	[26,27]
Enhanced oil recovery in oil and gas reservoirs	There is an economic incentive to inject CO_2 and recovered oil.	Transportation CO_2 for storage and handling is challenging.	[14, 28]
Storage in saline aquifers	Dissolved CO_2 react with metals and prepared carbonate precipitates.	Conversion rate is less and unpredictable.	[29]

Deep ocean storage	Compare to land, the sea part is 70% bigger	Injecting large amount of CO_2 may affect the seawater chemistry and marine life.	[30,31]
In-situ carbonation	It offers economic advantages over other processes, because no chemical plant is required, as it involves simple conversion to carbonate when reacted with rock.	Limited knowledge of rock, and conversion methods are still nucleated.	[13]

2. Approaches to mitigate global climate change through CO_2 utilization

Atmospheric CO_2 utilization can be done in two ways i.e. direct use of CO_2 and conversion of CO_2 to chemicals [32]. In the past years, CO_2 has been utilized in soft drink, food, fire-extinguishers, propellant, or as a fluid/solvent in various processes like drying-cleaning, separation, packaging, etc. [33]. However, emission rates are too high and there is still as need for new areas to store CO_2 such as feedstock to biofuel [34], carbonized cement materials [35], direct utilization of CO_2 via microalgae [34], CO_2 utilization in production of some special chemical products, energy products such as methanol, Dimethyl carbonate, Dimethyl ether etc. (see fig 1) [13, 32]. However, CO_2 still has certain disadvantages as a chemical reactant due to its inert, highly stable, non-reactive, and low Gibbs free energy properties [36]. Each potential conversion of CO_2 requires energy input; therefore, the life cycle assessment of such utilization processes or concepts must be applied to ensure a neutral or even negative carbon emission.

While considering GC mechanism and growth of global fuel demands, many research groups have tried conversion of fuels to organic molecules by sunlight [37-40]. Fuels that are obtained from utilizing solar light are called solar fuel. This is not a new mechanism but rahter a natural phenomenon where, solar energy and CO_2 are captured by plants to convert complex molecules such as glucose [37], which are then used as feedstock for production of biofuels. During the process, water splits into the oxygen and hydrogen, where oxygen is available for us to breathe and hydrogen combines with carbon to produce sugar [41]. Therefore, from an energy point of view, the synthesis of organic molecules represents a potential way to store hydrogen in the form of solar energy to chemical bond [42]. In the process, multiple electrons and protons are required to produce useful products such as methane or methanol [40]. There are several ways to reduce CO_2 with use of solar energy, and these methods can be divided into four major categories namely:

1. Homogeneous photo reduction by a molecular catalysis
2. Heterogeneous photo-electrochemical reduction by a semiconducting photo cathode

3. Electrochemical reduction by an electrolyzer powered by a photovoltaic device
4. Enzymatic photo-induced electrochemical reaction

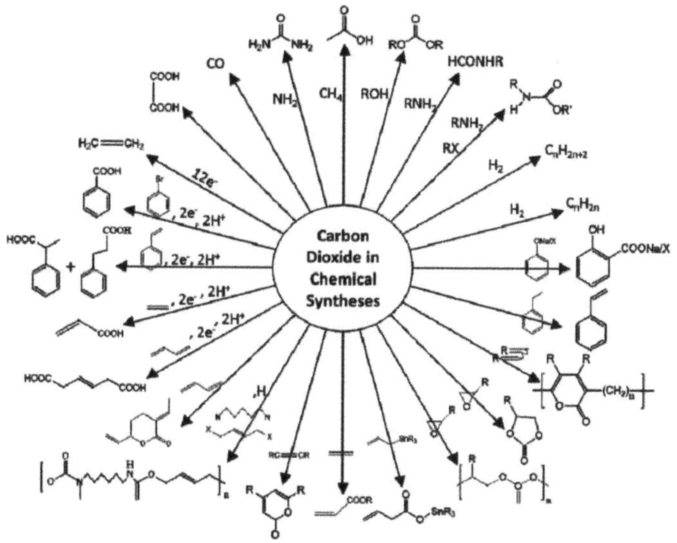

Figure 1. Utilization of CO_2 to synthesize various chemicals [Adapted Figure from 13].

2.1 Homogeneous photo reduction by a molecular catalysis with Case Study

A homogeneous CO_2 photoreduction system consists of a molecular catalyst; light absorber, sacrificial electron donor, and/or electron relay [43]. In the mechanism of photocatalytic reduction photosensitizer (P) and donor (D) are mostly involved to accomplish the reactions, where, P convert to P*(excited state) after absorbing radiation in the ultraviolet or visible region. After that, the excited state is reductively quenched by a sacrificial donor (D) generating a singly reduced photosensitizer (P^-) and oxidized donor ($D^{\cdot+}$). The P^- now transfers its electron to the catalyst to generate a reduced catalyst species. In most cases photosensitizer and catalyst have the same material with coordinated transition metal complexes. Ruthenium(II) trisbipyridine ($[Ru(bipy)_3]^{2+}$) is the most often applicable transition metal complex due to its high visible light absorption capacity and photo stability [44]. The electron transfer mechanism has been described as bellow equations.

$$P + hv \rightarrow P^*, \quad (1)$$

$$P^* + D \rightarrow P^- + D^{\cdot}, \qquad (2)$$

$$P^- + \text{catalyst} \rightarrow P + \text{catalyst}^-, \qquad (3)$$

$$\text{catalyst}^- + CO_2 \rightarrow \text{Products} + \text{catalyst}. \qquad (4)$$

Case Study: Homogeneous photocatalytic reduction of CO_2 to methanol and formate using ruthenium(II) trisphenanthroline as chromophore and pyridine as catalyst [45]. In order to produce formic acid, formaldehyde and methanol form CO_2, two, four and six electron were needed respectively to complete the reaction (Refer Eq. 5-7) [40, 46, 47]. However, multiple electron and proton transfer is necessary to produce methanol compared to formic acid and carbon monoxide (Refer Eq. 8) [43].

$$CO_2 + 2H^+ + 2e^- \rightarrow HCOOH, \qquad (5)$$

$$CO_2 + 4H^+ + 4e^- \rightarrow HCHO + H_2O, \qquad (6)$$

$$CO_2 + 6H^+ + 6e^- \rightarrow CH_3OH + H_2O, \qquad (7)$$

$$CO_2 + 2H^+ + 2e^- \rightarrow CO + H_2O. \qquad (8)$$

Thus, in the reaction mechanism, proper electron transfer is necessary which can be achieved by chromophore, pyridine and sacrificial donor system [48]. In the reaction, ascorbic acid can be used as a sacrificial reductant (electron supply), using visible light irradiation at 470 ± 20 nm. During the process of excitation, both chromophore $[Ru(phen)_3]^{2+}$ and pyridine shows metal-to-ligand charge transfer (MLCT) in the 400-500 nm region which yield long lived MLCT states that transiently localize the electron on one of the ligands (phen) and the hole on the metal center $[Ru^{III} (phen)_2 (phen^{\cdot -})]^{2+}$ [49]. However, the ratio of pyridine to chromophore is an important factor to generate the necessary six electrons. It was found that the turnover number during the process is 0.9 electron per $[Ru(phen)_3]^{2+}$, thus, pyridine was presented at large extend (1:200) to obtained 31±3 μM after 6h of irradiation [45]. Additionally, the presence of potassium salts (KCl) can enhance the yield of formate and methanol compared to electrolyte-free solutions. This is because ion-pairing between carboxylate functions with alkali and alkali earth metal cations can stabilize the transition states involving CO_2 reduction in transition-metal complexes coordinating CO2 ligands [50, 51]. Although photocatalytic reduction of CO_2 to solar fuel production has been achieved, much more progress is needed before it can become a practical commercial process as different solvent, electron donors, photosensitizers, and light sources have been deployed by various groups [40], but conversion ratio are still under consideration. More mechanistic work must be done in order to increase the stability and rates.

2.2 Heterogeneous photoelectrochemical reduction by a semiconducting photo cathode with case study

A heterogeneous CO_2 photoreduction system consists of a p-type or n-type semiconducting electrodes, which is utilized for CO_2 reduction [52-53], and an electrolyte. The semiconducting electrodes are containing band gap and thus they are different from a metallic electrode (with no band gap). Many semiconductors have been utilized as electrodes for CO_2 reduction and they are categorized depending on the charge densities of electrons and holes [54-61]. Among them, semiconductors with donor impurities that provide electrons to the lattice are n-type semiconductors with electrons as the majority charge carrier. On the other hand, semiconductors with acceptor impurities that remove electrons from the lattice are p-type semiconductors. In which, p-type semiconducting electrodes can act as photocathodes for photo assisted CO_2 reduction and it is classified in three major classes. (1) Direct heterogeneous CO_2 reduction by a biased semiconductor photo cathode; (2) heterogeneous CO_2 reduction by metal particles on a biased semiconductor photocathode and (3) heterogeneous CO_2 reduction by a molecular catalyst attached to the semiconductor photocathode surface.

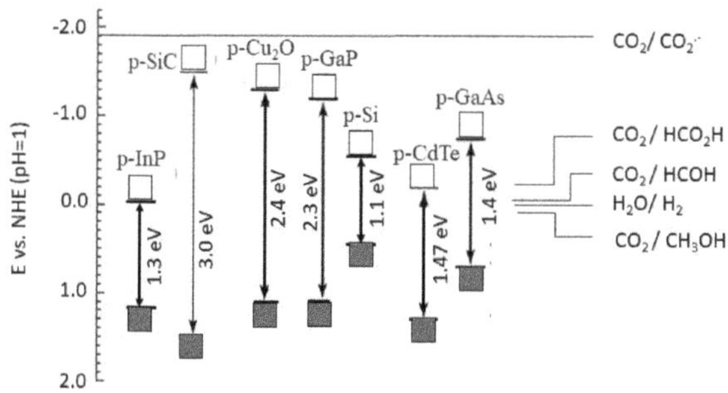

Figure 2. Conduction band (white square) and valance band (gray square) potentials of several p-type semiconductors at pH=1 versus a normal hydrogen electrode (NHE), with the potentials of several CO_2 and water redox potential of the homogeneous catalysts [54, 55, 58, 60-61].

Figure 2 shows band edge position versus an NHE for several common p-type semiconductor electrode with CO_2 reduction potential for different products at pH=1. The value of conduction band for CO_2 to $CO_2^{\cdot-}$, with single electron reduction product is much higher than other products such as HCOOH, HCHO and CH_3OH. Moreover, pH, co-catalyst and surfaces of electrodes are affecting the values of the band gaps [43]. Additionally, aqueous and non-aqueous solvents such as acetonitrile, dimethylformamide, dimethyl sulfoxide and methanol have been utilized for direct CO_2 photoelectochemical reduction [62, 63]. Among them, methanol has been utilized as solvent for CO_2 in Rectisol process due to its higher solubility [64-65]. In water hydration of CO_2 is different than solvents such as formation of bicarbonate (HCO_3^-) and carbonate (CO_3^{2-}) at pH 7-8 and >11.5, respectively [43, 66]. In order to produce solar fuel, high pressure reduction of CO_2 on p type semiconducting photocathode p-GaAs and p-GaP was performed to produce HCOOH or HCHO [67, 68]. However CH_3OH selectivity is still a challenge, it is obtained by increase concentration of carbonic acid, higher pressure and presence of Na_2CO_3, Na_2SO_4 and $HClO_4$ with water medium [67, 69]. The main product for CO_2 reduction in most non-aqueous solvents on these semiconductor surfaces is CO with varying Faradaic efficiency, except in methanol with a p-GaAs photocathode. In that p-type semiconductors such as p-Si, p-GaP, p-InP, and p-GaAs, CdTe is notable for its low CO_2 reduction (CO) onset potential and high quantum efficiency [43, 70]. Often, metal nano particles along with photocathode can improve the performance. The Au, Ag, Pd, Cu, and Ni have been utilized with p-InP photocathode to form methanol from CO_2, whereas for particulate-Pb/p-InP, the major product is HCOOH [71].

***Case Study*:** Photo electrochemical reduction of CO_2 to methanol using a single crystal catalyzed p-GaP and p-GaAs based semiconductor electrode [67]. In the system of a high pressure irradiation cell, an anode was bright platinum, glassy carbon or carbon road and the cathode was a p-GaP single crystal. A tungsten halogen lamp or Xe lamps with a maximum output of 7 suns have been utilized in these experiments. In the experiments, CO_2 reduction to formic acid and other organic products was achieved on negatively biased p-type semiconductor electrodes, which absorb in the visible light. Highest faradaic yield of reduction products, 80% was obtained with a p-GaP cathode at a cathodic current of -1.00V with 0.5M Na_2CO_3 under 8.5 atm pressure. While (-780) to (-1050) potential, maximum conversion of CH_3OH form HCOOH with nearly 37% yield. With the increased rate of the electric charge passed and lower potential, conversion rate decreases. Under elevated CO_2 pressure, it was possible to reached 80% Faradaic yields with the presence of a buffer such as $LiHCO_3$ and $NaHCO_3$. In that case adsorption of CO_2 on the surface of the semiconductor electrode such as GaP increased. While presence of $HClO_4$ (0.1 M) and CO_2 at atmosphere pressure 3 atm, conversion of CH_3OH was 61.0 μmol. With increase electric charge and pressure to 10 atm, formation of HCOOH was increased with decreased formation of CH_3OH. Thus it can be conclude that the Faradaic yield decreases when the reaction is carried out at a more negative potential. Good selectivity has also been observed at a catalyzed p-InP photoelectrochemical cell for the production of formic acid. While, these materials have shown the highest selectivity for the production of methanol only at exceptionally high potentials. Under improved experimental conditions, near 100% faradaic efficiency of CH_3OH was obtained with p-GaP semiconductor at potentials greater than 300 mV below the standard potential of -0.52V vs SCE at pH of 5.2 [67].

2.3 Electrochemical reduction by an electrolyzer powered by photovoltaic device with case study

A photochemical solar collector (device) was built, to assess the capability of various polycrystalline semiconductors for the photo assisted reduction of carbon dioxide and water under sunlight [72]. There are only few examples where PV-powered commercial electrolyzers have been used for CO_2 reduction to energy products rather than hydrogen generation [73, 74]. The idea to power an electrolyzer by a PV device was first proposed by Bard & Fox for the water splitting electrolyzer [75]. In the electrolyzer system, generated power from a photovoltaic generator has been supplied to electrolyzer (See Fig. 3). Electrolysis involves the conversion of reactants to products using a series of redox reactions in which one species loses electrons and is oxidized on the anode while another species gains electrons and is reduced on the cathode. In general case of direct conversion by direct methanol fuel cells (DMFC), CO_2 is converted to CO without the use of an external hydrogen source and followed by the product such as methanol [76]. On the cathode side, the reactions are much more complex as CO_2 can be reduced to a variety of products depending on the number of electrons donated (Refer Eq. 9). While on the anode side, water undergoes a four electron oxidation (Refer Eq. 10). This yields an overall reaction in a CO_2 electrolyzer which is the combination of both water and CO_2 to form high value electro-fuels (Refer Eq. 11). The general equation for this reaction is shown below.

$$\text{Reduction: } CO_2 + nH^+ + ne^- \rightarrow CH_xO_y + H_2O, \tag{9}$$

$$\text{Oxidation: } 2H_2O \rightarrow O_2 + 4H^+ + 4e^-, \tag{10}$$

$$\text{Overall: } CO_2 + nH_2O \rightarrow CH_xO_y + O_2. \tag{11}$$

The conversion can be affected by voltage, current, temperature, and gas output pressure parameters. An idealized diagram of CO_2 electrolysis device, including photovoltaic compartment is shown in Fig. 3. Typically CO_2 electrolysis is performed in modified two-compartment electrochemical cell that contains both anode and cathode materials in individual chambers separated by an ion conducting membrane [77]. The membrane is crucial in separating reaction products preventing the oxidation of the cathode products on the anode during operation. A voltage is applied between both electrodes to execute reaction. The equilibrium potential is fixed for the relevant oxidation and reduction reaction to occurred.

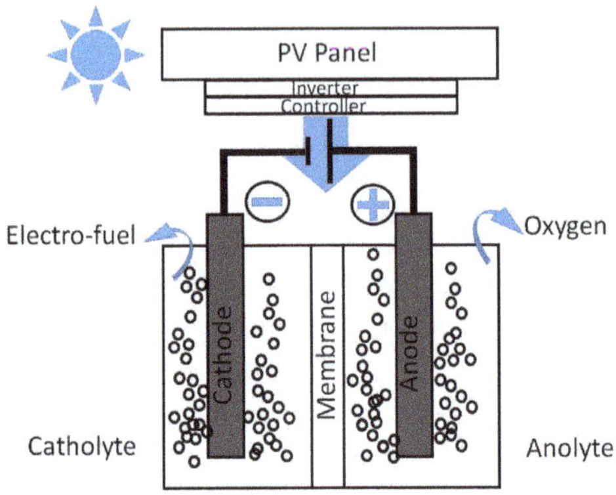

Figure 3. Schematic of CO_2 electrolyzer powered by photovoltaic device components.

Case Study: Catalytic conversion of CO and CO_2 into methanol through a photocell [78]. In the system redox reaction being composed through the reduction of CO or CO_2 into methanol and the oxidation of Everitt's salt (ES, $K_2Fe^{II}[Fe^{II}(CN)_6]$) to Prussian blue (PB, $KFe^{III}[Fe^{II}(CN)_6]$). In the electrolysis cell, n-type TiO_2 is used as anode and an ES-modified electrode (ES coated on pt plate) is used as cathode. While solar sell generating 5W power was utilized during the mechanism. The catalyst solution utilized in the system are prepared by dissolving a metal complex, methanol and 0.1M KCl. The reaction mechanism to convert methanol has been presented as per following equations (Refer Eq. 12, 13).

$$CO + 4ES + 4H^+ \leftrightarrow CH_3OH + 4PB + 4K^+, \quad (12)$$

$$CO_2 + 6ES + 6H^+ \leftrightarrow CH_3OH + 6PB + 6K^+ + H_2O. \quad (13)$$

In the reaction mechanism of equation 12 and 13, formation of methanol was obtained with transfer of 4 and 6 electron respectively. After the experiments, it was concluded that CO and CO_2 can be converted to methanol except at the cell voltage of 1.0V. The lack of formation of methanol at this voltage is ascribed the incomplete regeneration of ES in reaction. Meanwhile, ferrous complex with nitroso-R salt and tiron give high current efficiency of the methanol formation compared to cobalt(II)-tiron and iron(II)-phthalic acid complexes leading to no reduction of CO_2. Hence, the validity of a metal complex as the homogeneous catalyst is closely concerned with the coordination chemistry of the metal complex in the solution [79].

2.4 Enzymatic photoinduced electrochemical reaction with case study

Among the existing methods, the enzymatic method, offers a green and potent alternative for efficient CO_2 conversion due to its superior stereo-specificity and region/chemo-selectivity [80]. In nature, photoreaction can be seen in leafs, where glucose (organic molecule) is formed with the utilization of CO_2 [37]. In order to allow biological evolution to proceed efficiently, cells adopt six major routes (including the Calvin cycle, reductive citric acid cycle, reductive acetylCoA route, 3-hydroxypropionate cycle, 3-hydroxypropionate/4-hydroxybutyrate cycle and dicarboxylate/4-hydroxybutyrate cycle) for presenting the CO_2 metabolic process [81]. In all six major routes, the CO_2 fixation/conversion reaction is particularly important, in which oxidoreductases (i.e., formate dehydrogenase (FDH), CO_2 reductase, CODH, remodeled nitrogenase, etc.), synthases and lyases (i.e., carbonic anhydrase) play crucial roles in accelerating the reaction rate as well as create appropriate physicochemical microenvironments to suppress the denaturation of enzymes [40, 82]. An oxidoreductase is an enzyme that catalyzes the transferring of electrons from one molecule (the reductant or electron donor) to another (the oxidant or electron acceptor) and a lyase is an enzyme that catalyzes the breaking of chemical bonds and then generates a new double bond or a new ring structure. Equation 14 and 15 represents the formation of formate and CO through FDH reductase. During the redox reaction, NADPH/NADP+ or NADH/NAD+ is employed as an essential cofactor [40].

$$CO_2 + 2e^- + H^+ \rightarrow HCOO^-, \quad (14)$$

$$CO_2 + 2e^- + 2H^+ \rightarrow CO + H_2O, \quad (15)$$

$$CO_2 + H_2O \rightarrow H^+ + HCO_3^-. \quad (16)$$

A similar system has been deployed to synthesize a solar fuel from CO_2 with integrating several kinds of enzymes (including the enzyme that directly fixes/converts CO_2 and other enzymes that conduct the subsequent reactions) to implement multi-enzyme reactions. In natural photosynthesis of higher green plants and oxygenic photosynthetic cyanobacteria, the photosynthesis reaction consists of two photosystems such as Photosystem I and II [37]. Photosystem I performs the photoreduction of $NADP^+$ to NADPH, whereas photosystem II produces oxygen through water oxidation. The photosystem II in the photosynthesis protein acts as an oxidation catalyst of water [83, 84]. Through combining both reaction, conversion of an artificial leaf can be achievable.

In ordered to convert the ethanol end product, multi enzyme system such as FDH, aldehyde dehydrogenase (AlDH) and alcohol dehydrogenase (ADH) have been utilized (Fig. 4). However, equation 16 represent the lyase reaction. In which, CO_2 convert to bicarbonate through changing structure/elimination. Carbonic anhydrase, existing in a mammals, plants, algaes and bacteria is a typical lyase, which is mainly responsible for the (inter) conversion between CO_2 and bicarbonate to maintain the acid–base balance in blood and other tissues.

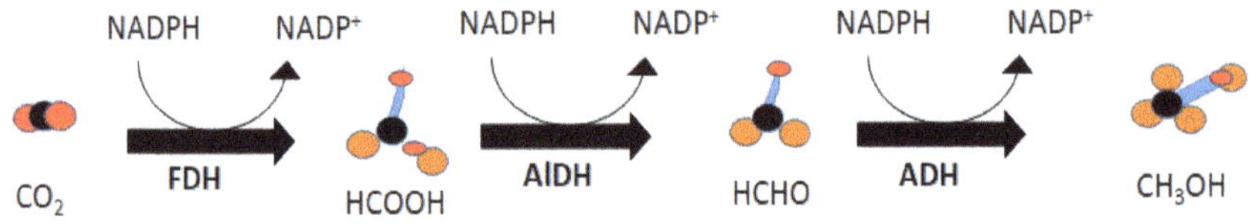

Fig. 4. CO_2 reduction to methanol in water promoted by FDH, AlDH and ADH.

Case Study: Photoinduced enzymatic conversion of CO_2 gas to solar fuel on cellulose nanofiber film [40]. In the multi-enzymatic system (FDH, AlDH and ADH induced on fiber), porphyrin is utilized as photosensitizer and dendrimer is utilized as carrier of electron/proton. The carbon capture on 2,2,6,6-tetramethylpiperidine-1-oxyl (TEMPO) oxidase cellulose nano fiber and dendrimer was successfully developed and now the utilization through photoinduced enzymatic system was developed successfully [85-86]. In the system, tetrakis-(4-carboxyphenyl)porphyrin (TCPP) have been deployed on TEMPO oxidized cellulose nano fiber (TOCNF) film through chemical bonding along with amine-terminated fourth generation poly(amido amine)(PAMAM) dendrimer. Apart from dendrimer and TCPP, multi enzyme systems and (b-nicotinamide adenine dinucleotide 2 0-phosphate-reduced tetrasodium salt hydrate (NADPH)) have been deployed on TOCNF film. To initiate the photo reaction, laser light (488 nm) has been utilized in the system and successful stepwise conversion such as formic acid from adsorbed CO_2, formaldehyde from converted formic acid and methanol from converted formaldehyde was achieved with 50 wt%, 80 wt% and 90 wt%, respectively on NADPH/enzyme-loaded Den–TCPP–TOCNF films (See Fig. 5). During the reaction, water molecule worked as an electron donor to TCPP+ and it was oxidized to O_2 molecules. The dendrimer played the role of an electron carrier to transfer electrons from TCPP to NADPH and converted to $NADP^+$. In the reaction mechanism of Fig. 5, the photo energy gained by the photosensitizer (TCPP) was transferred to electrons, which were then transferred to the electron carrier (dendrimer) and NADPH, and the electrons from NADPH were provided for the enzymatic reaction for the conversion of CO_2.

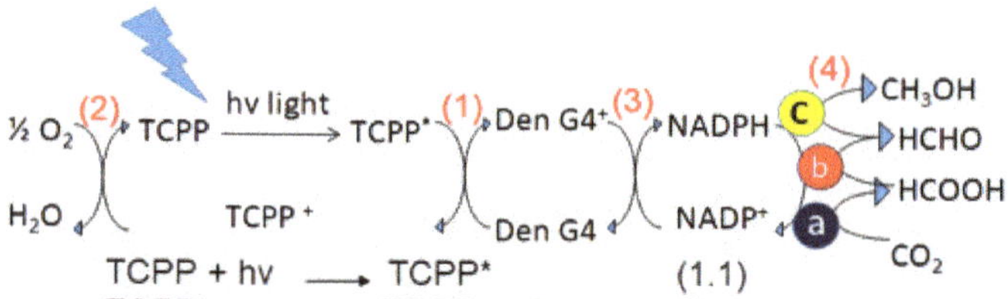

$$TCPP + h\nu \longrightarrow TCPP^{*} \quad (1.1)$$
$$TCPP^{*} \longrightarrow TCPP^{+} + 2e^{-} + H^{+} \quad (1.2)$$
$$Den\text{-}NH3^{+} + 2e^{-} + H^{+} \longrightarrow Den\text{-}NH_2 \quad (1.3)$$
$$TCPP^{*} + Den\text{-}NH_3^{+} \longrightarrow TCPP^{+} + Den\text{-}NH_2 \qquad (1)$$

$$TCPP^{+} + 2e^{-} + H^{+} \longrightarrow TCPP \quad (2.1)$$
$$TCPP^{+} + 2e^{-} + H^{+} \longrightarrow TCPP \quad (2.2)$$
$$H2O \longrightarrow \tfrac{1}{2}(o_2) + 4e^{-} + 2H^{+} \quad (2.3)$$
$$2\,TCPP^{+} + H2O \longrightarrow 2TCPP \qquad (2)$$

$$Den\text{-}NH_2 \longrightarrow Den\text{-}NH3^{+} + 2e^{-} + H^{+} \quad (3.1)$$
$$NADP^{+} + 2e^{-} + H^{+} \longrightarrow NADPH \quad (3.2)$$
$$Den\text{-}NH_2 + NADP^{+} \longrightarrow Den\text{-}NH3^{+} + NADPH \qquad (3)$$

$$1NADPH \longrightarrow NADP^{+} + 2e^{-} + H^{+} \quad (4.1)$$
$$2NADPH \longrightarrow NADP^{+} + 4e^{-} + 2H^{+} \quad (4.2)$$
$$3NADPH \longrightarrow NADP^{+} + 6e^{-} + 3H^{+} \quad (4.3)$$

$$2e^{-} + H^{+} + CO_2 \longrightarrow HCOOH \quad (4.1)$$
$$2e^{-} + H^{+} + HCOOH \longrightarrow HCHO \quad (4.2)$$
$$4e^{-} + 2H^{+} + CO_2 \longrightarrow HCHO \quad (4.2)$$
$$2e^{-} + H^{+} + HCHO \longrightarrow CH_3OH \quad (4.3)$$
$$6e^{-} + 3H^{+} + CO_2 \longrightarrow CH_3OH \quad (4.3) \qquad (4)$$

Fig. 5. Photoinduced enzymatic conversion and reaction mechanism of solar fuel on the NADPH/enzyme-loaded Den–TCPP–TOCNF films with immobilized a. FDH; b. AlDH and c. ADH enzymes.

3. Challenges to solar fuel productions

Virtually every approach under consideration for the photo-electrochemical reduction of CO_2 to fuels by either of ways requires catalysts to facilitate the formation with suitable yields. Development of a more efficient cathode can provide economic competitiveness of CO_2 conversion devices by enhancing the energetic efficiency, selectivity, and deactivation of cathode. The utilization of CO_2 based on chemical, photochemical and electrochemical technologies has great application potential. Thus, it is expected that integrating these routes with the enzymatic route may increase selectivity and productivity [80]. However, the integration of all technologies has challenges.

Currently, the main drawback of this process is the very low quantum yield that keeps the technology in the lab scale. Therefore, most likely, research effort will be focused the design of more efficient photoactive materials [87]. Nevertheless, there are some additional barriers to overcome such as analysis of the product, dealing with impurities, stability of materials and design of major photoreactors [88]. The detection limit of the analytical instruments are quite challenging due to formation of lower concentration of solar fuels. Recently, the optimization of the different methods of analysis has been approached by a combination of GC and HPLC measurements [89]. In which, CO_2 is a serious drawback for determining the composition of the products in the gas phase. Recently, many claims for high rate of conversions are suspicious of being affected by undetected impurities. This is because of the lower conversion rate making big differences in the product yields. This is a major reason to follow the GC way to obtain atom economy rather than just focus on yield numbers [90]. The stability of materials is obviously a crucial parameter which affects the engineering aspects but also the economy of the processes. Most of the case degradation/deformation was found which resulted to decrease the conversion ratio. Especially for the case of enzymes, stability for few hours makes big difference for mass utilization [80]. This is a paramount aspect for the practical development of this technology because any progress in enhancing the efficiency may render unsuccessful if the photo catalysts remain active for only a few hours. Finally the design of photoreactors is important to mitigate conversion related challenges. Cloudy days influence the intensity and continuity of the solar lights resulting in big differences in product yield. In addition, considering the limited rates of photocatalytic CO_2 reduction, recirculation of the unconverted products will be most likely required and accordingly should be considered in the photoreactor design. All these aspects have been important implications for both the economy and efficiency of the overall process.

4. Future perspectives

Photo-electrochemical conversion for CO_2 capture, sequestration and utilization provides a green and promising approach to reduce global warming and climate change [91]. Among the existing approaches usually consist of complicated reaction processes or expensive additives such as a noble metals. After the use of noble metals, bio-degradability, reusability and cost effectiveness are challengeable [92] which could be solved by bio-degradable products such as enzymes and bio-polymers. Among all methods, enzymatic conversation is successfully utilized for solar fuel production.

However, significant scientific and technical advances are still required for the large-scale utilization of CO_2 for solar fuel productions. Thus, considerably more research effort will be stimulated in attempts to discover new low-cost and low-energy input approaches for the regeneration and reuse of enzyme materials [80]. Meanwhile, the utilization of CO_2 based on chemical, photochemical and electrochemical technologies has great application potential. Thus, it is expected that integrating these routes with the enzymatic route may increase selectivity and productivity.

Based on the literature survey, the aspects which could help the improvement of conversions are cost effectiveness and efficiency of products to viable applicable in industrial productions. Developing an 'artificial leaf' that collects energy in the same way as a natural leaf is potentially the solution to the problems of sustainability of energy [93-96]. To avoid intermittency of solar energy, it is necessary to design systems that directly capture CO_2 and convert it into liquid solar fuels, which can be easily stored. There are several research groups namely, Joint Center on Artificial Photosynthesis (JCAP), Photosynthetic Antenna Research Center (PARC), Argonne-Northwestern Solar Energy Research (ANSER) Center, The [French] Alternative Energies and Atomic Energy Commission (CEA), Max Planck Gesellschaft, Water Oxidation Catalysts for Bio-Inspired Photoelectrochemical Cells, Negishi Artificial Photosynthesis Group and Solar Fuels Laboratory Nangang, working on artificial photosynthesis developments [97]. Although it has been explored for many years, much more effort should still be devoted to excavating facile and low-energy routes for CO_2 conversion by the use of cost-effective technologies.

Acknowledgement

Authors appreciate the Ministry of Science and Technology (MOST) of Taiwan (R.O.C.) under grant number MOST 106-3113-E-007-002 for the financial support.

References

[1] P.T. Anastas, M.M. Kirchhoff, Origins, current status, and future challenges of green chemistry, Acc. Chem. Res., 35 (2002) 686–694. https://doi.org/10.1021/ar010065m

[2] P. Anastas, N. Eghbali, Green Chemistry: Principles and Practice, (2009) 301–312.

[3] D. Patel, S. Kellici, B. Saha, Green Process Engineering as the Key to Future Processes, Processes, 2 (2014) 311–332. https://doi.org/10.3390/pr2010311

[4] I.T. Horvth, P.T. Anastas, Innovations and Green Chemistry, Chem. Rev., 107 (2007) 2169–2173. https://doi.org/10.1021/cr078380v

[5] T. V. Ramachandra, B.H. Aithal, K. Sreejith, GHG footprint of major cities in India, Renew. Sustain. Energy Rev., 44 (2015) 473–495. https://doi.org/10.1016/j.rser.2014.12.036

[6] K. Riahi, D. P. Van Vuuren, E. Kriegler, J. Edmonds, B. C. O'neill, S. Fujimori, W. Lutz, The shared socioeconomic pathways and their energy, land use, and greenhouse gas emissions implications: an overview, Global Environmental Change, 42 (2017) 153-168. https://doi.org/10.1016/j.gloenvcha.2016.05.009

[7] F.A. Rahman, M.M.A. Aziz, R. Saidur, W.A.W.A. Bakar, M.R. Hainin, R. Putrajaya, N.A. Hassan, Pollution to solution: Capture and sequestration of carbon dioxide (CO_2) and its utilization as a renewable energy source for a sustainable future, Renew. Sustain. Energy Rev., 71 (2017) 112–126. https://doi.org/10.1016/j.rser.2017.01.011

[8] M. Meinshausen, N. Meinshausen, W. Hare, S.C.B. Raper, K. Frieler, R. Knutti, D.J. Frame, M.R. Allen, Greenhouse-gas emission targets for limiting global warming to 2 degrees C., Nature, 458 (2009) 1158–1162. https://doi.org/10.1038/nature08017

[9] J. Rogelj, J. Nabel, C. Chen, W. Hare, K. Markmann, M. Meinshausen, M. Schaeffer, K. Macey, N. Höhne, Copenhagen Accord pledges are paltry, Nature, 464 (2010) 1126-1128. https://doi.org/10.1038/4641126a

[10] L.C. Lau, K.T. Lee, A.R. Mohamed, Global warming mitigation and renewable energy policy development from the Kyoto Protocol to the Copenhagen Accord - A comment, Renew. Sustain. Energy Rev. 16 (2012) 5280–5284. https://doi.org/10.1016/j.rser.2012.04.006

[11] J. Rogelj, M. Den Elzen, T. Fransen, H. Fekete, H. Winkler, R. Schaeffer, F. Sha, K. Riahi, M. Meinshausen, Perspective : Paris Agreement climate proposals need boost to keep warming well below 2° C, Nat. Clim. Chang. 534 (2016) 631–639.

[12] J. Gibbins, H. Chalmers, Carbon capture and storage, Energy Policy. 36 (2008) 4317–4322. https://doi.org/10.1016/j.enpol.2008.09.058

[13] P. Li, S.Y. Pan, S. Pei, Y.J. Lin, P.C. Chiang, Challenges and perspectives on carbon fixation and utilization technologies: An overview, Aerosol Air Qual. Res. 16 (2016) 1327–1344. https://doi.org/10.4209/aaqr.2015.12.0698

[14] J.D. Figueroa, T. Fout, S. Plasynski, H. McIlvried, R.D. Srivastava, Advances in CO_2 capture technology-The U.S. Department of Energy's Carbon Sequestration Program, Int. J. Greenh. Gas Control. 2 (2008) 9–20. https://doi.org/10.1016/S1750-5836(07)00094-1

[15] D.Y.C. Leung, G. Caramanna, M.M. Maroto-Valer, An overview of current status of carbon dioxide capture and storage technologies, Renew. Sustain. Energy Rev. 39 (2014) 426–443. https://doi.org/10.1016/j.rser.2014.07.093

[16] M. Kanniche, R. Gros-Bonnivard, P. Jaud, J. Valle-Marcos, J.M. Amann, C. Bouallou, Pre-combustion, post-combustion and oxy-combustion in thermal power plant for CO2 capture, Appl. Therm. Eng. 30 (2010) 53–62. https://doi.org/10.1016/j.applthermaleng.2009.05.005

[17] M.K. Mondal, H.K. Balsora, P. Varshney, Progress and trends in CO_2 capture/separation technologies: A review, Energy. 46 (2012) 431–441. https://doi.org/10.1016/j.energy.2012.08.006

[18] M.M. Hossain, H.I. de Lasa, Chemical-looping combustion (CLC) for inherent CO_2 separations-a review, Chem. Eng. Sci. 63 (2008) 4433–4451. https://doi.org/10.1016/j.ces.2008.05.028

[19] C.-H. Yu, A Review of CO_2 Capture by Absorption and Adsorption, Aerosol Air Qual. Res. (2012) 745–769. https://doi.org/10.4209/aaqr.2012.05.0132

[20] Z. Yong, V. Mata, A.E. Rodrigues, Adsorption of carbon dioxide at high temperature—a review, Sep. Purif. Technol. 26 (2002) 195–205. https://doi.org/10.1016/S1383-5866(01)00165-4

[21] H. Yang, Z. Xu, M. Fan, R. Gupta, R.B. Slimane, A.E. Bland, I. Wright, Progress in carbon dioxide separation and capture: a review., J. Environ. Sci. (China). 20 (2008) 14–27. https://doi.org/10.1016/S1001-0742(08)60002-9

[22] D. Aaron, C. Tsouris, Separation of CO2 from Flue Gas: A Review, Sep. Sci. Technol. 40 (2005) 321–348. https://doi.org/10.1081/SS-200042244

[23] P. Bernardo, E. Drioli, G. Golemme, Membrane gas separation: 1 review of state of the art, Ind. Chem. Eng. 48 (2009) 4638–63. https://doi.org/10.1021/ie8019032

[24] P. Babu, P. Linga, R. Kumar, P. Englezos, A review of the hydrate based gas separation (HBGS) process forcarbon dioxide pre-combustion capture, Energy. 85 (2015) 261–279. https://doi.org/10.1016/j.energy.2015.03.103

[25] K. Maqsood, A. Mullick, A. Ali, K. Kargupta, S. Ganguly, Cryogenic carbon dioxide separation from natural gas: A review based on conventional and novel

emerging technologies, Rev. Chem. Eng. 30 (2014) 453–477. https://doi.org/10.1515/revce-2014-0009

[26] S. J. Baines, R. H. Worden, Geological storage of carbon dioxide. Geological Society, London, Special Publications, 233(2004) 1-6. https://doi.org/10.1144/GSL.SP.2004.233.01.01

[27] J.Q. Shi, S. Durucan, CO_2 storage in underground coal seams, Oil & Gas Science and Technology Rev., 60 (2005) 1–18. https://doi.org/10.2516/ogst:2005037

[28] S. Thomas, Enhanced oil recovery-an overview, Oil & Gas Science and Technology Rev.,63 (2008) 9–19. https://doi.org/10.2516/ogst:2007060

[29] M. Bentham, G. Kirby, CO_2 Storage in Saline Aquifers, Oil & Gas Science and Technology Rev. 60 (2005) 559–567. https://doi.org/10.2516/ogst:2005038

[30] B. A. Seibel, P. J. Walsh, Potential impacts of CO_2 injection on deep-sea biota. Science, 294 (2001) 319-320. https://doi.org/10.1126/science.1065301

[31] V. Anderson, S. Woodhouse, O. F. Graff, J. S. Gudmundson, Hydrates for deep ocean storage of CO_2. In Proceedings of the Fifth International Conference on Gas Hydrates,13, 16, (2005) 1135-1139.

[32] E. Alper, O. Yuksel Orhan, CO_2 utilization: Developments in conversion processes, Petroleum. 3 (2016) 109–126. https://doi.org/10.1016/j.petlm.2016.11.003

[33] C.H. Huang, C.S. Tan, A review: CO_2 utilization, Aerosol Air Qual. Res. 14 (2014) 480–499. https://doi.org/10.4209/aaqr.2013.10.0326

[34] E. C. Francisco, D. B. Neves, E. Jacob-Lopes, T. T. Franco, (2010). Microalgae as feedstock for biodiesel production: carbon dioxide sequestration, lipid production and biofuel quality." Journal of Chemical Technology and Biotechnology 85 no. 3 (2010) 395-403. https://doi.org/10.1002/jctb.2338

[35] S.-Y. Pan, K.J. Shah, Y.-H. Chen, M.-H. Wang, P.-C. Chiang, Deployment of Accelerated Carbonation Using Alkaline Solid Wastes for Carbon Mineralization and Utilization Toward a Circular Economy, ACS Sustain. Chem. Eng. (2017) acssuschemeng.7b00291. doi:10.1021/acssuschemeng.7b00291.

[36] X. Xiaoding, J.A. Moulijn, Mitigation of CO_2 by Chemical Conversion: Plausible Reactions and Promising Products, Energy & Fuels. 10 (1996) 305–325. https://doi.org/10.1021/ef9501511

[37] J. Barber, Photosynthetic energy conversion: natural and artificial, Chem. Soc. Rev. 38 (2009) 185–96. https://doi.org/10.1039/B802262N

[38] A. Listorti, J. Durrant, J. Barber, Artificial photosynthesis: Solar to fuel, Nat. Mater. 8 (2009) 929–930. https://doi.org/10.1038/nmat2578

[39] H.B. Gray, Powering the planet with solar fuel, Nat. Chem. 1 (2009) 7. https://doi.org/10.1038/nchem.141

[40] K.J. Shah, T. Imae, Photoinduced enzymatic conversion of CO $_2$ gas to solar fuel on functional cellulose nanofiber films, J. Mater. Chem. A. 5 (2017) 9691–9701. https://doi.org/10.1039/C7TA01861D

[41] J. Barber, P.D. Tran, From natural to artificial photosynthesis., J. R. Soc. Interface. 10 (2013) 20120984. https://doi.org/10.1098/rsif.2012.0984

[42] Alberts, B., Johnson, A., Lewis, J., Raff, M., Roberts, K., & Walter, P. (2002). Catalysis and the Use of Energy by Cells.

[43] B. Kumar, M. Llorente, J. Froehlich, T. Dang, A. Sathrum, C.P. Kubiak, Photochemical and photoelectrochemical reduction of CO_2., Annu. Rev. Phys. Chem. 63 (2012) 541–69. https://doi.org/10.1146/annurev-physchem-032511-143759

[44] J.I. Goldsmith, W.R. Hudson, M.S. Lowry, T.H. Anderson, S. Bernhard, Discovery and high-throughput screening of heteroleptic iridium complexes for photoinduced hydrogen production, J. Am. Chem. Soc. 127 (2005) 7502–7510. https://doi.org/10.1021/ja0427101

[45] D.J. Boston, C. Xu, D.W. Armstrong, F.M. Macdonnell, Photochemical reduction of carbon dioxide to methanol and formate in a homogeneous system with pyridinium catalysts, J. Am. Chem. Soc. 135 (2013) 16252–16255. https://doi.org/10.1021/ja406074w

[46] E.B. Cole, P.S. Lakkaraju, D.M. Rampulla, A.J. Morris, E. Abelev, A.B. Bocarsly, Using a One Electron Shuttle for the Multi- electron Reduction of CO 2 to Methanol, J. Amercian Chem. Soc. (2010) 11539–11551.

[47] K. Ma, O. Yehezkeli, E. Park, J.N. Cha, Enzyme Mediated Increase in Methanol Production from Photoelectrochemical Cells and CO2, ACS Catal. 6 (2016) 6982–6986. https://doi.org/10.1021/acscatal.6b02524

[48] G. Seshadri, C. Lin, A.B. Bocarsly, A new homogeneous electrocatalyst for the reduction of carbon dioxide to methanol at low overpotential, J. Electroanal. Chem. 372 (1994) 145–150. https://doi.org/10.1016/0022-0728(94)03300-5

[49] J. A. Turner, A Realizable Renewable Energy Future, Sci., 285 (1999) 687–689. https://doi.org/10.1126/science.285.5428.687

[50] L. M. Teesch, J. Adams, Intrinsic interactions between alkaline earth metal ions and peptides: a gas-phase study, J. Am. Chem. Soc. (1990) 4110–4120. https://doi.org/10.1021/ja00167a003

[51] Y. Matsubara, D.C. Grills, Y. Kuwahara, Thermodynamic Aspects of Electrocatalytic CO2 Reduction in Acetonitrile and with an Ionic Liquid as Solvent or Electrolyte, ACS Catal. 5 (2015) 6440–6452. https://doi.org/10.1021/acscatal.5b00656

[52] C. Janáky, D. Hursán, B. Endrődi, W. Chanmanee, D. Roy, D. Liu, N.R. de Tacconi, B.H. Dennis, K. Rajeshwar, Electro- and Photoreduction of Carbon Dioxide: The Twain Shall Meet at Copper Oxide/Copper Interfaces, ACS Energy Lett. 1 (2016) 332–338. https://doi.org/10.1021/acsenergylett.6b00078

[53] D.H. Apaydin, E. Tordin, E. Portenkirchner, G. Aufischer, S. Schlager, M. Weichselbaumer, K. Oppelt, N.S. Sariciftci, Photoelectrochemical Reduction of CO 2 Using Third-Generation Conjugated Polymers, ChemistrySelect. 1 (2016) 1156–1162. https://doi.org/10.1002/slct.201600326

[54] A.J. Nozik, Photoelectrochemistr Y : Conversion, Ann. Rev. Phys. Chem. 29 (1978) 189–222. https://doi.org/10.1146/annurev.pc.29.100178.001201

[55] H. Ishida, K. Tanaka, T. Tanaka, Electrochemical CO_2 reduction catalyzed by ruthenium complexes [Ru(bpy)2(CO)2]2+ and [Ru(bpy)2(CO)Cl]+. Effect of pH on the formation of CO and HCOO-, Organometallics. 6 (1987) 181–186. https://doi.org/10.1021/om00144a033

[56] Y. Nakamura, R. Hinogami, S. Yae, Y. Nakato, Photoelectrochemical reduction of CO_2 at a metal-particle modified p-Si electrode in non-aqueous solutions, Adv. Chem. Conversions Mitigating Carbon Dioxide. 114 (1998) 565–568. https://doi.org/10.1016/S0167-2991(98)80822-7

[57] R. Hinogami, Y. Nakamura, S. Yae, Y. Nakato, An Approach to Ideal Semiconductor Electrodes for Efficient Photoelectrochemical Reduction of Carbon Dioxide by Modification with Small Metal Particles, J. Phys. Chem. B. 102 (1998) 974–980. https://doi.org/10.1021/jp972663h

[58] E.E. Barton, D.M. Rampulla, A.B. Bocarsly, Selective Solar-Driven Reduction of CO_2 to Methanol Using a Catalyzed, J. Am. Chem. Soc. 130 (2008) 6342–6344. https://doi.org/10.1021/ja0776327

[59] J.M. Saveant, Molecular catalysis of electrochemical reactions. Mechanistic aspects, Chem. Rev. 108 (2008) 2348–2378. https://doi.org/10.1021/cr068079z

[60] M.G. Walter, E.L. Warren, J.R. McKone, S.W. Boettcher, Q. Mi, E.A. Santori, N.S. Lewis, Solar Water Splitting Cells, Pubs.Acs.Org. (2010) 6446–6473.

[61] J.M. Smieja, C.P. Kubiak, Re(bipy-tBu)$(CO)_3$Cl-improved catalytic activity for reduction of carbon dioxide: IR-spectroelectrochemical and mechanistic studies, Inorg. Chem. 49 (2010) 9283–9289. https://doi.org/10.1021/ic1008363

[62] N.S. Spinner, J.A. Vega, W.E. Mustain, Recent progress in the electrochemical conversion and utilization of CO_2, Catal. Sci. Technol. Catal. Sci. Technol. (2012) 19–28.

[63] R. Kuriki, O. Ishitani, K. Maeda, Unique Solvent Effects on Visible-Light CO_2 Reduction over Ruthenium(II)-Complex/Carbon Nitride Hybrid Photocatalysts, ACS Appl. Mater. Interfaces. 8 (2016) 6011–6018. https://doi.org/10.1021/acsami.5b11836

[64] S. Kaneco, H. Katsumata, T. Suzuki, K. Ohta, Photoelectrochemical reduction of carbon dioxide at p-type gallium arsenide and p-type indium phosphide electrodes in methanol, Chem. Eng. J. 116 (2006) 227–231. https://doi.org/10.1016/j.cej.2005.12.014

[65] P. Luis, Use of monoethanolamine (MEA) for CO_2 capture in a global scenario: Consequences and alternatives, Desalination. 380 (2016) 93–99. https://doi.org/10.1016/j.desal.2015.08.004

[66] M. Mikkelsen, M. Jørgensen, F.C. Krebs, The Teraton Challenge. A Review of Fixation and Transformation of Carbon Dioxide, Energy Environ. Sci. 3 (2010) 43–81. https://doi.org/10.1039/B912904A

[67] B. Aurian-Blajeni, M. Halmann, J. Manassen, Electrochemical measurement on the photoelectrochemical reduction of aqueous carbon dioxide on p-Gallium phosphide and p-Gallium arsenide semiconductor electrodes, Sol. Energy Mater. 8 (1983) 425–440. https://doi.org/10.1016/0165-1633(83)90007-2

[68] J. A. Herron, J. Kim, A. A. Upadhye, , G. W. Huber, C. T. Maravelias, A general framework for the assessment of solar fuel technologies." Energy & Environmental Science 8, no. 1 (2015): 126-157. https://doi.org/10.1039/C4EE01958J

[69] K. Hara, Electrocatalytic Formation of CH_4 from CO_2 on a Pt Gas Diffusion Electrode, J. Electrochem. Soc. 144 (1997) 539. https://doi.org/10.1149/1.1837445

[70] E. Kecsenovity, B. Endrődi, P.S. Tóth, Y. Zou, R.A.W. Dryfe, K. Rajeshwar, C. Janáky, Enhanced Photoelectrochemical Performance of Cuprous Oxide/Graphene Nanohybrids, J. Am. Chem. Soc. 139 (2017) 6682–6692. https://doi.org/10.1021/jacs.7b01820

[71] Z. Guo, T. Liu, W. Li, C. Zhang, D. Zhang, Z. Pang, Carbon Supported Oxide-Rich Pd-Cu Bimetallic Electrocatalysts for Ethanol Electrooxidation in Alkaline Media Enhanced by Cu/CuOx, Catalysts. 6 (2016) 62. https://doi.org/10.3390/catal6050062

[72] B. Parida, S. Iniyan, R. Goic, A review of solar photovoltaic technologies, Renew. Sustain. Energy Rev. 15 (2011) 1625–1636. https://doi.org/10.1016/j.rser.2010.11.032

[73] K. Ogura, M. Yamada, M. Nakayama, N. Endo, Electrocatalytic Reduction of CO_2 to Worthier Compounds on a Functional Dual-Film Electrode with a Solar Cell as the Energy Source, Adv. Chem. Conversions Mitigating Carbon Dioxide. 114 (1998) 207–212. https://doi.org/10.1016/S0167-2991(98)80745-3

[74] M. Halmann, M. Ulman, B. Aurian-Blajeni, Photochemical solar collector for the photoassisted reduction of aqueous carbon dioxide, Sol. Energy. 31 (1983) 429–431. https://doi.org/10.1016/0038-092X(83)90145-7

[75] A. Bard, M.A. Fox, Artificial Photosynthesis: Solar Splitting of Water to Hydrogen and Oxygen, J. Chem. Inf. Model. 53 (2013) 1689–1699.

[76] A. Goeppert, M. Czaun, J.-P. Jones, G.K. Surya Prakash, G.A. Olah, Recycling of carbon dioxide to methanol and derived products – closing the loop, Chem. Soc. Rev. 43 (2014) 7995–8048. https://doi.org/10.1039/C4CS00122B

[77] M. Rahimnejad, A. Adhami, S. Darvari, A. Zirepour, S.E. Oh, Microbial fuel cell as new technol ogy for bioelectricity generation: A review, Alexandria Eng. J. 54 (2015) 745–756. https://doi.org/10.1016/j.aej.2015.03.031

[78] K. Ogura, Catalytic Conversion of Carbon Monoxide and Carbon Dioxide into Methanol with Photocells, J. Electrochem. Soc. 134 (1987) 2749. https://doi.org/10.1149/1.2100281

[79] K. Ogura, H. Uchida, Electrocatalytic reduction of carbon dioxide to methanol. Part 5. Relationship between the ability of metal complexes to engage in homogeneous catalysis and their co-ordination chemistry. Journal of the Chemical Society, Dalton Transactions, 6 (1987), 1377-1380. https://doi.org/10.1039/dt9870001377

[80] J. Shi, Y. Jiang, Z. Jiang, X. Wang, X. Wang, S. Zhang, P. Han, C. Yang, Enzymatic conversion of carbon dioxide, Chem. Soc. Rev. Chem. Soc. Rev. 44 (2015) 5981–6000. https://doi.org/10.1039/C5CS00182J

[81] N. V. D. Long, J. Lee, K.-K. Koo, P. Luis, M. Lee, Recent Progress and Novel Applications in Enzymatic Conversion of Carbon Dioxide, Energies. 10 (2017) 473. https://doi.org/10.3390/en10040473

[82] B.R. Crable, C.M. Plugge, M.J. McInerney, A.J.M. Stams, Formate formation and formate conversion in biological fuels production., Enzyme Res. 2011 (2011) 532536. https://doi.org/10.4061/2011/532536

[83] Y. Amao, Solar fuel production based on the artificial photosynthesis system, ChemCatChem. 3 (2011) 458–474. https://doi.org/10.1002/cctc.201000293

[84] T. Noji, K. Kawakami, J.R. Shen, T. Dewa, M. Nango, N. Kamiya, S. Itoh, T. Jin, Oxygen-Evolving Porous Glass Plates Containing the Photosynthetic Photosystem II Pigment-Protein Complex, Langmuir. 32 (2016) 7796–7805. https://doi.org/10.1021/acs.langmuir.6b02106

[85] K.J. Shah, T. Imae, A. Shukla, Selective capture of CO $_2$ by poly(amido amine) dendrimer-loaded organoclays, RSC Adv. 5 (2015) 35985–35992. https://doi.org/10.1039/C5RA04904K

[86] K.J. Shah, T. Imae, Selective Gas Capture Ability of Gas-Adsorbent-Incorporated Cellulose Nanofiber Films, Biomacromolecules. 17 (2016) 1653–1661. https://doi.org/10.1021/acs.biomac.6b00065

[87] J.H. Montoya, L.C. Seitz, P. Chakthranont, A. Vojvodic, T.F. Jaramillo, J.K. Nørskov, Materials for Solar Fuels and Chemicals, Nat. Mater. 16 (2017) 70-81. https://doi.org/10.1038/nmat4778

[88] A. Víctor, D. P. Serrano, J. M. Coronado, Current Challenges of CO_2 Photocatalytic Reduction Over Semiconductors Using Sunlight. In From Molecules to Materials (171-191). Springer International Publishing (2015).

[89] J. Hong, W. Zhang, J. Ren, R. Xu, Photocatalytic reduction of CO_2: a brief review on product analysis and systematic methods, Anal. Methods. 5 (2013) 1086–1097. https://doi.org/10.1039/c2ay26270c

[90] C.-J. Li, B.M. Trost, Green chemistry for chemical synthesis., Proc. Natl. Acad. Sci. U. S. A. 105 (2008) 13197–13202. https://doi.org/10.1073/pnas.0804348105

[91] S. Bensaid, G. Centi, E. Garrone, S. Perathoner, G. Saracco, Towards artificial leaves for solar hydrogen and fuels from carbon dioxide, ChemSusChem. 5 (2012) 500–521. https://doi.org/10.1002/cssc.201100661

[92] S. Lata, P.K. Singh, S.R. Samadder, Regeneration of adsorbents and recovery of heavy metals: a review, Int. J. Environ. Sci. Technol. 12 (2015) 1461–1478. https://doi.org/10.1007/s13762-014-0714-9

[93] J. H. Alstrum-Acevedo, M. K. Brennaman, T. J. Meyer, Chemical approaches to artificial photosynthesis. 2. Inorganic Chemistry, 44,20 (2005), 6802-6827. https://doi.org/10.1021/ic050904r

[94] D. Gust, T.A. Moore, A.L. Moore, Solar fuels via artificial photosynthesis, Acc.Chem.Res. 42 (2009) 1890–1898. https://doi.org/10.1021/ar900209b

[95] D.G. Nocera, The artificial leaf, Acc. Chem. Res. 45 (2012) 767–776. https://doi.org/10.1021/ar2003013

[96] Y. Tachibana, L. Vayssieres, J. R. Durrant, Artificial photosynthesis for solar water-splitting. Nature Photonics, 6,8 (2012) 511-518. https://doi.org/10.1038/nphoton.2012.175

[97] T.A. Faunce, Future Perspectives on Solar Fuels, Mol. Sol. Fuels. (2012) 506–528.

Chapter 8

'Surface-Modification' and 'Composite-Engineering' of Metal Chalcogenide Electrodes for Solar Hydrogen Production

Alka Pareek[1], Pramod H. Borse[1*]

[1]International Advanced Research Centre for Powder Metallurgy and New Materials (ARC International), Balapur PO, Hyderabad, Telangana, 500 005, India

Email: phborse@arci.res.in

Abstract

Solar energy is the future fuel and an apt solution for energy related concerns. Photoelectrochemical (PEC) cells can channelize solar energy directly into chemical energy and provide a useful fuel in the form of hydrogen. Exploring an efficient semiconductor material for such purpose is an essential prospect. This chapter highlights the importance of Cd chalcogenides in this technology. CdS is the most studied material for PEC research, possess perfect band gap, and band edge position as required for the desired photoanode material in the PEC cell. The efficiency of CdS photoanode can be improved specially by; (i) tuning of the electronic band structure of electrode lattice *via.* doping, and by (ii) electrode surface modification by utilizing metal-oxide nanoparticles or by loading of co-catalyst. Effect of modification of CdS photoanodes, with earth abundant transition metal hydroxide co-catalysts, on the PEC performance is reviewed.

Keywords

Solar, Hydrogen Energy, Surface Modification, Electrode, Photoelectrochemical, Chalcogenides

Contents

1. Introduction

1.1 Solar hydrogen energy by photoelectrochemical (PEC) cells

Existing and speculated energy demand for human being is a challenging issue in view of predicted scarcity of fossil fuel and related oil economy. Thus, it is necessary to shift from fossil fuel energy to sustainable renewable energy resources. Hydrogen is predicted to be one of the solutions, as it exists in most abundant resources in the universe in the form of water, biomass, natural gas, coal and oil [1, 2]. Hydrogen can be generated from renewable sources to generate energy for human utilization. Presently 95 % of hydrogen is produced from all vanishing natural resources as natural gas, those are non-renewable sources. Therefore, it is highly desirable to generate hydrogen from renewable resources those can yield clean, efficient and affordably energy for industry as well as for domestic purposes.

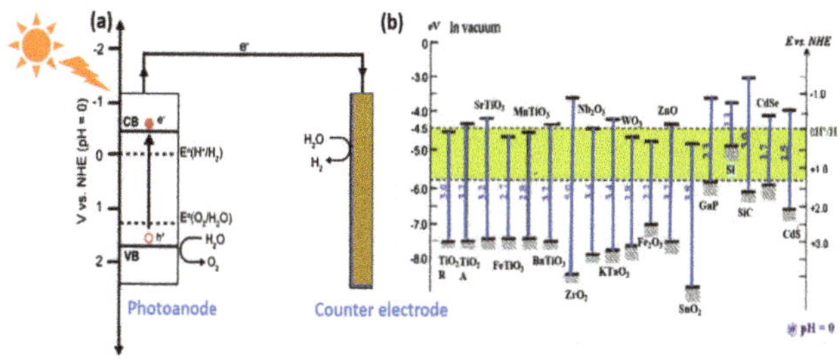

Figure 1 (a) Schematic of a PEC cell; (b) Band edge position of different known semiconductors with respect to the reduction and oxidation potential of water.

Photoelectrochemical (PEC) cell offers a promising technology to produce hydrogen that can be driven directly by solar energy utilization. PEC cell utilizes photons (light energy) to facilitate a chemical reaction for the splitting of water (H_2O) into hydrogen (H_2) and oxygen (O_2) gases [3]. Fig. 1(a) shows the principle working of a PEC cell. In a typical setup, the photoelectrochemical cell consists of a photo-anode, an electrolyte and a passive metal counter electrode. The photo-anode absorbs light and generate excitons *i.e.* electron (e^-) and hole (h^+) pair. These holes are extracted by electrolyte (responsible for oxidation) whereas electrons travel to the counter electrode where reduction occurs [4]. Generally, photons with energy greater than the semiconductor band gap can be absorbed by the semiconductor, creating electron-hole pairs, which are split by the electric field in the space-charge region between the semiconductor and the electrolyte. This process can be easily understood by the Eq. 1, 2 and 3.

When photons are incident on the photoanode material, e^- and h^+ are produced. Electrons are transferred to cathode via external circuit whereas water is oxidized at the **anode**, during the reaction [5, 6]:

$$\mathbf{H_2O + 2h^+ \rightarrow 2\ H^+ + 1/2O_2} \tag{1}$$

At the cathode**,** H^+ ions are reduced to form hydrogen gas by the following reaction:

$$\mathbf{2H^+ + 2\ e^- \rightarrow\ H_2} \tag{2}$$

Ideally, when a semiconductor is immersed in an electrolyte of a redox couple (A/A$^-$), an electron flows until the Fermi level of the semiconductor (E_F) and redox potential of electrolyte ($-qE^0(A/A^-)$) reaches equilibrium [7]. In a representative PEC cell, an *n*-type semiconductor is related to the O_2/H_2O potential; and a *p*-type semiconductor it related to the H^+/H_2 potential. In the case of an *n*-type semiconductor, the electrode will possess a positive charge due to the dopant atoms/defect states in the semiconductor, whereas the solution will have an excess negative charge. This positive charge on the semiconductor mainly appears at the depletion region, while the negative charge covers a very narrow part of the *Helmholtz layer*. Similarly, a *p*-type semiconductor induces a negative charge in depletion width whereas positive charge at the ***Helmholtz layer***.

The difference between the Fermi level of the semiconductor and the redox potential of an electrolyte leads to the band bending at the semiconductor/electrolyte interface as shown in Fig. 2. The band bending can be controlled by selection of a suitable redox couple. The region of band bending is known as depletion layer or space charge layer, which typically, has the width of around 0.1-1 μm. During operation of the PEC cell, this space charge layer assists in the separation of photogenerated charge carriers at the interface. When a photoanode is illuminated, Fermi level of semiconductor shifts to a negative potential, which results in generation of an output photovoltage in a PEC cell (ΔV in Fig. 2). The maximum photovoltage that can be achieved for a *semiconductor/electrolyte* pair is defined as the difference between the Fermi level of semiconductor and redox potential of the electrolyte [8]:

$$\mathbf{\Delta V = E_F - E_{\frac{A}{A^-}}} \tag{3}$$

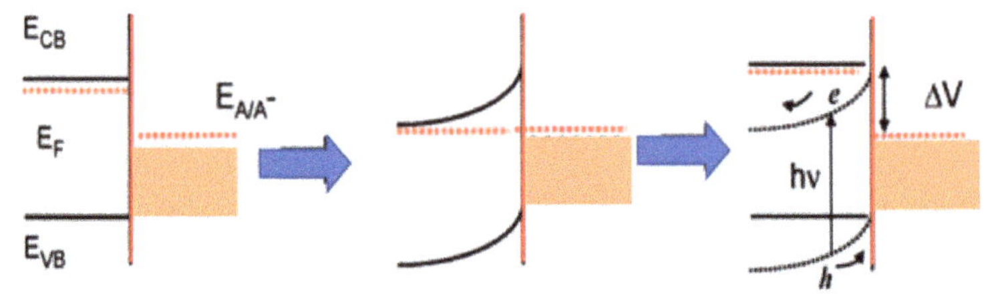

Figure 2 Schematic showing band diagram of semiconductor/electrolyte interface under dark (before and after attaining thermodynamic equilibrium) and under illuminated conditions.

The concept of PEC water splitting for hydrogen production has been known for decades and was first demonstrated in 1972 [9]; the world-record solar-to-hydrogen (STH) efficiency of 12.24 % and over 18 % of efficiency has been achieved for *p/n* type GaAs-/GAInP$_2$ and multijunction photoanode, *respectively*, until 1998 [10-11]. Recently a heterojunction PEC cell recorded 14 % STH in 2015 [12], however such high performances are always linked with poor stability or economic aspects.

1.2 Different Configuration of PEC cell

The PEC cell research involves various configurations with respect to the role of electrodes, types, geometries, type of electrolytes *etc.* Details of various configuration used in PEC Cell research are discussed here [13].

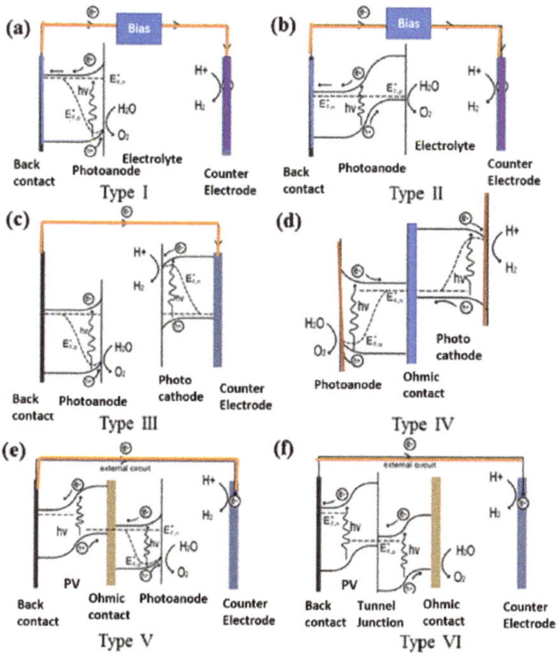

Figure 3. Band energetics in different configurations i.e. Type I – Type VI PEC cell.

(A) Type I

In this type of PEC reactor design, one electrode is kept as semiconductor and the other as a metal. The semiconductor electrode can be either a photoanode (*n*-type) or a photocathode (*p*-type). The band energetics of the Type I PEC cell is shown in Fig. 3(a) for an *n*-type semiconducting photoanode. External bias is applied in such cases to achieve efficient charge separation of the charge carriers.

(B)Type II

This configuration involves employing heterojunction electrodes. Primary semiconducting electrode (photoanode/photocathode) is modified with one or more semiconductors to improve the optical properties of the anode and for efficient charge separation at the interface. Fig. 3(b) shows band energetics of a Type II configuration PEC cell.

(C)Type III

This approach is mostly named as tandem cells where photoanode and photocathode are connected in series as shown in Fig. 3(c). The oxidation of water occurs at the photoanode, whereas reduction of water molecule occurs at the photocathode.

(D)Type IV

This configuration is a *wireless setup* of Type III configuration as shown in Fig. 3(d). Here, *electrode-hole* separation is carried out through a transparent conducting oxide substrate. It is important to note that in such configuration (where both the photoanode and photocathode are photoactive), the bandgap of the semiconductors should be such that light transmitted by one should be absorbed by the other.

(E) Type V

PV-PEC cell setup is used in this configuration. The PEC cell in such approach is biased by using a photovoltaic cell such as Si solar cells, III-V materials, hybrid perovskites and DSSC's as shown in Fig. 3(e). The benchmark efficiency of 12.4 % was achieved using this integrated system [10].

(F) Type VI

In this configuration photoelectrodes are not used instead a shielded PV junction is employed as shown in Fig. 3(f). That's why these cells are known as a buried PV cell. The highest reported Solar to Hydrogen conversion efficiency for such system is 30 % [14].

1.3 Significance of Electrode Material in PEC cell

The key aspect in a PEC cell is the choice of appropriate photoanode or photocathode material. An ideal material in a PEC cell requires appropriate electronic, optical properties. Especially it requires following characteristics:

(1) The bandgap of material should lie in between 1.23 eV to 3.0 eV, especially for solar active materials.

(2) Proper band edge positions those straddle the redox level of water so as to split water into O_2 and H_2

(3) It should be stable in electrolyte

(4) Nontoxic

(5) Cost effective

The specific requirement of a proper band gap lying in the range of 1.23 eV to 3.0 eV is important due to the fact that solar spectrum consist of 5% of UV (300-400 nm), 43 % of visible light (400-700 nm) and 52 % of infrared (700-2500 nm). Consequently, for an efficient utilization of the solar light photons the bandgap of a semiconductor should correspond to the visible light of solar spectrum. Further, since the water reduction potential lies at 0 V/NHE and oxidation potential at 1.23 V/NHE, the minimum bandgap of a semiconductor should be 1.23 eV. Therefore, desirable bandgap of a semiconductor lies in the range of 1.23 eV-3.0 eV.

The band edge position of the semiconductor is also very important parameter to predict their performance in the PEC cell. The conduction band edge position should be more negative than the reduction potential of water, whereas the valence band edge should be more positive than the oxidation potential of the water molecule (band edges of various semiconductors are shown in Fig. 1(b)). Subject to the fulfillment of band edge conditions, the photogenerated charge carriers will have sufficient energy to split water molecule. There is vigorous research going on for the identification of a suitable material for PEC application. This includes various compound semiconductors mostly covering transition metal oxides, nitrides and chalcogenides. Metal oxides, such as TiO_2 and ZnO are the most common materials used as the photoelectrodes in PEC cells due to their corrosion resistance, eco-friendly behavior and suited band edge alignments with respect to the water redox potentials [15]. However, their large bandgap (~ 3 eV) restrict their use for visible part of the solar spectrum, thus limiting the efficiency of these cells. Few small band gap metal oxides also exist, those capably utilize the visible light photon of solar spectrum *viz.* Fe_2O_3 (2.2 eV) and WO_3 (2.6 eV), but they suffer from a low conductivity, short carrier diffusion length and high over potential [16]. *Alternatively*, nitride semiconductors exhibit substantial corrosion resistance and have a bandgap ranging from 0.7 eV (InN) to 6.0 eV (AlN) that allows the bandgap engineering through alloying [17]. GaN based alloys are prime candidates for PEC cell applications. An addition of In to the GaN (E_g = 3.4 eV) lowers the bandgap thus making the $In_xGa_{1-x}N$ an excellent match to the solar spectrum, while shifting both the conduction and valence band edges closer to the redox potentials [18]. However, these semiconductors involve high fabrication costs, less stability and restricted large scale viability. Similar is the case of Si related compounds. Silicon is an earth abundant, relatively stable and suitable band gap material so it is regarded as an attractive candidate for the tandem cell photocathode [19]. But fabrication cost is too high in these semiconductors. In recent years, there has been considerable interest in the use of direct band gap II-VI binary and ternary compound semiconductors for PEC applications. The most famous semiconductors belonging to this group are Cadmium Chalcogenides (CdS, CdSe & CdTe) [20]. They comprise of the one of the most important class of semiconductors in photoelectrochemical research. Table 1 clearly shows a better PEC performance of CdSe and CdTe as compared to CdS but later shows high stability in polysulphide electrolyte as compared to the former. In addition, ease of the synthesis of CdS by low cost chemical methods is another lucrative feature [21].

Table 1.Comparision showing the output parameters of electrode made from CdX (X= Te, Se, S) in a typical PEC cell, as reported in literature.

S.No.	Cell parameters	CdTe	CdSe	CdS
1.	V_{oc} (open-circuit voltage)	433 mV	290 mV	210 mV
2.	I_{sc} (short-circuit current)	188 mA	163 μA	63 μA
3.	FF (Fill factor)	0.45	0.53	0.56
4.	η (Efficiency)	5.1%	3.4%	1.0%
5.	Stability in polysulphide solution (1 M S: 1 M NaOH : 1 M Na_2S)	Less	Moderate	High
6.	Lattice structure	HCP Wurtzite	HCP Wurtzite	FCC/HCP

1.4 Cadmium sulphide as an electrode material in PEC cell

Cadmium Sulphide (CdS) is one of the most studied photoanode materials for PEC cells since it has a suitable band gap of ~2.42 eV, high index of refraction, high absorption coefficients, long lifetimes, important optical properties, ease of fabrication and numerous device applications [22]. Different relevant properties of CdS are summarized in Table 2. It is a compound semiconductor possessing sulphur vacancies and a high electron affinity, where the lattice defects impart CdS *n*-type conductivity. Historically, the single crystalline CdS photoelectrochemical solar cell were first reported by Gerischer [23] demonstrating a power conversion efficiency of 1.5%. In the past, most of the studies reported in the literatures are on single crystals. The change in surface crystallinity has been noted as a reason for its instability and consequent low efficiency.

However, systems of commercial interest use polycrystalline thin films because of the cost considerations. Unfortunately, because of its inherent defects present in the polycrystalline materials, a considerable portion of the photogenerated charge carriers is lost due to recombination. There are numerous reports on CdS nanostructures for its application in PEC and the solar cells. Dongre *et al* [24] reported the deposition of CdS nano wire films by a simple chemical bath deposition process followed by chemical etching. The nanowires displayed the widths in the range of 50–150 nm and the lengths of the order of a few micrometers.

Table 2. Different Physical properties of CdS.

Parameters	**CdS (Wurtzite)**
Relative density	4.92
Molecular weight	144.46 (g/mol)
Lattice parameters	a = 4.136 Å, c = 6.713 Å
Direct band gap	2.42eV (300K)
Effective mass of electrons mn*/mc	0.153 - 0.171
Effective mass of holes mp*/mc	0.7 light holes,5 heavy holes
Thermal conductivity	0.20 W $K^{-1}cm^{-1}$(*c*-axis)
Dielectric constant	8.64
Reflective index	2.3 (=2μm), 2.26 (=14μm)
Electron mobility	~ 400 cm^2 $V^{-1}s^{-1}$
Hole mobility	15 $cm^2V^{-1}s^{-1}$

Nanowires with the optical band gap of 2.48 eV were reported to be more efficient in PEC cells as compared to the nano particulate electrodes. Further, Flower-like CdS film had been reported by the same group [25] using the CBD method, followed by an electroless chemical etching technique. They found that the higher temperature induced a dense flower like microstructures that uniformly covers the substrate, while lower temperature produced interconnected short nanorod type surface structures. The efficiency of the heterojunction solar cells was found to be improved by using flower-like nanostructures of CdS photoelectrode as a window material.

Gao *et al.* [26] reported the formation of branched single crystal CdS nano wires. This branched CdS nanostructure is prepared by a simple surfactant-directing method, which uses readily available reagents and provides a convenient route to obtain high-yield *single crystal branched nano wires.* These branched nano wires with average diameter of about 40 nm and length up to several micrometers, exhibited a strong emission at about 700 nm, as confirmed from photoluminescence measurements. Recently Patel *et al.* reported on chemical bath deposited CdS with highest reported I_{sc} of 99 μA/cm^2 [27]. Fig. 4(a) shows highly aligned CdS nanorods on the conducting substrate produced by a simple chemical method. Photocurrent density of 0.4 mA/cm^2 is recorded for these CdS nanorods (Fig. 4 (b)). Zyod *et al.* reported CdS film deposition by electrochemical deposition, where a highest I_{sc} of 0.18 mA/cm^2 was observed with the stability for 200 min [28]. Till now highest the I_{sc} of 224 μA and 240 μA/cm^2 are reported by Yadav *et al.* and Pareek *et al.* for the bare CdS thin film under AM 1.5 solar simulator and a maximum incident photon to current conversion efficiency (IPCE) of 5 % [29-31]. This photocurrent is still quite low to realize CdS as an efficient photoanode for photoelectrochemical cells. There are various strategies adopted to modify the properties of CdS and enhance its PEC performance, those will be discussed in a later section of this chapter.

1.5 Modified CdS electrodes in PEC cells

1.5.1 Doped CdS system

Doping can be defined as an intentional introduction of impurity atoms in the host lattice to control the fundamental properties *viz.* structural, optical, electrical and thermal properties of an intrinsic semiconductor. Doping is widely used to tailor the band gap of the semiconductor, improve their electrical conductivity, and tune their optical and magnetic properties. A dopant is a foreign element which is incorporated in the host lattice. Fig. 4(c) shows the formation of metastable states in the band gap of the semiconductor by doping of Cu metal-ion. Such metastable states modify the electronic bandgap of the semiconductor and improve their electrical properties. Various groups studied doped CdS systems to maximize the PEC performance of CdS in PEC cell. Bagdare *et al.* [32] deposited thin films of $Zn_xCd_{(1-x)}S$ with varying Zn concentration, $0.1 \leq x \leq 0.9$ on stainless steel (SS) and amorphous glass substrate by simple and convenient chemical bath deposition (CBD) technique. The blue shift in the optical transmission spectra was found with an increase in the x in $Zn_xCd_{(1-x)}S$ in the range of $0.1 \leq x \leq 0.9$. An interesting change in the film morphology was oserved, from flake to spherical particle structure, with increase in the Zn concentration. These films were successfully utilized for (PEC) cell application. The charge transfer resistance (R_{ct}) determined from Nyquist plots were found to decrease initially with an increase in the x, minimum for $x = 0.5$ and increased thereafter. Correspondingly PEC performance was also found to be synchronously dependent on the concentration of Zn. The maximum short circuit current density (J_{sc}) and open circuit voltage (V_{oc}) of 457 A/cm^2 and 389 mV were respectively found for $x = 0.5$ under 10 mW/cm^2 of illumination. Zhou *et.al.* [33] reported a chemical bath deposition of thin films of $Cd_{1-x}Zn_xS$ ($x \leq 10\%$) and their application in the CdTe solar cells. The best cell efficiency was reported to be 15.7%. The use of low zinc concentration in the $Cd_{1-x}Zn_xS$ layer and *inter-diffusion* between Cd_{1-}

$_{x}Zn_{x}S$ and CdTe layers were assumed to be the reason for achieving a high V_{oc} and FF for the solar cell.

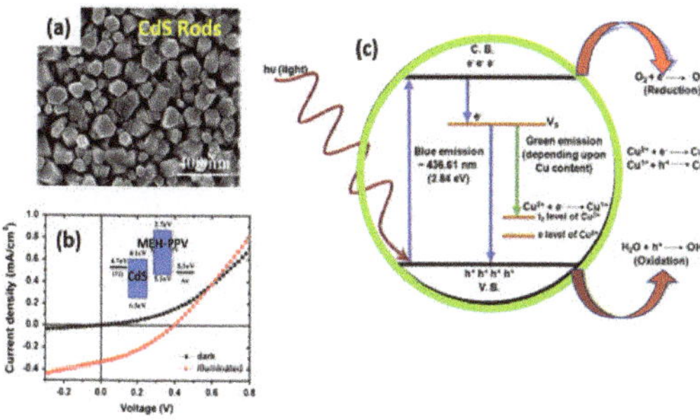

Figure 4. (a) FESEM image of CdS nanorods; (b) Schematic showing formation of meta-stable states in the forbidden gap of semiconductor by doping with copper and; (c) corresponding PEC performance.

Xia *et al.* [34] reported the high efficiencies in $SnO_2/Cd_{1-x}Zn_xS$ /CdTe cells. $Cd_{1-x}Zn_xS$ films suitable for application in photovoltaic devices have been produced by annealing CdS films in vaporous zinc chloride. The incorporation of Zn in CdS was found to be dependent on the annealing temperature and duration, reaching a concentration as high as 50% (x~0.5) and producing a blueshift of 0.3eV in the band gap energy. CdTe solar cells fabricated with $Cd_{1-x}Zn_xS$ showed substantially higher short-circuit current than the other solar cell made with CdS as the window layer. The optimal concentration of Zn was found to be around 5%.

There are quite a few reports on doped electrodes [35, 36], one of an important one is discussed here. Raviprakash *et al.* [37] reported on deposition of $Cd_xZn_{(1\_x)}S$ (x = 0, 0.2, 0.4, 0.6, 0.8, and 1) thin films by the chemical spray pyrolysis technique using a less used combination of chemicals. They observed that the inclusion of Cd into the structure of ZnS improved the crystallinity of the films. The value of lattice constant '*a*' and '*c*' have been observed to vary with composition from 0.382 to 0.415 nm and 0.625 to 0.675 nm, *respectively*. The band gap of the thin films varied from 3.32 to 2.41 eV as composition varied from x = 0.0 to 1.0. It was observed that the presence of lower amount of cadmium results in remarkable changes in the optical band gap of ZnS. The light absorption capability of semiconductor and the position of the conduction band are crucial factors influencing its photo activity. Iacomi *et al.* deposited Mn doped CdS nanocrystalline thin films [38]. They reported a *blue shift* in the band gap of CdS with Mn doping. It could be explained due to size effect of the small grains and Mn-substitution in the CdS lattice. Akintunde *et al.* studied aluminum and chlorine doping in CdS in the range of 0.0002 wt% to 0.01 wt % [39]. They observed increase in the thickness of the CdS with indium concentration which leads to decrease in optical transmittance. After Cl doping the electrical conductance and optical transmittance of Al doped CdS thin films were found to improve. Shadia *et al.* reported on In doped CdS thin film by spray pyrolysis deposition in the temperature range of 360°C to 490°C [40]. Table 3 compares the PEC performance of different doped CdS systems.

Table 3.Comparison of the PEC performance of CdS electrode doped with various metal-ions [ref. 40-43], in all case source of AM 1.5 solar simulator (P: 100 mW/cm^2) was used.

Sr.No.	Dopant	PEC Performance	Ref.
1.	Undoped	Photocurrent- 137 μA/cm^2	40
2.	Zn	Photocurrent- 817 μA/cm^2	41
4.	Cu	Photocurrent- 515 μA/cm^2	42
5.	Pb	Power efficiency-0.24 %	43

1.5.2 Metal oxide modified system

Surface modification of n-type or p-type semiconductor with metal oxides is a well-known technique to improve the efficiency and stability of semiconductors. In n-type Si solar cells Al_2O_3 is used for passivation of surface states which in turn improved the overall cell efficiency. In PEC cells, modifying anode with stable metal oxides improves photocurrent as well as imparts longevity to the anode. These metal-oxide coatings reduce the electron-hole recombination process by passivating the surface states of semiconductor and improving its performance. In some reports, it has been validated that TiO_2 modified anodes has shown a higher electron lifetime than the unmodified one. Such coatings also act as protective layer against photocorrosion of low bandgap semiconductors. An important point to be noted here is that thickness of metal oxide should be less than 2 nm so that it may allow the tunneling of charges through the transparent coating. Campet *et al.* published a research paper exclusively on the protection of photoanode against corrosion by metal oxide led surface modification [44]. They discussed on the criteria for choosing different protective coatings over semiconductors: (1) Protection by a non-conducting oxide coating (electrolyte-insulator–semiconductor structure, EIS structure): Three semiconductors were considered under this category viz. TiO_2 (3 eV), Al_2O_3 (4 eV) and Nb_2O_5. Since these are wide band gap semiconductors and valence band edge is well below the CdS so the only method for charge transfer is by tunneling mechanism. Yang *et al.* discussed various metal-oxide coatings on semiconductor electrode for improvement in efficiency and stability of low band gap semiconductors [45]. Our group worked extensively on metal-oxide (TiO_2, Nb_2O_5) modified CdS photoanodes and demonstrated the importance of using a linking agent to attach the nanoparticles over the CdS surface. The efficiency and stability of semiconductor was found to be drastically improved after modification [46-49].

In this work, in order to explore the effect of modification of CdS photoanode with metal oxide, The Nb_2O_5 and TiO_2 nanoparticles (size range~ 2-5 nm) were synthesized using hydrothermal methodology. Fig. 5(a) shows the process of making modified electrodes using a linking agent. Spray deposited CdS films were first modified with linking agent and then Nb_2O_5 nanoparticles were attached to its surface. As shown in Fig. 5(b) metal-oxide nanoparticle modified CdS electrode show remarkable improvement in the PEC performance of CdS photoanodes.

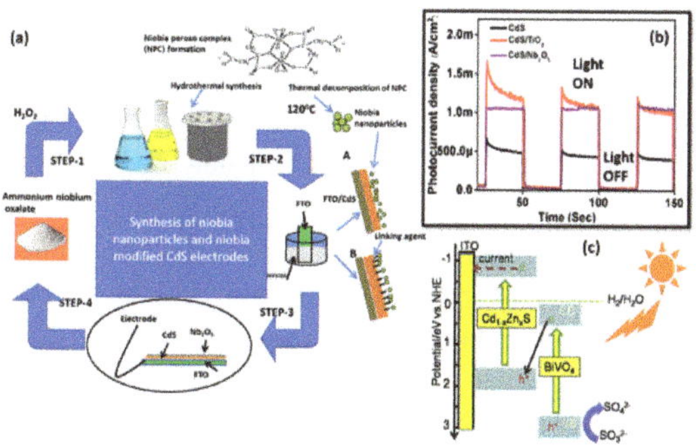

Figure 5. (a) Schematic of nanoparticle synthesis of oxide nanoparticle and further modification of CdS photoanode by nanoparticles using a specific linking agent; (b) Chronoamperometry performance of Nb_2O_5 and TiO_2 nanoparticle modified CdS films; (c) Charge transfer mechanism in metal oxide modified low band gap electrode materials.

This improvement is attributed to the passivation of the CdS surface by TiO_2 nanoparticles. Passivation of surface states reduces the recombination process and leads to improvement in output photocurrent. Jiang *et al.* explained the possible mechanism of the charge transfer when a metal oxide having conduction band edge positive than base semiconductor (Fig. 5(c)). In such cases, there is a possibility that the electrons from $BiVO_4$ conduction band goes to CdS valence band and recombine with the hole that in turn suppresses the *electron-hole* recombination in CdS [50].

1.5.3 Co-catalyst modified system

Among various methods explored to improve the performance of PEC cell, *co-catalyst* loading is another most promising way. However, most semiconductors have appropriate band edge positions required for the photoinduced redox reaction of water. But presence of overpotential acts as a barrier for hydrogen or oxygen evolution reaction. Overpotential reduces the net output photovoltage produced during the water splitting reaction. This could be easily understood by studying the energy losses in the PEC cell. The energy loss in the PEC cell can be understood by following equation [51]:

$$\Delta G_{loss} = \Delta G_{transport} + e\,(U_a + U_b + IR) \qquad (4)$$

Where, ΔG_{loss}- sum of energy losses in cell, $\Delta G_{transport}$ – energy loss due to charge transport, U_a- overpotential at anode, U_b-overpotential at cathode, *IR*- potential loss.

So, it is clear from the above reaction that in order to improve the PEC performance one needs to reduce the potential loss across the cell. The *co*-catalyst loading over the semiconductor reduces the barrier height for gas evolution reaction, which in turn lowers the overpotential at the interface and thus improves the efficiency of the cell. On the contrary, in the absence of such *co*-catalysts photo-generated holes participate in the anodic oxidation process of semiconductors to ultimately degrade it. Use of *co*-catalysts on the CdS surface, increases the rate of oxygen evolution at photoanode surface that consumes holes and thus the holes cannot participate in the photocorrosion reaction. Previously, enormous efforts have been put with respect to loading of Pt, Ru, RuO_2 and other inert-metals on the semiconductor surface to improve the efficiency and to contain the problem of the photo-corrosion in the low-band gap materials. However, these chemically *inert metals* are known to be very expensive and thus make such technology economically unviable and unfeasible. Consequently, there is a huge demand of the cost-effective *co*-catalysts those can improve the efficiency and stability of the semiconducting photoanodes in a PEC cell. It is important to note that, to reduce the activation energy for water oxidation, various earth-abundant *co*-catalysts *viz.* Co-Pi, $Co(OH)_2$, Co_2O_3, Co-Pi, $Ni(OH)_2$, NiO, $NiFeO_2$, MoS_2 have been known; to be utilized over the different semiconductors like hematite, Ta_3N_4, CdS [52-57], CdS, TiO_2, ZnO *etc.* Our group has studied effect of various transition metal (Ni, Mn, Cu, Fe, Co) hydroxides and oxide nanoparticle coating on the CdS surface [55] and studied thereupon the improvement in the PEC performance due to these systems.

2. Deposition of CdS thin films

All the above-discussed work is discussed with respect to the deposition of CdS films on transparent conducting substrates. We have dominantly utilized the nanostructured CdS films developed by one of simple and economic technique. So it becomes important to discuss the film deposition methodology at this stage. Nanostructure CdS thin films were deposited using spray pyrolysis deposition technique owing to its tremendous advantage over the other deposition processes, *viz.* scalability of process, ease of doping, moderate operation conditions as temperatures (100 $^{\circ}$C - 500 $^{\circ}$C) and cost-effectiveness with respect to equipment cost and energy consumed during deposition process. This method controllably yields the desired thickness, crystallinity, composition, possibility of multilayer deposition and possibility of its usability for variety of substrates. Preciously equal amounts of solution of $CdCl_2$ (0.1M) and $(NH_2)_2CS$ (0.1M) were dissolved in double distilled water to achieve homogenous precursor. Fluorine doped tin oxide; FTO (Pilkington TCO-15) with resistivity 12 Ωcm was used as substrate. The distance between nozzle and substrate was maintained at 23±2 cm throughout the experiment. Nanostructuring was carried out by varying the flow rate from 2 ml/min to 11 ml/min and deposition time from 1 min to 7 min. Temperature of the substrate was maintained at 350 $^{\circ}$C throughout the experiments. The photograph of as-deposited films is shown in Fig. 6(a).

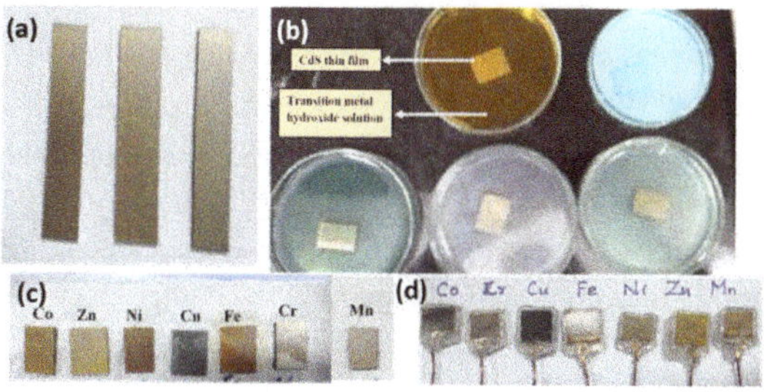

Figure 6. Photographs of (a)as deposited CdS thin films by spray pyrolysis; (b) Modification of the CdS thin film by metal hydroxides using "chemical impregnation method"; (c) Metal hydroxide modified films; (d) Metal hydroxide modified electrodes.

3. Synthesis of metal oxide and hydroxides

The modification of CdS surface by ***M***-OH (*M=Ni, Co, Mn, Fe, Cu, Zn*) nanoparticles were carried out by chemical impregnation method [54]. Accordingly, the $Ni(OH)_2$ nano-colloidal solution was prepared by addition of NaOH in 0.1 M $Ni(NO_3)_2$ aqueous solution. This yielded light green colored colloid of $Ni(OH)_2$. In case of pink colored $Co(OH)_2$ colloid and other hydroxides, similar procedure was followed. The optimal CdS films were immersed in these colloidal solutions for 1 hour, then washed with the de-ionized water, and dried carefully to obtain hydroxide modified CdS films (***M***-OH) as shown in Fig. 6(b). Fig. 6(c) and Fig. 6(d) shows metal hydroxides modified CdS films and electrodes, *respectively*. The photoelectrodes were fabricated from as deposited films. Films of area 1 cm^2 were cut from the deposited films. Silver paste is used to make an ohmic contact over the film surface. Thus, formed electrode assembly was finally sealed with a leak proof adhesive and dried.

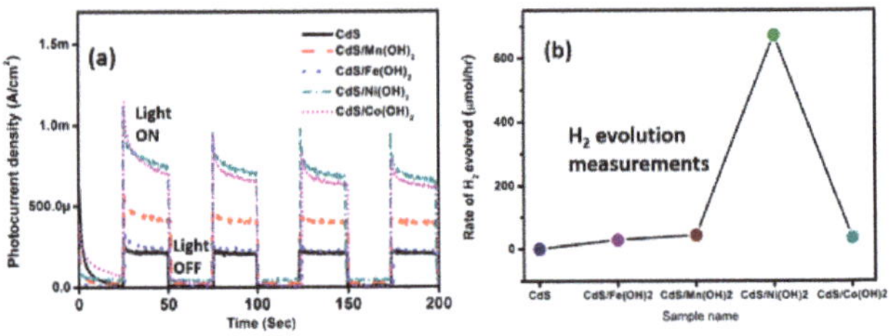

Figure 7. (a) Chronoamperometric measurements (b) hydrogen evolution measurements of hydroxide modified electrodes.

Photoelectrochemical measurements were carried out for selected stable electrodes (after 1 hour of impregnation) as shown in Fig. 7 (a). PEC measurements were carried out using two-electrode cell (or 3- electrode cell, wherever mentioned) with the photoanode as a working electrode and graphite (/platinum) as a counter electrode in a quartz PEC reactor. In case of 3-electrode cell, SCE (standard calomel electrode) was used as reference electrode. The electrolyte was made using 0.01 M Na_2S & 0.02 M Na_2SO_3 to minimize the photo-corrosion. The solar simulator with AM 1.5 filter (Newport) having irradiance of $80mW/cm^2$ was used as light source. Gas chromatography was used for hydrogen evolution measurements. Fig. 7(a) shows the chronoamperometric curves of bare and metal hydroxide modified CdS photoanodes. It can be clearly seen that the modified electrodes show better performance with respect to bare CdS. Highest photocurrent density of 680 $\mu A/cm^2$ is observed for $Ni(OH)_2$ modified CdS films that is around 3.4 times higher than bare CdS photoanodes (210 $\mu A/cm^2$). $Co(OH)_2$ modified CdS films demonstrates similar improvement in photocurrent. CdS/$Mn(OH)_2$ shows an improved photocurrent density of 0.42 mA/cm^2. Fig. 7(b) reveals the H_2 evolution rate at graphite counter electrode at 0.2 V/SCE bias. There is a drastic improvement in the catalytic properties of CdS after different hydroxide modifications. $Ni(OH)_2$ modified CdS films shows highest rate of hydrogen evolution of 600 μmol/hr. with respect to bare CdS photoanodes (1.17 μmol/hr). These results clearly validate the point that loading of *co*-catalysts improved the efficiency of CdS photoanodes in a PEC cell.

4. Summary

CdS is the most studied PEC material with optimum band gap for utilization of visible light and perfect band edge position for the splitting of water molecule. But low *charge separation* efficiency in CdS photoanodes reduces the overall efficiency for PEC cell. To improve the performance of CdS photoanodes ***three mechanisms*** are discussed. *Firstly*, doping increases the conductivity of the semiconductors and creates sub-states in the band gap of semiconductors which enhances the absorbance and reduces the recombination process. *Second* approach to improve the performance of CdS photoanode in a PEC cell is *metal-oxide nanoparticles coating*. Metal-oxide passivates the surface states and hence reduces the recombination process. Such coatings also act as a protective layer against the photocorrosion. *Thirdly*, *co*-catalyst loading on the surface of the CdS enhances the oxygen evolution reaction and hence improves the PEC performance and stability of the cell. This discussion validates the point that CdS is a promising material for PEC application but still there is a long way to realize it for practical applications.

Acknowledgement

The authors thank the support of the Director, ARCI, India.

References

[1] M. Gratzel, Photoelectrochemical cells, Nature, 414 (2001) 338. https://doi.org/10.1038/35104607

[2] Pramod H. Borse, "Hydrogen from water" *Sustainable utilization of natural resources*, Edited by P. Mondal, A.K. Dalai, Taylor & Francis Group, 6000 Broken Sound Parkway NW, Suite 300, Boca Raton, FL 33487-2742, CRC Press 2017, pp 441–457, Print ISBN: 978-1-4987-6183-3.

[3] M. Graetzel, Solar Energy Conversion by Dye-Sensitized Photovoltaic Cells, Inorg. Chem., 44 (2005) 6841. https://doi.org/10.1021/ic0508371

[4] G. Wang, Y. Ling, H. Wang, X. Lu, Y. Li, Chemically modified nanostructures for photoelectrochemical water splitting, J Photochem. Photobio. C: Photochem. Rev. 19 (2014) 35. https://doi.org/10.1016/j.jphotochemrev.2013.10.006

[5] D.W. Jing, L.J. Guo, L.A. Zhao, X.M. Zhang, H.A. Liu, M.T. Li, S.H. Shen, G.J. Liu, X.W. Hu, X.H. Zhang, K. Zhang, L.J. Ma, P.H. Guo, Efficient solar hydrogen production by photocatalytic water splitting: From fundamental study to pilot demonstration, Int. J. Hydrogen Energy 35 (2010) 7087. https://doi.org/10.1016/j.ijhydene.2010.01.030

[6] V. Lionel, Solar Hydrogen production, John Wiley & Sons (Asia) Pte Ltd, Singapore, 2009.

[7] C. Jiang, S.J.A. Moniz, A. Wang, T. Zhang and J. Tang, Photoelectrochemical devices for solar water splitting – materials and challenges, Chem. Soc. Rev. 46 (2017) 4645-4660. https://doi.org/10.1039/C6CS00306K

[8] M.G. Walter, E.L. Warren, J.R. McKone, S.W. Boettcher, Q. Mi, E.A. Santori,and N.S. Lewis, Solar Water Splitting Cells, Chem. Rev. 110 (2010) 6446–6473. https://doi.org/10.1021/cr1002326

[9] A. Fujishima, K. Honda, Electrochemical Photolysis of Water at a Semiconductor Electrode, Nature, 238 (1972) 37-38. https://doi.org/10.1038/238037a0

[10] O. Khaselev, J.A. Turner, A monolithic photovoltaic-photoelectrochemical device for hydrogen production via water splitting, Science, 280 (1998) 425-427. https://doi.org/10.1126/science.280.5362.425

[11] G. Peharz, F. Dimroth, U. Wittstadt, Solar hydrogen production by water splitting with a conversion efficiency of 18%, Int. J. Hydrogen Energy, 32 (2007) 3248-3252. https://doi.org/10.1016/j.ijhydene.2007.04.036

[12] M.M. May, H.J. Lewerenz, D. Lackner, F. Dimroth, T. Hannappel, Efficient direct solar-to-hydrogen conversion by in situ interface transformation of a tandem structure, Nat Commun., 6 (2015) 8286. https://doi.org/10.1038/ncomms9286

[13] C. Jiang, S.J.A. Moniz, A. Wang, T. Zhang and J. Tang, Photoelectrochemical devices for solar water splitting – materials and challenges, Chem. Soc. Rev. 46 (2017) 4645-4660. https://doi.org/10.1039/C6CS00306K

[14] J.Y. Jia, L.C. Seitz, J.D. Benck, Y. J. Huo, Y.S. Chen, J.W.D. Ng, T. Bilir, J.S. Harris and T.F. Jaramillo, Solar water splitting by photovoltaic-electrolysis with a solar-to hydrogen efficiency over 30%, Nat. Commun. 7 (2016) 13237. https://doi.org/10.1038/ncomms13237

[15] J.Z. Zhang, Metal oxide nanomaterials for solar hydrogen generation from photoelectrochemical water splitting, MRS BULLETIN, 36 (2011) 49-55. https://doi.org/10.1557/mrs.2010.9

[16] K. Sivula, F.L. Formal and M. Graetzel, Solar Water Splitting: Progress Using Hematite (α-Fe_2O_3) Photoelectrodes, Chem. Sus. Chem., 4 (2011) 432–449. https://doi.org/10.1002/cssc.201000416

[17] J. Wu, W. Walukiewicz, K.M. Yu, J. W. Ager III, E.E. Haller, H. Lu, W.J. Schaff, Y. Saito, and Y. Nanishi, Unusual properties of the fundamental band gap of InN, Appl. Phys. Lett. 80 (2002) 3967. https://doi.org/10.1063/1.1482786

[18] J. Wu, W. Walukiewicz, K.M. Yu, J.W. Ager III, E.E. Haller, H. Lu, W.J. Schaff, Small band gap bowing in $In_1\grave{A}_xGa_xN$ alloys, Appl. Phys. Lett, 80 (2002) 4741. https://doi.org/10.1063/1.1489481

[19] R.N. Dominey, N.S. Lewis, J.A. Bruce, D.C. Bookbinder and M.S. Wrighton, Improvement of photoelectrochemical hydrogen generation by surface modification of *p*-type silicon semiconductor photocathodes, J. Am. Chem. Soc., 104 (1982) 467–482. https://doi.org/10.1021/ja00366a016

[20] A.A. Yadav, E.U. Masumdar, Optical and electrical transport properties of spray deposited $CdS_{1-x}Se_x$ thin films, J. Alloys Compd. 505 (2010) 787. https://doi.org/10.1016/j.jallcom.2010.06.141

[21] Mark S. Wrighton, Jeffrey M. Bolts, Andrew B. Bocarsly, Michael C. Palazzotto, and Erick G. Walton, Stabilization of n-type semiconductors to photoanodic dissolution: II–VI and III–V compound semiconductors and recent results for n-type silicon, Journal of Vacuum Science and Technology 15 (1978) 1429. https://doi.org/10.1116/1.569801

[22] R.K. Pandey, Shikha Mishra, Sanjay Tiwari, P. Sahu, B.P. Chandra, Comparative study of performance of CdTe, CdSe and CdS thin "films-based photoelectrochemical solar cells, Solar Energy Materials & Solar Cells 60 (2000) 59-72. https://doi.org/10.1016/S0927-0248(99)00063-X

[23] R. Memming, Solar energy conversion by photoelectrochemical processes, Electrochim. Acta, 25 (1980) 77. https://doi.org/10.1016/0013-4686(80)80054-5

[24] H. Gerischer, J. Gobrecht, On the power characteristics of electrochemical solar cells. Ber Bunsenges. Phys. Chem. 80 (1976) 327–333. https://doi.org/10.1002/bbpc.19760800412

[25] J.K. Dongre,V. Nogriya, M. Ramrakhiani, Structural, optical and photoelectrochemical characterization of CdS nanowire synthesized by chemical bath deposition and wet chemical etching, App. Surf. Sci. 255 (2009) 6115–6120. https://doi.org/10.1016/j.apsusc.2009.01.064

[26] J.K. Dongre, M. Ramrakhiani, Synthesis of flower-like CdS nanostructured films and their application in photoelectrochemical solar cells, J. Alloys. Compds. 487 (2009) 653–658. https://doi.org/10.1016/j.jallcom.2009.08.031

[27] F. Gao, Q. Lu, Single Crystalline Cadmium Sulfide Nanowires with Branched Structure, Nanoscale Res Lett 4 (2009) 371–376. https://doi.org/10.1007/s11671-009-9256-3

[28] S.B. Patil, A.K. Singh, Effect of complexing agent on the photoelectrochemical properties of bath deposited CdS thin films, Applied Surface Science 256 (2010) 2884–2889. https://doi.org/10.1016/j.apsusc.2009.11.043

[29] Ahed Zyoud, Iyad Saadeddin, Sahar Khurduj, Mumen Marie, Zafer M. Hawash, Maryam I. Faroun, Guy Campet, Dae Hoon Park, Hikmat S. Hilal, J. Electroana. Chem. 707 (2013) 117–121. https://doi.org/10.1016/j.jelechem.2013.08.022

[30] A.A. Yadav, E.U. Masumdar, Photoelectrochemical investigations of cadmium sulphide (CdS) thin film electrodes prepared by spray pyrolysis, J. Alloys Compd. 509 (2011) 5394–5399. https://doi.org/10.1016/j.jallcom.2011.02.061

[31] A. Pareek, R. Dom, P. H. Borse, Fabrication of large area nanorod like structured CdS photoanode for solar H_2 generation using spray pyrolysis technique, Inter. J. Hydrogen Energy 38 (2013) 36-44. https://doi.org/10.1016/j.ijhydene.2012.10.057

[32] P.B. Bagdare, S.B. Patil, A.K. Singh, Phase evolution and PEC performance of ZnxCd(1-x) S nanocrystalline thin films deposited by CBD, J. Alloys Compds. 506 (2010) 120–124. https://doi.org/10.1016/j.jallcom.2010.06.152

[33] J. Zhou, X. Wu, G. Teeter, B. To, Y. Yan, R. G. Dhere, T. A. Gessert, CBD-Cd1−xZnxS thin films and their application in CdTe solar cells, Phys. Stat. Sol. (b), 241 (2004) 775–778. https://doi.org/10.1002/pssb.200304218

[34] W. Xia, J.A. Welt, H. Lin, H.N. Wu, M.H. Ho, C. W.Tang, Fabrication of Cd1−xZnxS films with controllable zinc doping using a vapor zinc chloride treatment, Sol. Energy Mater. Sol. Cells, 94 (2010) 2113-2118. https://doi.org/10.1016/j.solmat.2010.06.037

[35] F. Chen, W. Qiu, X. Chen, L. Yang, X. Jiang, M. Wang, H. Chen, Large-scale fabrication of CdS nanorod arrays on transparent conductive substrates from aqueous solutions, Sol. Energy 85 (2011) 2122–2129. https://doi.org/10.1016/j.solener.2011.05.020

[36] J. Kaur, M. Sharma, O.P. Pandey, Structural and optical studies of undoped and copper doped zinc sulphide nanoparticles for photocatalytic application, Superlat. Micro. 77 (2015) 35-53.

[37] Y. Raviprakash, K.V. Bangera, G.K Shivakumar, Preparation and characterization of Cd x Zn 1− x S thin films by spray pyrolysis technique for photovoltaic applications, Solar Energy 83 (2009) 1645-1651. https://doi.org/10.1016/j.solener.2009.06.004

[38] F. Iacomi, I. Salaoru, N. Apetroaei, A. Vasile, C.M. Teodorescu, D. Macovei, J. Optpelectr. Physical characterization of CdMnS nanocrystalline thin films grown by vacuum thermal evaporation, J. Optoelec. Adv. Mater., 8 (2006) 266-270.

[39] J.A. Akintunde, Dual impurity doping of buffer solution cadmium sulphide thin films: electrical and optical properties, J. Mater. Sci.: Mater. Electr., 11(2000) 503-508. https://doi.org/10.1023/A:1008920602583

[40] S. J. Ikhmayies, R.N. Ahmad-Bitar, A study of the optical bandgap energy and Urbach tail of spray-deposited CdS:In thin films, App. Surf. Sci. 256 (2010) 3541–3545. https://doi.org/10.1016/j.apsusc.2009.12.104

[41] A. Pareek, R. Dom and P.H. Borse, Fabrication of large area nanorod like structured CdS photoanode for solar H_2 generation using spray pyrolysis technique, Inter. J. Hydrogen Energy 38 (2013) 36 -44. https://doi.org/10.1016/j.ijhydene.2012.10.057

[42] A. Pareek, R. Thotakuri, R. Dom, H.G. Kim and P.H. Borse, Nanostructure Zn Cu co-doped CdS chalcogenide electrodes for opto-electric-power and H_2 generation, Inter. J. Hydrogen Energy 42 (2017) 125-132. https://doi.org/10.1016/j.ijhydene.2016.10.029

[43] M.A. Barote, S.S. Kamble, A.A. Yadav, R.V. Suryavanshi, L.P. Deshmukh, E.U. Mazumdar, Thickness dependence of $Cd_{0.825}$ $Pb_{0.175}$ S thin film properties, Materials Letters 78 (2012) 113–115. https://doi.org/10.1016/j.matlet.2012.03.018

[44] G. Campet, J.P. Manaud, C. Puprichitkun, Z.W. Sun, P. Salvador, Protection of photoanodes against photo-corrosion by surface deposition of oxide films: criteria for choosing the protective coating, Active and Passive Elec. Comp.13 (1989) 175-189. https://doi.org/10.1155/1989/78914

[45] R. Liu, Z. Zheng, J. Spurgeon, B. S. Brunschwig, X. Yang, Enhanced Photoelectrochemical Water-Splitting Performance of Semiconductors by Surface

Passivation Layers, Energy Environ. Sci.,7 (2014) 2504-2517. https://doi.org/10.1039/C4EE00450G

[46] A. Pareek, R. Purbia, P. Paik, N.Y. Hebalkar, H.G. Kim, P.H. Borse, Stabilizing effect in nano-titania functionalized CdS photoanode for sustained hydrogen generation Int. J. Hydrogen Energy, 39 (2014) 4170-4180. https://doi.org/10.1016/j.ijhydene.2013.12.185

[47] A. Pareek, P. Paik and P.H. Borse, Nanoniobia Modification of CdS Photoanode for an Efficient and Stable Photoelectrochemical Cell, Langmuir, , 30 (2014) 15540-15549. https://doi.org/10.1021/la503713t

[48] A. Pareek, P. Paik and P.H. Borse, Fabrication of a highly efficient and stable nano-modified photoanode for solar H_2 generation, RSC Adv. 3 (2013) 19905-19908. https://doi.org/10.1039/c3ra42632g

[49] A. Pareek, P. Paik and P.H. Borse, Characterization of Nano-Titania Modified CdS/Polysulfide Electrolyte Interface by Utilizing Mott-Schottky and Electrochemical Impedance Spectroscopy, Electroanalysis, 26 (2014) 2403 – 2407. https://doi.org/10.1002/elan.201400241

[50] A. Pareek, A. Gopalakrishnan and P.H. Borse, Efficiency and stability aspects of CdS photoanode for solar hydrogen generation technology, Journal of Physics: Conference Series 755 (2016) 012006. https://doi.org/10.1088/1742-6596/755/1/012006

[51] J. Jiang, M. Wang, R. Li, L. Ma, L. Guo, Fabricating CdS/$BiVO_4$ and $BiVO_4$/CdS heterostructured film photoelectrodes for photoelectrochemical applications, Int. J. Hydrogen Energy, 38 (2013) 13069-13076. https://doi.org/10.1016/j.ijhydene.2013.03.057

[52] T. Bak, J. Nowotny, M. Rekas, C.C. Sorrell, Photo-electrochemical hydrogen generation from water using solar energy. Materials-related aspects, Int. J. Hydrogen Energy 27 (2002) 991–1022. https://doi.org/10.1016/S0360-3199(02)00022-8

[53] G. Wang, Y. Ling, X. Lu, T. Zhai, F. Qian, Y. Tong and Y. A. Li, A mechanistic study into the catalytic effect of $Ni(OH)_2$ on hematite for photoelectrochemical water oxidation, Nanoscale 5 (2013) 4129. https://doi.org/10.1039/c3nr00569k

[54] J. Li, F. Meng, S. Suri, W. Ding, F. Huang and N. Wu, Photoelectrochemical performance enhanced by a nickel oxide–hematite *p–n* junction photoanode, Chem. Commun. 48 (2012) 8213-8215. https://doi.org/10.1039/c2cc30376k

[55] A. Pareek, P. Paik and P.H. Borse, Role of Transition Metal-Hydroxide (M-OH_x, M=Mn, Fe, Ni, Co) Co-catalyst Loading : Efficiency and Stability of CdS Photoanode, Mater. Res. Soc. Symp. Proc. Vol. 1776 (2015). https://doi.org/10.1557/opl.2015.363

[56] A. Pareek, P. Paik and P.H. Borse, Stable hydrogen generation from Ni- and Co-based co-catalysts in supported CdS PEC cell, Dalton Trans., 45 (2016)11120–11128. https://doi.org/10.1039/C6DT01277A

[57] A. Pareek, P. Paik and P.H. Borse, Ultrathin MoS_2–MoO_3 nanosheets functionalized CdS photoanodes for effective charge transfer in photoelectrochemical (PEC) cells, J. Mater. Chem. A 5 (2017) 1541-1547. https://doi.org/10.1039/C6TA09122A

Chapter 9

Enhanced Hydrogen Storage Properties of Hydrothermally Synthesized TiO_2 Nanotube-Multiwall Carbon Nanotube Nanocomposite

Manoj C. Raj [1,2], Thillai Sivakumar Natarajan [1,2], Rajesh J Tayade [1]*, Hari C. Bajaj [1,2] *

[1]Inorganic Materials and Catalysis Division, CSIR-Central Salt and Marine Chemicals Research Institute (CSIR-CSMCRI), Gijubhai Badheka Marg, Bhavnagar - 364 002, Gujarat, India.

[2]Academy of Scientific and Innovative Research (AcSIR), CSIR-Central Salt and Marine Chemicals Research Institute (CSIR-CSMCRI), Gijubhai Badheka Marg, Bhavnagar 364 002, Gujarat, India.

Fax: +91 278 2567562 / 2566970; Tel.: +91 278 2471793

Email address: hcbajaj@csmcri.res.in, tayade@csmcri.res.in,

Abstract

Titanium dioxide (TiO_2)-carbon based composite materials have gained greater attention because of their eco-friendly nature, higher adsorption capacity and enhanced photocatalytic activity. Among the TiO_2-carbon composite materials, TiO_2-multiwall carbon nanotube (TNT-MWCNT) composite materials are attractive for hydrogen storage application due to the presence of two tubular structures with high surface area. The present study focus on the hydrogen uptake studies of hydrothermally synthesized TiO_2 nanotube-multiwall carbon nanotube nanocomposite by *in-situ* addition of MWCNT. Subsequently characterized by powder X-ray diffraction (PXRD), transmission electron microscope (TEM), CHNS analysis, and nitrogen adsorption-desorption isotherm analysis. Hydrogen uptake studies revealed that 0.25 wt% MWCNT@TNT nanocomposite exhibited enhanced H_2 uptake (132 cc/g) than other composite and bare MWCNT (36 cc/g) and TNT (54 cc/g) respectively. The increment in the hydrogen uptake capacities of the composite materials was attributed to the enhancement in the surface area as well as micro pore volume by multi walled carbon nanotube incorporation.

Keywords

Multiwall Carbon Nanotube, TiO_2 Nanotube, Hydrothermal, Hydrogen Storage, Nanocomposite

Contents

1. Introduction

Depletion of fossil fuel and the escalating risk of global warming necessitated new energy concept for the future mankind. Hydrogen is a promising alternative fuel compared to hydrocarbon because of its abundance, carbon free and mass energy density. However, the storing of hydrogen is difficult, therefore conventional methods such as storing of hydrogen in compressed gas cylinders at high pressure and cryogenic storage of liquid hydrogen has been developed. However there is a concern over the conventional system like poor efficiency, safety issues and they are economically not viable. These processes has heightened the researcher interest in the development of feasible method to efficiently trap hydrogen [1,2]. Physisorption of hydrogen in porous materials at cryogenic temperature is a feasible method for hydrogen storage in mobile applications. It has the advantage of possessing complete reversibility, fast adsorption-desorption kinetics and low heat of adsorption due to weak van der Waals interactions. The US

Department of Energy (DOE) has set gravimetric and volumetric storage targets for onboard hydrogen storage for the year 2020 as 5.5 wt% of hydrogen or 40 g/L near-ambient temperatures and applicable pressure [3-5].

Since the discovery of carbon nanotubes (CNT) by Lijima et al., tremendous research efforts have been carried out in CNTs owing to their higher surface area, and its unique electronic, optical and mechanical properties [6-12] etc. Subsequently hydrogen storage capacity of CNT has been enhanced by making composite material with metal and metal oxide nanoparticles [13-15]. Among the metal oxide, TiO_2 based CNT composite material has been developed for hydrogen storage due to its unique properties. Sami-ullah et al. studied the hydrogen adsorption study on carbon nanotubes impregnated TiO_2 nanorods and nanoparticles from room temperature to 8-18 atm pressure and exhibited higher H_2 uptake than the pristine CNTs [16]. The enhanced hydrogen capacity of CNT-TiO_2 nanoparticles is explained by initial binding of hydrogen on TiO_2 and its subsequent spillover in CNT–TiO_2 nanoparticles. Further the discovery of CNT attracted the researchers to develop metal oxide nanotubular materials. Among the nanotubular structure titanium dioxide (TiO_2) nanotubes is a promising material due to its high surface area and high adsorptive capacity compared to TiO_2 nanoparticle [17-21]. Bavykin et al. synthesized multilayered titanium dioxide nanotubes and reported a hydrogen uptake of 3.8 wt% at 6 bar; revealed hydrogen is occupied at interstitial cavities between the layers in the walls of TiO_2 nanotubes [22]. Therefore the making of CNT-TNT composite materials will lead to higher hydrogen adsorption capacity, however to the best of our knowledge no reports have been reported on synthesis of CNT-TNT composite and its hydrogen storage application.

Further the optimization of the pore radius, enhanced adsorption potential due to the overlapping force fields of the opposite micropore walls could make the adsorption of hydrogen molecule more feasible even at room temperature. Thus, optimal storage can be achieved on maximization of the micropore volume [23, 24]. To retain hydrogen in the tunnels and cages of porous solids, one needs strong interactions between the walls of the adsorbent and the H_2 molecules, provided that the pores are sufficiently large to accommodate H_2 molecules. Therefore tuning of the pore size and enhancement in micropores in TNTs can be obtained by incorporation of other microporous materials such as activated multiwalled carbon nanotubes into titanium nanotube, which also have sufficient hydrogen sorption capacities.

The present work discusses the hydrogen sorption capacity of multiwall carbon nanotube-TiO_2 nanotube (MWCNT@TNT) composite materials, synthesized by hydrothermal method with in-situ addition of MWCNT. Hydrogen adsorption capacities of synthesized composite materials are performed at 77 K up to 4000 KPa.

2. Experimental

2.1 Chemicals and materials

Multi-wall carbon nanotubes (MWCNT, > 95 %, outer diameter (OD = 60-100nm), inner diameter (ID= 5-10 nm) and length = 0.5 - 500 μm) and anatase TiO_2 (AT, >99%) were purchased from Sigma Aldrich. Sodium hydroxide (NaOH), nitric acid (HNO_3), and hydrochloric acid (HCl) were purchased from S.D. Fine Chemicals Ltd, Mumbai. All the chemicals used in our experiments were of analytical grade. The double distilled water was used to prepare the experimental solutions.

2.2 Pretreatment of MWCNT

The raw MWCNT (MWCNT Bare) was suspended in concentrated HNO_3 (25 ml), sonicated and refluxed at 353 K for 20 h. After 20 h, the reaction mixture was centrifuged and continuously washed with copious amount of distilled water until the pH of the reaction mixture reached neutral. Then dried in an oven at 383 K for 12 h and powdered. Successively the powdered MWCNT was mixed with 50 % HNO_3 solution, followed by refluxing at 403 K for 24 h. Thereafter MWCNT was centrifuged and repeatedly washed with distilled water to decrease the pH to neutral, dried at 343 K for 12 h. The modified MWCNT was labeled as MWCNT Act.

2.3 Synthesis of MWCNT@TiO_2 nanotube composites

MWCNT@TiO_2 nanotube composite was prepared by hydrothermal method (Fig. 1). MWCNT@TNT composite with different percentage of MWCNT loaded (0.25, 0.5, 0.75 and 1 wt%) samples was synthesized by adding calculated amount of MWCNT into a 50 mL of 10 N NaOH solution containing 1.2 g of anatase TiO_2 (AT), sonicated and transferred to a teflon lined stainless steel autoclave. The autoclave was heated in an oil bath at 403 K for 48 h under autogenous pressure with stirring (250 rpm). After 48 h, the autoclave was cooled to room temperature and the formed MWCNT@TNT was washed with copious amount of distilled water until the pH of the filtrate was less than 7. Afterward, synthesized MWCNT@TNT was treated with 0.1 M HCl solution under stirring at room temperature for 12 h. Subsequently, the MWCNT@TNT was washed with distilled water until it was free from chloride ion (silver nitrate test). Then the synthesized MWCNT@TNT was dried at 343 K overnight and then calcined at 523 K for 2 h under N_2 atmosphere with the ramp rate of 2 °C/min. The synthesized MWCNT@TNT composite materials were denoted as 0.25wt% MWCNT@TNT, 0.5wt%MWCNT@TNT, 0.75wt%MWCNT@TNT and 1wt% MWCNT@TNT respectively. Similarly the pristine titanium dioxide nanotubes (TNT) was synthesized without addition of MWCNT and designated as TNT Bare.

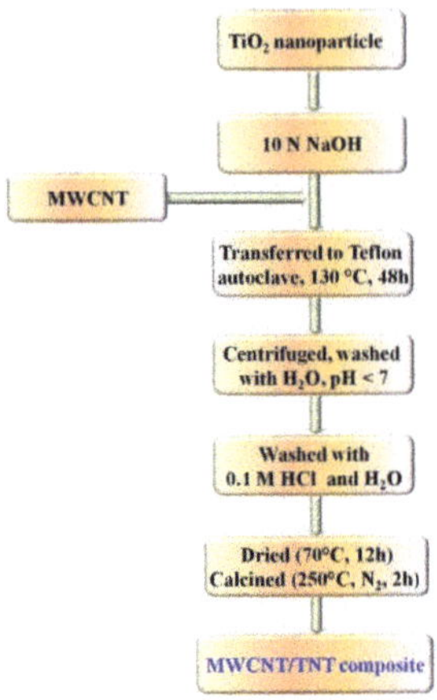

Figure 1. Flow chart of MWCNT@TNT composite synthesis.

3. Characterization

The powder X-ray diffraction (PXRD) patterns of the synthesized nanocomposite materials were recorded using PHILIPS X'pert MPD diffractometer at ambient temperature in the 2θ range of 2° - 80° with the scan speed of 0.1°sec^{-1} using CuK$_{\alpha 1}$ (λ=1.54056 Å) radiation. The elemental compositions of the sample were determined by CHNS analysis. The tubular morphologies of MWCNT, TNT and MWCNT@TNT samples were observed by transmission electron microscopy (TEM, JEOL JEM-2010 electron microscope) analysis with the operating voltage of 200 kV. Prior to analysis a small amount of the sample was dispersed in ethanol by sonication and a drop of resultant suspension was placed on a carbon coated copper (Cu) grid, dried and used for TEM analysis.

The BET surface area, Langmuir surface area, pore volume, pore width of the synthesized materials were determined in a static volumetric gas adsorption system (Micromeritics Instrument Corporation, USA, model ASAP 2020) using N_2 adsorption-desorption isotherm at 77.4 K up to 101 KPa pressure. Prior to adsorption measurements the samples were activated by heating at 373 K under vacuum (5×10^{-6} mmHg) for 4 h with the heating rate of 1 K min^{-1}. Surface area and pore diameter of composite materials were obtained using BET (Brunauer –Emmett-Teller) equation and Barett Joyner Halenda (BJH) desorption isotherm. Pore volume was obtained by single point adsorption method and micropore volume was obtained by density functional theory (DFT) using CO_2 at 273 K temperature.

Hydrogen adsorption measurements were carried out at 77.4 K up to 4000 KPa in an automated high pressure gas adsorption system (BELSORP-HP, BEL Japan, Inc.). Prior to sorption isotherm measurement, the sample was activated at 373 K under vacuum (1.2 x 10^{-2} Pa) at the heating rate of 1 K min^{-1}. After activation, the samples were allowed to cool down to room temperature. The quantity of the sample was determined from the weight difference of the sample before and after activation. Before sorption measurement the temperature and vacuum was maintained for 6 h and analysis was performed out at 77.4 K by immersing the sample in liquid nitrogen Dewar with automatic liquid nitrogen level controller. The dead volume of the material was obtained using helium. The errors in the measurements of hydrogen uptake values were within the range of $\pm$ 0.5%.

4. Results and discussion

4.1 Powder X-ray diffraction analysis

The PXRD patterns of MWCNT Act, TNT Bare, and MWCNT@TNT composite materials are shown in Fig. 2. PXRD results reveal that MWCNT incorporated TNT samples showed the same diffraction patterns as of TNT Bare (Fig. 2). This confirms that the structure of TNT was not affected by MWCNT incorporation. As the percentage loading of MWCNT increases the composite materials loses its crystallinity. All the PXRD peak of TNT Bare and MWCNT@TNT composite material at 2θ values of 25.3°, 36.6°, 48.0°, 54.5°, 62.6°, 68.0°, and 75.5° are indexed to planes of anatase phase of TiO_2 such as (101), (004), (200), (105), (204), (116), and (107) (JCPDS-00-021-1272) respectively.

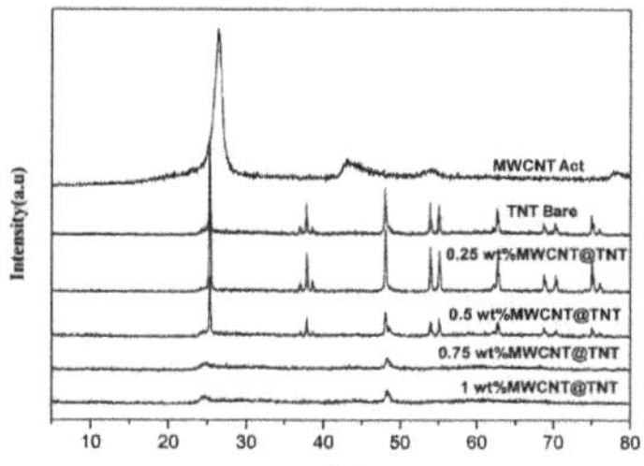

Figure 2. Powder XRD pattern of MWCNT Act, TNT bare & MWCNT@TNT composites.

This confirms the retaining of the anatase phase of TiO_2 after nanotube formation. Moreover, no significant shift in the XRD peaks of composite material compared with bare TNT indicated that MWCNT does not enter into the lattice to substitute Ti partly. It is further observed that nanocomposites yield no peak at the 2θ positions of 26.0° and 43.48°, which would be characteristic of CNTs. This implies that the diffraction peak of MWCNTs at 26.0° might overlap with the main peak of anatase TiO_2 at 25.3° [25]. Moreover, the crystallinity of MWCNTs is much lower than that of TiO_2, causing shielding of MWCNTs peaks by TiO_2 peaks. The lower residual peak intensity of the composite material may be due to the uptake of MWCNT inside the pores which reduces the scattering contrast between the pore wall and pore [26]. From the above results, it is concluded that the *in-situ* addition of MWCNT does not restrict the formation of TNT and was further confirmed by surface area and TEM analysis.

4.2 CHNS analysis

The elemental compositions of composite material are depicted in Table 1. The result demonstrated that the weight percentage loading of MWCNT are comparable to the calculated value of MWCNT percentage. From the aforementioned results it can be concluded that the MWCNT was successfully loaded into TNT.

Table 1. CHN analysis of MWCNT and its composites.

Sample	%C	%H
MWCNT Bare	99.53	0.326
MWCNT Act	96.53	0.316
0.25wt% MWCNT@TNT	0.57	1.405
0.5wt% MWCNT@TNT	0.61	0.828
0.75wt% MWCNT@TNT	0.76	0.524
1wt% MWCNT@TNT	1.40	0.959

4.3 Nitrogen adsorption-desorption isotherms

The N_2 adsorption- desorption isotherm of MWCNT Bare and MWCNT Act, TNT Bare and MWCNT@TNT composite materials are shown in Fig. 3 and Fig. 4 respectively. The results demonstrated that according to the IUPAC classification N_2 isotherm of all samples is a typical type IV isotherm. A hysteresis loop is observed at relative pressure of $P/P^0 \sim 0.43$ for TNT and its composites indicating the capillary condensation which is intermediate between the H1 (P/P^0 0.5 to 0.8) and H3 (P/P^0-0.8) type [27-29].

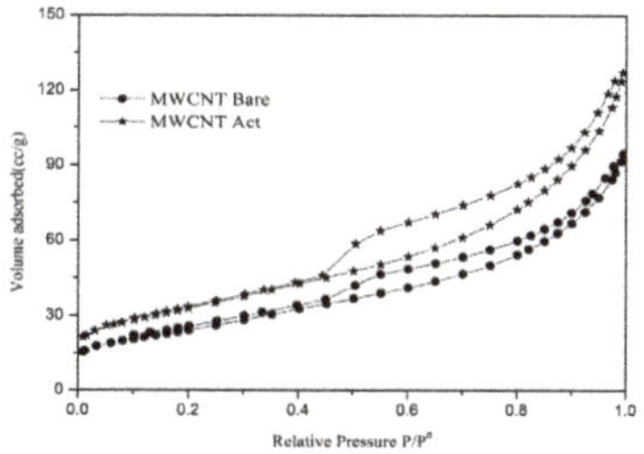

Figure 3. Nitrogen adsorption-desorption isotherm of MWCNT Bare and MWCNT Act.

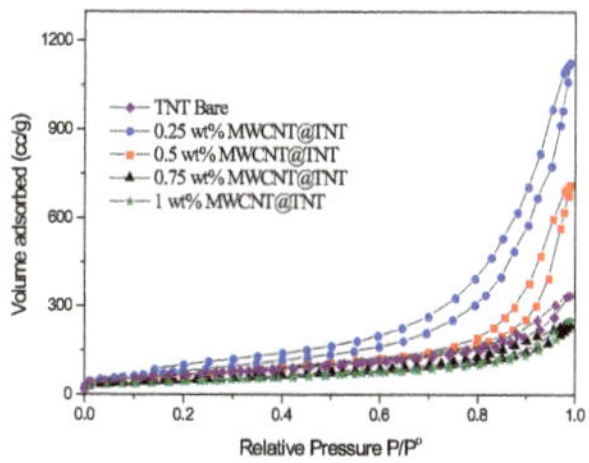

Figure 4. Nitrogen adsorption-desorption isotherm of TNT Bare and MWCNT@TNT composites.

The specific surface area (SSA) was determined by Brunauer-Emmet-Teller (BET) measurements in the relative pressure (P/P^0) range of 0.05 - 0.3, the total pore volume is calculated by Gurvitsch equation at pressure P/P^0 of 0.95. The results revealed that MWCNT Bare has a surface area of 84 m^2/g and it was increased to 117 m^2/g after acid activation (Table 2). The enhancement in the surface area of activated MWCNT (MWCNT Act) presumably due to the removal of metal impurities and the opening of the tube tips which makes the inner cavities accessible for adsorption. The pore volume and pore diameter also increased after acid treatment. TNT Bare has a surface area of 174 m^2/g and the surface area was enhanced to 299 m^2/g in 0.25 wt% MWCNT@TNT composite material. It obviously shows the successful synthesis of MWCNT/TNT composite materials and further increase in the MWCNT concentration leads to decrease in the surface area. However surface area analysis further demonstrated that increase in the surface area was observed after MWCNT loading and all composite materials shows higher surface area than TNT Bare. It confirms the successful loading of MWCNT into TNT during the hydrothermal treatment. Moreover there was no trend for obtained pore volume. The total pore volume is not related to hydrogen adsorption at 77 K. Micropore specially ultramicropore in the range of < 0.7 nm contribute a lot to the enhanced hydrogen adsorption due to favorable interaction [30]. The enhanced micropore has been studied by density functional method using CO_2 as adsorbate.

Table 2. BET and Langmuir surface area (SA), pore volume, pore width of MWCNT Bare, MWCNT Act, TNT Bare and its composites.

Sample Name	BET SA (m^2/g)	Langmuir SA (m^2/g)	Pore volume (cm^3/g)	Pore width (nm)
MWCNT Bare	84	114	0.13	5.88
MWCNT Act	117	163	0.17	6.0
TNT Bare	174	238	0.314	7.39
0.25 wt% MWCNT@TNT	299	426	1.417	6.79
0.5 wt% MWCNT@TNT	249	349	0.879	8.92
0.75 wt% MWCNT@TNT	238	301	0.52	12.7
1 wt% MWCNT@TNT	156	210	0.39	14.4

4.4 TEM analysis

Fig. 5 shows the TEM images of anatase TiO_2, synthesized TNT, MWCNT Act and MWCNT@TNT composite materials. The result demonstrated that anatase TiO_2 nanoparticles (Fig. 5a) were spherical in shape. Fig. 5b clearly shows the tubular structure of synthesized TNT and clearly shows the opening end of TNT. Further anatase TiO_2 nanoparticles were not observed around the TNT. It validates that anatase TiO_2 nanoparticle was completely transformed into TNT after hydrothermal treatment with NaOH. Further the selected area electron diffraction (SAED) pattern of TNT (Fig. 5c) show that the circular rings were corresponding to the *(hkl)* planes of the anatase phase such as (101) and (200) respectively. Similar results were reported by Tayade et al. [31,19] and Kasuga et al. [32,17] in the hydrothermal synthesis of TiO_2 nanotube. Fig. 5d demonstrates the opening end of MWCNT Bare and it clearly shows that the MWCNT was successfully opened after acid treatment. It further corroborates the enhancement in the surface area of MWCNT after acid treatment (Table 2). Moreover HRTEM image of MWCNT (Fig. 5e) evidently exhibit the multiwall nature of the carbon nanotube. TEM image of 0.25 wt% MWCNT@TNT sample (Fig. 5f) demonstrate the uniform tubular structure of composite materials and it further proves that addition of MWCNT during the synthesis does not restrict the formation of TNT. Whereas MWCNT was not observed in the TEM image of 0.25 wt% MWCNT@TNT, this may be due to the low percentage loading of MWCNT. However, surface area and PSD analysis (Fig. 4 and Fig. 7) clearly demonstrates the enhancement in the surface area and formation of extra pores were observed after addition of MWCNT. It confirms that MWCNT was successfully loaded into the TNT samples. TEM images of 1 wt% MWCNT@TNT composite material (Fig. 5g) clearly shows the presence of both the MWCNT and TNT in synthesized samples. However, it was observed that higher percentage loading of MWCNT leads to decrease in surface area.

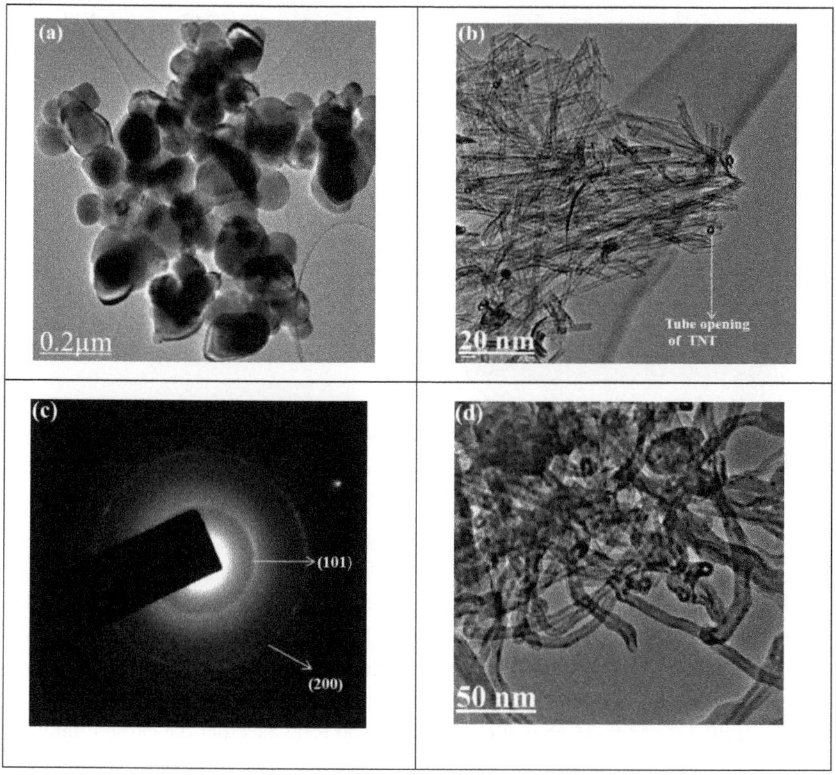

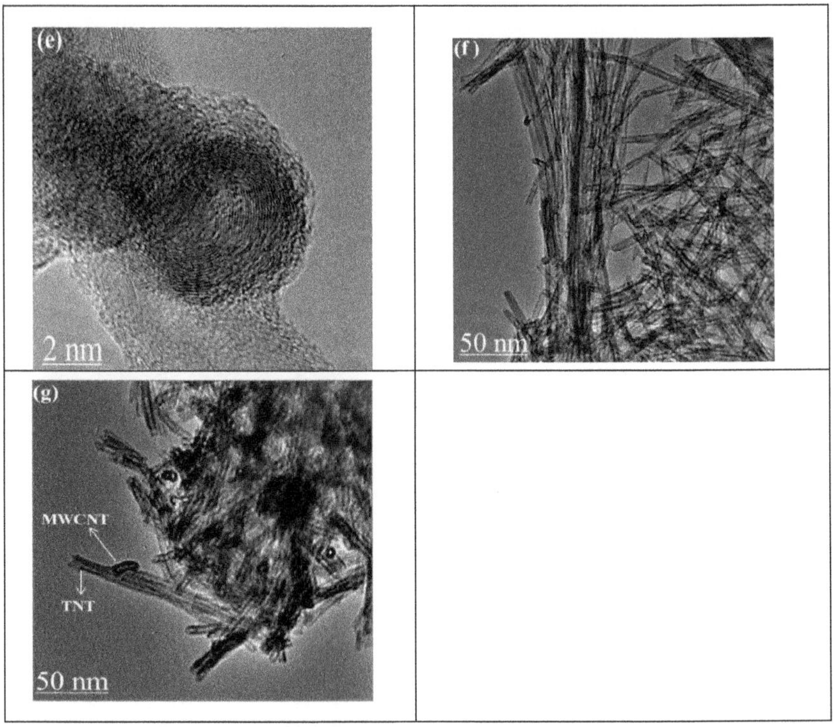

Figure 5. TEM images of (a) AT, (b) TNT Bare, (c) SAED pattern of (b), (d) MWCNT Act, (e) HRTEM of MWCNT-Act, (f) 0.25%MWCNT@TNT and (g) 1wt% MWCNT@TNT.

4.5 Hydrogen adsorption analysis

The high pressure hydrogen adsorption isotherms were measured for MWCNT Bare, MWCNTAct, TNT Bare, and its composites at 77.4 K up to 4000 KPa. The hydrogen adsorption isotherm is depicted in Fig. 6 and the hydrogen uptake values are given in Table 3. The results demonstrated that hydrogen uptake of MWCNT samples was enhanced from 21 cc/g to 36 cc/g after activation and it reveals that the carbon impurities are removed and the channels inside the nanotubes are accessible for hydrogen adsorption.

Further the incorporation of MWCNT into TNT enhanced the hydrogen uptake capacity of composite materials. The hydrogen uptake capacities of four composite materials 0.25,

0.5 0.75 and 1 wt% MWCNT@TNT are 132, 102, 54, and 52 cc/g respectively (Table 3). Whereas in the case of TNT Bare it was 54 cc/g . As compared to TNT Bare, 40 % increase in the H_2 uptake was observed for 0.25 wt% MWCNT@TNT composite material.

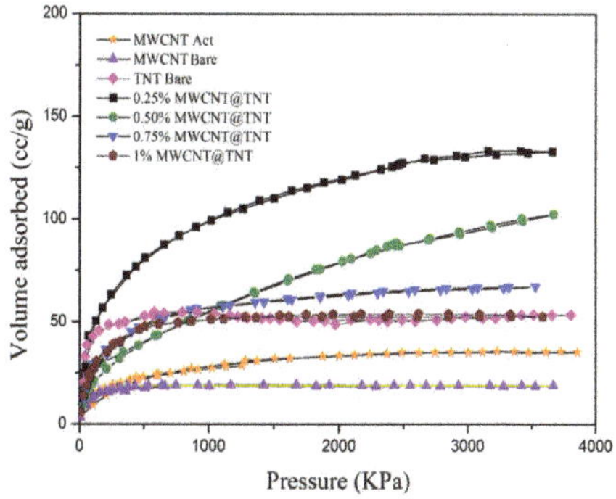

Figure 6. Hydrogen adsorption isotherm of MWCNT Bare, MWCNT Act, TNT Bare and MWCNT@TNT composites.

Table 3. Hydrogen adsorption studies of MWCNT Bare, MWCNT Act, and MWCNT@TNT composites.

Sample	H_2 uptake (cc/g)	H_2 uptake (Wt %)
MWCNT Bare	21	0.1875
MWCNT Act	36	0.321
TNT Bare	54	0.4821
0.25wt% MWCNT@TNT	132	1.178
0.5wt% MWCNT@TNT	102	0.910
0.75wt% MWCNT@TNT	54	0.5
1wt% MWCNT@TNT	52	0.46

Moreover, a gradual decrease in the hydrogen uptake with increasing the amount of MWCNT incorporation has been observed. The higher loading of MWCNT might be blocking the pores and also less van der Waal interaction between the hydrogen molecule and walls of the composite material resulting in lesser hydrogen uptake. From the aforementioned results, it concludes that 0.25 wt% of MWCNT loading is the optimum percentage for higher hydrogen uptake using present reaction condition.

4.6 Density functional theory (DFT) method

The enhancement in the hydrogen adsorption capacity of 0.25wt% MWCNT@TNT compared to TNT Bare may be due to the formation of micropore volume. During the reaction the multiwalled carbon nanotube rupture into small fragments and the formation of micropore volume in the framework is possible. The formation of micropores mainly ultra microporores in the range of < 0.7 nm has been analyzed by DFT method using CO_2 at 273 K. The characterization of ultra microporous (micropores of molecular dimensions) materials by nitrogen adsorption is difficult, because the filling of pores with diameter from 0.5 to 1 nm occur at relative pressure of 10-7 to 10-5 where the rates of diffusion and adsorption are very slow. This phenomenon is related to the fact that the nitrogen is diatomic and possesses a quadruple moment, which leads to specific fluid-wall interactions and adsorption cannot take place in the pores of molecular dimensions [33,34]. Therefore CO_2 adsorption at 273 K was preferred over N2 at 77.4 K for accessing the micropores because at 273 K and higher absolute pressure (P0 = 26200 Torr) CO_2 can access ultra micropores, which are not accessible to nitrogen at 77.4 K [35]. Due to higher diffusion rate, equilibrium is achieved faster as compared to nitrogen adsorption at 77 K thereby dramatic decrease in analysis time i.e., 3-5 h for CO_2 versus 30-50 h using N_2.

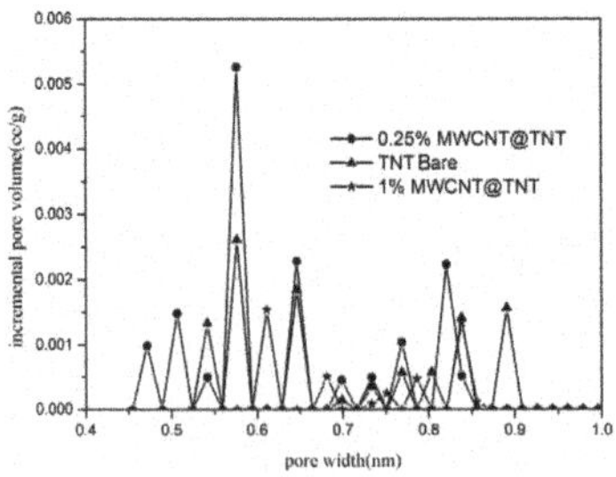

Figure 7. DFT method using CO_2 at 273 K.

Pore size distribution of TNT bare, 0.25 wt% MWCNT@TNT and 1wt% MWCNT@TNT were calculated by DFT method using CO_2 as adsorbate at 273 K and results were shown in Figure 7. The result reveals that both TNT bare and 0.25 wt%

MWCNT@TNT showed a sharp peak in the range 0.52 - 0.6 and 0.62 - 0.66 nm. However 0.25 wt% MWCNT@TNT composite material shows a sharp peak at 0.576 nm, which is due to the presence of TNT and additional two minor peaks were observed at 0.471 and 0.506 nm respectively. The additional peaks were not observed in TNT, it may be due to the generation of additional pores by incorporation of MWCNT into TNT which resulted in the reduction of pore size and increase in the micropore volume of the composite material. There was no additional pores observed for 1wt% MWCNT@TNT which indicates that higher MWCNT loading blocked the micropores. Hence 0.25 wt% loading is the optimum value of MWCNT value and showed enhanced hydrogen adsorption capacity using present reaction conditions.

5. Conclusions

MWCNT@TNT composite materials was successfully synthesized by hydrothermal methods and characterized by various physicochemical techniques. The tubular structure of TNT and MWCNT@TNT composite material was confirmed by TEM analysis and surface area analysis. The hybrid composite TNT with low percentage of MWCNT loading showed a significant increase in hydrogen adsorption capacity at 77.4 K due to the enhanced surface area and formation of micropores respectively. The result demonstrated that 0.25wt%MWCNT@TNT disclosed higher adsorption capacity using present reaction conditions and the order is 0.25wt%MWCNT@TNT > 0.5wt% MWCNT@TNT > 0.75wt% MWCNT@TNT ~ TNT > 1wt% MWCNT@TNT > MWCNT Act > MWNCT Bare respectively. The result further demonstrated that hydrogen sorption capacity decreased as the percentage loading of MWCNT was higher than the optimum value (0.25wt %).

Acknowledgements

Authors are thankful to CSIR, New Delhi, India, for funding through Network Project on HYDEN (CSC- 0122). We also thankful to Analytical Science Discipline and Centralized Instrumentation Facility of the institute for analytical support. MCR and TSN thanks to AcSIR for enrolment in Ph.D.

References

[1] M. Paik Suh, H.J. Park, T.K. Prasad, D. Lim, Hydrogen Storage in Metal–Organic Frameworks, Chem. Rev. 112 (2012) 782-835. https://doi.org/10.1021/cr200274s

[2] L. Schlapbach, A. Zuttel, Hydrogen-storage materials for mobile applications, Nature. 414 (2001) 353-358. https://doi.org/10.1038/35104634

[3] M. Bastos-Neto, C. Patzschke, M. Lange, J. Mollmer, A. Moller, S. Fichtner, C. Schrage, D. Lassig, J. Lincke, R. Staudt, H. Krautscheid, R. Glaser, Assessment of hydrogen storage by physisorption in porous materials, Energy Environ. Sci. 5 (2012) 8294-8303. https://doi.org/10.1039/c2ee22037g

[4] Information on http://www1.eere.energy.gov/hydrogenandfuelcells/storage/current_technology.html

[5] S. Satyapal, J. Petrovic, C. Read, G. Thomas, G. Ordaz, The U.S. Department of Energy's National Hydrogen Storage Project: Progress towards meeting hydrogen-powered vehicle, Catal. Today. 120 (2007) 246-256. https://doi.org/10.1016/j.cattod.2006.09.022

[6] S. Iijima, Helical microtubules of graphitic carbon, Nature. 354 (1991) 56-58. https://doi.org/10.1038/354056a0

[7] A.C. Dillon, K.M. Jones, T.A. Bekkedahl, C.H. Kiang, D.S. Bethune, M.J. Heben, Storage of hydrogen in single-walled carbon nanotubes, Nature. 386 (1997) 377-379. https://doi.org/10.1038/386377a0

[8] G. Gundiah, A. Govindaraj, C.N.R. Rao, Hydrogen storage in carbon nanotubes and related materials, J. Mater. Chem. 13 (2003) 209-213. https://doi.org/10.1039/b207107j

[9] S.J. Yang, J.Y. Choi, H.K. Chae, J.H. Cho, K.S. Nahm, C.R.. Park, Preparation and Enhanced Hydrostability and Hydrogen Storage Capacity of CNT@MOF-5 Hybrid Composite, Chem. Mater. 21 (2009) 1893-1897. https://doi.org/10.1021/cm803502y

[10] K.P Prasanth, P.Rallapalli, M.C. Raj, H.C. Bajaj, R.V. Jasra, Enhanced hydrogen sorption in single walled carbon nanotube incorporated MIL-101 composite metal–organic framework, Int. J. Hydrogen Energ. 36 (2011) 7594-7601. https://doi.org/10.1016/j.ijhydene.2011.03.109

[11] M. Endo, T. Hayashi, Y.A. Kim, M. Terrones, M.S. Dresselhaus, Applications of carbon nanotubes in the twenty-first century, Phil. Trans. R. Soc. Lond. A. 362 (2004) 2223-2238. https://doi.org/10.1098/rsta.2004.1437

[12] M. Trojanowicz, Analytical applications of carbon nanotubes: a review, Trends Analyt Chem. 25 (2006) 480-489. https://doi.org/10.1016/j.trac.2005.11.008

[13] M. T. Yildirim, S. Ciraci, Titanium-Decorated Carbon Nanotubes as a Potential High-Capacity Hydrogen Storage Medium, Phys. Rev. Lett. 94 (2005) 175501-175504. https://doi.org/10.1103/PhysRevLett.94.175501

[14] E. Liu, J. Wang, J. Li, C. Shi, C. He, X. Du, N. Zhao, Enhanced electrochemical hydrogen storage capacity of multi-walled carbon nanotubes by TiO_2 decoration, Int. J. Hydrogen Energ. 36 (2011) 6739-6743. https://doi.org/10.1016/j.ijhydene.2011.02.128

[15] A. Lueking, R.T. Yang, Hydrogen Spillover from a Metal Oxide Catalyst onto Carbon Nanotubes—Implications for Hydrogen Storage, J Catal. 206 (2002) 165-168. https://doi.org/10.1006/jcat.2001.3472

[16] S. Rather, N. Mehraj-ud-din, R. Zacharia, S.W. Hwang, A.R. Kim, K.S. Nahm, Hydrogen storage of nanostructured TiO_2-impregnated carbon nanotubes, Int. J. Hydrogen Energ. 34 (2009) 961 - 966. https://doi.org/10.1016/j.ijhydene.2008.09.089

[17] T. Kasuga, M. Hiramatsu, A. Hoson, T. Sekino, K. Niihara, Titania Nanotubes Prepared by Chemical Processing , Adv. Mater. 11 (1999) 1307 - 1311. https://doi.org/10.1002/(SICI)1521-4095(199910)11:15%3C1307::AID-ADMA1307%3E3.0.CO;2-H

[18] D.V. Bavykin, J.M. Friedrich, F.C. Walsh, Protonated titanates and TiO_2 nanostructured materials: synthesis, properties, and applications, Adv. Mater. 18 (2006) 2807 - 2824. https://doi.org/10.1002/adma.200502696

[19] T.S Natarajan, K. Natarajan, H.C. Bajaj, R.J. Tayade, Enhanced photocatalytic activity of bismuth-doped TiO_2 nanotubes under direct sunlight irradiation for degradation of Rhodamine B dye, J. Nanopart. Res. 15 (2013) 1669-1687. https://doi.org/10.1007/s11051-013-1669-3

[20] P. Roy, S. Berger, P. Schmuki, TiO_2 Nanotubes: Synthesis and Applications Angew. Chem. Int. Edit. 50 (2011) 2904-2939. https://doi.org/10.1002/anie.201001374

[21] S. Tohru, Inorganic and Metallic Nanotubular Materials Recent Technologies and Applications: T. Kijima (Ed.): Inorganic and Metallic Nanotubular Materials. @ Springer-Verlag Berlin Heidelberg 2010 ,Top. App. Phys. 117(2010) 17–32

[22] D.V. Bavykin, A.A. Lapkin, P.K. Plucinski, J.M. Friedrich, F.C. Walsh, Reversible Storage of Molecular Hydrogen by Sorption into Multilayered TiO_2

Nanotubes, J. Phys. Chem. B. 109 (2005) 19422-19427. https://doi.org/10.1021/jp0536394

[23] P.B.S. Rallapalli, M.C. Raj, D.V. Patil, K.P. Prasanth, R.S. Somani, H.C Bajaj, Activated carbon @ MIL-101(Cr): a potential metal-organic framework composite material for hydrogen storage, Int. J. Energy Res, 37 (2013) 746–753. https://doi.org/10.1002/er.1933

[24] H.C. Bajaj, R.S. Somani, P.B.S. Rallapalli, D.V. Patil, K.P. Prasanth, M.C. Raj, R.S. Thakur, M. John, B.L. Newalkar, N.V. Choudary, US Patent 9,433,919 (2016).

[25] Y.H. Tseng, C.Y. Yen, M.Y. Yen, and C.C.M. Ma, Effects of the acid pretreated multi-walled carbon nanotubes on the photocatalytic capacity of TiO_2/multi-walled carbon nanotubes nanocomposites, Micro and Nano Letters, 5 (2010) 1–6. https://doi.org/10.1049/mnl.2009.0094

[26] M. Kruk, M. Jaroniec, C.H. Ko, R. Ryoo, Characterization of the Porous Structure of SBA-15, Chem. Mater. 12 (2000) 1961 - 1968. https://doi.org/10.1021/cm000164e

[27] S.H, Lim, J. Luo, Z. Zhong, W. Ji, J. Lin, Room-Temperature Hydrogen Uptake by TiO_2 Nanotubes, Inorg, Chem. 44 (2005) 4124 - 4126. https://doi.org/10.1021/ic0501723

[28] F. Rouquerol, J. Rouquerol, K Sing, Adsorption by Powders and Porous Solids Principles, Methodology and Applications, second ed., Academic Press London: 1999

[29] S.J. Gregg, K.S.W. Sing, Adsorption, Surface Area and porosity, first ed., Academic Press London:1982.

[30] K.M. Thomas, Hydrogen adsorption and storage on porous materials, Catal Today. 120 (2007) 389-398. https://doi.org/10.1016/j.cattod.2006.09.015

[31] R.J. Tayade, D.L. Key, Synthesis and Characterization of Titanium Dioxide Nanotubes for Photocatalytic Degradation of Aqueous Nitrobenzene in the Presence of Sunlight, Mater. Sci. Forum. 657 (2010) 62-74. https://doi.org/10.4028/www.scientific.net/MSF.657.62

[32] T. Kasuga, M. Hiramatsu, A. Hoson, T. Sekino, K. Nihara, Formation of Titanium Oxide Nanotube, Langmuir. 14 (1998) 3160-3163. https://doi.org/10.1021/la9713816

[33] M. Thommes. Physical Adsorption Characterization of Ordered and Amorphous Mesoporous Materials. Nanoporous Materials: Science And Engineering editors, London: Imperial College Press, 2004, pp. 317-364. https://doi.org/10.1142/9781860946561_0011

[34] A.V. Neimark, K.S.W. Sing, & Thommes, M, Handbook of Heterogeneous Catalysis, second ed., Weinheim: Wiley-VCH, 2008.

[35] M. Armandi, B. Bonelli, K. Cho, R. Ryoo, E. Garrone, Study of hydrogen physisorption on nanoporous carbon materials of different origin, Int. J. Hydrogen Energ. 36 (2011) 7937-7943. https://doi.org/10.1016/j.ijhydene.2011.01.049

Chapter 10

Silver Phosphate Based Photocatalysis: A Brief Review from Fundamentals to Applications

Alaka Samal,[1,2] Ayonbala Baral,[1,3] Dipti P. Das[1,2*]

[1]Academy of Scientific and Innovative Research, New Delhi, India

[2]Colloids and Material Chemistry, CSIR- Institute of Minerals and Materials Technology, Bhubaneswar - 751013, Odisha, India

[3]Hydro & ElectroMetallurgy, CSIR- Institute of Minerals and Materials Technology, Bhubaneswar - 751013, Odisha, India

[*]E-mail: das_dipti77@yahoo.co.in

Abstract

Attributable to the superior visible light active nature and high efficiency, silver phosphate (Ag_3PO_4) has attracted gigantic attention for decomposition of organic contaminants and fuel production. The photoresponsivity of Ag_3PO_4 hugely depends upon the morphology, method of fabrication, formation of hybrids and photocorrosive nature of it. The cause of high activity, activity based on morphology and various methods of synthesis of Ag_3PO_4 based photocatalysts to improve the stability of Ag_3PO_4 for applications towards energy and environment is the crux of the matter in this review. Important applications including photocatalytic pollutant degradation, O_2/H_2 production, and bacterial degradation are also addressed. Finally, summary and outlooks on the challenges and future perspectives of this emerging photocatalyst are presented.

Keywords

Photocatalysis, Dye Degradation, Water Splitting, Visible Light, Ag_3PO_4, Heterostructure, Electronic Structure, Pollutants

Contents

1. Introduction

In the twenty-first century the diminution of fossil fuel and the environmental pollution has raised much global anxiety. From the introduction of photocatalysis, much research has been devoted towards the search of semiconductors which can harvest the plentiful available solar light as the nature does by photosynthesis. For the last some decades, photocatalysis technique is used for various applications *Ca.* water splitting to produce hydrogen and oxygen fuel, organic pollutant degradation and decontamination of bacteria causing hazardous diseases from water (Scheme **1**). Owing to the multitasking capacity of various semiconductors, the photocatalysis process is the most green and easy technology for the coming generation. For this, researchers from the world are trying to develop new and extra ordinary catalysts which are active under visible light, highly recyclable, having high quantum efficiency and are easy to prepare.

Scheme 1. Different applications of photocatalysis.

In order to use around 45 % of the visible light available in the solar light, narrow band gap photocatalysts are needed. The narrow band gap photocatalysts should have a band gap between 2.0 eV to 2.8 eV, so that they can be efficiently used for any photocatalytic reaction under visible light and also they should not face charge carrier recombination. Along with the narrow band gap a photocatalyst should possess suitable redox potential and strong absorption of light. Unfortunately, most of the semiconductors at present lack some of these expectations and are not able to meet all those conditions. So, the invention of materials with narrow band gaps such as Fe_2O_3, Cu_2O, $BiVO_4$, CdS, $CoTiO_3$, NiTiO3 etc. were done. But, these narrow band gap photocatalysts usually suffer from low efficiency. For example, $BiVO_4$ demonstrates a low quantum yield of only 9% at 450 nm irradiated light wavelength [1]. Non-oxide semiconductors, which could have high quantum yields, unluckily suffer from photocorrosive nature. Again sulphides and nitrides experienced to be self-oxidative in a photoelectrochemical setup [2,3]. Doping with cations (e.g. V, Mn, or Cr) and anions (e.g. N, S, or P) has also been commonly explored as a strategy to improve absorption of visible light, with an aim of introducing donor or acceptor levels into the bulk band gap of a host oxide that remains chemically inert. This approach suffers from a loss of performance due to the introduction of abundant recombination centres for the photoexcited charge carriers [4]. It is also said that the synthesis of nanoporous metal phosphates with different morphologies at all times is very attractive due to their unique advantages of having abundant active sites for reactions with high surface areas and fast interfacial transport of protons/electrons by decreasing the diffusion path length through the porous structure [5–8]. Considering the metal phosphates particularly, in the field of heterogeneous catalysis several metal phosphates devoted for various application oriented reactions. Like vanadium phosphate phase [9], Iron phosphates, other important oxidation catalysts, also showed high activity in a variety of oxidation reactions [10–12]. Zirconium and titanium phosphates are the most studied members of solid acids [11-12]. Their electric behaviors, on the other hand, have been widely investigated. Tian *et al.* found that an air electrode manufactured from mesoporous zirconium phosphate exhibited remarkable electrocatalytic activity for oxygen reduction reaction [13]. Due to the relatively high specific surface area and richness in surface hydroxyl group, mesoporous transition metal phosphates materials have been frequently employed as adsorbents for radionuclide materials [14] and heavy metal ions [15]. Our previous work is also based on a metal phosphate ($Co_3(PO_4)_2$) which was applied for the production of hydrogen by water splitting photocatalytically. [16] Yamauchi et al. also focused on the synthesis of nanoporous nickel, aluminum, and zirconium phosphates and studied their superior energy storage application [17]. Ho et al. synthesized copper phosphate microflowers for the artificial solar-to-fuel conversion system [18]. Thus, the metal phosphates are extensively used in photocatalytic applications currently.

Since the introduction of better-quality visible light active Ag_3PO_4 by Yi et al. [19] lots of research was focused towards Ag_3PO_4 based photocatalysis (Scheme 2). It is a yellow colored material having superior visible light activity. Photodegradation of organic pollutants is mainly due to very deep position of the valence band which produces the strong oxidative ability of photogenerated holes (h^+). Ag_3PO_4 has very high quantum efficiency towards O_2 evolution and is dozens of time quicker than $BiVO_4$ and $TiO_{2-x}N_x$, towards the decomposition of methylene blue due to its very high photooxidation property indeed [19]. Additionally, the most motivating fact is that this unique photocatalyst can achieve a quantum efficiency of up to 90% at wavelengths greater than 420 nm, which is significantly higher than the previous reported values [19]. So, these findings could be expected to solve the global energy crisis and environment problems with huge availability of solar light and hence, attracting number of researchers towards itself. But, unfortunately the consumption of a large amount of noble metal and the low structural stability of pure Ag_3PO_4 and photocorrosive behaviour in presence of light strongly limits its practical environmental applications in spite of its promising attributes. Since then, many efforts have been dedicated for further improvement and optimization of its photoelectric and photocatalytic properties and stability. It is a defy to uphold both the photocatalytic property and photostability of the Ag_3PO_4 under the reaction condition of light as metallic silver forms within the reaction condition ($Ag^+ + e^- \rightarrow Ag^0$) in presence of photon. So, many efforts were made by different research groups for tuning the properties Ag_3PO_4 which can ultimately upgrade its activity as well as stability. In this review, a clear summary of research progress on silver phosphate based photocatalyst and their applications were described along with the structure, properties, and theoretical aspects of Ag_3PO_4.

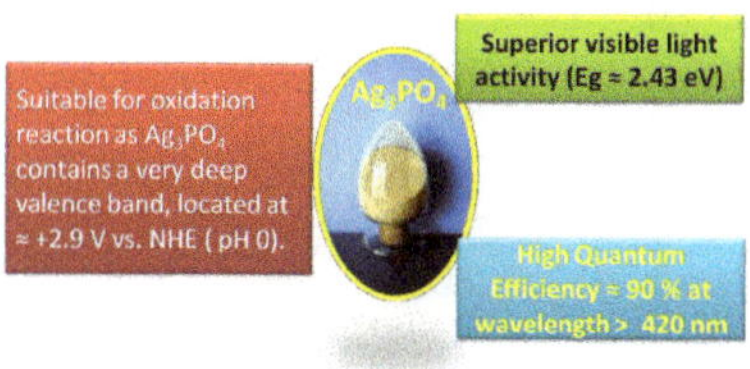

Scheme 2. The superior properties of Ag_3PO_4 as a photocatalyst.

1.1 Pristine Ag_3PO_4 as photocatalyst

No matter how much the efficiency of the miraculous Ag_3PO_4, it also has so many cons like other semiconductors used as photocatalyst. The band diagram of Ag_3PO_4 is shown in Scheme 3. The conduction band minima of the Ag_3PO_4 is aligned at 0.45 V, NHE which is incapable for the proton reduction reaction as the conduction band minima is more positive than the hydrogen reduction potential 0.0 V, NHE. Also, it cannot produce superoxide radicals due to the same reason of having positive conduction band position. Another promising limitation of Ag_3PO_4 is that it undergoes photocorrosion in presence of light and yields metallic silver as mentioned earlier. So many researchers have tried to understand the fundamentals of its ability and also they wanted to improve the activity by modifying the phases or else scheming a heterostructure with silver phosphate to avoid the photocorrosion. This part is well described further in the article.

Besides owing an extraordinary photoactivity silver phosphate tends to photodegradation in the presence of light and this limits its practical application towards energy and environment. The authors also studied the photocorrosion property of pure Ag_3PO_4 where black metallic silver comes out of the photocatalyst after the photoreaction. For which reusability of the photocatalyst cannot be studied. As can be seen from Scheme 3, the position of top of the valence band (E_{VB}) Ag_3PO_4 is at 0.45 V *vs.* NHE at pH 0. As described by Yi et al. [19] the electrode potential of Ag/Ag_3PO_4 lies between the hydrogen reduction potential and $Ag/AgNO_3$ (0.80 V *vs.* NHE at pH 0). For which Ag_3PO_4 cannot produce H_2 through water splitting. The main issue regarding the practical use of Ag_3PO_4 is that it undergoes self reduction under the irradiation of light. Again due to the more positive conduction band the photocatalyst cannot reduce dissolved oxygen to superoxide radical ($^{\bullet}O_2^-$) which is a useful radical to decompose any organic pollutants. But due to its deep VB position the holes can directly oxidize the organic pollutants efficiently and also the VB can produce hydroxyl radical which is another highly oxidizing radical to flash decomposition of pollutants. So, to make Ag_3PO_4 stable is the ultimate goal while retaining its photocatalytic activity.

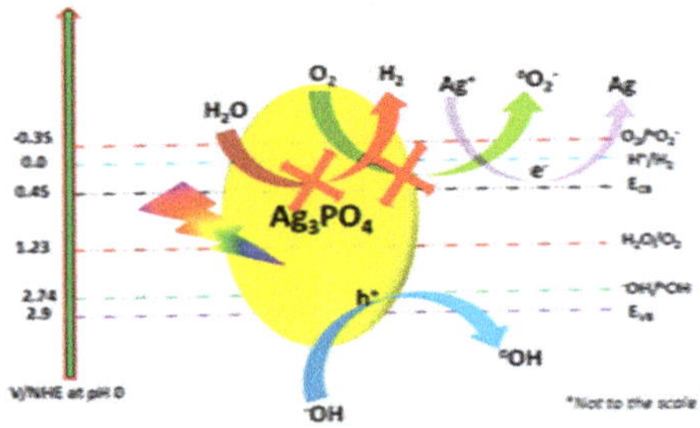

Scheme 3. Band diagram of Ag_3PO_4 showing CB and VB.

2. Electronic understanding of Ag_3PO_4

As discussed in the previous section, Ag_3PO_4 is a promising photocatalyst in the direction of both oxygen evolution and photodegradation of organic contaminants. As a result, many researchers tried to analyse the origin of stupendous activity of it by computational analysis. Recently, Kahk et al. and Ma et al. described the structure of Ag_3PO_4. It is known to crystallize in the cubic structure with space group P4-3n [20,21]. The crystal structure consists of isolated, regular PO_4 tetrahedral forming a body-centred cubic lattice. As shown in Fig. 1, the Ag atom experiences 4-fold coordination by four O atoms. The P atoms have 4-fold coordination surrounded by four O atoms, while the O atoms have 4-fold coordination surrounded by three Ag atoms and one P atom [20,21]. Kahk and group described that the calculated lattice parameter of Ag_3PO_4 obtained from hybrid DFT calculations is 6.072 Å and the value of the lattice parameter was determined from X-ray diffraction to be 6.004 Å. Ma and group deduced the polyhedron configuration of Ag_3PO_4. Fig. 1(c) consists of tetrahedral PO_4 and AgO_4. They mentioned that one PO_4 tetrahedron and three tetrahedral AgO_4 are combined with each other through the corner oxygen and the absolute cause of the photocatalytic activity of Ag_3PO_4 is explained as PO_4^{3-} ions having a large negative charge (an inductive effect) which maintains a large dipole in the Ag_3PO_4, which results in the distortion of tetrahedral AgO_4. Again, they precisely described that as PO_4^{3-} possessing a large electron cloud overlapping prefers to attract holes and repel electrons, which helps the e^-/h^+ separation which thereby remains as another cause of the extraordinary activity. Similarly, Botelho et al. studied the structural behavior by means of X-ray diffraction, Rietveld refinement, and Raman spectroscopy to model the cubic Ag_3PO_4 structure (Fig. 1(d))[20].

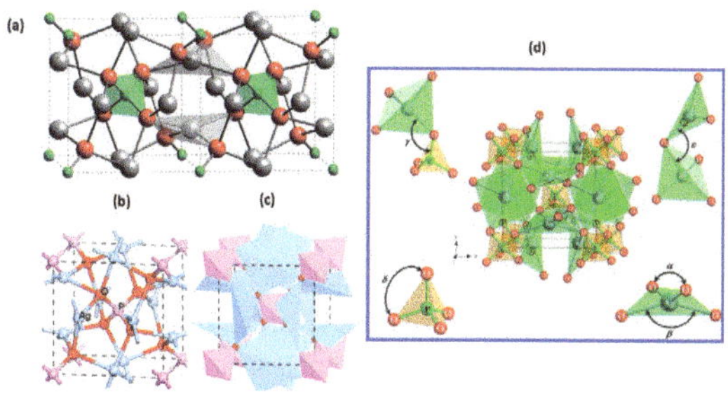

Figure 1. (a)The cubic crystal structure of Ag_3PO_4. The local coordination environments of P (tetrahedral) and Ag (distorted tetrahedral) are shown. Reprinted with permission. Copyright Kahk et al., 2014, RSC Publishing group.[21] Unit-cell structure of cubic Ag_3PO_4, showing (b) ball and stick and (c) polyhedron configurations. Red, purple, and blue spheres represent O, P, and Ag atoms, respectively. Reprinted with permission. Copyright Ma et al., 2011, ACS Publications.[20] (d) Schematic representation of the cubic Ag_3PO_4 structure, illustrating both tetrahedral [AgO_4] and [PO_4] clusters. Reprinted with permission. Copyright Botelho et al., 2015, ACS Publications [20].

To look into the origin of photocatalytic activation of Ag_3PO_4, an electronic structure is needed to explain. Ag_3PO_4 having an indirect band-gap of 2.36 eV as well as a direct transition of 2.43 eV, which was deduced from the ultraviolet-visible diffuse reflectance spectrum [19]. Thus, Ag_3PO_4 is able to absorb irradiation with a wavelength shorter than 530 nm, well extending into the visible region.

Again theoretical studies have been carried out for the illustration of electronic structure and purpose of understanding the excellent high photocatalytic activity of Ag_3PO_4 by using first-principle of density functional theory (DFT) (Fig. 2). Ma et al. [20] used first-principle DFT combined with the LDA+U formalism to show that Ag_3PO_4 has a large dispersion of conduction band, which facilitates the separation of charge carriers. Moreover, high concentration of Ag vacancies in Ag_3PO_4 lattice has a significant effect on the separation of electron–hole pairs and optical absorbance in the visible-light region. On basis of the above studies and results, Liu et al. [22] used hybrid density functional method to more precisely get the electronic structure of Ag_3PO_4 photocatalyst (Fig. 3). The group got great results with a band gap of 2.43 eV, which agrees well with the experimental result. The conduction bands are credited to Ag5s and 5p states, while the valence bands mainly consist of O2p and Ag4d states. The VBM potential was 2.67 eV (*Vs.* NHE), which indicates an adequate driving force for water oxidation or pollutants degradation. Xu et al. [23] determined the stability and mechanisms of the electron transfer, as well as the photocatalytic efficiency of the composite by the interaction between the Ag_3PO_4 surface and graphene (GR). The Ag_3PO_4/GR geometry and separation characterize the strength of the interfacial interaction. Kahk et al. [21] represented high-resolution X-ray photoelectron spectra of Ag_3PO_4, together with results from theoretical calculations using hybrid DFT. Their group discussed the agreement between the theoretical and experimental results and also shown that hybrid DFT calculations predict the structural, electronic and optical properties of Ag_3PO_4 with good accuracy. Additionally, they presented an interpretation of the detailed electronic structure of Ag_3PO_4 from the perspective of molecular orbital (MO) theory (Fig. 4). In the report, they have rationalized electronic structure of Ag_3PO_4 by using an approach based on molecular orbital (MO) theory which reveals the highly localized nature of bonding in these materials. In case of Ag_3PO_4, as the covalent interactions between the phosphorus and the oxygen atoms are far stronger than those between silver and oxygen, they have constructed a molecular orbital diagram for Ag_3PO_4, starting from that of an isolated $(PO_4)^{3-}$ unit. For the $(PO_4)^{3-}$ unit, it is sufficient to consider only sigma interactions between the orbitals on P and O. The four oxygen atoms contribute two orbitals of sigma symmetry each: the 2s, and one of the three 2p. Under a tetrahedral environment these yield two triply degenerate ligand group orbitals (LGOs) of t_2 symmetry, and two LGOs of a1 symmetry. These overlap with the P 3s (a_1) and P 3p (t_2) orbitals to form twelve molecular orbitals (MOs) as shown in Fig. 4.

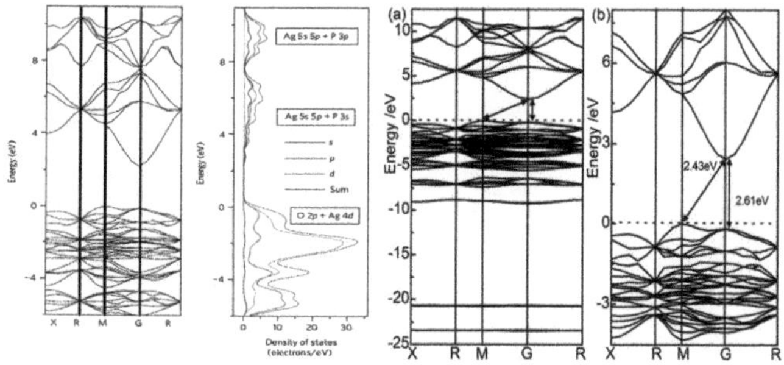

Figure 2. (a) Band structure of Ag_3PO_4 calculated using the PBE0 approach. (b) A magnified view of band structure near the Fermi level. Reprinted with permission. Copyright Liu et al., 2011, AIP Publications.[22]

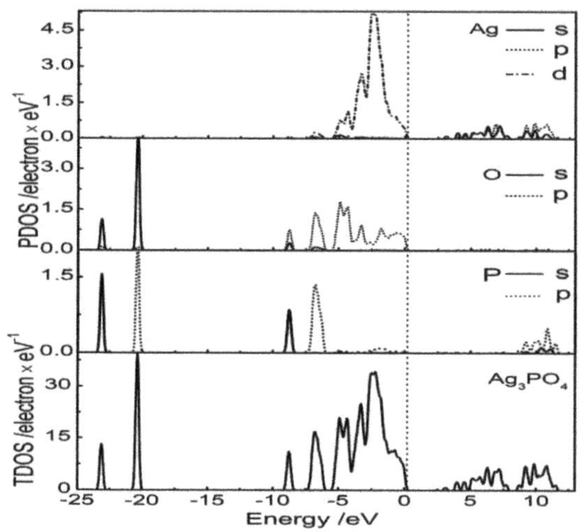

Figure 3. TDOS and PDOS of Ag_3PO_4 using PBE0 approach. Reprinted with permission. Copyright Liu et al., 2011, AIP Publications.[22]

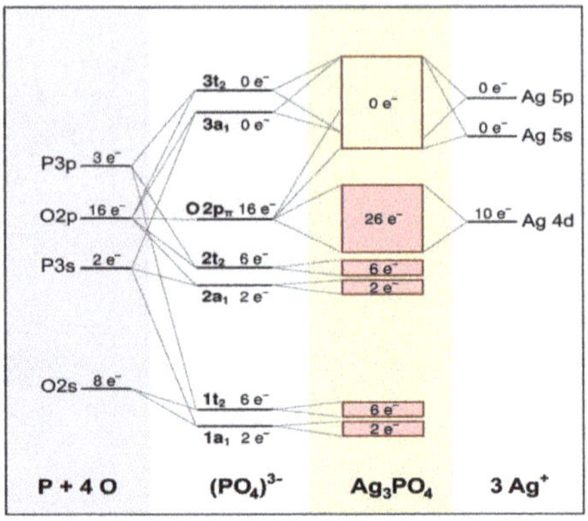

Figure 4. A molecular orbital (MO) diagram of Ag_3PO_4. Reprinted with permission. Copyright Kahk et al., 2014, RSC Publishing group.[21]

3. Ag_3PO_4 based composite for environmental applications

To harvest photons in the visible region, many narrow band gap metal oxides or chalcogenides have been coupled with TiO_2 to fabricate visible-light photocatalysts, which exhibit visible light photocatalytic activity to a certain extent. Such a strategy is also applied to modify Ag_3PO_4 photocatalyst to enhance its photocatalytic activity and/or improve its stability. The previous investigations showed that the photocatalytic activity of Ag_3PO_4 can be enhanced by Ag nanoparticles deposition on Ag_3PO_4. Because, the Ag_3PO_4 decomposition to metallic silver provides the platform to capture the photogenerated electrons and thus prevent the recombination of electron-hole pairs within the Ag_3PO_4 samples at the initial stage of photocatalytic reactions. However, the photoactivity decreases with increasing Ag contents due to the formation of Ag layers on the surface of Ag_3PO_4 that shield light absorption, inhibit the transfer of holes from the valance band of Ag_3PO_4 to the interface between photocatalyst and solution and also hinder the contact of pollutant molecules with Ag_3PO_4 and accordingly, the photocatalytic activity deteriorates gradually.[24] This deterioration of photocatalytic activity due to photocorrosion largely limits its practical application as a recyclable highly efficient photocatalyst. It is found that the Ag/Ag_3PO_4 heterocubes synthesized by reacting Ag_3PO_4 cubes with glucose in an aqueous ammonia solution exhibit higher photocatalytic activities than pure Ag_3PO_4 cubes for degradation of organic contaminants under visible-light irradiation [25]. The stability improvement of Ag_3PO_4 by covering Ag^0 nanoparticles on the surface of Ag_3PO_4 is attributed to the localized surface plasmon resonance (LSPR) effects of silver nanoparticles and a large negative charge of PO_4^{3-}

ions, [26,27] which effectively inhibit the reducibility of Ag^+ ions in the Ag_3PO_4 lattice. Ag_3PO_4 can also be rejuvenated from weak photocatalytically active Ag as a recyclable highly efficient photocatalyst by oxidizing Ag with H_2O_2 under a PO_4^{3-} ion atmosphere [28]. However, these methods have so many cons form a practical application perspective. Thus, the fabrication of Ag_3PO_4 based composite photocatalysts with high photocatalytic activity and excellent stability as well as lower Ag usage for their large scale applications is desirable.

Yao et al. synthesized Ag_3PO_4/TiO_2 visible light photocatalyst by depositing of Ag_3PO_4 nanoparticles onto the TiO_2 (P25) surface photocatalyst [29]. Their results show that the Ag_3PO_4/TiO_2 heterostructured photocatalyst shows enhanced activity and is much more stable than unsupported Ag_3PO_4. The enhanced activity is attributed to the electron hole effective separation and the larger surface area of the Ag_3PO_4/TiO_2 composite, while the enhanced stability is ascribed to the chemical adsorption of Ag^+ cations in Ag_3PO_4 and O^- anions in TiO_2. Moreover, the silver weight percentage of the photocatalyst decreases from 77% to 47%, significantly reducing the cost of Ag_3PO_4 based photocatalysts for the Ag_3PO_4/TiO_2 composite [29]. The UV photocatalytic activity of Ag_3PO_4/TiO_2 composite heterostructures was comparable to that of Ag_3PO_4 nanoparticles surfaces. Not only the stability, but also the reusability of the Ag_3PO_4/TiO_2 heterostructure catalysts was substantially enhanced as compared to that of the Ag_3PO_4 nanoparticles and TiO_2 nanobelts alone. These results were attributed to the improved charge separation of the photogenerated electrons and holes under UV light at the Ag_3PO_4/TiO_2 interface and/or surfactant-like function of the nanobelts in stabilizing the Ag_3PO_4 nanoparticles. Ag_3PO_4/TiO_2 composite heterostructures appear to be more desirable in long-term applications because of their photocatalytic activity as well as the enhanced chemical stability [30].

Considering that the VB level of Ag_3PO_4 is appreciably lower than that of TiO_2 with +2.7 V (*Vs.* NHE) and Ag_3PO_4 can be severed as an appropriate sensitizer for TiO_2. Lee et al. fabricated the novel heterojunction structures of Ag_3PO_4-core/TiO_2-shell by covering the Ag_3PO_4 nanoparticles with polycrystalline TiO_2 by sol-gel method. The prepared Ag_3PO_4/TiO_2 composites show notably enhanced photocatalytic activity in decomposing gaseous 2-propanol and evolving CO_2 compared to bare Ag_3PO_4 and TiO_2. They explained the unusually high visible-light photocatalytic activity of Ag_3PO_4/TiO_2 composite originates from the unique relative band positions of the two semiconductors [31].

Besides Ag_3PO_4/TiO_2 composite heterostructures, AgX/Ag_3PO_4 (X=Cl, Br, I) heterocrystals have also attracted much attention due to the excellent photocatalytic activity. Bi and co-workers have reported that the AgX/Ag_3PO_4 (X=Cl,Br,I) heterocrystals prepared by *in-situ* ion-exchange method embodied some advantages compared to the single Ag_3PO_4 and reported to be more promising and fascinating visible-light-driven photocatalyst than pure Ag_3PO_4 [32]. The $AgBr/Ag_3PO_4$ hybrid synthesized using an in situ anion-exchange method displayed much higher photocatalytic activity than single AgBr or Ag_3PO_4, as well as high stability under visible light irradiation. The high stability was attributed to the formed $Ag@AgBr/Ag_3PO_4@Ag$ plasmonic system, which effectively retains its activity due to the efficient transfer of photoinduced electrons [33].

Similarly, Ag_3PO_4-based composite photocatalysts including Fe_3O_4/Ag_3PO_4 (magnetic separable) [34], $In(OH)_3/Ag_3PO_4$ (enhancing absorption by tuning surface electric property) [35], Ag_3PO_4/carbon nanotube-stabilized pickering emulsion (enhancing activity by surface-chemical design of novel micro-reaction system) [36], Ag_3PO_4-graphene [37], $Ag@(Ag_2S/Ag_3PO_4)$ (facilitating migration of charge carriers) (enhancing activity *via* synergistic effect of Ag and Ag_2S) [38], AgX/Ag_3PO_4 (improving stability *via* core-shell structure) [39] and so forth have been successfully synthesized and studied. Each of the synthesized photocatalysts shows exclusive characters (described in the brackets behind) as well as common features shared with others like the promotion of charge carrier separation, increase of surface area and so forth.

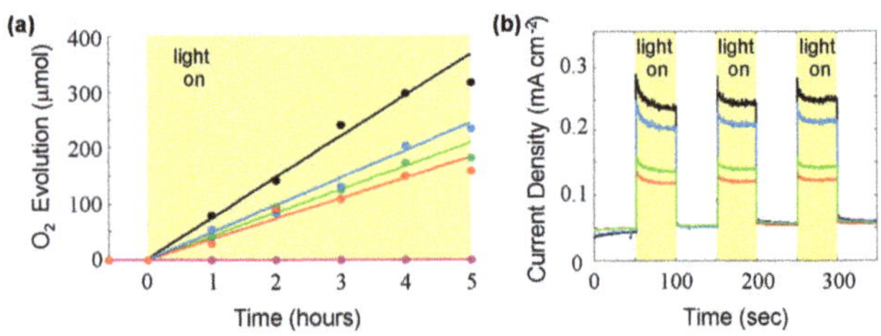

Figure 5. Photocatalytic O_2 evolution under visible light irradiation (wavelength > 420 nm) over bare Ag/AgBr, Ag_3PO_4, Ag_3PO_4/RGO, Ag_3PO_4/Ag/AgBr, and Ag_3PO_4/Ag/AgBr/RGO (from bottom to top) (a) and transient photocurrent responses of electrodes functionalized with Ag_3PO_4-based materials in the same order (bottom to top) as in panel (b) (Measurements proceeded in a 0.01 mol/L Na_2SO_4 aqueous solution under visible light irradiation (wavelength > 420 nm, I0=64 mW/cm^2) at 0.5 V (Vs. SCE) bias. Reprinted with permission. Copyright Hou et al., 2012, ACS Publications.[40]

Hou et al. prepared graphene-supported $Ag_3PO_4/Ag/AgBr$ *via* photoassisted deposition–precipitation method [40]. This was followed by subsequent hydrothermal treatment. He also found that, compared to the bare Ag_3PO_4 powder the O_2-evolution rate of the nano-composite was two times under the irradiation of visible light. Compared with unsupported Ag_3PO_4/Ag/AgBr, graphene supported bare Ag_3PO_4 and Ag/AgBr, it also performed improved activity (Fig. 5). The depletion of the conduction band electrons of Ag_3PO_4, downshift of the Ag_3PO_4 valence band influenced by silver and charge transferring onto the graphene support were responsible for the enhanced activity (Fig. 6).

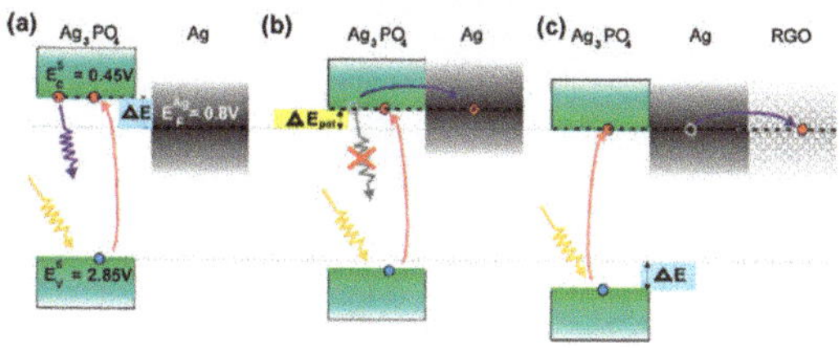

Figure 6. Idealised model of the synergistic increase of photocatalytic activity of Ag_3PO_4 upon functionalization with Ag/AgBr and RGO. Reprinted with permission. Copyright Hou et al. 2013, ACS Publications. Reprinted with permission. Copyright Hou et al., 2012, ACS Publications.[40]

Another special case, in terms of the preparation method which is well investigated by Yu et al. [41] Ag_3PO_4–polyacrylonitrile (PAN) hetero-nanofibers were successfully fabricated through electro-spinning technique (Fig. 7). The Necklace-like Ag_3PO_4-PAN (Fig. 8) exhibited excellent photocatalytic activities for the degradation of organic contaminants under visible light irradiation. By tuning the mass ratio (between Ag_3PO_4 and PAN) and applied voltage, the morphology of the product can be changed correspondently. Since, electrospun polyacrylonitrile is used in the clothing production; this research implies a possibility to make the clothing material photocatalytically self-cleanable.

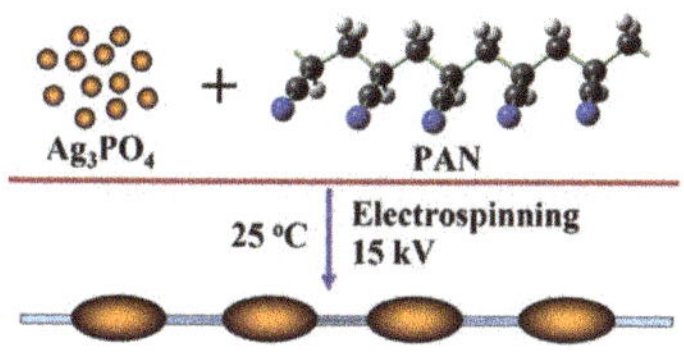

Figure 7. A schematic illustration of the formation process of the Ag_3PO_4–PAN necklace-like nanofibers prepared by electrospinning. Reprinted with permission. Copyright Yu et al., 2013, RSC publishing group.[41]

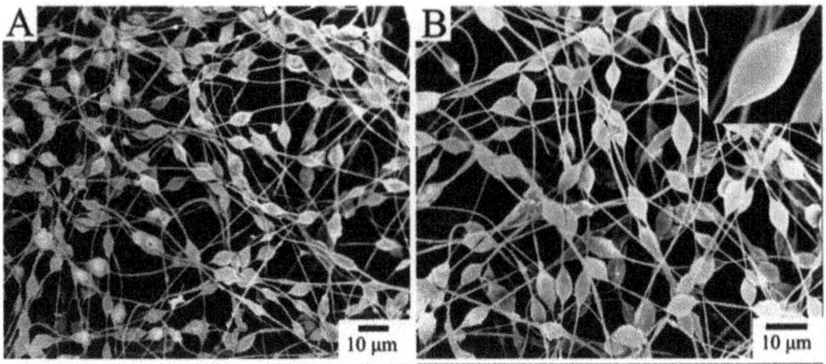

Figure 8. Schematic illustration of formation process of Ag_3PO_4–PAN necklace-like nanofibers prepared by electrospinning (a) and SEM images of products (b, c). Reprinted with permission. Copyright Yu et al., 2013, RSC publishing group. [41]

As mentioned later, examples also exist in the field whereby Ag_3PO_4 is coupled with either a metal or semiconductor (or both) and used to degrade organic contaminants/dyes. In all cases, the modified heterojunction can significantly outperform Ag_3PO_4 alone. The working theory suggests that electron/hole migration from semiconductor to semiconductor prevents recombination, and thus either the electron or hole has a significantly longer lifetime. Wang et al. [42] synthesized $AgBr/Ag_3PO_4$ hybrid microstructures by a facile method. These hybrids exhibited the enhanced photocatalytic performance for degradation of rhodamine B (RhB) under the visible light illumination. Additionally, the effect of AgBr on the recyclability of the hybrid was investigated by them. Consequently, the photo-corrosion was compensated which showed the same degradation rate during the second cycle as the first cycle. They described the charge carrier transfer that when the visible light illuminated, the photogenerated holes in an Ag_3PO_4 particle quickly transferred to an AgBr particle, while the photogenerated electrons migrated to the Ag_3PO_4 particle, promoting the separation of photogenerated carriers in the photocatalytic system.

Fan et al. [43] designed for the first time photocatalytic composite, Ag_3PO_4/g-C_3N_4 to investigate the activities in converting CO_2 to fuels under simulated solar light. It was observed that the loading of Ag_3PO_4 remarkably promoted the photocatalytic activity of g-C_3N_4 on CO_2 photoreduction, proving that the system works in the Z-scheme way. Shalom et al. [44] showed the facile synthesis of an efficient silver phosphate/graphitic carbon nitride (Ag_3PO_4/g-C_3N_4) photocatalyst for oxygen production and pollutant degradation by using electrostatically driven assembly and ion-exchange processes. The composite materials demonstrate a sheet-like C_3N_4 structure, decorated with different

Ag_3PO_4 particles sizes. So, a Z-scheme photocatalytic mechanism for the degradation of RhB and O_2 evolution is proposed by the group. They elaborated that in the beginning, the system behaves similar to an ordinary heterojunction and electrons could pass from the CB of ECN to Ag_3PO_4 or they are generated in Ag_3PO_4 as such. Song et al. reported the magnetic $Ag_3PO_4/TiO_2/Fe_3O_4$ heterostructured nanocomposite. The nanocomposite was found to exhibit markedly enhanced photocatalytic activity, cycling stability and long-term durability in the photodegradation of acid orange 7 (AO7) under visible light. Moreover, the antibacterial film prepared from $Ag_3PO_4/TiO_2/Fe_3O_4$ nanocomposite presented excellent bactericidal activity and recyclability toward *Escherichia coli* (*E. coli*) cells under visible-light irradiation (Fig. 9) [45]. Reunchan et al., using hybrid density-functional calculations, studied the possibilities for n-type and p-type doping in Ag_3PO_4 by doping sulphur and silicone in Ag_3PO_4. It was found that sulfur substituted for phosphorous (SP) has relatively low formation energy (high solubility) and acts as a shallow donor in any examined growth conditions (Fig. 10). Whereas, substitutional silicon at phosphorous site (SiP) is a deep acceptor and its solubility is low, indicating that p-type conductivity is unlikely to occur by Si doping [46]. Chen et al. investigated PANI/Ag/Ag_3PO_4 composite obtained by *in-situ* depositing Ag_3PO_4 nanoparticles on the surface of the prepared PANI and studied the best photocatalytic activity towards Rhodamine B degradation by 20 wt. % PANI/Ag/Ag_3PO_4 composite, which is approximately 4 times higher than that of pure Ag_3PO_4 [47]. The interfacial electric field, formed at the interface of Ag_3PO_4 and PANI can promote the separation efficiency of photogenerated electrons and holes of Ag_3PO_4. Meanwhile, the direction of the electric field may also have contributed to the improvement of the photocatalytic stability of the PANI/Ag_3PO_4 composite. The relevant mechanism is schematically shown in Fig. 11. The long chain of leucoemeraldine base PANI is connected to each other by the –NH- bonds and the N atoms in the outer layer are very easy to adsorb Ag^+ because of its strong negative lone-pair electrons. Then, the adsorbed Ag^+ oxidizes –C-NH- bond to –N-C- bond due to the weak reduction ability of leucoemeraldine base PANI, resulting in the formation of Ag^0. PO_4^{3-} will react with the adsorbed Ag^+ and Ag_3PO_4 particles is then *in-situ* formed on the long chain of PANI. The design and synthesis of highly efficient visible-light-driven photocatalysts through a facile, environmentally friendly, and economical method have become a key aim in the photocatalytic field. In this study, they prepared reduced graphene oxide grafted Ag_3PO_4 (RGO/Ag_3PO_4) composites with enhanced photocatalytic activity by the *in-situ* deposition of Ag_3PO_4 nanoparticles on the surface of RGO sheets. The as prepared photocatalysts were tested towards the photocatalytic degradation of RhB dye.[48] Yang et al. reported a novel Ag@Ag_3PO_4@ZnO ternary heterostructures synthesized through a three-step approach [49]. They fabricated Ag nanorods *via* a modified polyol method which serve as the substrates for subsequent Ag_3PO_4 deposition in an aqueous solution containing mainly $[Ag(NH_3)_2]^+$ and HPO_4^{2-} (I). They described the PVP plays a key role as the bridging molecule to effectively decrease the high interfacial energy between Ag nanorods and Ag_3PO_4 nanoparticles. Also, the authors prepared the samples at different stages in which, the Ag nanorods serve as non-planar substrates with smooth surface are 10 μm in length and 150 nm in diameter. When they adjusted the reactant concentrations to 0.15 M (standard condition), the deposited Ag_3PO_4 crystals are polycrystalline nanoparticles arranged continuously along the Ag nanorods with a proper density. Furthermore, the three components of Ag, Ag_3PO_4 and ZnO possess a good match in their energy band structure, for example, their band-edge positions: ZnO (CB = -0.6 eV, VB = 2.6 eV), Ag_3PO_4 (CB = 0.45 eV, VB = 2.88 eV) and Ag (E_f Ag = 0.4 eV *Vs.* normal hydrogen electrode (NHE)). Therefore, they illustrated in the article that due to the coaxial structure and the matched energy band-edge of the three components, photo-generated electrons can readily transfer to and concentrate on the surface of Ag nanorods, while the holes transport to the valence band of ZnO, thus decreasing the recombination rate of the photo-induced charge carriers.

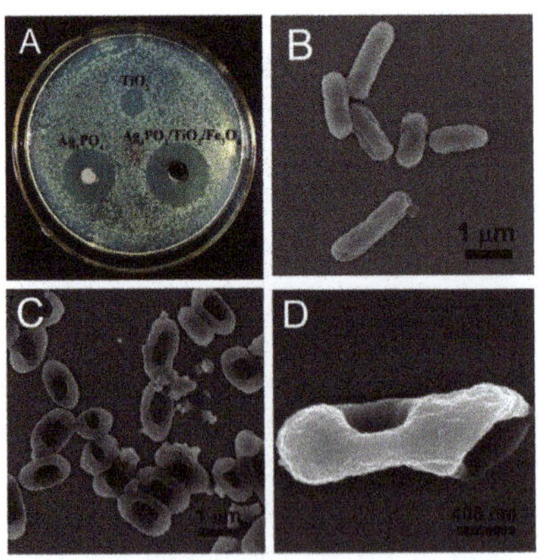

Figure 9. (A) Photographs of antibacterial results on E. coli for TiO_2, Fe_3O_4, TiO_2/Fe_3O_4, Ag_3PO_4, Ag_3PO_4/Fe_3O_4, and $Ag_3PO_4/TiO_2/Fe_3O_4$ samples. (B,C) FE-SEM images of E. coli (B) before and (C) after being killed on $Ag_3PO_4/TiO_2/Fe_3O_4$ under visible-light irradiation. (D) Enlarged view of a damaged E. coli cell. Reprinted with permission. Copyright Xu et al., 2014, ACS Publications.[45]

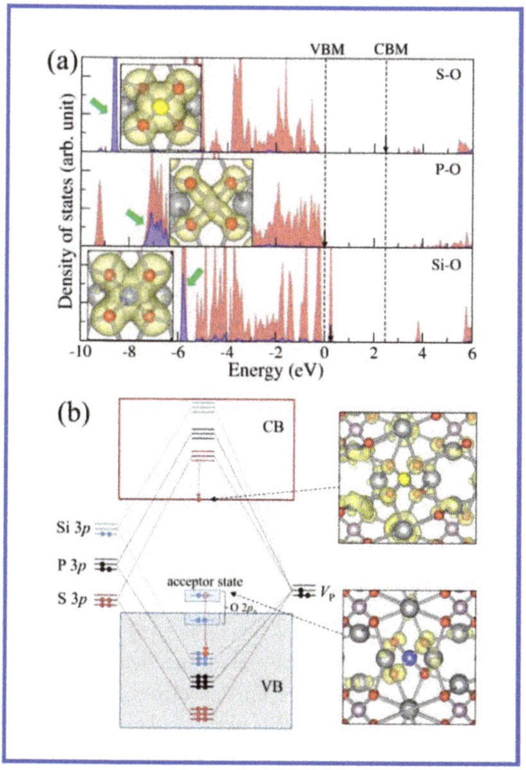

Figure 10. (a) Projected density of states (PDOS) of S, P and Si 3p orbitals (blue shaded) and 2p orbitals (red shaded) of the four nearest O atoms for P S, P P and P Si in the neutral charge state. The insets show the charge distribution of the bonding state associated with S-O, P-O and Si-O bonds, which are indicated by the solid green arrows. The highest occupied state is schematically shown by the vertical black arrows. (b) Schematic diagram for the interaction between 3p orbitals of S, P or Si with the p orbitals of P vacancy (P V) in P S , P P and P Si in Ag_3PO_4. The solid dots represent the electron occupation. The charge distributions of the bottom of the conduction band in P S and of the acceptor state which is located above the valence band (VB) in P Si are also shown. The isosurfaces in all cases are set to about 90% of the maximum of the respective charge densities. Reprinted with permission. Copyright Umezawa et al., 2015, ACS Publications.[46]

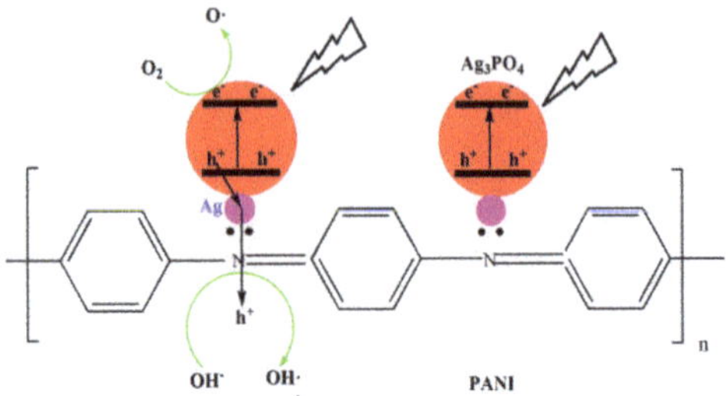

Figure 11. Proposed mechanism for the charge transfer of the photogenerated electrons and holes by Ag_3PO_4 in the PANI/Ag/Ag_3PO_4 composite under visible light illumination. Reprinted with permission. Copyright Bu et al., 2014, ACS Publications. [47]

Zhai et al. also used carbon compounds to help improve the photooxidation of water by Ag_3PO_4 (Fig. 12) [36]. By using multi-wall carbon nanotubes (MWNTs), the authors were able to double the amount of oxygen produced by Ag_3PO_4, and also greatly improve stability. The MWNTs could shuttle photogenerated electrons away from Ag_3PO_4 due to their relatively large work functions (4.30 to 4.95 eV) and excellent electronic conductivity as a result from specific p-conjugated structures, thus being able to prevent photodegradation [50-52]. Such electron transfer from Ag_3PO_4 to MWNTs was evidenced by Pd^{2+} reduction to Pd on the MWNTs, not yet found directly on the Ag_3PO_4, or in the absence of Ag_3PO_4. It should be noted that the contribution from plasmonic effects of photoreduced Ag nanoparticles on this electron shuttling could be actually ruled out because; (1) Ag nanoparticles were not formed in this system as evidenced by the great improvement of stability of Ag_3PO_4; (2) plasmonic effects usually cause low quantum efficiency which is not likely to lead to the photodeposition of Pd on MWNTs in 20 min irradiation [53-54].

Silver phosphate can also be utilized for valence band charge transfer when combined with $SrTiO_3$, demonstrated by Guan et al. (Fig. 13) [55]. The composite was synthesized using a facile hydrothermal method, whereby nanosized $SrTiO_3$ particles were grown on Ag_3PO_4 polyhedrons. By controlling the molar ratio of $SrTiO_3$ to Ag_3PO_4, the authors were able to enhance the photoactivity for oxygen evolution up to 75%. Interestingly, the composite was much less active in comparison to that synthesized *via* a hydrothermal method, which was ascribed to poor dispersion and bad interfacial contact between the two semiconductors. The authors attributed the increase in activity to an electron–hole transfer mechanism, decreasing recombination and increasing charge separation. The $SrTiO_3$–Ag_3PO_4 composite demonstrated greater stability as evidenced by a larger TON (2.04) in comparison to bare Ag_3PO_4 (1.69). Despite this, $SrTiO_3$ has a more negative conduction band than that of Ag_3PO_4. Therefore, electrons cannot be transferred from $SrTiO_3$ to Ag_3PO_4 and the composite still suffers from bulk photocorrosion (unless a scavenger is used).

Whilst, it is evident that electron reduction of silver cations has serious photocorrosion issue in Ag_3PO_4, several photocatalysts have the ability to accept electrons due to their lower conduction band. For example, Li et al. very recently demonstrated that deposited Ag_3PO_4 nanoparticles on (040) facets of large $BiVO_4$ crystals could efficiently degrade methylene blue (MB), how the CB of $BiVO_4$ lies at a more positive position than the CB of Ag_3PO_4 and reasons of its excellent stability [56]. Very recently, FeOOH was shown to have favorable conduction band positioning (+0.5 V *Vs.* NHE) and also shown to have significant enhancement in the photooxidation ability of $BiVO_4$ when deposited as a catalyst, boosting IPCE (incident photon-to-electron conversion) from 9% up to 60% in the visible light range [57-58]. Considering the position of the CB of Ag_3PO_4 (*Ca.* +0.4 V *Vs.* NHE), the same strategy could also be applied, so that electrons are shuttled effectively from Ag_3PO_4 to FeOOH and thereby increasing hole lifetimes leading to efficient water photooxidation. In terms of regeneration, Wang and co-workers used sodium phosphate and hydrogen peroxide to rejuvenate Ag to Ag_3PO_4, showing that it can be easily recycled [59]. Alternatively, Yi et al. developed two strategies to anodically oxidise photoreduced Ag back to Ag_3PO_4; here they definitively show ions in the reactant solution can serve as a phosphate source and replenish the photocatalyst with successive rejuvenations showing no loss of photocurrent [60]. Moniz et al. have recently reviewed successes based on $BiVO_4$ architectures with reported STH conversion efficiencies of over 4.5% [61]. Considering $BiVO_4$ was discovered over a decade before Ag_3PO_4 and both have similar CB positions, it is envisaged that with continued development silver phosphate has the potential to achieve commercial success, provided problems such as photodegradation are solved.

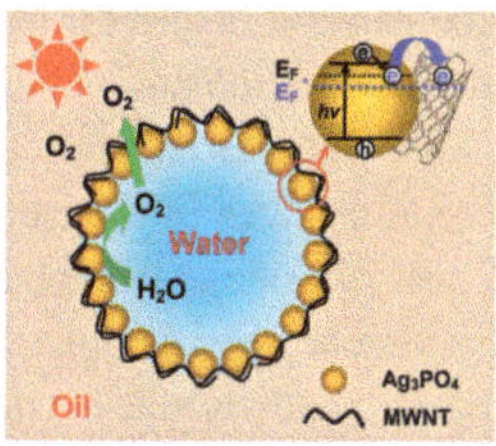

Figure 12. Schematic of a highly efficient Pickering emulsion-based photocatalytic system formed by self-assembling Ag_3PO_4–MWNT nanohybrid at the water/oil interface. Reprinted with permission. Copyright Zhai et al. 2013, ACS Publications.[36]

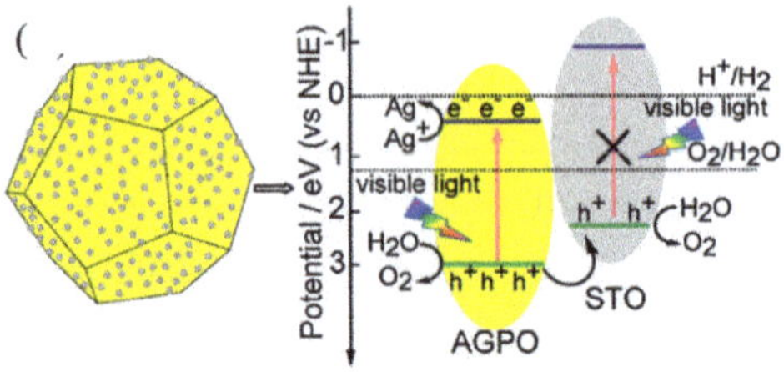

Figure 13. Schematic diagram of band structure and expected charge separation of $SrTiO_3/Ag_3PO_4$ composite under visible-light irradiation. Reprinted with permission. Copyright 2014, Guan et al. ACS Publications.[55]

To improve upon the stability and photoactivity of Ag_3PO_4, various systems coupled with silver phosphate as described. Our group also tried to augment prominently the stability and reactivity of Ag_3PO_4 by designing Z-scheme heterostructure with RGO (Fig. 14) without any mediator third party [62]. A visible light driven, direct Z-scheme reduced graphene oxide–Ag_3PO_4 (RGO–Ag_3PO_4) heterostructure was synthesized by means of a simple one-pot photoreduction route by varying the amount of RGO under visible light illumination. Furthermore, total organic carbon (TOC) analysis confirmed 97.1% mineralization of organic dyes over RGO–Ag_3PO_4 in just five minutes under visible light illumination. The use of different quenchers in the photomineralization suggested the presence of hydroxyl radicals ($^{\bullet}OH$), superoxide radicals ($^{\bullet}O_2^-$), and holes (h^+), which play a significant role in the mineralization of organic dyes. Additionally, clean hydrogen fuel generation was also observed with excellent reusability. Unlikely to the previous reports on water oxidation over Ag_3PO_4, it is the first report on hydrogen evolution from Ag_3PO_4 hybrids. The RGO–Ag_3PO_4 heterostructure has a high H_2 evolution rate of 3690 $\mu molh^{-1}g^{-1}$, which is 6.15 times higher than that of RGO. Also we have studied the photostability and reusability of the heterostructure for practical applications by both dye mineralization and H_2 production experiments. The production of H_2 increased progressively with illumination time without any deactivation, even after four cycles (Fig. 15a). To ascertain the structural changes of the catalysts after 12 h of illumination during water splitting, HRTEM analysis of the photocatalyst after the fourth run was carried out (Fig. 15b). It is clearly visible from the micrograph that there was no change in the catalyst as far as morphology and particle size is concerned.

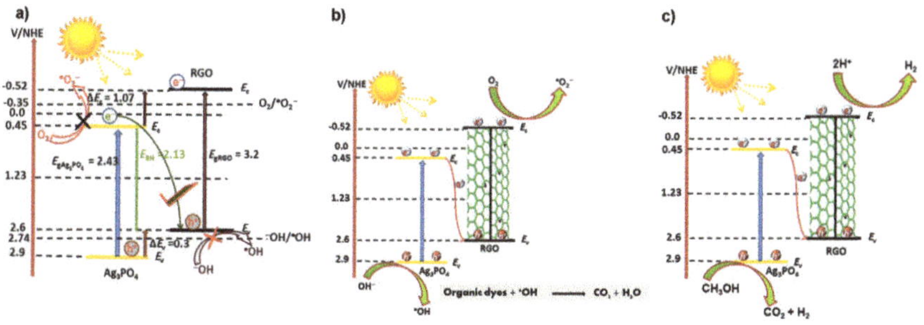

Figure 14. a) Schematic representation of e–h+ transfer process. Representative possible Z-scheme mechanism for b) mineralization of dyes and c) H_2 evolution over the xRGO–Ag_3PO_4 heterostructure under visible light illumination. Reprinted with permission. Copyright 2016, Samal et al. Wiley-VCH. [62]

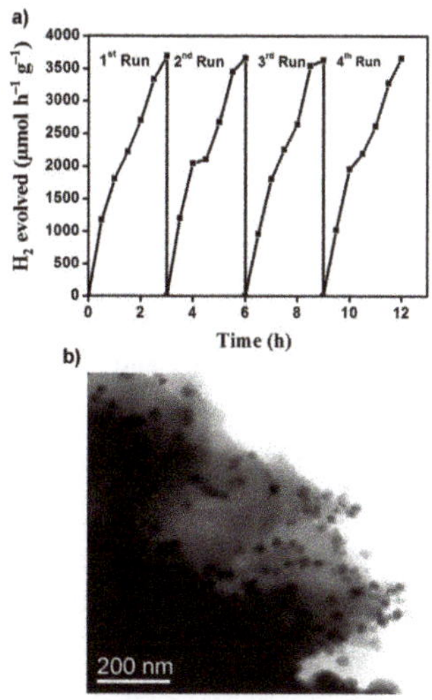

Figure 15. a) Reusability test for water splitting experiments up to the fourth run. b) TEM micrograph of 4RGO–Ag_3PO_4 nanocomposite employed after fourth run. Reprinted with permission. Copyright 2016, Samal et al. Wiley-VCH. [62]

To briefly summaries, we have reviewed various strategies that have been employed to improve the photocatalytic properties of Ag_3PO_4 for organic contaminant degradation (Table 1). Although, many significant improvements in developing Ag_3PO_4 based photocatalysts have been made up till now, the current performance is still far from satisfactory for practical applications and more improvement is needed.

Table 1. Recent studies on Ag_3PO_4 based photocatalysts and corresponding organic pollutant degradation efficiency.

Photocatalyst	Organic pollutant degradation efficiency[Ref]
RGO-Ag_3PO_4 Z-scheme heterostructure	20 mg/L of concentration of organic dyes got degraded upto 100% in 5 min of visible light irradiation (our work)[62]
Ag_3PO_4 nanocrystals	90% MB degraded within 80 min, 2 and 4.6 times higher that of Ag_3PO_4 microcrystals and N-doped TiO_2-P25 [63]
Graphene/nanosized-Ag_3PO_4	90% MB degraded within 10 min [64]
Ag_3PO_4 porous microcubes	99% RhB degraded within 24 min, 2 times higher that of solid Ag_3PO_4 samples [65]
Various Ag_3PO_4 crystals (branch, tetrapod, nanorod, and triangular prism)	100% MB and 98% RhB degraded on branched Ag_3PO_4 within 30 min and 35 min [66]
Dendritic Ag_3PO_4	Completely RhB degradation in 3 min, about 500 times higher that of N-doped TiO_2 [67]
Ag_3PO_4 rhombic dodecahedrons {110} and cubes {100}	Rhombic dodecahedrons: completely MO and RhB degradation in 4 min and 3 min, around 560 and 360 times faster than that of commercial N-doped TiO_2, respectively. Photocatalytic activities with the order rhombic dodecahedrons{110} 4 cubes {100} [68]
Ag_3PO_4 tetrahedron {111}	Photocatalytic activities with the order tetrahedrons {111} 4 rhombic dodecahedrons {110} 4 cubes {100} (MB, MO, and RhB degradation) [69]
Trisoctahedral Ag_3PO_4 enclosed by {221} and {332} facets	Completely degrade RhB: 3 min (Ag_3PO_4 trisoctahedrons), 8 min (Ag_3PO_4 cubes) [70]
Ag/Ag_3PO_4	Completely degrade RhB: 4 min (Ag/Ag_3PO_4 heterocrystals), 8 min (Ag_3PO_4 cubes) [71]

Necklace-like Ag nanowire/Ag_3PO_4 cubes	Completely degrade RhB: 2 min (Ag nanowire/Ag_3PO_4 cubes), 8 min (Ag_3PO_4 cubes) Completely degrade MO: 8 min (Ag nanowire/Ag_3PO_4 cubes), 14 min (Ag_3PO_4 cubes) [72]
Ag_3PO_4/AgBr/Ag	3.8 and 3.2 times MO and MB degradation rates those of Ag_3PO_4, respectively [73]
Ag_3PO_4–graphene	Nearly 100% RhB degraded in 2 min, about 2 times faster than that of Ag_3PO_4 polyhedra [74]
Graphene oxide–Ag_3PO_4	9.3 times MB degradation rates that of Ag_3PO_4 nanocrystals [75]
Ag_3PO_4/nitridized-$Sr_2Nb_2O_7$	70 times isopropanol decomposition rate of pure Ag_3PO_4 [76]
Bi^{3+} doped Ag_3PO_4	7.3 times MO degradation rate that of undoped Ag_3PO_4 [77]

4. Ag_3PO_4 photoactivity based on morphology

It is commonsensical in the field of photocatalysis, the morphology (size, shape and kind of exposed facets etc.) of the photocatalysts has a control on the efficiency and activity. So, in case of Ag_3PO_4/based materials also this morphology engineering has a significant effect. Since, Ag_3PO_4 is a promising photocatalyst, lots of researches have been conducted to the morphology tuning. Such novel shapes as branch, tetrapod, nanorod, triangular prism [66] pine tree [78] and porous microcubes [79] are successfully obtained. Generally, those preparation methods involve precipitation (with the aid of Ag^+–ligand complex, organic additives or templates) [79,68], chemical or electrochemical oxidation of metallic Ag into Ag^+ (subsequently captured by PO_4^{-2}) [80], ligand assisted anion exchange process [81] and hydrothermal synthesis [82] etc. Except for modulating the inner conditions of the reactions, those outside factors like ultrasound were also demonstrated to be influential to the morphology [66]. Wang et al. [83] first managed to synthesize uniform tetrapod-like Ag_3PO_4 microcrystals (T-Ag_3PO_4) with a simple hydrothermal method without adding any template or surfactant. He used the precursor was phosphoric acid while they tuned the pH value of the reaction system by urea.

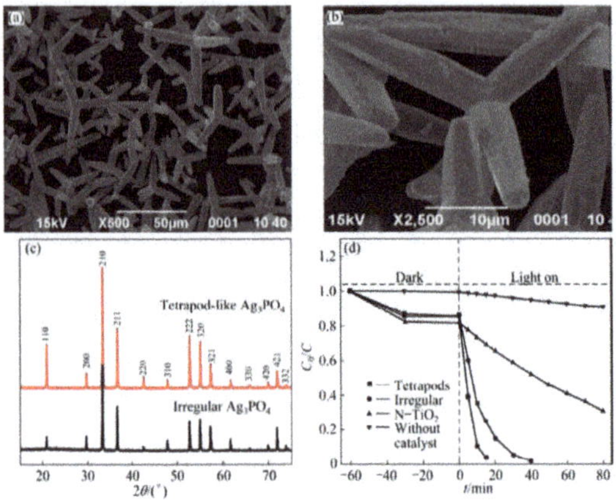

Figure 16. SEM images of T-Ag_3PO_4 at different magnifications (a,b), XRD patterns of tetrapod-like and irregular Ag_3PO_4 (c) and degradation of RhB with tetrapod-like Ag_3PO_4, irregular Ag_3PO_4 and N-doped TiO_2 under visible-light (λ>420 nm) (d). Reprinted with permission. Copyright 2013, Wang et al. RSC Publishing group.[83]

By varying the amount of the urea, reaction time and temperature, the morphology of the product was successfully tuned by the researchers. As the SEM images (Fig. 16(a) and (b)) shown, four arms of the tetrapod are cylindrical microrods with an average diameter of 5 μm and a length of 15–30 μm. And the XRD patterns showed (Fig. 16(c)), the intensity ratios of (110)/(200) and (222)/(321) for T-Ag_3PO_4 are 2.9 and 1.6, which are remarkably higher than those (0.56 and 0.9) of the irregular counterpart, respectively.

From the XRD pattern, the high exposing rate of the (110) facet is demonstrated by the researchers too. They described the origin for higher intensity ratio of (110)/(200) is credited to the high surface energy of (110). The activity of the product was verified in RhB degradation (Fig. 16(d)). In comparison with N-doped TiO_2, the Ag_3PO_4 samples exhibited higher catalytic activities, while the T-Ag_3PO_4 possessed the highest activity. The highest activity of T-Ag_3PO_4 was resulted from the higher surface energy of (110) facets than those of (200) facets.

While considering a special example with the preparation method, Jiao et al. [70] fabricated various shaped Ag_3PO_4 microcrystals based on the heteroepitaxial growth procedure, in which different seeds were added into the reaction system before precipitation happened. This is a procedure well recognized in the field of nano fabrication, in which the nucleation and crystal growth are separated in terms of space and time [84]. Fig. 17 [70] illustrates the synthesis procedure along with the SEM images of the products. In point of fact, the structure multiplicity of the crystal nucleus will lead to drastically different shaped products, the usage of pure single sorts of seeds are beneficial to forming pure, uniform shaped crystals as well as to obtaining the desired morphology. Though, in the literature no such comparison is reported between individuals in terms of activity, nearly all of them showed an enhanced reaction rate with regard to the spherical counterpart and N-doped TiO_2. Considering the large amount of the researches, some of the reliable and representative instances are listed in Table 1 with regard to the morphology, preparation method and photocatalytic activity.

Dong et al. interestingly synthesized Ag_3PO_4 microcrystals with different morphologies, including tetrahedrons with round edge sand corners, short tetrapods, polyhedrons, and dendritic long tetrapods *via* simple and green routes [85]. When the group increased the concentration of KH_2PO_4 to 0.6 mol L^{-1}, the number of short tetrapods increased basically. They showed clearly in the morphological description that tetrahedral crystal nucleus with exposed {111} facets are in the centre of four short arms and the four arms with length of 1 μm preferentially grow along the [110] direction. And when glacial acetic acid was added into system the tetrapods with longer dendritic arms formed as main product. Every arm is like one dendrite with symmetrical sawtooth and one fillister runs through the whole dendrite. They described the formation of dendritic tetrapods is attributed to the decrease of nuclei number in the initial reaction stage because of the introduction of glacial acetic acid. It was revealed that Ag_3PO_4 tetrahedral, tetrapods and dendritic long tetrapods could absorb visible light with wavelengths shorter than 530 nm, whereas the adsorption edge of polyhedrons is at 510 nm. The authors examined the photocatalytic activity of Ag_3PO_4 products and N-doped TiO_2 nanoparticles at the first cycle. They also seen that even though the ambient temperature was as low as 13 °C, all the Ag_3PO_4 products presented excellent photocatalytic activity and tetrahedral Ag_3PO_4 with round edges and corners had the highest activity. Bi et al. fabricated concave trisoctahedral Ag_3PO_4 microcrystals enclosed with {221} and {332} facets based on the heteroepitaxial growth procedure, which exhibit much higher photocatalytic activities than cubic Ag_3PO_4 and commercial N-doped TiO_2 [68].

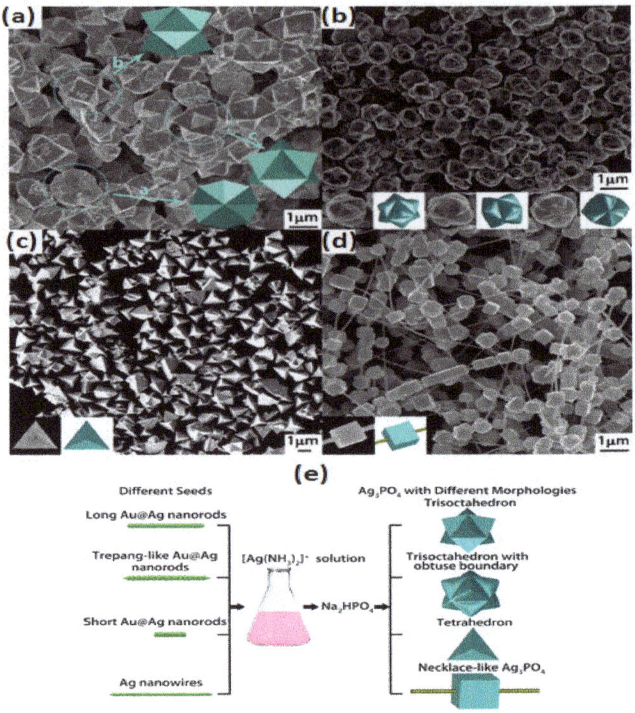

Figure 17. Schematic illustration of growth process of Ag_3PO_4 with different morphologies fabricated by seed-mediated method using different seeds (a) and SEM images of four kinds of products (b-e). Reprinted with permission. Copyright 2013, Jiao et al. RSC Publishing group.[70]

Jiao et al. [70] fabricated concave trisoctahedral Ag_3PO_4 microcrystals enclosed by {221} and {332} facets based on the heteroepitaxial growth procedure, which exhibit much higher photocatalytic activities than cubic Ag_3PO_4 and commercial N-doped TiO_2 as shown in Fig. 18. Fig. 18(a) shows a SEM image of a single trisoctahedral Ag_3PO_4 and they used red lines to accurate the edges that form the outline of the trisoctahedral Ag_3PO_4. A model of trisoctahedral projected from the {110} direction has been shown in Fig. 18(b), and the four edge-on facets are indicated by the black arrows. Besides, their group compared the photocatalytic behaviors of trisoctahedral Ag_3PO_4 for the degradation of RhB under visible-light irradiation with cubic Ag_3PO_4 and N-doped TiO_2.

They observed that except N-doped TiO_2, both these Ag_3PO_4 photocatalysts showed excellent photocatalytic activities for the RhB degradation reaction Fig. 18(d). But, the trisoctahedral one is more active as compared to cubic Ag_3PO_4 photocatalysts. Furthermore, the photocatalytic behaviors of TOH Ag_3PO_4 for the degradation of RhB under visible-light irradiation were explored. For comparison, the performances of cubic Ag_3PO_4 and N-doped TiO_2 were also investigated. As shown in Fig. 18(e), except for N-doped TiO_2, both these Ag_3PO_4 photocatalysts exhibited excellent photocatalytic activities for the RhB degradation reaction. Also they explained that although owning much larger dimensions than cubes (about 500 nm), the trisoctahedral Ag_3PO_4 microcrystals exhibited higher photocatalytic activity than cubic submicro-crystals.

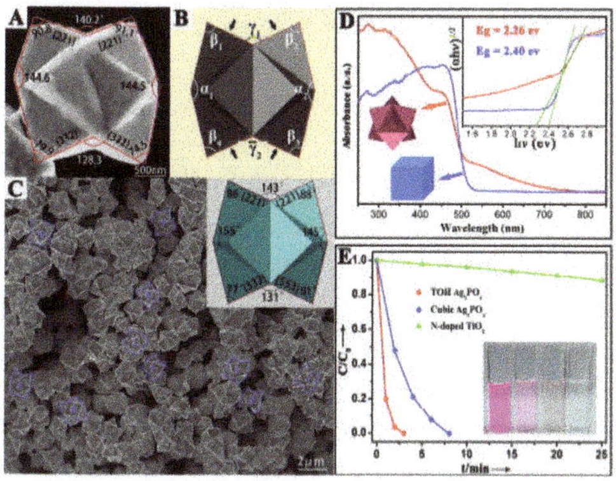

Figure 18. (a) SEM image and (b) model of a TOH Ag_3PO_4 microcrystal viewed along the <110> direction. (c) SEM images of TOH Ag_3PO_4 and the inset is an ideal model fabricated by the average values of ten TOH microcrystals whose edges are accurated by blue lines. (d) UV-vis diffusive reflectance spectra and the inset show photographs and plots of $(\alpha hv)1/2$ vs hv. (e) photocatalytic activities of TOH Ag_3PO_4 for RhB degradation under visible-light irradiation. Reprinted with permission. Copyright 2013, Jiao et al. RSC Publishing group.[70]

Ye et al. have developed a facile and general route for high-yield fabrication of single-crystalline Ag_3PO_4 rhombic dodecahedrons with only {110} facets exposed and cubes bounded entirely by {100} facets [68]. Among them, the Ag_3PO_4 rhombic dodecahedrons exhibited the highest photocatalytic activity, as they could completely degrade MO dye in 4 min under visible-light irradiation. In contrast, the cubes decomposed MO in 14 min, while the spherical Ag_3PO_4 particles required ~28 min. They further studied the surface structures and surface energies of Ag_3PO_4 {100} and {110} planes through density functional theory (DFT) calculations as shown in Fig. 19.

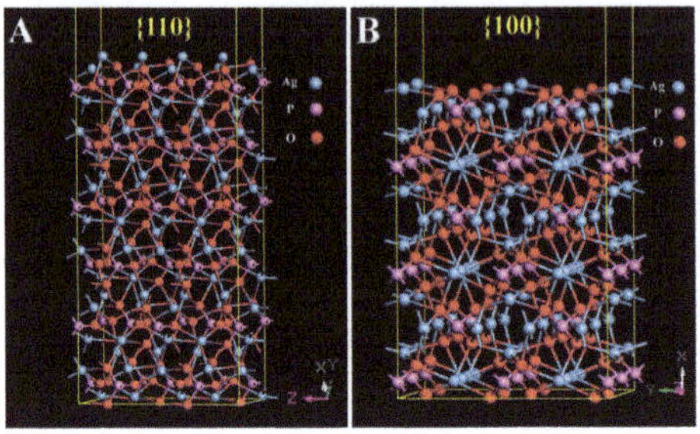

Figure 19. Relaxed geometries for the (A) {110} and (B) {100} surfaces of Ag_3PO_4 based on a 192-atom slab model. The vacuum region was set to the same thickness as Ag_3PO_4. Reprinted with permission. Copyright 2011, Bi et al. ACS Publications.[68]

Wang et al. investigated a facile soft-chemical method for the synthesis of Ag_3PO_4 crystals with various new morphologies like branch, tetrapod, nanorod, triangular prism (Fig. 20). Also, they studied the morphological effect on the photocatalytic activity of the obtained Ag_3PO_4 crystals [66]. They have studied the photocatalytic activity of branched Ag_3PO_4 and nanorod-shaped Ag_3PO_4 photocatalysts exhibit higher photocatalytic activities than the irregular spherical Ag_3PO_4 particles for MB degradation reaction. In particular, the branched Ag_3PO_4 exhibits the highest photocatalytic activity among these samples, as it could completely degrade MB dye in 30 min under visible light irradiation. In contrast, about 78% of the initial MB molecules were decomposed by the irregular spherical Ag_3PO_4 particles within 30 min.

Teng et al. also found that the surface morphology and the crystal structure of the product were significantly affected by the amount of glycine added, the reaction time and the reaction temperature. In particular, glycine was found to be the most vital factor to control the growth of Ag_3PO_4 as can be evidenced from the Fig. 21. First, they have investigated the effect of the amount of glycine added to the products [86]. Furthermore, the method for degradation of Rhodamine B (RhB) was used as a probe reaction to investigate the photocatalytic activity of the etched Ag_3PO_4 under visible light irradiation. Ag_3PO_4 sample prepared at 2:1 has the highest degradation efficiency.

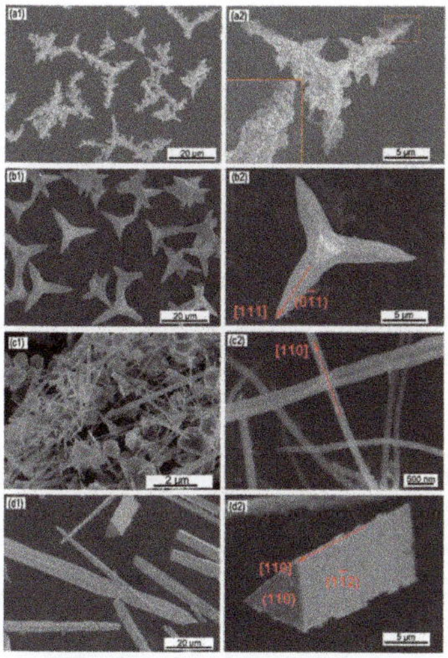

Figure 20. SEM images of branched (a), tetrapod (b), nanorod-shaped (c) and triangular-prism-shaped (d) Ag_3PO_4 crystal. Reprinted with permission. Copyright 2013, Dong et al. RSC Publishing group.[66]

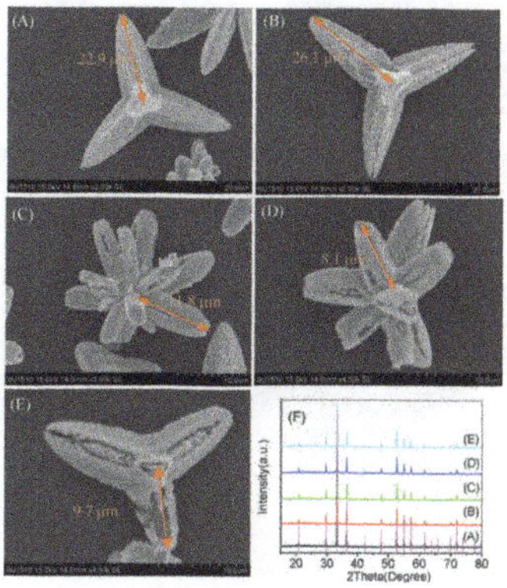

Figure 21. SEM images (A–E) and XRD patterns (F) of the samples prepared at different molar ratios of glycine/Ag(I): (A) 0.5 : 1, (B) 1 : 1, (C) 2 : 1, (D) 3:1, and (E) 4:1. Reaction temperature: 20 °C; stirring time upon addition of phosphoric acid: 40 min. Reprinted with permission. Copyright 2014, Wang et al. RSC Publishing group.[86]

Martin et al. [87] investigated the photooxidation of water using faceted Ag_3PO_4. The group wonderfully performed theoretical calculations to predict the optimum morphology for solar energy conversion by probing the surface energies of three primary low index facets of Ag_3PO_4: {100}, {110} and {111}. Fig. 22, shows the different morphologies confirmed by SEM micrographs. TEM tilt studies also they have shown in figure to confirm the tetrahedron morphology. The group also found that in comparison to rhombic dodecahedron {110} and cubic {100} structures, tetrahedral crystals show an extremely high activity for water photooxidation, with an initial oxygen evolution rate exceeding 6 mmol $h^{-1}g^{-1}$, 10 times higher than either {110} or {100} facets. So, this informs the morphology as well as facet engineering has a great impact on the photo-catalytic activity of Ag_3PO_4.

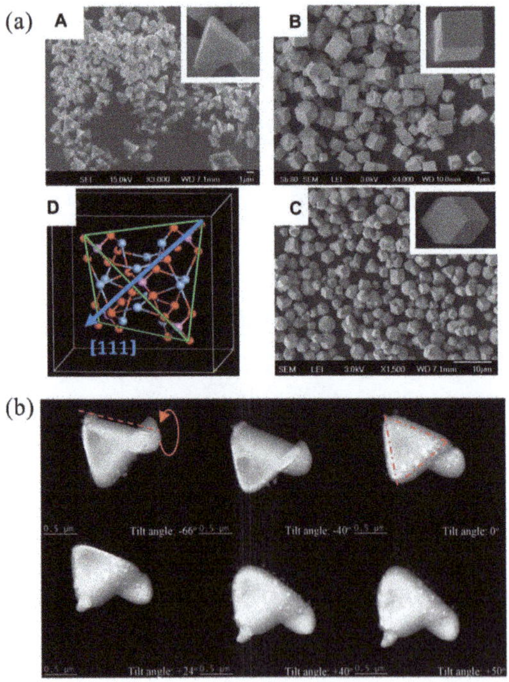

Figure 22. (a) SEM micrograph of Ag3PO4 crystals: (A) tetrahedron, (B) cubic, (C) rhombic dodecahedron, and (D) Materials Studio visualisation of a tetrahedron. (b) TEM tilt studies. A tetrahedron indicated by dotted lines is rotated on an axis from -66 to +50o. Reprinted with permission. Copyright 2013, Martin et al. RSC Publishing group.[87]

5. Summary and outlook

Silver phosphate is one of the most active photocatalysts recorded to date since its first application in photocatalysis in 2009, boasting quantum yields larger than all commonly known oxidation photocatalysts such as TiO_2, $BiVO_4$, and WO_3. In this review, we have demonstrated that Ag_3PO_4 can be augmented in terms of both water photooxidation and decontamination including bacterial disinfection. The effective control on morphology and facet engineering affect the extraordinary activity of Ag_3PO_4 in a positive way which further gives a new light in the path of photocatalytic science. Particular faceted crystals have been shown to be more active due to preferential charge transfer to specific crystal planes. The construction of a heterojunction, e.g. metal–semiconductor, semiconductor–semiconductor, or more purposefully, a p–n junction, is evidently beneficial for either photooxidation or organic decomposition. By transferring the charge carriers from either conduction or valence bands, by using interfaces such as RGO/C_3N_4/fullerenes, it is possible to both improve electron–hole separation and reduce photocorrosion. This is a very important aspect as experimental studies show the active species for degradation and oxidation is in most cases VB holes and a two-electron CB reduction. Unfortunately, scanty literatures are available towards the description of chemistry of water photooxidation efficiency of Ag_3PO_4. It would appear that photocatalyst stability remains as a major problem of Ag_3PO_4 in both water photooxidation and organic decomposition. To solve the problem, a suitable and readily available inorganic electron acceptor, e.g. metal ions such as Ce^{4+}, Fe^{3+} etc., could be used to accept electrons from the CB of Ag_3PO_4. The more promising strategy is to build a junction structure where an electron acceptor e.g. carbon-based, Ag metal or (reduced) graphene oxide, is used to stabilize the photocatalyst as demonstrated by several preliminary studies, leading to the subsequent stable and efficient reduction and oxidation reactions. Subsequently, there is an urgent need for comprehensive experimental mechanistic studies to fully understand the nature behind enhanced stability created by combining Ag_3PO_4 with carbonaceous materials. In addition, the photocatalyst can also be chemically or photoelectrochemically rejuvenated *in-situ* using hydrogen peroxide or electrical bias respectively. Although, the outlook is promising, there are some important considerations which must be taken into account before commercialization. It is clear that a combination of crystal facet control, and heterojunction construction will be vital future components for both water splitting and water treatment. In the context of this review, Ag_3PO_4 is currently one of the most efficient photocatalysts to date, but an easy way to synthesize and the cost is highly sought after. By solving these problems, a practical device can be fabricated to obtain a commercially feasible system for solar fuel generation and water purification.

Acknowledgements

Authors are very much thankful to Mr. S.K. Mishra, Acting Director, CSIR-IMMT for giving permission to publish the review. AS is thankful to CSIR for the grant of Senior Research Fellowship (SRF).

References

[1] A. Kudo, K. Omori, H. Kato, A Novel Aqueous Process for Preparation of Crystal Form-Controlled and Highly Crystalline $BiVO_4$ Powder from Layered Vanadates at Room Temperature and Its Photocatalytic and Photophysical Properties, J. Am.Chem. Soc. 121 (1999) 11459-11467. https://doi.org/10.1021/ja992541y

[2] F.E. Osterloh, Inorganic Materials as Catalysts for Photochemical Splitting of Water, Chem. Mater. 20 (2008) 35-54. https://doi.org/10.1021/cm7024203

[3] F.E. Osterloh, Inorganic nanostructures for photoelectrochemical and photocatalytic water splitting, Chem. Soc. Rev. 42 (2013) 2294–2320. https://doi.org/10.1039/C2CS35266D

[4] M.A. Henderson, A surface science perspective on TiO_2 photocatalysis, Surf. Sci. Rep. 66 (2011) 185–297. https://doi.org/10.1016/j.surfrep.2011.01.001

[5] P.M. Forster, J. Eckert, J.S. Chang, S.E. Park, G. Ferey, A.K. Cheetham, Hydrogen Adsorption in Nanoporous Nickel(II) Phosphates, J. Am. Chem. Soc. 125 (2003) 1309-1312. https://doi.org/10.1021/ja028341v

[6] Y.C. Liao, C.H. Lin, S.L. Wang, Direct White Light Phosphor: A Porous Zinc Gallophosphate with Tunable Yellow-to-White Luminescence, J. Am. Chem. Soc. 127 (2005) 9986-9987. https://doi.org/10.1021/ja0512879

[7] A.S. Arico, P. Bruce, B. Scrosati, J.M. Tarascon, W.V. Schalkwijk, Nanostructured materials for advanced energy conversion and storage devices, Nat. Mater. 4 (2005) 366-377. https://doi.org/10.1038/nmat1368

[8] M.E. Davis, Ordered Porous Materials for Emerging Applications, Nature 417 (2002) 813-821. https://doi.org/10.1038/nature00785

[9] G.J. Hutchings, Vanadium phosphate: a new look at the active components of catalysts for the oxidation of butaneto maleic anhydride, J. Mater. Chem. 14 (2004) 3385–3395. https://doi.org/10.1039/b404610m

[10] E. Muneyama, A. Kunishige, K. Ohdan, M. Ai, Reduction and Reoxidation of Iron Phosphate and Its Catalytic Activity for Oxidative Dehydrogenation of Isobutyric Acid, J. Catal. 158 (1996) 378–384. https://doi.org/10.1006/jcat.1996.0039

[11] Y. Wang, X.X. Wang, Z. Su, Q. Guo, Q.H. Tang, Q.H. Zhang, H.L. Wan, SBA-15-supported iron phosphate catalyst for partial oxidation of methane to formaldehyde, Catal. Today 93–95 (2004) 155–161. https://doi.org/10.1016/j.cattod.2004.06.037

[12] R.H. Lin, Y.J. Ding, L.F. Gong, W.D. Dong, J.H. Wang, T. Zhang, Efficient and stable silica-supported iron phosphate catalysts for oxidative bromination of methane, J. Catal. 272 (2010) 65–73. https://doi.org/10.1016/j.jcat.2010.03.011

[13] X.Y. Tian, W. He, J.J. Cui, X.D. Zhang, W.J. Zhou, S.P. Yan, X.N. Sun, X.X. Han, S.S. Han, Y.Z. Yue, Mesoporous zirconium phosphate from yeast biotemplate, J. Colloid Interface Sci. 343 (2010) 344–349. https://doi.org/10.1016/j.jcis.2009.11.037

[14] [X.S. Li, A.R. Courtney, W. Yantasee, S.V. Mattigod, G.E. Fryxell, Templated synthesis of mesoporous titanium phosphates for the sequestration of radionuclides, Inorg. Chem. Commun. 9 (2006) 293–295. https://doi.org/10.1016/j.inoche.2005.11.014

[15] A. Dutta, A.K. Patra, A. Bhaumik, Porous organic–inorganic hybrid nickel phosphonate: Adsorption and catalytic applications, Micropor. Mesopor. Mater. 155 (2012) 208–214. https://doi.org/10.1016/j.micromeso.2012.01.017

[16] A. Samal, S. Swain, B. Satpati, D.P. Das, B.K. Mishra, 3 D $Co_3(PO_4)_2$ -Reduced Graphene Oxide Flowers for Photocatalytic Water Splitting: A Type II Staggered Heterojunction System, ChemSusChem 9 (2016) 3150-3160. https://doi.org/10.1002/cssc.201601214

[17] M. Pramanik, R.R. Salunkhe, M. Imura, Y. Yamauchi, Phosphonate-Derived Nanoporous Metal Phosphates and Their Superior Energy Storage Application, ACS Appl. Mater. Interf. 8 (2016) 9790–9797. https://doi.org/10.1021/acsami.6b01012

[18] J. Wang, T. Zhu, G.W. Ho, Nature-Inspired Design of Artificial Solar-to-Fuel Conversion Systems Based on Copper Phosphate Microflowers, ChemSusChem 9 (2016) 1 – 5. https://doi.org/10.1002/cssc.201501673

[19] Z. Yi, J. Ye, N. Kikugawa, T. Kako, S. Ouyang, H. Stuart-Williams, H. Yang, J. Cao, W. Luo, Z. Li, An orthophosphate semiconductor with photooxidation properties under visible-light irradiation, Nat. Mat. 9 (2010) 559–564. https://doi.org/10.1038/nmat2780

[20] (a) X. Ma, B. Lu, D. Li, R. Shi, C. Pan, Y. Zhu, Origin of Photocatalytic Activation of Silver Orthophosphate from First-Principles, J. Phys. Chem. C 115 (2011) 4680–4687. (b) G. Botelho, J.C. Sczancoski, J. Andres, L. Gracia, E. Longo, Experimental and Theoretical Study on the Structure, Optical Properties, and Growth of Metallic Silver Nanostructures in Ag_3PO_4, J. Phys. Chem. C 119 (2015) 6293–6306. https://doi.org/10.1021/jp512111v

[21] J.M. Kahk, A.D.L. Sheridan, B.A.B. Kehoe, C.D.O. Scanlon, B.J. Morgan De, F.G.W. Watsonc, D.J. Payne, The electronic structure of silver orthophosphate: experiment and theory, J. Mater. Chem. A 2 (2014) 6092-6099. https://doi.org/10.1039/C3TA14191H

[22] J.J. Liu, X.L. Fu, S.F. Chen, Y.F. Zhu, Electronic structure and optical properties of Ag_3PO_4 photocatalyst calculated by hybrid density functional method, Appl. Phys. Lett. 99 (2011) 191903-191905. https://doi.org/10.1063/1.3660319

[23] L. Xu, W.-Q. Huang, L.-L. Wang, G.-F. Huang, P. Peng, Mechanism of Superior Visible-Light Photocatalytic Activity and Stability of Hybrid Ag_3PO_4/Graphene Nanocomposite, J. Phys. Chem. C 118 (2014) 12972–12979. https://doi.org/10.1021/jp5034273

[24] W.G. Wang, B. Cheng, J.G. Yu, G. Liu, W.H. Fan, Visible-Light Photocatalytic Activity and Deactivation Mechanism of Ag_3PO_4 Spherical Particles, Chem. Asian J. 7 (2012) 1902–1908. https://doi.org/10.1002/asia.201200197

[25] Y.P. Bi, H.Y. Hu, S.X. Ouyang, Z.B. Jiao, G.X. Lu, J.H. Ye, Selective Growth of Metallic Ag Nanocrystals on Ag_3PO_4 Submicro-Cubes for Photocatalytic Applications, Chem. Eur. J. 18 (2012) 14272–14275. https://doi.org/10.1002/chem.201201435

[26] Y.P. Liu, L. Fang, H.D. Lu, L.J. Liu, H. Wang, C.Z. Hu, Highly efficient and stable Ag/Ag_3PO_4 plasmonic photocatalyst in visible light, Catal. Commun. 17 (2012) 200–204. https://doi.org/10.1016/j.catcom.2011.11.001

[27] Y.P. Liu, L. Fang, H.D. Lu, Y.W. Li, C.Z. Hu, H.G. Yu, One-pot pyridine-assisted synthesis of visible-light-driven photocatalyst Ag/Ag_3PO_4, Appl. Catal. B 115-116 (2012) 245–252. https://doi.org/10.1016/j.apcatb.2011.12.038

[28] H. Wang, Y.S. Bai, J.T. Yang, X.F. Lang, J.H. Li, L. Guo, A Facile Way to Rejuvenate Ag_3PO_4 as a Recyclable Highly Efficient Photocatalyst, Chem. Eur. J. 18 (2012) 5524–5529. https://doi.org/10.1002/chem.201103189

[29] W.F. Yao, B. Zhang, C.P. Huang, C. Ma, X.L. Song, Q.J. Xu, Synthesis and characterization of high efficiency and stable Ag_3PO_4/TiO_2 visible light photocatalyst for the degradation of methylene blue and rhodamine B solutions, J. Mat. Chem. 22 (2012) 4050–4055. https://doi.org/10.1039/c2jm14410g

[30] R.Y. Liu, P.G. Hu, S.W. Chen, Photocatalytic activity of Ag_3PO_4 nanoparticle/TiO_2nanobelt heterostructures, Appl. Surf. Sci. 258 (2012) 9805–9809. https://doi.org/10.1016/j.apsusc.2012.06.033

[31] S.B. Rawal, S.D. Sung, W.I. Lee, Novel Ag_3PO_4/TiO_2 composites for efficient decomposition of gaseous 2-propanol under visible-light irradiation, Catal. Commun. 17 (2012) 131–135. https://doi.org/10.1016/j.catcom.2011.10.034

[32] Y.P. Bi, S.X. Ouyang, J.Y.Cao, J.H. Ye, Facile synthesis of rhombic dodecahedral AgX/Ag_3PO_4 (X = Cl, Br, I) heterocrystals with enhanced photocatalytic properties and stabilities, Phys. Chem. Chem. Phys. 13 (2011) 10071–10075. https://doi.org/10.1039/c1cp20488b

[33] J. Cao, B.D. Luo, H.L. Lin, B.Y. Xu, S.F. Chen, Visible light photocatalytic activity enhancement and mechanism of $AgBr/Ag_3PO_4$ hybrids for degradation of methyl orange, J. Hazar. Mat. 217 (2012) 107–115. https://doi.org/10.1016/j.jhazmat.2012.03.002

[34] G. Li, L. Mao, Magnetically separable Fe_3O_4–Ag_3PO_4 sub-micrometre composite: facile synthesis, high visible light-driven photocatalytic efficiency, and good recyclability, RSC Adv. 2 (2012) 5108–5111. https://doi.org/10.1039/c2ra20504a

[35] J. Guo, S. Ouyang, H. Zhou, T. Kako, J. Ye, $Ag_3PO_4/In(OH)_3$ Composite Photocatalysts with Adjustable Surface-Electric Property for Efficient Photodegradation of Organic Dyes under Simulated Solar-Light Irradiation, J. Phy. Chem. C 117 (2013) 17716–17724. https://doi.org/10.1021/jp4062972

[36] W. Zhai, G. Li, P. Yu, L. Yang, L. Mao, Silver Phosphate/Carbon Nanotube-Stabilized Pickering Emulsion for Highly Efficient Photocatalysis, J. Phys. Chem. C 117 (2013) 15183–15191. https://doi.org/10.1021/jp404456a

[37] X. Yang, H. Cui, Y. Li, J. Qin, R. Zhang, H. Tang, Fabrication of Ag_3PO_4-Graphene Composites with Highly Efficient and Stable Visible Light Photocatalytic Performance, ACS Catal. 3 (2013) 363–369.

[38] J. Tang, W. Gong, T. Cai, T. Xie, C. Deng, Z. Peng, Q. Deng, Novel visible light responsive Ag@(Ag_2S/Ag_3PO_4) photocatalysts: synergistic effect between Ag and

Ag_2S for their enhanced photocatalytic activity, RSC Adv. 3 (2013) 2543–2547. https://doi.org/10.1021/cs3008126

[39] I. Lee, M.A. Albiter, Q. Zhang, J. Ge, Y. Yin, F. Zaera, New nanostructured heterogeneous catalysts with increased selectivity and stability, Phys. Chem. Chem. Phys. 13 (2011) 2449–2456. https://doi.org/10.1039/C0CP01688H

[40] Y. Hou, F. Zuo, Q. Ma, C. Wang, L. Bartels, P. Feng, Ag_3PO_4 Oxygen Evolution Photocatalyst Employing Synergistic Action of Ag/AgBr Nanoparticles and Graphene Sheets, J. Phys. Chem. C 116 (2012) 20132–20139. https://doi.org/10.1021/jp303219j

[41] H.C Yu, Z.G Jiao, H.Y Hu, G.X Lu, J.H Ye, Y.P. Bi, Fabrication of Ag_3PO_4–PAN composite nanofibers for photocatalytic applications, CrystEngComm 15 (2013) 4802–4805. https://doi.org/10.1039/c3ce00073g

[42] B. Wang, X. Gu, Y. Zhao, Y. Qiang, A comparable study on the photocatalytic activities of Ag_3PO_4, AgBr and AgBr/Ag_3PO_4 hybrid microstructures, App. Surf. Sci. 283 (2013) 396– 401. https://doi.org/10.1016/j.apsusc.2013.06.121

[43] Y. He, L. Zhang, B. Teng, M. Fan, New Application of Z-Scheme Ag_3PO_4/g-C_3N_4 Composite in Converting CO_2 to Fuel, Environ. Sci. Technol. 49 (2015) 649-656. https://doi.org/10.1021/es5046309

[44] X. Yang, H. Tang, J. Xu, M. Antonietti, M. Shalom, Silver Phosphate/Graphitic Carbon Nitride as an Efficient Photocatalytic Tandem System for Oxygen Evolution, ChemSusChem 8 (2015) 1350 – 1358. https://doi.org/10.1002/cssc.201403168

[45] J. Xu, Z. Da Gao, K. Han, Y. Liu, and Y. Yan Song, Synthesis of Magnetically Separable $Ag_3PO_4/TiO_2/Fe_3O_4$ Heterostructure with Enhanced Photocatalytic Performance under Visible Light for Photoinactivation of Bacteria, ACS Appl. Mater. Interf. 6 (2014) 15122–15131. https://doi.org/10.1021/am5032727

[46] P. Reunchan, N. Umezawa, Sulfur and silicon doping in Ag3PO4, J. Phys. Chem. C 119 (2015) 2284–2289. https://doi.org/10.1021/jp509715b

[47] Y. Bu and Z. Chen, Role of Polyaniline on the Photocatalytic Degradation and Stability Performance of the Polyaniline/Silver/Silver Phosphate Composite under Visible Light, ACS Appl. Mater. Interf. 6 (2014) 17589–17598. https://doi.org/10.1021/am503578s

[48] B. Chai, J. Li, Q. Xu, Magnetically Separable $ZnFe_2O_4$–Graphene Catalyst and its High Photocatalytic Performance under Visible Light Irradiation, Ind. Eng. Chem. Res. 53 (2014) 8744–8752. https://doi.org/10.1021/ie4041065

[49] C. Jin, G. Liu, L. Zu, Y. Qin, J. Yang, Preparation of Ag@Ag_3PO_4@ZnO ternary heterostructures for photocatalytic studies, J. Coll. Interf. Sci. 453 (2015) 36–41. https://doi.org/10.1016/j.jcis.2015.03.066

[50] H. Ago, T. Kugler, F. Cacialli, W.R. Salaneck, M.S. Shaffer, A.H. Windle, R.H. Friend, Work Functions and Surface Functional Groups of Multiwall Carbon Nanotubes, J. Phys. Chem. B 103 (1999) 8116–8121. https://doi.org/10.1021/jp991659y

[51] M. Shiraishi, M. Ata, Work function of carbon nanotubes, Carbon 39 (2001) 1913–1917. https://doi.org/10.1016/S0008-6223(00)00322-5

[52] H. Tang, C.M. Hessel, J. Wang, N. Yang, R. Yu, H. Zhao, D. Wang, Two-dimensional carbon leading to new photoconversion processes, Chem. Soc. Rev.43 (2014) 4281–4299. https://doi.org/10.1039/C3CS60437C

[53] Z. Bian, T. Tachikawa, P. Zhang, M. Fujitsuka, T. Majima, Au/TiO_2 Superstructure-Based Plasmonic Photocatalysts Exhibiting Efficient Charge Separation and Unprecedented Activity, J. Am. Chem. Soc. 136 (2013) 458–465. https://doi.org/10.1021/ja410994f

[54] L. Liu, P. Li, B. Adisak, S. Ouyang, N. Umezawa, J. Ye, R. Kodiyath, T. Tanabe, G.V. Ramesh, S. Ueda, H. Abe, Gold photosensitized SrTiO3 for visible-light water oxidation induced by Au interband transitions, J. Mater. Chem. A 2 (2014) 9875–9882. https://doi.org/10.1039/c4ta01988a

[55] X. Guan, L. Guo, Cocatalytic Effect of $SrTiO_3$ on Ag_3PO_4 toward Enhanced Photocatalytic Water Oxidation, ACS Catal. 4 (2014) 3020–3026. https://doi.org/10.1021/cs5005079

[56] C. Li, P. Zhang, R. Lv, J. Lu, T. Wang, S. Wang, H. Wang, J. Gong, Selective Deposition of Ag_3PO_4 on Monoclinic $BiVO_4$(040) for Highly Efficient Photocatalysis Small 9 (2013) 3951–3956.

[57] S.B. Rawal, A.K. Chakraborty, W.I. Lee, FeOOH/TiO_2 Heterojunction for Visible Light Photocatalyst Heterojunction of FeOOH and TiO_2 for the Formation of Visible Light Photocatalyst, Bull. Korean Chem. Soc. 30 (2009) 2613–2616. https://doi.org/10.5012/bkcs.2009.30.11.2613

[58] J.A. Seabold, K.-S. Choi, Efficient and Stable Photo-Oxidation of Water by a Bismuth Vanadate Photoanode Coupled with an Iron Oxyhydroxide Oxygen Evolution Catalyst, J. Am. Chem. Soc. 134 (2012) 2186–2192. https://doi.org/10.1021/ja209001d

[59] H. Wang, Y. Bai, J. Yang, X. Lang, J. Li, L. Guo, A Facile Way to Rejuvenate Ag_3PO_4 as a Recyclable Highly Efficient Photocatalyst, Chem. Eur. J. 18 (2012) 5524–5529. https://doi.org/10.1002/chem.201103189

[60] Z. Yi, R.L. Withers, Y. Liu, A two-step approach towards solar-driven water splitting, Electrochem. Commun. 13 (2011) 28–30. https://doi.org/10.1016/j.elecom.2010.11.004

[61] S.J.A. Moniz, S.A. Shevlin, D.J. Martin, Z.X. Guo, J. Tang, Visible-light driven heterojunction photocatalysts for water splitting – a critical review, Ener. Environ. Sci. 8 (2015) 731–759. https://doi.org/10.1039/C4EE03271C

[62] A. Samal, D.P. Das, K.K. Nanda, B.K. Mishra, J. Das, A. Dash, Reduced Graphene Oxide–Ag_3PO_4 Heterostructure: A Direct Z-Scheme Photocatalyst for Augmented Photoreactivity and Stability, Chem. Asian J. 11 (2016) 584 – 595. https://doi.org/10.1002/asia.201501286

[63] D. Cao-Thang, N. Thanh-Dinh, F. Kleitz, D. Trong-On, Large-scale synthesis of uniform silver orthophosphate colloidal nanocrystals exhibiting high visible light photocatalytic activity, Chem. Commun. 47 (2011) 7797–7799. https://doi.org/10.1039/c1cc12014j

[64] Q. Xiang, D. Lang, T. Shen, F. Liu, Graphene-modified nanosized Ag_3PO_4 photocatalysts for enhanced visible-light photocatalytic activity and stability, Appl. Catal. B 162 (2015) 196–203. https://doi.org/10.1016/j.apcatb.2014.06.051

[65] Q. Liang, W. Ma, Y. Shi, Z. Li and X. Yang, Hierarchical Ag_3PO_4 porous microcubes with enhanced photocatalytic properties synthesized with the assistance of trisodium citrate, CrystEngComm 14 (2012) 2966–2973. https://doi.org/10.1039/c2ce06425a

[66] P. Dong, Y. Wang, H. Li, H. Li, X. Ma, L. Han, Shape-controllable synthesis and morphology-dependent photocatalytic properties of Ag_3PO_4 crystals, J. Mater. Chem. A 1 (2013) 4651–4656. https://doi.org/10.1039/c3ta00130j

[67] Y. Bi, H. Hu, Z. Jiao, H. Yu, G. Lu, J. Ye, Two-dimensional dendritic Ag_3PO_4 nanostructures and their photocatalytic properties, Phys. Chem. Chem. Phys. 14 (2012) 14486–14488. https://doi.org/10.1039/c2cp42822a

[68] Y. Bi, S. Ouyang, N. Umezawa, J. Cao, J. Ye, Facet Effect of Single-Crystalline Ag_3PO_4 Sub-microcrystals on Photocatalytic Properties, J. Am. Chem. Soc. 133 (2011) 6490–6492. https://doi.org/10.1021/ja2002132

[69] B. Zheng, X. Wang, C. Liu, K. Tan, Z. Xie, L. Zheng, High-efficiently visible light-responsive photocatalysts: Ag_3PO_4 tetrahedral microcrystals with exposed {111} facets of high surface energy, J. Mater. Chem. A 1 (2013) 12635–12640. https://doi.org/10.1039/c3ta12946b

[70] Z. Jiao, Y. Zhang, H. Yu, G. Lu, J. Ye, Y. Bi, Concave trisoctahedral Ag_3PO_4 microcrystals with high-index facets and enhanced photocatalytic properties, Chem. Commun. 49 (2013) 636–638. https://doi.org/10.1039/C2CC37324F

[71] Y. Bi, H. Hu, S. Ouyang, Z. Jiao, G. Lu, J. Ye, Selective Growth of Metallic Ag Nanocrystals on Ag_3PO_4 Submicro-Cubes for Photocatalytic Applications, Chem. Eur. J. 18 (2012) 14272–14275. https://doi.org/10.1002/chem.201201435

[72] Y. Bi, H. Hu, S. Ouyang, Z. Jiao, G. Lu, J. Ye, Selective growth of Ag_3PO_4 submicro-cubes on Ag nanowires to fabricate necklace-like heterostructures for photocatalytic applications, J. Mater. Chem. 22 (2012) 14847–14850. https://doi.org/10.1039/c2jm32800c

[73] W.S. Wang, H. Du, R.-X. Wang, T. Wen, A.-W. Xu, Heterostructured Ag_3PO_4/AgBr/Ag plasmonic photocatalyst with enhanced photocatalytic activity and stability under visible light, Nanoscale 5 (2013) 3315–3321. https://doi.org/10.1039/c3nr00191a

[74] X. Yang, H. Cui, Y. Li, J. Qin, R. Zhang, H. Tang, Fabrication of Ag_3PO_4-Graphene Composites with Highly Efficient and Stable Visible Light Photocatalytic Performance, ACS Catal. 3 (2013) 363–369. https://doi.org/10.1021/cs3008126

[75] S. Bai, X. Shen, H. Lv, G. Zhu, C. Bao, Y. Shan, Assembly of Ag_3PO_4 nanocrystals on graphene-based nanosheets with enhanced photocatalytic performance, J. Coll. Interf. Sci. 405 (2013) 1–9. https://doi.org/10.1016/j.jcis.2013.05.023

[76] J. Guo, H. Zhou, S. Ouyang, T. Kako, J. Ye, An Ag_3PO_4/nitridized $Sr_2Nb_2O_7$ composite photocatalyst with adjustable band structures for efficient elimination of gaseous organic pollutants under visible light irradiation, Nanoscale 6 (2014) 7303–7311. https://doi.org/10.1039/C4NR00537F

[77] S. Zhang, S. Zhang and L. Song, Super-high activity of Bi^{3+} doped Ag_3PO_4 and enhanced photocatalytic mechanism, Appl. Cat. B 152 (2014) 129–139. https://doi.org/10.1016/j.apcatb.2014.01.020

[78] J.K. Liu, C.X. Luo, J.D. Wang, X.H. Yang, X.H. Zhong, Controlled synthesis of silver phosphate crystals with high photocatalytic activity and bacteriostatic activity, CrystEngComm 14 (2012) 8714–8721. https://doi.org/10.1039/c2ce25604e

[79] Q. Liang, W. Ma, Y. Shi, Z. Li, X. Yang, Hierarchical Ag_3PO_4 porous microcubes with enhanced photocatalytic properties synthesized with the assistance of trisodium citrate, CrystEngComm 14 (2012) 2966–2973. https://doi.org/10.1039/c2ce06425a

[80] H. Hu, Z. Jiao, H. Yu, G. Lu, J. Ye, Y. Bi, Facile synthesis of tetrahedral Ag_3PO_4 submicro-crystals with enhanced photocatalytic properties, J Mater. Chem. A 1 (2013) 2387–2390. https://doi.org/10.1039/c2ta01151d

[81] Y.S. Xu, W. Zhang, Morphology-controlled synthesis of Ag_3PO_4 microcrystals for high performance photocatalysis, CrystEngComm 15 (2013) 5407–5411. https://doi.org/10.1039/c3ce40172c

[82] R. Que, High-yield synthesized silver orthophosphate nanowires and their application in photoswitch, Front. Optoelectr. China 4 (2011) 176–180. https://doi.org/10.1007/s12200-011-0160-y

[83] J. Wang, F. Teng, M. Chen, J. Xu, Y. Song, X. Zhou, Facile synthesis of novel Ag_3PO_4 tetrapods and the {110} facets-dominated photocatalytic activity, CrystEngComm 15 (2013) 39–42. https://doi.org/10.1039/C2CE26060C

[84] S.E. Skrabalak, Y. Xia, Pushing Nanocrystal Synthesis toward Nanomanufacturing, ACS Nano 3 (2009) 10–15. https://doi.org/10.1021/nn800875p

[85] L. Dong, P. Wang, S. Wang, P. Lei, Y. Wang, A simple way for Ag_3PO_4 tetrahedron and tetrapod microcrystals with high visible-light-responsive activity, Mater. Lett. 134 (2014) 158–161. https://doi.org/10.1016/j.matlet.2014.07.094

[86] L. Wang, N. Li, Q. Zhang, S. Lou, Y. Zhao, M. Chen, F. Teng, An innovative glycine complexing approach to silver phosphate myriapods with improved photocatalytic activity, CrystEngComm 16 (2014) 9326-9330. https://doi.org/10.1039/C4CE01296H

[87] D.J. Martin, N, Umezawa, X. Chen, J. Ye, J. Tang, Facet engineered Ag_3PO_4 for efficient water photooxidation, Ener. Environ. Sci. 6 (2013) 3380-3386. https://doi.org/10.1039/c3ee42260g

Chapter 11

Shape-Control Synthesis and Photocatalytic Applications of CeO_2 to Remediate Organic Pollutant Containing Wastewater: A Review

Kinjal J. Shah [a,b,*], Pen-Chi Chang [a,b,*]

[a] Graduate Institute of Environmental Engineering, National Taiwan University, Taiwan

[b] Carbon Cycle Research Center, National Taiwan University, Taipei City 10673, Taiwan

Email: kinjalshah8@gmail.com (K. J. Shah); pcchiang@ntu.edu.tw (P. C. Chang).

Abstract

Shape-control ceria (CeO_2) nanomaterial has drawn more attention in recent years due to their excellent physicochemical properties and applications. The method of shape-controlling synthesis of metal nano-catalysts are determined by the necessities of the application. Present review focused on the brief description of current research activities and their emphasis on the synthesis approaches of shape-controlled of CeO_2 nanostructures and their wastewaters treatment applications. Most of the favorable strategies of shape-controlled synthesis of CeO_2 nanomaterial applicable for photocatalysis degradation of organic pollutant containing wastewater treatments are divided into the following three sections: (i) the use of a shaped CeO_2 nanomaterial; (ii) doped CeO_2 nanomaterials with other materials; and (iii) CeO_2 as a dopant in many materials. In the last part of review, we have provided challenges associated with shape-control synthesis of CeO_2 nanomaterials.

Keywords

Ceria (CeO_2) Nanomaterial, Photocatalysis, Shape-controlled Synthesis, Dye Degradation

Contents

1. Introduction

The World Health Organization (WHO) estimated that 25% of the diseases facing humans today are occurring due to environment pollutions including air, soil and water [1]. All of them are essential substance for all life on earth, and a precious resource for human civilization. Recently, the rapidly growing global population and the higher living standards continuously drive up the demand on natural resources. Among them, pressure has been building up on water supplies from unconventional water sources (e.g., storm water, contaminated fresh water, brackish water, wastewater and seawater) rather than conventional natural (fresh) water sources, especially in water stressed regions [2]. Thus, clean and affordable water is considered one of the most basic humanitarian goals, and remains a major global challenge for the 21st century. Therefore, to protect the environment and fulfill the needs of the pure water for drinking and other daily life applications, the removal of pollutants from waste water is very essential. In general, several approaches, such as adsorption [3], biological treatment [4], membrane-based separations [5], and chemical treatments [6-7] have been used for waste water treatment. Currently, advance oxidation processes along with combination of various conventional methods has shown promising results for the waste water treatment [8]. During the process of advance oxidation, sequences of reactions are taking place and break down the macromolecule (pollutant) into less harmful and smaller substances by oxidation through hydroxyl radical (·OH). Those things can be achieved by ultrasonic cavitation, electron beam irradiation, photocatalytic oxidation, Fenton's reaction and strong oxidizing species like ozone and hydrogen peroxide [9]. Among them, photocatalytic oxidation has the potential to facilitate the utilization of natural sunlight source to obtained cleaner and greener production apart from UV light source. Therefore, it is advisable to put research efforts for development of novel photo catalyst, in particular under visible light irradiation to degrade pollutants.

For the development of novel photo catalyst, semiconductor based materials, specially metal oxides are the most reliable materials for water/wastewater treatment, due to their inexpensive, nontoxic, higher efficient to transfer multielectrons, tunable and regenerable nature [8, 10,11]. Moreover, photocatalysis through semiconductor materials have their own benefits such as complete degradation of pollutants, no addition of other chemicals in the process, cost effective approaches, use of solar/UV light and ambient operating conditions. A wide variety of semiconductor materials i.e. SnO_2, ZnO, TiO_2, In_2O_3, Fe_2O_3, NiO, CeO_2 and others have been utilized as photocatalytic degradation of organic pollutants [12-13]. The value of band gap positions (with compared to normal hydrogen electrode (NHE)) of listed semiconductors, which are highly responsible for photocatalytic degradation of pollutants have been presented in Figure 1. The value of conduction band for water splitting and oxidation of organic molecules comes under the range of most of the semiconductor metal oxides. Moreover, pH, co-catalyst and surfaces of electrodes are affecting the values of band gaps [14]. Apart from band gap regulations, optical, electronic, and structural properties of the material have to be considered carefully, which includes morphological controls, shape of the crystal, surface sensitization, phase and compositions to develop highly efficient photocatalyst.

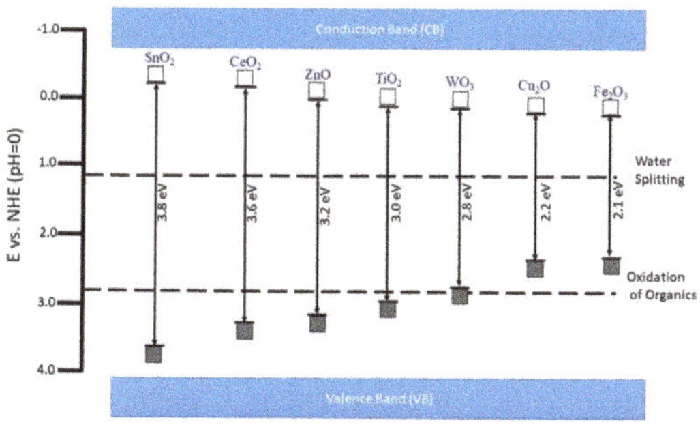

Figure 1. Conduction band (CB=white square) and valance band (VB=gray square) potentials (eV) of various semiconductors (metal oxide) at pH=0 versus a normal hydrogen electrode (NHE).

While considering the semiconductor materials, cerium is the most reactive rare earth metals found in the Earth's crust with a form of Ce-carbonate, -phosphate, -silicate, and -(hydr)oxide [15]. Cerium dioxide, ceria has received much attention in the global nanotechnology market due to their highly stable, inexpensive, and excellent capability to give active oxygen [16]. It has been proven as a promising material which exists in different nanostructures including nanorods, nanowires, nanotubes and nanoflakes. A number of synthesis methods can be used to prepare CeO_2 nanomaterials, such as precipitation method, sol–gel method, hydrothermal or solvothermal method, electrochemical deposition, chemical vapor deposition, flame spray pyrolysis, surfactant assisted method, and so on [16-18]. All of the above stated preparation method; basic need is to increase oxygen defects which are responsible for highly catalytic activity in CeO_2. Thus, much interest has been paid in the synthesis techniques to develop better shape and catalytically activity. However, only a few authors have written about the correlation between synthesis techniques and photocatalytic activity of CeO_2.

In this perspective, a general, comprehensive review on the photocatalytic application of CeO_2 in wastewater treatment is presented with detailed shape-controlled synthesis techniques of CeO_2 nanomaterials. In the first half part, we have discussed the synthesis techniques and associate mechanisms of controllable synthesis of CeO_2 nanomaterials. In the second part, we present their applications in polluted wastewater treatments. That has been divided into three parts, shaped CeO_2, dopend-CeO_2 and CeO_2 loaded in materials for polluted wastewater treatments. Finally, we give an outlook on the challenges in the development of controlled synthesis and applications of CeO_2 nanomaterials.

2. CeO_2: material properties

Cerium (electronic configuration: [Xe] $4f^1$ $5d^1$ $6s^2$) exhibits a variable electronic structure (dual valence states of +3 and +4) at lower temperature because a very small amount of energy is necessary to change the relative occupancy [15, 19]. In which, cerium is exhibiting a unique stability in the tetravalent ceric state (Ce^{4+} with [Xe] $4f^0$) as compared with trivalent cerous state (Ce^{3+} with [Xe] $4f^1$). This is different than other lanthanides containing trivalent stable states. This feature allows for easy separation of cerium from other rare-earth elements through oxidation (forming CeO_2). Meanwhile, nanostructured CeO_2 materials have received tremendous attention because of their unique properties which are derived from their size, shape, orientation, and high surface area.

The CeO_2 nanocrystal has a fluorite crystal structure (face centered cubic- fcc). It consists of a simple cubic oxygen sub-lattice with the cerium ions occupying alternate cube centers [20]. In the structure, each cerium cation is coordinated by eight nearest neighbor oxygen anions, while each oxygen anion is coordinated by four nearest neighbor cerium cations. In which, cerium is at the center of the tetrahedron whose corners are occupied by oxygen atoms. In CeO_2, two kinds of defects, namely, intrinsic and extrinsic defects have been found. The presence of intrinsic defect may be due to thermal disorderness in crystal lattice, which can be formed by redox reaction between the solid and surrounding atmosphere. The extrinsic defects in crystal may be formed by introducing foreign dopant or impurities. Apart from these, the stable known defects are presence due to oxygen vacancies [15]. This defect is observed due to reversible transition in the oxidation state of two cerium ions from Ce^{3+} to Ce^{4+}, which may generate neutral oxygen vacancies in ceria (equation 1). In which, a neutral species $1/2O_2$ (g) is formed, if an oxygen ion (O^{2-}) leaves ceria lattice. The two electrons left behind were trapped at two cerium sites, i.e. they become localized at two cerium sites. The higher Ce^{3+} concentration of total cerium that exists, the more defects form.

$$O^{2-} + 2Ce^{4+} \leftrightarrow 2Ce^{3+} + \frac{1}{2}O_2(g), \quad (1)$$

CeO_2 nanocrystal can accommodate high oxygen deficiency by the substitution of lower valent elements on the cation sub lattice. Due to this property, higher oxygen ion conductivity is observed. Meanwhile, the reduction limit of non-stoichiometric CeO_2 is Ce_2O_3, where all cerium ions are found in a Ce^{3+} oxidation state. The electronic band structure of Ce_2O_3 bears resemblance to that of partially reduced CeO_2.

There are three (100), (110) and (111) low index lattice planes existing on the surface of the CeO_2 nanocrystal (See Figure 2). The stability based on density functional theory of all three planes are different and follows the sequence (111) > (110) > (100), while the activity follows the opposite order [20]. Thus, there are more oxygen vacancies on the (110) and (100) planes, as the energy required to form oxygen vacancies on the (111) surface is greater. Meanwhile, considering the shape control formation through CeO_2 nanocrystal, it can be concluded that nanorods containing (110) and (100) planes, and nanocubes containing (100) planes have higher oxygen vacancies. The increased diffusion rate of oxygen vacancies in the CeO_2 nanocrystal causes increased catalytical activity. Thus the activity of CeO_2 nanocrystal directly relates to the number of oxygen vacancies in the crystal, which are affected by shape, size or crystal, doping elements, temperature and others parameters. By considering above phenomenon, decreasing size by nanotechnology, formation of more oxygen vacancies may help for higher catalytical valuation in CeO_2.

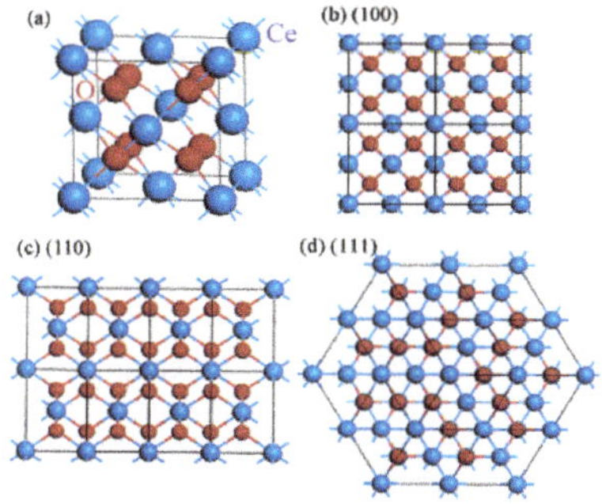

Figure 2. (a) Unit (face centred crystal) cell of the CeO_2 structure. (b-d) The (100) [or (200)], (110), and (111) planes of the CeO_2 structure. (Reprinted with permission from ref 20 @ 2003 American Chemical Society.

3. Shape controlled synthesis methods for CeO_2

In ordered to obtain nanocrystal structure, size and morphology control in the formation process is essentials and it can be achieved by crystal growth process by nucleation and growth process. The control of the size can be achieved by controlling rate of nucleation and growth process [21]. In which, the big nanoparticles can form at lower nucleation rate with lower concentration of nuclei. Alternatively, higher nucleation rates can result in higher concentration of nuclei, and smaller particles will be formed. There are many parameters such as concentration of precursor, temperature, setting time, solvent, pH, pressure and many, can affect the rate of nucleation and growth process [21-22]. With controlling the above parameters, aim of controlled crystal size has been achieved. The surface energies play an important part in the controlled synthesis of different shapes and that can be controlled by appropriate additives such as ligands, surfactants, ions and polymers [23].

In various crystalline forms such as cubic, hexagonal, monoclinic, orthorhombic, tetragonal, astriclinic, mix crystal framework have been synthesize by sol-gel, microwave assisted, ultrasonic irradiation, electrochemical deposition, precipitation, hydrothermal, solvothermal methods [18, 24]. Meanwhile, specialized structure such as hollow spheres, nanotubes, nanorods, nanowires have been synthesized by soft/hard template-directed synthesis, emulsion, sol gel, hydrothermal, solvothermal synthesis and so on techniques (See Table 1).

Table 1. Details of morphology, synthesis procedure and properties of CeO_2 nanomaterials.

CeO_2 nanomaterials with morphology	Synthesis methods	BET (m^2g^{-1}) Or particle size(nm)	Ref.
Nanoparticles	Precipitation	18-22 nm	[26]
Hierarchical CeO_2 Nanocrystal	Sol-gel method	65 m^2g^{-1}	[27]
CeO_2 Nanoparticles, Nanocubes, Nanorods	Microwave assisted method	NA	[35]
Polyhedral shape of CeO_2	Hydrothermal process	6nm	[37]
Three-Dimensional CeO_2 Nanocrystals	Hydrothermal process	5nm and 6.8 nm	[38]
CeO_2 nanorods	Solvothermal method	40–50 nm	[40]
CeO_2 hollow nanospheres	Hydrothermal.	30-300 nm	[48]
CeO_2 mesoporous spheres	Solvothermal	216 m^2g^{-1}	[41]
CeO_2 nanospindles	Solvothermal synthesis	127 m^2g^{-1}	[39]
CeO_2triangular microplates	Hydrothermal method	300-500 nm	[49]
CeO_2 spherical nanoparticles	Spray pyrolysis	62 m^2g^{-1}	[42]
Tadpole shape CeO_2	Sol–gel synthesis	1.2 nm	[29]
Mesoporous CeO_2 nanoflowers	Solution synthesis	95.7 m^2g^{-1}	[50]
CeO_2 powder	Microemulsion technique	250 m^2g^{-1}	[51]
CeO_2 nanoparticles	Sonochemical method	Crystal size 5 nm	[31]
CeO_2 nanoparticles	Electrochemical deposition method	10-30 nm	[45]

3.1 Precipitation method

The growth of nanoparticles, nanocrystals and regular shaped nano materials have been synthesized under this method. It is a simple and easily handle method to prepare CeO_2 nanomaterials and thus it is widely used in industrial applications. During the process, basic compound of CeO_2 has been precipitate by heat treatment [25]. During the process, pH, temperature, concentration of the aqueous solution, O_2 content in atmosphere, precipitating agent and its amount, and set time are influencing the shape and morphology of the products [25]. In which, ae higher particle size can be obtained by increasing the temperature or decreasing the oxygen content. However, aggregation was observed after the heating treatment, which cannot be control during the process and resulted in an un-uniform morphology with different sizes. Ketzial and Nesaraj [26] studied the precipitation of CeO_2 nanoparticles by precipitation method and found that the presence of the poly ethylene glycol and poly vinyl pyrollidone, surfactants reduced the size of the CeO_2 nanoparticles by decreasing chances of aggregation during calcine process.

3.2 Sol-gel method

The method used for the fabrication of metal oxides, especially oxides for semiconductor materials. The process involves conversion of monomers into a colloidal solution (*sol*) that acts as the precursor for an integrated network of discrete particles. Xiao et al. [27] has synthesized hierarchical CeO_2 nanocrystal by sol-gel method, Ferreira et al. [28] have synthesized CeO_2 nanomaterials with native cassava starch as size limiting chelating agent and Yu et al. [29] have synthesize uniform size CeO_2 nanocrystal with spherical, wire and tadpole shape. The solvent plays an essential role in the precursor material transformation. By changing the parameters such as solubility, ionic properties, reactant and temperature can provide high quality CeO_2 nanomaterials [18]. During presence of native cassava starch chelating agent, oxygen vacancies induced due to presence of structural defects caused by reduction in the valence of Ce^{4+} to Ce^{3+}.

3.3 Ultrasonic irradiation method

It is very difficult to synthesis CeO_2 nanomaterials with only the ultrasonic irradiation method. Thus, it is usually combined with other methods to synthesize CeO_2 nanomaterials, such as sonochemical synthesis. In which, the chemical reaction takes place in presence of ultrasound. During sonication, ultrasonic waves radiate through the solution causing an alternating high and low pressure in the liquid media. It has been well recognized that ultrasonic irradiation caused cavitation in a liquid medium where formation, growth and implosive collapse of bubbles occur and where many chemical reactions are taking place [30]. Jamshidi et al. fabricated CeO_2 nanomaterials with surfactant, poly vinyl pyrrolidon and polyethylene glycol through the sonochemical method [31]. Under sonochemical synthesis, collision between two CeO_2 nanomaterials increased, but with surfactant, agglomeration rate decrease and uniform products are obtained. This kind of method has a higher potential to obtained an optimized surface area and, therefore, better potential for further applications.

3.4 Microwave assisted synthesis method

Microwave synthesis can provide a simple and fast route to the preparation of nanomaterials. Under microwave irradiation conditions, organic reactions can be accelerated and selectivity of the ensuing products can be obtained easily because of the advantages associated with this techniques such as instant and rapid heating, higher temperature homogeneity, low cost, pressure control and selective heating [32]. The use of the microwave assisted heating technique has become an essential tool in all areas of synthetic organic chemistry, nanoparticle synthesis, solvent-free reactions and water-mediated reactions [18]. For the synthesis of CeO_2 nanomaterials, microwave synthesis is also considered as an effective and fast method [33-35]. Yang et al. synthesized rod like and sphere like CeO_2 by microwave assisted reaction of $Ce(NO_3)_3{\cdot}6H_2O$ in the presence of NaOH [33]. Moreover, Tao et al. synthesized nanoparticles, nanocubes and nanorods from $Ce(NO_3)_3{\cdot}6H_2O$ tert-butylamine, oleic acid and ethylene glycol materials [35]. Under the microwave treatment, the energy of selective dielectric heating rapidly transferred directly to the reactants, leading to higher effective temperature. Based on this concept, with increasing the irradiation time, the shape of the materials are changing. Tao et al. have synthesized nanoparticles, nanocubes and nanorods with an irradiation time of 2, 10, 30 min respectively [35]. The morphologies changed from nanoparticles to nanorods and the sizes increased when the irradiation time increased from 2 to 30 min, and it clearly showed that the nanorods and nanocubes grew by the cost of CeO_2 nanocrystal. Due to the possibility to synthesize various shape control nanomaterials at lower cost, microwave assisted synthesis became of interest for industrial scale production of nanomaterials.

3.5 Hydrothermal method

Hydrothermal synthesis is a method to synthesize nanoparticles, especially single crystal. The shape and particle shapes depends on the solubility of minerals in hot water under high pressure [36]. In ordered to maintain the pressure and temperature, an autoclave has been utilized. The growth of crystals have been controlled by varying aqueous solution, ionic properties and reactivity of reactants with solution, temperature or pressure. The reaction is promoted under high pressure and temperature and the product can be separated after the reaction. Materials having higher vapor pressure near their melting points can be grown by this method. However, the intermediate observation of product and expense of the autoclave are the negative point of this method [18]. Through this method, CeO_2 nanocrystal with polyhedral shape and three dimensional shapes, nanorods, nanocubes, uniform size particles have been synthesized. Wang and Feng synthesized polyhedral shape CeO_2 nanoparticles with particle size 3-10 nm [37]. While three dimensional CeO_2 nanocrystal developed by Tan et al [38]. In above both cases, octahedron with {111} were obtained under different synthesis conditions. Beside reaction, the common solution being utilized is NaOH. The concentration of NaOH, reaction time and pressure for reaction changes the shape of the particles. Thus, it is concluded that the controlled shape can be obtained by controlling the reaction conditions. Moreover, the presence of surfactant, polymer and anions like Cl^-, NO_3^-, SO_4^{2-} and PO_4^{3-}, have affect the shape of the final particles/crystals. These anions have a selective interaction with faces of the CeO_2, which leads to the growth of nanocrystal with varying morphologies. CeO_2 triangular microplates and CeO_2 hollow also synthesized under this preparation method.

3.6 Solvothermal method

This method is very similar to the hydrothermal method but the precursor solution is not aqueous. This procedure utilizes sol-gel and hydrothermal processes together. Thus, this method allows the precise control over the size, shape distribution and crystallinity of nanomaterials. CeO_2 nanospindles [39], CeO_2 nanorods [40], CeO_2 mesoporous spheres [41] so on can be synthesize with the solvothermal method with glycerin, ethylenediamine anhydrous and formic acid/ethanol act as a mass transfer medium in the reaction, respectively. Moreover, the organic reagents play a key role in determining the growth and stability of the nanocrystal. Thus, a capping agent is introduced for the system. Surfactant are often used as capping reagent to prepare CeO_2 nanomaterials with different size and shape. The CeO_2 nanocrystal preferably grow as octahedral shape but at a low ratio of the organic ligands, generation of nanocubes are higher. However, if the ratio is higher, generation of octahedral or smaller crystals are higher.

3.7 Spray pyrolysis

In general, pyrolysis is a thermochemical decomposition of organic materials at elevated temperature. During the spray pyrolysis, a thin film is deposited by spraying solution on heated surfaces. Due to the thermal heat, reactants convert to product. For CeO_2 particles prepared with this process, the morphologies of hollow and porous were frequently observed rather than solid spherical particles because of the fast evaporation rate and lack of time for solute diffusion and particle densification. Shih et al. synthesized CeO_2 nanomaterials with use of different precursor such as $Ce(NO_3)_3 \cdot 6H_2O$, $Ce(NH_4)_2(NO_3)_6$ and $Ce(CH_3COO)_3 \cdot 1.5H_2O$ [42]. They observed that the solubility of $Ce(NH_4)_2(NO_3)_6$ and $Ce(NO_3)_3 \cdot 6H_2O$ increase with increasing temperature. This is possibly due to the relatively strong hydrophilic properties of both cerium complexes comparing to

$Ce(CH_3COO)_3 \bullet 1.5H_2O$. Due to different solubility of these three precursors, volume precipitation, partial volume and surface precipitation and surface precipitation dominates the spray pyrolysis process to form main morphologies of spherical solid, hollow (closed and open pores) and porous (closed pores) of the ceria particles from $Ce(NH_4)_2(NO_3)_6$ and $Ce(CH_3COO)_3 \bullet 1.5H_2O$ and $Ce(NO_3)_3 \cdot 6H_2O$ precursors, respectively. Based on the result, by choosing suitable precursors, different morphological ceria particles can be obtained for different applications.

3.8 Emulsion and microemulsion technique

The microemulsion technique promises to be a versatile preparation method which enables one to control the particle properties such as mechanisms of particle size control, geometry, morphology, homogeneity and surface area [43]. Surfactants are the principal agent in emulsion and microemulsion techniques. Emulsions are stable dispersions of immiscible liquids but they are not thermodynamically stable. While, microemulsions are thermodynamically stable. These two techniques are often used to prepare CeO_2 nanomaterials with different shapes, such as rods, wires, plates and so on. The common surfactant used for CeO_2 nanomaterials synthesis is cetyltrimethylammonium bromide (CTAB) and octadecylamine. Kockrick et al. synthesized CeO_2 nanoparticles using n-haptane, marlophen NP 5 as surfactant and $Ce(NO_3)_3$ system [44]. The micelle and particle size in the range of 5–12 nm were controlled by varying the molar water to surfactant ratio. Increase of the concentration of surfactant resulted in decrease of the size by decreasing the possibility of agglomeration. But after increasing the surfactant, increase particle size due to charged formation appeared.

3.9 Electrochemical deposition method

Electrochemical deposition is a process by which a desired coating of metal, oxide, or salt can be deposited onto the surface of a conductor substrate by simple electrolysis of a solution containing the desired metal ion or its chemical complex. This method is mostly utilized to prepared doped CeO_2 nanomaterials. The current density mostly affect the properties of the nanomaterials. With increasing the current density, formation of nanocrystals increase. Wang and Sun synthesized CeO_2 nanomaterials with help of cathode, working electrode of Ru-Ti mesh and platinum as anode [45]. The experiment is performed in the galvanostatic mode at current densities of 100 A/m^2 and at room temperature. The results indicate that the as-prepared powders after being treated at 650 °C (calcined) are nanocrystalline with the cubic fluorite structure and spherically in shape with size of 10-30 nm.

3.10 Flame spray pyrolysis method

A flame spray pyrolysis method produces metal oxide powders from highly volatile gaseous metal chlorides that are decomposed/oxidized in hydrogen-oxygen flames to form nano-oxide powders. In this procedure, precursor solutions are first atomized into droplets, which are subsequently pyrolyzed to produce solid particles, usually in the hot reaction zone. There are three types of atomizer namely, ultrasonic, mechanical, and electrospray have been used to spray the precursor into droplets. Oh and Kim synthesized CeO_2 nanomaterials from $Ce(NO_3)_3{\cdot}6H_2O$, ethanol and diethylene glycol butyl ether materials [46]. In which, ethanol was chosen as the solvent for its good solubility and diethylene glycol butyl ether was used to suppress the liquid evaporation. During the process, ceria particles with sizes below 10 nm formed by precursor evaporation and subsequent gas-phase growth has been reported in flame spray pyrolysis.

3.11 Soft-template and hard-template directed synthesis

This synthesis method relies on the use of reversible non-covalent bonding interactions between molecular building blocks in order to pre-organize them into a certain relative geometry as a prelude to covalent bond formation. The soft template-directed synthesis often refers to the Ostwald ripening mechanism [47]. The assistance of surfactants can promote the self-assembly of CeO_2 nanoparticles, resulting in the formation of CeO_2 nanomaterials with different morphologies. Structure such as rods, plates, cubes, flowers, spindles etc. have been synthesized through soft-template directed synthesis. While, CNT, nanorods, carbon sphere kind of materials have been synthesized by hard-template directed synthesis [18]. CTAB, benzyl alcohol, polyvinylpyrrolidone, polyethylene glycol are the major surfactant materials have been utilized in soft-template directed synthesis techniques.

As mentioned above all techniques suggest that CeO_2 nanomaterials with different shape controlled has been synthesized by controlled experimental conditions. Moreover, to enhance the amount of oxygen vacancies, decrease valance region and thermal stability, the need of doping with some elements such as transition metals in the CeO_2 nanomaterials arises. The doped CeO_2 nanomaterials have been synthesized from above mention methods, among them, hydrothermal/solvothermal treatment can provide better results. Sometime, CeO_2 nanomaterials prepared by a doping method have higher crystallinity but their size distribution is poor. The preparation of CeO_2 nanomaterials is often based on trial and error experience. Thus, many possibilities exist to develop new methods or controlled conditions to develop a variety of CeO_2 nanomaterials in the future.

4. Photocatalytic applications

4.1 Photocatalysis with CeO_2 nano materials

Semiconductor metal oxide photocatalysts such as CeO_2 enable the generation of reactive oxidants that help in the photochemical oxidation of organic compounds namely, rhodamine B dye, methylene blue, methyl orange, azodyes (acid orange 7), acidic black 10B etc. In some cases, surface loading of metal induced CeO_2 called doped- CeO_2 or CeO_2 loaded called CeO_2 nanomaterials as dopant on supported materials are necessary for extending the photo response in the visible light region. Meanwhile, CeO_2 nanomaterials can uptake and release oxygen via the transformation between Ce^{3+} and Ce^{4+}, thus it can be used in many catalytic reactions and display excellent catalytical properties. Table 2 summarized controlled size CeO_2 nanomaterials with its preparation methods and photocatalytic efficiency. Summarized all shapes of CeO_2 nanomaterials are involved for dye degradation applications for water remediation.

Rhodamine B dye degraded by using CeO_2 nanocubes, mesoporous CeO_2 nanowires, CeO_2 microspheres, lamellar CeO_2, mono-dispersed CeO_2 nanocubes and so on types of CeO_2 nanomaterials. Among them, most of the CeO_2 nanomaterials have been synthesized by hydrothermal or solvothermal mechanisms. Liu et al. confirmed that the photodegradation experiments of RhB dyes, CeO_2 nanocubes exposed with {100} facets, exhibit excellent photocatalytic performance [52] and the smaller size CeO_2 nanocubes exhibit better photocatalytic activity such as 35% against 25% of dye degradation. Yu et al. [53] synthesized CeO_2 solid particle, porous particle and jar particles through electrospray ionization techniques. The photo degradation of rhodamine B dye with CeO_2 solid particle, porous particle and jar particles are in the ordered of porous particle > jar particles > solid particle with 70.5, 60 and 56%, respectively. In which, the specific surface area of particles are in ordered of jar (19.2 m^2/g) > porous particle (16.7 m^2/g) > solid particles (10.4 m^2/g). Thus, it can be concluded that the surface area is not only responsible for dye degradation but shape of the particles also have equal importance. With having jars with semi closed structure, it is difficult to reached UV light with compared to porous open structure.

Methylene blue dye degrades by using CeO_2 nanocubes, nanocrystals, X-architecture CeO_2 and so on by precipitation, solvothermal and microwave assisted methods. Zheng et al. [54] synthesized X-architecture CeO_2 using the hydrothermal route assisted by polyvinylpyrrolidone. They found the photocatalytic activity of X-architecture CeO_2 is higher than traditional CeO_2 samples because of its irregular boundaries, which increased the surface area of traditional CeO_2 samples (27.2 m^2/g) to (42.7 m^2/g) of X-architecture

CeO_2. This phenomenon clearly suggests the active role of shape and surface area over methylene bleu dye degradation.

Table 2. Photocatalytic application of different shape assisted CeO_2 nanomaterials.

CeO_2 nanomaterials & morphology	Synthesis methods	Pollutant (% degradation)	Treatment conditions	Ref.
CeO_2 nanocubes	Facile hydrothermal method	Rhodamine B dye (35%)	Halogen lamp: 420 nm, 1000 W, t= 2h, C_o= 10 ml/L.	[52]
Mesoporous CeO_2 nanowires	Facile hydrothermal method	Rhodamine B dye (100%)	UV lamp: 365nm, 125W, t= 100 min, C_o= 10^{-5} $molL^{-1}$.	[57]
CeO_2 microspheres	Electrospray ionization technique	Rhodamine B dye (70.5%)	Mercury lamp: 365nm, 300W, t= 140min, C_o= 10ml/L	[53]
Lamellar CeO_2	Direct precipitation	Rhodamine B dye (100%)	Mercury lamp: UV range, 500 W, t= 100min, C_o= 10 ml/L.	[58]
CeO_2 nanocrystals	Precipitation method	Methylene Blue (93%)	Mercury lamp: 332 nm, 70W, t= 130 min, C_o= 5-40 ml/L.	[59]
Mono-dispersed CeO_2 nanocubes	Facile hydrothermal method	Rhodamine B dye (20 - 30%)	UV irradiation	[60]
CeO_2 microcrystals	Solvothermal method	Rhodamine B dye (60%)	Xe lamp: 300 W, t= 150 min, C_o= 100 ml dye.	[61]
One-dimensional CeO_2 materials	Microwave assisted hypothermal method	Methyl orange (98%)	UV light by mercury vapor lamp, 250W, t= 100 min, C_o= 1* 10^{-5} M dye.	[62]
CeO_2 nanorods	Microwave-assisted hydrothermal method	Methyl orange (95%)	Mercury lamp: 250W, t= 120 min, C_o= 1* 10^{-5} M dye.	[55]
X-architecture CeO_2	Hydrothermal route	Methylene Blue (90%)	UV light irradiation: 300 W; Mercury lamp: 500W, t= 120 min, C_o= 40 mg/L dye.	[54]
CeO_2 powder	Precipitation method	Azodyes acid orange 7(90%)	Halogen lamp: 420nm, 1000W, t= 11h min, C_o= 70 mg/L dye.	[63]
CeO_2 nanocrystalline	Hydration & Precip-itation method	Acidic Black 10B (96.9)	UV visible: t= 2h, C_o= 70 mg/L dye.	[56]
CeO_2 nanoparticles	Hydrothermal synthesis	Acridine orange (45.8)	Mercury lamp: 491 nm, 250 W, t= 170 min, C_o= 0.03 mmol dye.	[64]
One-dimensional CeO_2 nanotubes	Template-free hydrothermal treatment	Benzene (Comparable degradation)	UV lamps: 254 nm, 4 W, t= 22h, C_o= 250 ppm dye.	[65]

A similar way, Phuruangrat et al. [55] synthesized shape-controlled CeO_2 nanorods with microwave assisted hydrothermal method. They found NaOH concentration play an important role for synthesis of CeO_2 nanorods from CeO_2 nanoparticles. With concentration of 20 ml NaOH, only nanorods/nanowires with 400–600 nm long were detected in place of nanoparticles of 15-20 nm. After shape-controlled synthesis of such nanorods they are photocatalytically utilized for methyl orange degradation, 95% of degradation was observed with 120 min of UV irradiation. Apart from NaOH, ammonium bicarbonate (NH_4HCO_3) are being utilized as precipitant material for synthesize of CeO_2 nanocrystals [56]. The average particle size increases after increasing the calcined temperature from 8 nm to 34 nm. It can be utilized for photodegradation of Acidic Black 10B dye.

After considering all above facts, it can be concluded that the physicochemical properties and photocatalytic performance of CeO_2 are controlled by condition at preparation stage or calcination temperature. At higher temperature, agglomeration increases with growth of the crystal. Thus, nanorods, nanocube and nanotube have higher calcination/processing temperature to get proper shape. During dye degradation, the shape of the crystal, nanoparticles are considered after the active surface of the materials. As active surfaces always giving better performance than close or partially closed surfaces.

4.2 Photocatalysis with doped CeO_2

Several properties of semiconductor based photocatalysts materials have been varied by impurities and defects, so called doping materials. The synthesis of n- and p-type doped systems motivates the development of visible light photocatalysts materials. To accomplish this development, a comprehensive understanding of different doped materials and its effects on photocatalytic properties on CeO_2 materials have been listed in Table-3. Photocatalytic performance of CeO_2 is found to be lower as compared to more widely used ZnO and TiO_2 due to lower reduction potential of conduction band. However, doping of different transition metals such as Fe^{3+}, Ti^{4+}, V^{5+}, Mo^{6+} etc. have ability to enhance the absorption band shift into the visible light region [66]. This phenomenon was observed due to the occupation of the interstitial site or increasing the concentration of defects like oxygen vacancies, which shift the band gap absorption to longer wavelength. In most of the cases, doped-CeO_2 materials have been synthesized by precipitation, co-precipitation or hydrothermal methods. TiO_2 and ZnO are highly conductive compared to the CeO_2 material. In ordered to enhance conductivity of CeO_2, ZnO mixed CeO_2 nanoparticles synthesized by *in-situ* synthesis method has been reported by Shah et al. [67]. They found, rhodamine B dye degradation by ZnO-CeO_2 nano composite materials was 55.3% with compared to CeO_2 nanomaterials having 36.5%.

Table 3. Photocatalytic application of different shape assisted doped-CeO_2 nanomaterials.

CeO_2 nanomateri-als	Synthesis methods	Pollutant (% degradation)	Treatment conditions	Ref.
In-Situ synthesis of CeO_2/ZnO composite nanoparticles	Precipitation	Rhodamine B dye (36.5% with CeO_2, 55.3 % with CeO_2-ZnO)	UV lamp: 365 nm Intensity:2000 W/cm^2, t= 3h, C_o=25 ppm dye.	[67]
Fe^{3+} Doped mesopores CeO_2	Hydrotherma l	Rhodamine B dye (100%)	Natural light irradiation, C_o= 25 mg/L RhB solution,	[68]
Ag_3PO_4 /CeO_2	Hydrotherma l & *in situ* precipitation	Rhodamine B dye (88% with mercury lamp)	Mercury lamp: 356 nm, 300 W, t= 3h, C_o= 10 mg/L dye.	[69]
Co-doped CeO_2	Co-precipita-tion method	Methylene blue (88.9%)	UV irradiation: 365nm, 30W, t= 420 min, C_o= 15 ppm dye	[73]
Ti, Mn, Fe and Co- doped CeO_2	Co-precipita-tion method	Methylene blue)	Mercury light: <400 nm	[70]
ZnO/CeO_2 nanocomposite	Thermal decompose-tion	Methyl orange (95.9%), Methyl-ene blue (97.4%),	UV visible light: 250W, t= 150 min, C_o= 500 ml of solution,	[71]
Fluorine doped CeO_2 Nanocubes (F- CeO_2)	Solution combustion	Methylene blue (92.1%)	UV visible light: 300 W; fluorescent lamp, 50 W; t= 18 min , C_o= 10 mg/L dye	[74]
$BiVO_4/CeO_2$ composite	Precipitation & hydrother-mal method	Methylene blue (100%), Methyl orange (100%)	Halogen lamp: 185 mW cm^{-2}, t= 30 min , C_o= 7.5 mg/L dye	[75]
Fe-Ce mixed oxide	Precipitation	Phenol (13%), Methylene Blue (93%), Congo red (100%)	Solar light: intensity 100000 lux, C_o= 10 ppm for phenol, 100 ppm for dye	[72]
p-CuO/*n*-CeO_2	Ribbon-ignition method	Rhodamine B dye (86.2%)	Xenon lamp: 420 nm, 350W, t= 24 min , C_o= 10 mg/L dye	[76]

Meanwhile, Hao et al. [68] synthesized Fe^{+3} doped mesopores CeO_2 nanomaterials, which having complete rhodamine B dye degradation at natural light irradiation. This phenomenon was observed because 25% Fe-doped CeO_2 has a surface area of 144.7 m^2/g compared to CeO_2 nanomaterials having a surface area in the range of 108.9 m^2/g. A high BET surface area is helpful to enhance the adsorption capacity of organic dyes and thus to increase the photocatalytic efficiency. Not only transition elements or metal oxides but p-n type (p-type Ag_3PO_4) and (n-type CeO_2) composite also have significant effect on rhodamine B dye degradation [69]. Similarly, rhodamine B dye degradation was observed for the case of p-CuO doped in CeO_2 nanomaterials. Yue and Zhang synthesized Ti, Mn, Fe, Co doped CeO_2 particles via co-precipitation method [70]. The methylene blue dye degradation was observed in ordered of Co- > Mn- > Ti- > Fe- > CeO_2 nanoparticles. In which, the doping of the ceria nanoparticles with transition metals significantly improves their optical activity, which may be due to the enhancement of energy transfer. Rajendran et al. [71] studied photocatalytic decomposition of three component namely methyl orange, methylene blue and phenol with ZnO doped CeO_2 nanocomposite, synthesized by thermal decomposition method in presence of visible light. The degradation of dyes followed the order: 97.4% methylene blue > 95.9% methyl orange > 96.2% phenol. Pradhan and Parida [72] studied photocatalytic decomposition of three dyes namely congo red, methylene blue and phenol with mesoporous iron-cerium mixed oxides, synthesized by precipitation method. The degradation of dyes followed the order: 97.4% congo red > 93% methylene blue > 13% phenol.

4.3 Photocatalysis of dopent-CeO_2 materials on different supported materials

Table 4 displays the details of several CeO_2 nanomaterials used as dopant or loaded on supported materials. Generally, CeO_2 nanomaterials with a high surface area, which has more chance to contact with reactants, should display a better catalytic performance. Sometimes, nanoparticles, with higher surface area, smaller size, displayed poorer catalytic performance; whilst nanorods, with lower surface area and larger diameter, showed better catalytic performance. The main factor causing the unusually properties is the exposed planes. Liu and Huang developed carbon nano fiber loaded with Cu_2O nanoparticles, which are applicable for photocatalytic degradation of rhodamine B dye with above 85% of dye degradation [77]. Meanwhile, Kasinathan et al. degrade rhodamine B dye with 99.9% through doping of CeO_2 over TiO_2 nanocomposite [78]. Similar way, CeO_2 nanomaterials have been decorated over CdS nano rod and ZnS materials [79-80]. In both cases, rhodamine B dye degradation was achieved 96% and 73%, respectively. The synthesis method and experimental methods have been described in detail in Table 4. In all above cases, surface area, shape of the CeO_2 nanomaterials, type of CeO_2 planes and morphology take active participation for photocatalytic degradation of rhodamine B dye. This suggests the importance and significance for shaped control synthesis of CeO_2 nanomaterials for multiple applications.

Table 4. Photocatalytic application of different shape dopent-CeO_2 nanomaterials on different support materials.

CeO_2 nanomaterials with morphology	Synthesis methods	Pollutant (% degradation)	Treatment conditions	Ref.
Cu_2O nanoparticles supported on carbon nano fibers	Electrospinn-ing	Rhodamine B dye Phenol Pollutant (above 85 %)	Xenon lamp: 500 W, t= 40 min , C_o= 10 ppm phenol	[77]
CeO_2 doped TiO_2 nanocomposites	Hydrother-mal method	Rhodamine B dye (99.9 %)	UV visible light: 15W, t= 800 sec, C_o= 1mM dye.	[78]
ZnS decorated CeO_2 nanoparticles	Microwave method	Rhodamine B dye (73%)	UV lamp: 254 nm, 15W, t= 120 min, C_o= 0.1mM dye.	[79]
CeO_2 on CdS nanorod	Solvothermal method	Rhodamine B dye (96.68%)	UV spectra: t= 50 min, C_o= 40 mg/L dye.	[80]

5. Challenges and prospects of CeO_2 nanomaterials

This review described the shape-controlled synthesis and applicability of CeO_2 nanomaterials for dye degradation as a part of wastewater remediation. Different path and mechanisms have been described to synthesize CeO_2 nanomaterials systematically. The main parameter associated with CeO_2 nanomaterials, which is responsible for redox/photocatalytic activity i.e. presence of Ce^{3+} defects in the crystal lattice. The size, shape, morphology and structure of CeO_2 nanomaterials have been controlled by controlling the defects in the crystal lattice. However, there are many challenges associated with CeO_2 nanomaterials i.e. synthesize, shape controlled and application oriented. The controlled synthesis of CeO_2 nanomaterials are still challenging. Still more research are needed to navigate synthesize techniques for specific applications. Recently, most of the cases to synthesize CeO_2 nanomaterials are associated with hit and trial bases or by experience bases, in which agglomeration was form at higher temperature and sudden drop of surface area, shape and reduction of crystal defects. Similarly, calcination temperature and aging time also affects the oxidation state of CeO_2 nanomaterials. With aging, oxidation potential of the solution decreases, leading to the transformation of CeO_2 nanomaterials from mostly Ce^{4+} state to Ce^{3+} state. Many anti sintering materials such as zirconium and other earth metals have been utilized to control the temperature effects, but the semiconductor properties of CeO_2 nanomaterials were affected. Moreover, many surfactant materials and polymer materials have been utilized for controlling the shape of CeO_2 nanomaterials. However, controlling the shape and morphology is still a big challenge. Meanwhile, development of green chemistry for green synthesis and waste minimization are another aspect to be considering for the synthesis of CeO_2 nanomaterials. We have to develop more valuable, cost effectiveness and shaped control synthesis techniques for industrial applications. Apart from synthesis techniques, many applications are associated with CeO_2 nanomaterials based on the size and shape. The catalytic application of CeO_2 nanomaterials and their morphology development properties should be extended for more reactions. Therefore, further work will focus on green route synthesis of CeO_2 nanomaterials, control of morphology during synthesis and its consequences. In short, there are many chances to grow in this area i.e. for synthesis aspects, for application aspects or for cost effectiveness aspects. Research towards this aim will need a concerted effort of material scientists, chemists, physicists and engineers.

Acknowledgement:

Authors appreciate the Ministry of Science and Technology (MOST) of Taiwan (R.O.C.) under grant number MOST 106-3113-E-007-002 for the financial support.

References

[1] O.C. Nweke, W.H. Sanders, Modern environmental health hazards: A public health issue of increasing significance in Africa, Environ. Health Perspect. 117 (2009) 863–870. https://doi.org/10.1289/ehp.0800126

[2] B.G. Rahm, S.J. Riha, Evolving shale gas management: water resource risks, impacts, and lessons learned, Environ. Sci. Process. Impacts. 16 (2014) 1400–1412. https://doi.org/10.1039/C4EM00018H

[3] I. Ali, V.K. Gupta, Advances in water treatment by adsorption technology, Nat. Protoc. 1 (2007) 2661–2667. https://doi.org/10.1038/nprot.2006.370

[4] C.R. Woolard, R.L. Irvine, Biological treatment of hypersaline wastewater by a biofilm of halophilic bacteria, Water Environ. Res. 66 (1994) 230–235. https://doi.org/10.2175/WER.66.3.8

[5] L. Ma, X. Dong, M. Chen, L. Zhu, C. Wang, F. Yang, Y. Dong, Fabrication and water treatment application of carbon nanotubes (CNTs)-based composite membranes: A review, Membranes (Basel). 7 (2017). 1-21. https://doi.org/10.3390/membranes7010016

[6] W.S. Wan Ngah, M.A.K.M. Hanafiah, Removal of heavy metal ions from wastewater by chemically modified plant wastes as adsorbents: A review, Bioresour. Technol. 99 (2008) 3935–3948. https://doi.org/10.1016/j.biortech.2007.06.011

[7] F. Fu, Q. Wang, Removal of heavy metal ions from wastewaters: A review, J. Environ. Manage. 92 (2011) 407–418. https://doi.org/10.1016/j.jenvman.2010.11.011

[8] M. Muruganandham, R.P.S. Suri, S. Jafari, M. Sillanpaa, G.-J. Lee, J.J. Wu, M. Swaminathan, Recent Developments in Homogeneous Advanced Oxidation Processes for Water and Wastewater Treatment, Int. J. Photoenergy. 2014 (2014) 21. https://doi.org/10.1155/2014/821674

[9] M.A. Rauf, M.A. Meetani, S. Hisaindee, An overview on the photocatalytic degradation of azo dyes in the presence of TiO_2 doped with selective transition metals, Desalination. 276 (2011) 13–27. https://doi.org/10.1016/j.desal.2011.03.071

[10] V.K. Yemmireddy, Y.C. Hung, Using Photocatalyst Metal Oxides as Antimicrobial Surface Coatings to Ensure Food Safety—Opportunities and Challenges, Compr. Rev. Food Sci. Food Saf. 16 (2017) 617–631. https://doi.org/10.1111/1541-4337.12267

[11] V. Binas, D. Venieri, D. Kotzias, G. Kiriakidis, Modified TiO_2 based photocatalysts for improved air and health quality, J. Mater. 3 (2017) 3–16.

[12] W. Atul, G. Gaikward, M. Dondhe, N. Khaty, Removal of organic pollutant from water by heterogeneous photocatalysis: a review, Res J Chem Env. 17 (2013) 84–94.

[13] N. Muhd Julkapli, S. Bagheri, S. Bee Abd Hamid, Recent advances in heterogeneous photocatalytic decolorization of synthetic dyes, Sci. World J. 2014 (2014). https://doi.org/10.1155/2014/692307

[14] F. Nandjou, S. Haussener, Degradation in photoelectrochemical devices: review with an illustrative case study, J. Phys. D. Appl. Phys. 50 (2017) 124002. https://doi.org/10.1088/1361-6463/aa5b11

[15] J.T. Dahle, Y. Arai, Environmental geochemistry of cerium: Applications and toxicology of cerium oxide nanoparticles, Int. J. Environ. Res. Public Health. 12 (2015) 1253–1278. https://doi.org/10.3390/ijerph120201253

[16] T. Montini, M. Melchionna, M. Monai, P. Fornasiero, Fundamentals and Catalytic Applications of CeO_2-Based Materials, Chem. Rev. 116 (2016) 5987–6041. https://doi.org/10.1021/acs.chemrev.5b00603

[17] N. Rajput, Methods of preparation of nanoparticles – A REVIEW, International Journal of Advances in Engineering & Technology, 7, 4 (2015) 1806-1811.

[18] D. Zhang, X. Du, L. Shi, R. Gao, Shape-controlled synthesis and catalytic application of ceria nanomaterials, Dalt. Trans. 41 (2012) 14455. https://doi.org/10.1039/c2dt31759a

[19] L. Thi, T. Tuyen, D.Q. Khieu, H.T. Long, D.T. Quang, C. The, L. Trang, T.T. Hoa, N.D. Cuong, Monodisperse Uniform CeO_2 Nanoparticles : Controlled Synthesis and Photocatalytic Property, J. Phys. Chem. C. 2016 (2016) 10009–10016.

[20] Z. Wang, X. Feng, Polyhedral shapes of CeO_2 nanoparticles, J. Phys. Chem. B. 107 (2003) 13563–13566. https://doi.org/10.1021/jp036815m

[21] J. Polte, Fundamental growth principles of colloidal metal nanoparticles – a new perspective, CrystEngComm. 17 (2015) 6809–6830. https://doi.org/10.1039/C5CE01014D

[22] J. Chang, E.R. Waclawik, Colloidal semiconductor nanocrystals: controlled synthesis and surface chemistry in organic media, RSC Adv. 4 (2014) 23505–23527. https://doi.org/10.1039/C4RA02684E

[23] S. Razzaque, S.Z. Hussain, I. Hussain, B. Tan, Design and utility of metal/metal oxide nanoparticles mediated by thioether end-functionalized polymeric ligands, Polymers (Basel). 8 (2016). https://doi.org/10.3390/polym8040156

[24] P. Delgado-Martínez, C. Freire, R. González-Prieto, R. Jiménez-Aparicio, J.L. Priego, M.R. Torres, Synthesis, crystal structure, and magnetic properties of amidate and carboxylate dimers of ruthenium, Crystals. 7 (2017) 1–16. https://doi.org/10.3390/cryst7070192

[25] H. Yao, Y. Wang, G. Luo, Article A new size-controllable precipitation method to prepare CeO_2 nanoparticles in a membrane dispersion micro-reactor A new size-controllable precipitation method to prepare CeO_2 nanoparticles in a membrane dispersion micro-reactor, Ind. Eng. Chem. Res. 56 (2017) 4993-4999. https://doi.org/10.1021/acs.iecr.7b00289

[26] J.J. Ketzial, A.S. Nesaraj, Synthesis of CeO_2 nanoparticles by chemical precipitation and the effect of a sur- factant on the distribution of particle sizes, J. Ceram. Process. Res. 12 (2011) 74–79.

[27] H. Xiao, Z. Ai, L. Zhang, Nonaqueous Sol - Gel Synthesized Hierarchical CeO_2 Nanocrystal Microspheres as Novel Adsorbents for Wastewater Treatment, J. Phys. Chem. C. 113 (2009) 16625–16630. https://doi.org/10.1021/jp9050269

[28] N.S. Ferreira, R.S. Anglica, V.B. Marques, C.C.O. De Lima, M.S. Silva, Cassava-starch-assisted sol-gel synthesis of CeO_2 nanoparticles, Mater. Lett. 165 (2016) 139–142. https://doi.org/10.1016/j.matlet.2015.11.107

[29] T. Yu, J. Joo, Y. Il Park, T. Hyeon, Large-scale nonhydrolytic sol-gel synthesis of uniform-sized ceria nanocrystals with spherical, wire, and tadpole shapes, Angew. Chemie - Int. Ed. 44 (2005) 7411–7414. https://doi.org/10.1002/anie.200500992

[30] J.J. Hinman, K.S. Suslick, Nanostructured Materials Synthesis Using Ultrasound, Top. Curr. Chem. 375 (2017) 1–36. https://doi.org/10.1007/978-3-319-54271-3_3

[31] P. Jamshidi, M. Salavati-Niasari, D. Ghanbari, H.R. Shams, Synthesis, Characterization, Photoluminescence and Photocatalytic Properties of CeO_2 Nanoparticles by the Sonochemical Method, J. Clust. Sci. 24 (2013) 1151–1162. https://doi.org/10.1007/s10876-013-0605-0

[32] S. Nomanbhay, M.Y. Ong, A Review of Microwave-Assisted Reactions for Biodiesel Production, Bioengineering. 4 (2017) 57, 1-21.

[33] H. Yang, C. Huang, A. Tang, X. Zhang, W. Yang, Microwave-assisted synthesis of ceria nanoparticles, Mater. Res. Bull. 40 (2005) 1690–1695. https://doi.org/10.1016/j.materresbull.2005.05.014

[34] E. Kockrick, C. Schrage, A. Grigas, D. Geiger, S. Kaskel, Synthesis and catalytic properties of microemulsion-derived cerium oxide nanoparticles, J. Solid State Chem. 181 (2008) 1614–1620. https://doi.org/10.1016/j.jssc.2008.04.036

[35] Y. Tao, H. Wang, Y. Xia, G. Zhang, H. Wu, G. Tao, Preparation of shape-controlled CeO2 nanocrystals via microwave-assisted method, Mater. Chem. Phys. 124 (2010) 541–546. https://doi.org/10.1016/j.matchemphys.2010.07.007

[36] S. Kuśnieruk, J. Wojnarowicz, A. Chodara, T. Chudoba, S. Gierlotka, W. Lojkowski, Influence of hydrothermal synthesis parameters on the properties of hydroxyapatite nanoparticles, Beilstein J. Nanotechnol. 7 (2016) 1586–1601. https://doi.org/10.3762/bjnano.7.153

[37] Z. Wang, X. Feng, Polyhedral shapes of CeO_2 nanoparticles, J. Phys. Chem. B. 107 (2003) 13563–13566. https://doi.org/10.1021/jp036815m

[38] J.P.Y. Tan, H.R. Tan, C. Boothroyd, Y.L. Foo, C. Bin He, M. Lin, Three-dimensional structure of CeO_2 nanocrystals, J. Phys. Chem. C. 115 (2011) 3544–3551. https://doi.org/10.1021/jp1122097

[39] D. Zhang, F. Niu, T. Yan, L. Shi, X. Du, J. Fang, Ceria nanospindles: Template-free solvothermal synthesis and shape-dependent catalytic activity, Appl. Surf. Sci. 257 (2011) 10161–10167. https://doi.org/10.1016/j.apsusc.2011.07.010

[40] C. Sun, H. Li, H. Zhang, Z. Wang, L. Chen, Controlled synthesis of CeO_2 nanorods by a solvothermal method, Nanotechnology. 16 (2005) 1454–1463. https://doi.org/10.1088/0957-4484/16/9/006

[41] X. Liang, J. Xiao, B. Chen, Y. Li, Catalytically stable and active CeO_2 mesoporous spheres, Inorg. Chem. 49 (2010) 8188–8190. https://doi.org/10.1021/ic100795p

[42] S.-J. Shih, Y.-Y. Wu, C.-Y. Chen, C.-Y. Yu, Controlled Morphological Structure of Ceria Nanoparticles Prepared by Spray Pyrolysis, Procedia Eng. 36 (2012) 186–194. https://doi.org/10.1016/j.proeng.2012.03.029

[43] M.A. Malik, M.Y. Wani, M.A. Hashim, Microemulsion method: A novel route to synthesize organic and inorganic nanomaterials. 1st Nano Update, Arab. J. Chem. 5 (2012) 397–417. https://doi.org/10.1016/j.arabjc.2010.09.027

[44] E. Kockrick, C. Schrage, A. Grigas, D. Geiger, S. Kaskel, Synthesis and catalytic properties of microemulsion-derived cerium oxide nanoparticles, J. Solid State Chem. 181 (2008) 1614–1620. https://doi.org/10.1016/j.jssc.2008.04.036

[45] T. Wang, D.C. Sun, Preparation and characterization of nanometer-scale powders ceria by electrochemical deposition method, Mater. Res. Bull. 43 (2008) 1754–1760. https://doi.org/10.1016/j.materresbull.2007.07.008

[46] H. Oh, S. Kim, Synthesis of ceria nanoparticles by flame electrospray pyrolysis, J. Aerosol Sci. 38 (2007) 1185–1196. https://doi.org/10.1016/j.jaerosci.2007.09.007

[47] H. Chun Zeng, Ostwald Ripening: A Synthetic Approach for Hollow Nanomaterials, Curr. Nanosci. 3 (2007) 177–181. https://doi.org/10.2174/157341307780619279

[48] G. Shen, H. Liu, Q. Wang, Z. Wang, Y. Chen, Self-template hydrothermal synthesis of CeO 2 hollow nanospheres, J. Nanoparticle Res. 14 (2012).

[49] Z. Guo, F. Du, G. Li, Z. Cui, Synthesis and characterization of single-crystal $Ce(OH)CO_3$ and CeO_2 triangular microplates, Inorg. Chem. 45 (2006) 4167–4169. https://doi.org/10.1021/ic052189r

[50] J. Wei, Z. Yang, H. Yang, T. Sun, Y. Yang, A mild solution strategy for the synthesis of mesoporous CeO_2 nanoflowers derived from $Ce(HCOO)_3$, CrystEngComm. 13 (2011) 4950–4955. https://doi.org/10.1039/c1ce05324h

[51] A. Bumajdad, M.I. Zaki, J. Eastoe, L. Pasupulety, Microemulsion-based synthesis of CeO_2 powders with high surface area and high-temperature stabilities, Langmuir. 20 (2004) 11223–11233. https://doi.org/10.1021/la040079b

[52] Z. Liu, S. Guo, C. Hong, Z. Xia, Synthesis and photocatalytic properties of CeO_2 nanocubes, J. Mater. Sci. Mater. Electron. 27 (2016) 2146–2150. https://doi.org/10.1007/s10854-015-4004-1

[53] W. Yu, Q. Ma, C. Wang, X. Dong, J. Wang, G. Liu, Electrospray ionization preparation and photodegradation properties of CeO_2 microspheres with tunable morphologies, Mater. Express. 4 (2014) 435–440. https://doi.org/10.1166/mex.2014.1189

[54] X. Zheng, S. Huang, D. Yang, H. Zhai, Y. You, X. Fu, J. Yuan, X. Zhou, J. Wen, Y. Liu, Synthesis of X-architecture CeO_2 for the photodegradation of methylene blue under UV-light irradiation, J. Alloys Compd. 705 (2017) 131–137.

[55] A. Phuruangrat, T. Thongtem, S. Thongtem, Effect of NaOH on morphologies and photocatalytic activities of CeO_2 synthesized by microwave-assisted hydrothermal method, Mater. Lett. 193 (2017) 161–164. https://doi.org/10.1016/j.matlet.2017.01.105

[56] Y. Zhai, S. Zhang, H. Pang, Preparation, characterization and photocatalytic activity of CeO_2 nanocrystalline using ammonium bicarbonate as precipitant, Mater. Lett. 61 (2007) 1863–1866. https://doi.org/10.1016/j.matlet.2006.07.146

[57] Y. Shao, Y. Ma, Mesoporous CeO_2 nanowires as recycled photocatalysts, Sci. China Chem. 55 (2012) 1303–1307. https://doi.org/10.1007/s11426-012-4636-4

[58] P. Zhao, J. Song, M. Xia, Y. Wang, D. Meng, H. Wang, R. Li, Photocatalytic and electrochemical properties of CeO_2 with lamellar structure inherited from cerium oxalate complex, J. Porous Mater. 24 (2017) 1089–1094. https://doi.org/10.1007/s10934-016-0349-y

[59] H.R. POURETEDAL, A. KADKHODAIE, Synthetic CeO_2 Nanoparticle Catalysis of Methylene Blue Photodegradation: Kinetics and Mechanism, Chinese J. Catal. 31 (2010) 1328–1334. https://doi.org/10.1016/S1872-2067(10)60121-0

[60] Y.C. Zhang, M. Lei, K. Huang, C. Liang, Y.J. Wang, S.S. Ding, R. Zhang, D.Y. Fan, H.J. Yang, Y.G. Wang, A facile route to mono-dispersed CeO_2 nanocubes and their enhanced photocatalytic properties, Mater. Lett. 116 (2014) 46–49. https://doi.org/10.1016/j.matlet.2013.10.085

[61] G. Cheng, J. Xiong, F.J. Stadler, A facile polyol-mediated approach to tunable CeO_2 microcrystals and their photocatalytic activity, Powder Technol. 249 (2013) 89–94. https://doi.org/10.1016/j.powtec.2013.07.033

[62] A. Phuruangrat, S. Thipkonglas, T. Thongtem, S. Thongtem, Hydrothermal synthesis and characterization of Dy-doped MoO_3 nanobelts for using as a visible-light-driven photocatalyst, Mater. Lett. 195 (2017) 37–40. https://doi.org/10.1016/j.matlet.2017.02.053

[63] P. Ji, J. Zhang, F. Chen, M. Anpo, Study of adsorption and degradation of acid orange 7 on the surface of CeO_2 under visible light irradiation, Appl. Catal. B Environ. 85 (2009) 148–154. https://doi.org/10.1016/j.apcatb.2008.07.004

[64] S.B. Khan, M. Faisal, M.M. Rahman, K. Akhtar, A.M. Asiri, A. Khan, K.A. Alamry, Effect of Particle Size on the Photocatalytic Activity and Sensing Properties of CeO_2 Nanoparticles, Int. J. Electrochem. Sci. Int. J. Electrochem. Sci. 8 (2013) 7284–7297. www.electrochemsci.org.

[65] Z.-R. Tang, Y. Zhang, Y.-J. Xu, A facile and high-yield approach to synthesize one-dimensional CeO_2 nanotubes with well-shaped hollow interior as a photocatalyst for degradation of toxic pollutants, RSC Adv. 1 (2011) 1772. https://doi.org/10.1039/c1ra00518a

[66] Z. Li, W. Luo, M. Zhang, J. Feng, Z. Zou, Photoelectrochemical cells for solar hydrogen production: current state of promising photoelectrodes, methods to improve their properties, and outlook, Energy Environ. Sci. 6 (2013) 347–370. https://doi.org/10.1039/C2EE22618A

[67] N.H. Shah, K.R. Bhangaonkar, S.T. Mhaske, In-Situ Synthesis of CeO_2/ZnO Composite Nanoparticles and Its Application in Degradation of Rhodamine B Using Sonocatalytic & Photocatalytic Method, Int. J. Mater. Sci. Eng. 5 (2017) 16–27.

[68] S. Hao, J. Hou, P. Aprea, T. Lv, Photocatalytic activity under weak visible light of Fe3+ doped mesoporous CeO_2, Ind. Eng. Chem. Res. 53 (2014) 14617–14622. https://doi.org/10.1021/ie501585z

[69] W. Zhang, C. Hu, W. Zhai, Z. Wang, Y. Sun, F. Chi, S. Ran, X. Liu, Y. Lv, Novel Ag_3 PO_4 / CeO_2 p-n Hierarchical Heterojunction with Enhanced Photocatalytic Performance, 19 (2016) 673–679.

[70] L. Yue, X. Zhang, Structural characterization and photocatalytic behaviors of doped CeO 2 nanoparticles, 475 (2009) 702–705.

[71] S. Rajendran, M.M. Khan, F. Gracia, J. Qin, V.K. Gupta, S. Arumainathan, Ce^{3+}-ion-induced visible-light photocatalytic degradation and electrochemical activity of ZnO/CeO_2 nanocomposite, Sci. Rep. 6 (2016) 31641. https://doi.org/10.1038/srep31641

[72] G.K. Pradhan, K. Parida, Fabrication of iron-cerium mixed oxide: an efficient photocatalyst for dye degradation, Int. J. Eng. Sci. Technol. 2 (2011) 53–65. https://doi.org/10.4314/ijest.v2i8.63780

[73] J. Saranya, K.S. Ranjith, P. Saravanan, D. Mangalaraj, R.T. Rajendra Kumar, Cobalt-doped cerium oxide nanoparticles: Enhanced photocatalytic activity under UV and visible light irradiation, Mater. Sci. Semicond. Process. 26 (2014) 218–224. https://doi.org/10.1016/j.mssp.2014.03.054

[74] H. Miao, G.F. Huang, J.H. Liu, B.X. Zhou, A. Pan, W.Q. Huang, G.F. Huang, Origin of enhanced photocatalytic activity of F-doped CeO_2 nanocubes, Appl. Surf. Sci. 370 (2016) 427–432. https://doi.org/10.1016/j.apsusc.2016.02.122

[75] N. Wetchakun, S. Chaiwichain, B. Inceesungvorn, K. Pingmuang, S. Phanichphant, A.I. Minett, J. Chen, $BiVO_4/CeO_2$ nanocomposites with high visible-light-induced photocatalytic activity, ACS Appl. Mater. Interfaces. 4 (2012) 3718–3723. https://doi.org/10.1021/am300812n

[76] N. Wang, Y. Pan, T. Lu, X. Li, S. Wu, J. Wu, A new ribbon-ignition method for fabricating p-CuO/n-CeO_2 heterojunction with enhanced photocatalytic activity, Appl. Surf. Sci. 403 (2017) 699–706. https://doi.org/10.1016/j.apsusc.2017.01.232

[77] Y. Liu, Q. Huang, G. Jiang, D. Liu, W. Yu, Cu_2 nanoparticles supported on carbon nanofibers as a cost-effective and efficient catalyst for RhB and phenol degradation, J. Mater. Res. (2017) 1–11.

[78] K. Kasinathan, J. Kennedy, M. Elayaperumal, M. Henini, M. Malik, Photodegradation of organic pollutants RhB dye using UV simulated sunlight on

ceria based TiO_2 nanomaterials for antibacterial applications, Sci. Rep. 6 (2016) 38064. https://doi.org/10.1038/srep38064

[79] C.W. Raubach, L. Polastro, M.M. Ferrer, A. Perrin, C. Perrin, A.R. Albuquerque, P.G.C. Buzolin, J.R. Sambrano, Y.B. V De Santana, J.A. Varela, E. Longo, Influence of solvent on the morphology and photocatalytic properties of ZnS decorated CeO2 nanoparticles, J. Appl. Phys. 115 (2014). https://doi.org/10.1063/1.4880795

[80] S. Gu, Y. Chen, X. Yuan, H. Wang, X. Chen, Y. Liu, Q. Jiang, Z. Wu, G. Zeng, Facile synthesis of CeO_2 nanoparticle sensitized CdS nanorod photocatalyst with improved visible-light photocatalytic degradation of rhodamine B, RSC Adv. 5 (2015) 79556–79564.

Chapter 12

Synthesis, Characterization and Photocatalytic Study of Sm^{3+} Doped Mesoporous CeO_2 Nanoparticles

Sajith N. V.[a], Sheethu Jose[a], Soumya B. Narendranath[a], Pradeepan Periyat[b]*

[a]Department of Chemistry, Central University of Kerala, Kasaragod, India. 671314

[b]Department of Chemistry, University of Calicut, India. 673635

pperiyat@uoc.ac.in

Abstract

CeO_2 and Sm^{3+} (2.5, 10 and 15 wt%) doped CeO_2 nanoparticles have been synthesized by combining sol-gel and hydrothermal method. Synthesized samples were characterized by using Powder X-Ray Diffraction, FT-IR spectroscopy, X-Ray photoelectron spectroscopy, Transmission Electron Microscopy, UV-visible spectroscopy, TGA/DSC Analysis and BET surface area analysis. XRD pattern showed that as synthesized crystalline structures of CeO_2 nanoparticles are cubic fluorite type structure. TEM showed uniform particle size ranges from 10-20 nm and XPS confirmed the successful incorporation of Samarium. CeO_2 and Sm^{3+} doped CeO_2 have high surface area and their pore size distribution is in the mesoporous range. Photocatalytic activity of prepared sample was studied using methylene blue (MB) dye degradation. Photocatalytic study showed that 10 wt% Sm^{3+} doped CeO_2 sample has the highest catalytic activity among various sample synthesized.

Keywords

Cerium Dioxide, Photocatalysis, Methylene Blue, Samarium Doped, Dye Degradation

Contents

1. Introduction

Solar energy is an important source of renewable energy. Large magnitude of its free availability with 5% UV, 43% visible and 52% IR radiation makes it a highly appealing source of energy. In the present scenario researchers are greatly involved in the development of technologies to harvest solar energy using nano materials having potentiality to meet the present and future energy crisis. The nano-dimensions of a material bring on significant changes of its properties, like optical absorption, electronic conductivity, chemical reactivity, biocompatibility, in comparison with the macro-dimensions of the same material. With decrease in particle size, an increase of surface area is observed and the number of active sites present on the surface of the particle is higher. The enhanced surface area produces a considerable change of surface properties (*e.g.* energy, morphology). All these factors change the basic characteristics and the reactivity of the nanomaterials. Researchers all over the world are involved in utilizing nanomaterial fabrications for photocatalysis for the current environmental issues such as pollution, energy crisis *etc.* Photocatalysis proceeds over a semiconductor by the irradiation of light. It can be considered as one of the best methods for waste water treatment for the detoxification of organic pollutants especially dyes, whose mechanism has been mostly studied [1-6]. This is because semiconductors are inexpensive and have the potential to mineralize various organic compounds [7]. Metal oxides such as TiO_2 [8], ZnO [9], WO_3 [10] and CeO_2 are the most prominent semiconductors for photocatalysis [11] and their significant characteristics are the high surface area, desired band gap [12], suitable morphology [13] and high stability. The photocatalytic process involves the generation of electron-hole pairs upon irradiation with sunlight and the formation of reactive oxygen species (ROS) such as superoxide radical anion ($O_2^{\cdot -}$), hydroxyl radical ($OH^{\cdot}$), hydroperoxy radical ($OOH^{\cdot}$) *etc.* which in turn degrade or transform the harmful compounds to byproducts such as CO_2 and H_2O [14].

CeO_2 is a very active rare earth metal oxide having cubic fluorite structure in which each cerium site is surrounded by eight oxygen sites in face centered cubic arrangement and each oxygen site has a tetrahedron cerium site [15]. CeO_2 based materials have attracted much attention in catalytic systems for a variety of applications such as automobile Three Way Catalysis (TWCs) [16], Solid Oxide Fuel Cells (SOFC) [17], Polymer Exchange Membrane Fuel Cells (PEMFCs) [18], Oxygen sensors [19] and Diesel engines owing to its high oxygen transport capacity, resistance to chemical and photo corrosion, unique redox potential of Ce^{4+}/Ce^{3+} results fast shuttling of oxidation states ($Ce^{4+}$and Ce^{3+}) and high oxygen storage capacity. Most of the application of ceria are due to the presence of the large amount of oxygen vacancies and due to the presence of Ce^{+3}/Ce^{+4}redox couple [20]. Moreover, Cerium is the most abundant of the rare earth elements (about 0.0046 wt % of the Earth's crust) and is ecofriendly in nature. Properties such as wide band gap, stability and non-toxicity of CeO_2 shows similarity to other metal oxides such as TiO_2 and ZnO [21, 22]. The band gap energy of CeO_2 is 3.1 eV [23] and studies have proven that doping of ceria with trivalent elements change the structural and electronic properties of ceria. It is due to the increase in the conversion of Ce^{4+} to Ce^{3+} which increases active oxygen vacancies and results in improved photocatalytic activity.

Herein, we report Sm^{3+} doped CeO_2 nanoparticles with greater photocatalytic activity under direct sunlight. They were characterized with XRD, TG-DTA, FT-IR, and BET surface area analysis.

2. Experimental section

2.1 Materials used

Cerium nitrate hexahydrate ($Ce(NO_3)_3.6H_2O$) (Alfa Aesar, 99.99%), Samarium nitrate hexahydrate ($Sm(NO_3)_3.6H_2O$) (Alfa Aesar, 99.99%), 30% Ammonia solution (Merck, India), 10% HCl (Merck, India) and Methylene blue[$C_{16}H_{18}N_3SCl$] (Loba Chemie). All chemicals were in reagent grade and used no further purification.

2.2 Methods

0.2M $Ce(NO_3)_3.6H_2O$ solution in 500 mL was prepared with distilled water. 28-30% ammonia solution was added drop wise to the clear solution for the precipitation of $Ce(OH)_4$ until the pH became10 which indicated the complete precipitation of $Ce(OH)_4$. Whole solution was transferred into the centrifuging cell and centrifuged for 45 minutes at the rpm of 10,000. The sample was then washed with water until the solution become nitrate free. The following equation correspond to the chemical reaction involved,

$$Ce(NO_3)_3.6H_2O + H_2O \rightarrow Ce(OH)_4 \quad (1)$$

After washing, the precipitate was collected and dispersed in 1000 mL water. 10% HCl was added in small quantities to this solution in order to maintain pH 2. The sol was stable at pH range of 1.8-2. Again, it was stirred for 2 h. The stabilized sol was transferred into an autoclave and heated at 150 °C for 48 h. The heat-treated sol was transferred into a Petri dish and dried at 150 °C for 48 h to get CeO_2 nanoparticles.

Three different weight percentages (2.5, 10 and 15 wt%) of Sm^{3+} doped CeO_2 sol were prepared by adding a calculated quantity of Samarium Nitrate $Sm(NO)_3.6H_2O$ dissolved in 100 mL water. Samarium Nitrate solution is added to the sol drop wise with constant stirring. Stirring of the sol is continued for 2 days. The stabilized sol was transferred into an autoclave and heated at 150 °C for 48 h. The heat-treated sol was transferred into a Petri dish and dried at 150 °C for 48 h to get different weight percentage added Sm^{3+} doped CeO_2 nanoparticles.

2.3 Photocatalytic study

Photocatalytic activity of CeO_2 and Sm^{3+} doped CeO_2 was investigated by using methylene blue as a model system. Methylene blue is extensively used as redox indicator in analytical chemistry. It shows blue color when dissolved in water and turns colorless in the presence of a reducing agent [24]. 0.1g of photocatalyst was accurately weighed and transferred into a 100 mL beaker. 50 mL of 10^{-5} M methylene blue was added into the beaker. The solution was kept in the dark for 30 minutes with constant stirring so that adsorption/desorption equilibrium is attained. After that, it was exposed to the sunlight with constant stirring. Sample was collected at regular intervals and centrifuged. The absorbance of each sample was measured using a Shimadzu 2600 spectrophotometer. Data obtained from the UV-visible spectrophotometer was plotted in origin 8 software. Photocatalytic activities of CeO_2 as well as Sm^{3+} doped CeO_2 nanoparticles in three different ratios were investigated.

2.4 Characterization

The XRD patterns were recorded using a Rigaku Miniflex 600 diffractometer with Cu Kα (λ=1.54Å) radiation. Data were collected by step scanning over a 2θ range from 10 to 80° with a step size of 0.01. The resultant XRD patterns were analyzed by comparing them with the reported ones. The crystallite size was calculated by using the Debye-Scherrer equation:

$$\varphi = 0.9\lambda/\beta cos\theta \tag{3}$$

where φ is the crystallite size (nm), θ is the Braggs angle (radian) and β is the peak width at half maximum (radian). Fourier transform infrared spectral analysis was carried out by Perkin Elmer Spectrum 2 infrared spectrometer in the range of 4000 to 400 cm^{-1} with KBr as reference. Thermo gravimetric-differential thermal analysis (TG-DTA) were carried out by Perkin Elmer STA 6000 Thermal Analyser, Singapore from 50 to 600 °C in N_2 atmosphere at a constant heating rate of 10 °C/min. X-ray Photoelectron Spectroscopy (XPS) analysis was performed on a Thermo VG Scientific (East Grinstead, U.K.) Sigma Probe spectrometer. The instrument employs a monochromated Al Kα X-ray source. Particle size and shape were analyzed using Transmission Electron Microscopy (TEM) in JEOL/JEM 2100 transmission electron microscope. BET (Brunauer, Emmett and Teller) surface area and pore size measurements were carried out by nitrogen adsorption using a TriStar 3020 surface area analyzer. Measurements were performed after degassing the powder samples for 2 h at 120 °C.

Optical properties were measured by Shimadzu 2600 UV-visible spectrophotometer over a range of 200-800 nm. The band gap E is calculated by using the equation,

$$E=hc/\lambda_{in} \tag{4}$$

where, h is the Plank's constant, c is the velocity of light and λ_{int} is the wavelength.

3. Results and discussion

3.1 Powder X-ray diffraction analysis

Figure. 1 represents the XRD patterns of CeO_2 and Sm^{3+} doped CeO_2 nanoparticles. A set of diffraction peaks can be indexed to (111), (200), (220), (311), (222), (400), (331) and (420) corresponding to a cubic fluorite structured ceria phase with space group Fm-3m (JCPDF: 01-075-0151) with a lattice parameter value of 5.418 nm [25]. No additional peaks were observed in the XRD patterns which represent absence of secondary phase in the sample. There are no impurity peaks present in the XRD pattern of the Sm^{3+} doped

CeO_2 up to the dopant concentration of 100:10, this indicates the doped metal ion does not significantly affect the cubic fluorite phase of the CeO_2. A small blue shift in the peak associated with the (111) direction (from 28.4348 to 27.9316°) is observed in the case of 15 wt% Sm^{3+} doped sample shown in Figure 1(b). This may be explained by smaller Ce^{4+} ions (0.97 nm) being replaced by larger Sm^{3+} ions (1.079 nm) [26], thus increasing lattice parameters and crystallite size. The broad reflection peaks indicate the crystal size is small and it should be in the range of nanoscale. The crystallite size of all samples was calculated by using Scherrer equation and expressed in Table 1. The result reveals that the crystallite size varies in the range of 3-6 nm.

Table 1. Crystallite size of bare and Sm^{3+} doped CeO_2 nanoparticles.

Sm^{3+} doped CeO_2	
wt% of Sm^{3+}	Crystallite size (nm)
0	3.237
2.5	4.294
10	5.368
15	5.928

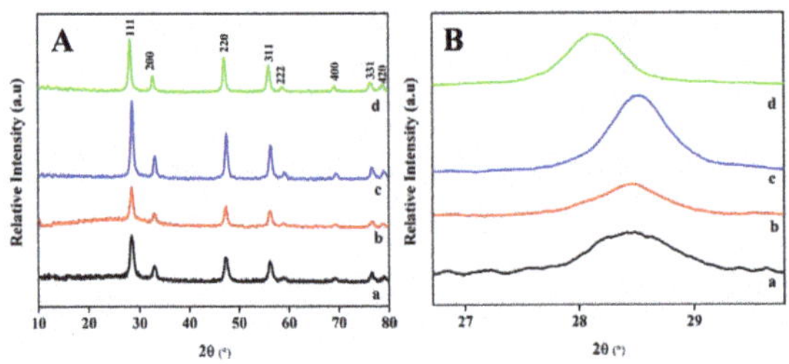

Figure. 1. A) XRD patterns of a) undoped CeO_2, b) 2.5wt%, c) 10wt% and d) 15wt% Sm^{3+} doped CeO_2 nanoparticles B) Enlarged portion from 2θ, 25-35°.

3.2 FT-IR spectroscopic analysis

The as synthesized CeO_2 nanoparticles give characteristic peaks in the range of 3300 cm^{-1}, 1632 cm^{-1}, 1402 cm^{-1}, and 553 cm^{-1} in the FT-IR spectrum. The broad band at 3300 cm^{-1} shown in the Figure. 2 is due to the OH vibrations of adsorbed water molecules. The band at 1632 cm^{-1} is due to the scissor bending of H-O-H [27, 28]. The band at 553 cm^{-1} (below 700 cm^{-1}) is due to the stretching vibration of Ce-O. The very intense band at 1402 cm^{-1} is attributed to Ce–O bonds [27-30].

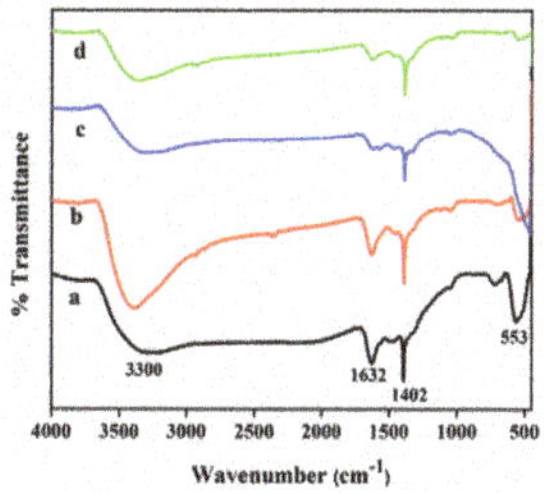

Figure. 2. FT-IR spectra of a) undoped CeO_2, b) 2.5wt%, c) 10wt% and d) 15wt% Sm^{3+} doped CeO_2 nanoparticles.

3.3 Thermogravimetric-Differential thermal analysis (TG-DTA)

The thermogram of bare CeO_2 and Sm^{3+} doped CeO_2 is represented in Figure 3. The weight percentage of the sample is decreasing with respect to temperature which indicating the thermal stability of the CeO_2 sample. There is no strong endothermic or exothermic peak which indicates the phase transition of the sample is absent in the temperature range OF 30-600 °C.

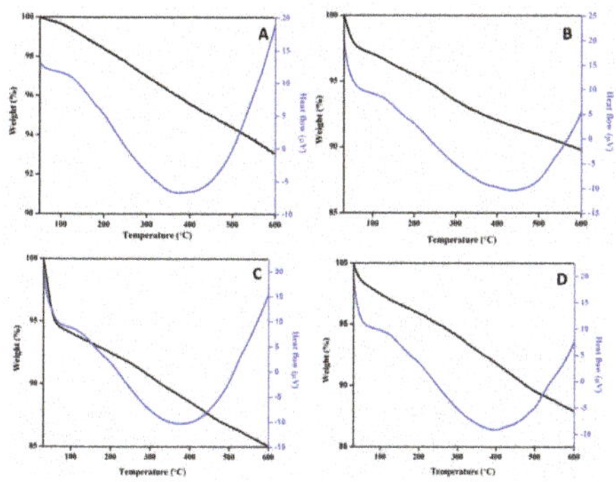

Figure 3. TG-DTA curves of A) undoped CeO_2, B) 2.5wt%, C) 10wt% and D) 15wt% Sm^{3+} doped CeO_2 nanoparticles.

3.4 X-ray photoelectron spectroscopy

The incorporation of samarium was confirmed with XPS. The XPS of bare cerium oxide and 10 wt% Sm doped cerium oxide are shown in Figure 4. Figure 4A and Figure 4B represent the Ce $3d_{3/2}$ and $3d_{5/2}$ spectra in bare CeO_2 and Sm^{3+} doped CeO_2 respectively. The peak at 916.5 eV is the satellite peak corresponds to Ce(IV). The peaks located at

888.84 and 882.5 eV correspond to Ce $3d_{5/2}$ levels. The binding energies correspond to 898.4, 900.7 and 907.3 eV correspond to $Ce3d_{3/2}$ levels [31]. The XPS of Sm(III) is represented in Figure 4C. The characteristic 3d peaks of Sm(III) appear at 1111.08 ($3d_{3/2}$) and 1083.6 eV ($3d_{5/2}$) which is in agreement with the reported values [32].

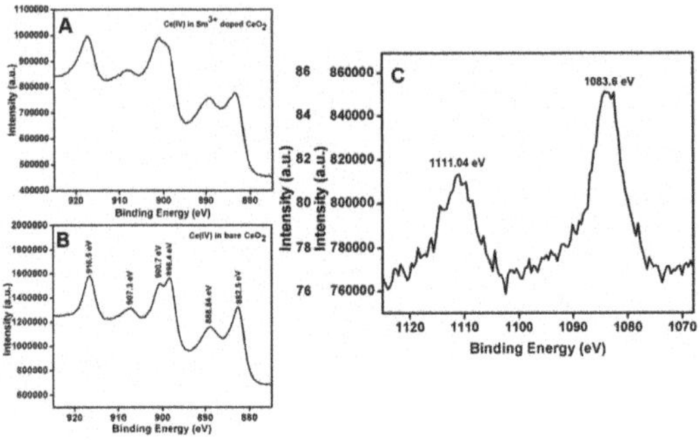

Figure 4. XPS of Ce 3d in A) Sm^{3+} doped CeO_2, B) Bare CeO_2 and C) XPS of Sm 3d in 10 wt% Sm doped CeO_2.

3.5 Transmission electron Mmicroscopy (TEM)

TEM analysis reveals particle shape, size and morphology. Figure 5 represents the TEM images of bare CeO_2 and 10 wt% Sm(III) doped CeO_2. The particles are uniformly distributed and have almost equal size. The particle sizes of bare and doped CeO_2 are found to be in the range of 10-20 nm.

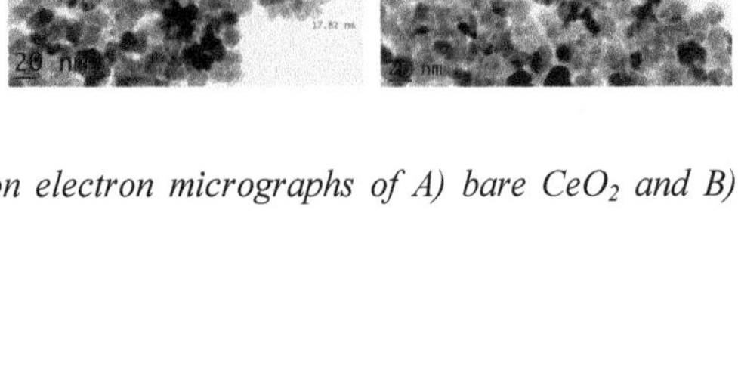

Figure 5. Transmission electron micrographs of A) bare CeO_2 and B) 10 wt% Sm^{3+} doped CeO_2.

3.6 BET-surface area analysis

Table 2. Physical properties from N2 adsorption and desorption studies of bare and Sm^{3+} doped CeO_2 nanoparticles.

wt% of Sm3+ doped	BET-Surface area (m^2/g)	Pore volume (cm3/g)	Pore diameter (nm)
0	65.9650 ± 0.2265	0.129950	7.87994
2.5	74.8435 ± 0.2172	0.109179	5.83508
10	64.4726 ± 0.1902	0.104895	6.50789
15	55.5602 ± 0.1664	0.099378	7.15459

BET analysis provides a specific surface area of materials by nitrogen multilayer adsorption as a function of relative pressure using BET surface area analyzer. The N_2 adsorption and desorption isotherms and BET surface area plots of bare CeO_2 and Sm^{3+} doped CeO_2 are shown in Table 2 and Figure 6. The adsorption and desorption isotherms of all samples show type IV behaviour with the typical hysteresis loop and this is the characteristics of mesoporous materials [12]. BJH analysis can also be employed to determine pore size and specific pore volume using adsorption and desorption techniques. This technique characterizes that pore size distribution is independent of the pore volume of the sample. The mesoporous nature of the samples is also evident from the adsorption and desorption pore size measurement using BJH method (Table 2).

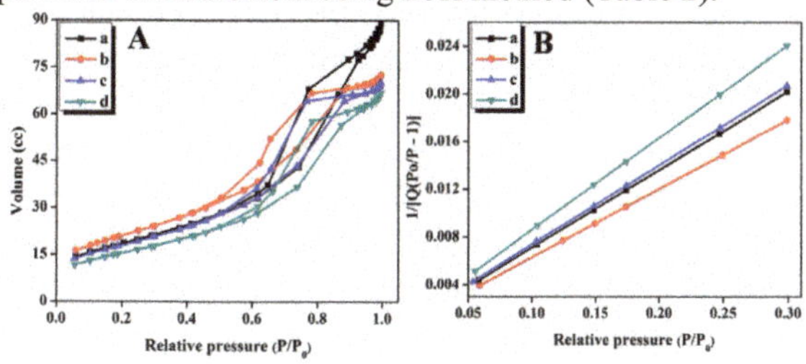

Figure. 6. A) N_2 adsorption isotherm and B) BET surface area plot of a) undoped CeO_2, b) 2.5 wt%, c) 10 wt% and d) 15 wt% Sm^{3+} doped CeO_2 nanoparticles.

3.7 UV-visible spectroscopy

Figure 7 shows the UV-visible absorbance spectra of bare CeO_2 and Sm^{3+}doped CeO_2 samples. The enlarged portion of the spectra from 200 to 450 nm is shown in the inset of Figure 7. All the samples have absorption ranges from 200 to 400 nm showing that CeO_2 and Sm^{3+} doped CeO_2 absorbing in the UV region.

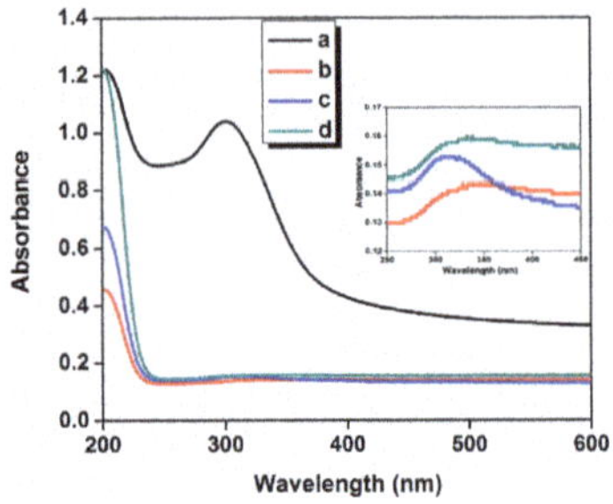

Figure 7. UV-visible absorption spectra of a) undoped CeO_2, b) 2.5wt%, c) 10wt% and d) 15wt% Sm^{3+} doped CeO_2 nanoparticles.

3.8 Photocatalytic study

The photocatalytic activity studies of CeO_2 and Sm^{3+} doped CeO_2 were carried out with aqueous methylene blue under natural sunlight. The absorbance of methylene blue corresponds to photocatalytic degradation was represented in Figure 8. Methylene has two characteristic absorption peaks at 614 and 612 nm. The photocatalytic activity was tested for a duration of 120 minutes with sample withdrawing intervals of 30 minutes. Dark adsorption/desorption activity was also tested with the system. The reaction mixture was stirred for 30 minutes in dark to attain the adsorption-desorption equilibrium. After irradiation with sunlight the absorbance of the dye was found to be decreased at each time interval for CeO_2 and Sm^{3+} doped CeO_2 photocatalyst. Among the various samples studied 10 wt% Sm^{3+} doped CeO_2 composition showed highest activity and it completely degraded the methylene blue dye within 120 minutes. From Figure 8 the absorbance spectra correspond to methylene blue degradation by 10 wt% Sm^{3+} doped CeO_2 (Figure 8C) shows that there is a marginal decrease in the absorbance at each time interval which is not very significant in bare CeO_2, 2.5 wt% Sm^{3+} doped CeO_2 and 15 wt% Sm^{3+}doped CeO_2. This points to the higher photocatalytic activity of 10 wt% Sm^{3+} doped CeO_2 nanoparticles towards methylene blue degradation. However, Sm^{3+} doping in CeO_2 enhanced the photo assisted dye degradation activity compared to bare CeO_2. The photocatalytic activity increases up to 10 wt% Sm^{3+} doping and further doping decreases the activity. Generally doping created disorders and defects in crystalline nanoparticles. These defects can be a trapping centre for electrons and holes which will effectively restrict the recombination of these photo generated charge carriers [33]. As a result, the photocatalytic activity will be enhanced. Here Sm^{3+} doping creates defects in CeO_2 which is indicated by the peak in PXRD patterns (Figure 1b). There are no noticeable peak shifts up to a doping concentration of 10 wt%, further increase in Sm^{3+} in CeO_2 lattice creates defects indicated by the marginal peak shift in PXRD pattern indicated in Figure 1a and Figure 1b. Thus, dopant concentration is critical in deciding the photocatalytic activity of CeO_2 nanoparticles.

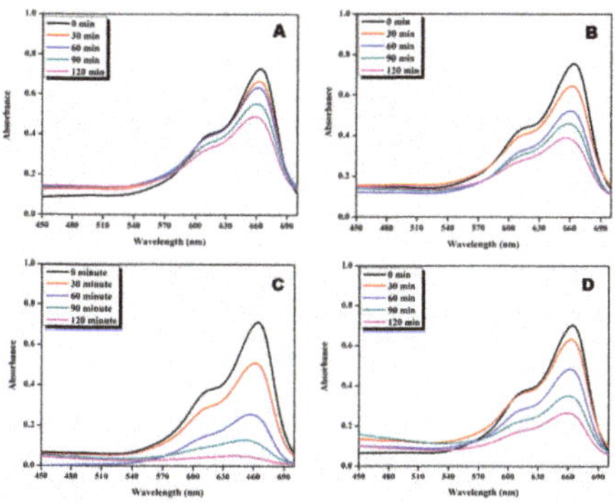

Figure 8. UV-vis absorption spectra of methylene blue using (A) undoped CeO_2, B) 2.5 wt%, C) 10 wt% and D) 15 wt% Sm^{3+} doped CeO_2 nanoparticles.

4. Conclusions

The present work describes sol-gel hydrothermal synthesis of CeO_2 and various weight percentages of Sm^{3+} doped CeO_2 nanoparticles. The synthesized nanoparticles were characterized with various techniques such as Powder X-Ray Diffraction, FT-IR spectroscopy, UV-visible spectroscopy, TGA/DSC Analysis and BET surface area analysis. The structural characterizations reveal that Sm3+ is doped into CeO_2 crystal structure. Peak shifts in the PXRD patterns show that doping imparts disorders in the crystal structure and the peak shift becomes noticeable with a doping concentration 15 wt% of Sm^{3+}. TG-DTA analysis confirms the thermal stability of the synthesized samples. TEM images showed uniform size distribution of highly crystalline particles and XPS revealed effective incorporation of Sm^{3+} in CeO_2 structure. Mesoporous behaviour of the CeO_2 and Sm^{3+} doped CeO_2 nanoparticles were proven by the typical Type IV hysteresis as shown from the BET measurements. The photocatalytic study of CeO_2 and Sm^{3+} doped CeO_2 under direct sunlight were examined using the methylene blue dye as model system. This study showed that doping enhances the photocatalytic activities of CeO_2 nanoparticles. The optimum doping concentration of Sm^{3+} in CeO_2 lattice is found to be 10 wt%, further doping creates more disorders in the system as indicated by the PXRD patterns which decreases the photocatalytic activity.

Acknowledgement

The author SNV acknowledges UGC for funding. The authors SJ and SBN acknowledge DST-SERB, India for financial assistance. The authors acknowledge Sophisticated Test and Instrumentation Centre, Kochi for TEM analysis.

References

[1] A. Mills, S. Le Hunte, An overview of semiconductor photocatalysis, J. Photochem. Photobiol. A: Chem. 108 (1997) 1-35. https://doi.org/10.1016/S1010-6030(97)00118-4

[2] M. Faisal, M. A. Tariq, M. Muneer, Photocatalysed degradation of two selected dyes in UV-irradiated aqueous suspensions of titania, Dyes Pigm. 72 (2007) 233-9. https://doi.org/10.1016/j.dyepig.2005.08.020

[3] M. M. Rahman, A. Jamal, S. B. Khan, M. Faisal, Characterization and applications of as-grown β-Fe_2O_3 nanoparticles prepared by hydrothermal method, J Nanopart Res. 13 (2011) 3789-99. https://doi.org/10.1007/s11051-011-0301-7

[4] D. Ravelli, D. Dondi, M. Fagnoni, A. Albini, Photocatalysis. A multi-faceted concept for green chemistry, Chem Soc Rev. 38 (2009) 1999-2011. https://doi.org/10.1039/b714786b

[5] S. Malato, P. Fernández-Ibáñez, M. I. Maldonado, J. Blanco, W. Gernjak, Decontamination and disinfection of water by solar photocatalysis: recent overview and trends, Catal Today. 147 (2009) 1-59. https://doi.org/10.1016/j.cattod.2009.06.018

[6] L-Y. Yang, S-Y. Dong, J-H. Sun, J-L. Feng, Q-H. Wu, S-P. Sun, Microwave-assisted preparation, characterization and photocatalytic properties of a dumbbell-shaped ZnO photocatalyst, J. Hazard. Mater. 179 (2010) 438-43. https://doi.org/10.1016/j.jhazmat.2010.03.023

[7] M. Stylidi, D. I. Kondarides, X. E. Verykios, Pathways of solar light-induced photocatalytic degradation of azo dyes in aqueous TiO_2 suspensions, Appl. Catal., B: Environmental. 40 (2003) 271-86. https://doi.org/10.1016/S0926-3373(02)00163-7

[8] A. Houas, H. Lachheb, M. Ksibi, E. Elaloui, C. Guillard, J. M. Herrmann. Photocatalytic degradation pathway of methylene blue in water, Appl. Catal., B: Environmental. 31 (2001) 145-57. https://doi.org/10.1016/S0926-3373(00)00276-9

[9] W. Shen, Z. Li, H. Wang, Y. Liu, Q. Guo, Y. Zhang. Photocatalytic degradation for methylene blue using zinc oxide prepared by codeposition and sol–gel methods, J. Hazard. Mater. 152 (2008) 172-5. https://doi.org/10.1016/j.jhazmat.2007.06.082

[10] S. W. Kim, H-K. Kim, H. W. Choi, D. H. Yoo, E. J. Kim, S. H. Hahn, Photocatalytic Activity of Metal-Inserted WO, Nano sci Nano technol. 13 (2013) 7053-5. https://doi.org/10.1166/jnn.2013.7733

[11] O. M. Ishchenko, V. Rog, G. Lamblin, D. Lenoble, TiO_2- and ZnO-Based Materials for Photocatalysis: Material Properties, Device Architecture and Emerging Concepts. In: W Cao, editor. Semiconductor Photocatalysis - Materials, Mechanisms and Applications, InTech (2016). https://doi.org/10.5772/62774

[12] P. Periyat, S.C. Pillai, D. E. McCormack, J. Colreavy, S. J. Hinder. Improved high-temperature stability and sun-light-driven photocatalytic activity of sulfur-doped anatase TiO_2, J. Phys. Chem. C. 112 (2008) 7644-52. https://doi.org/10.1021/jp0774847

[13] [13] Y. Liu, L. Tian, X. Tan, X. Li, X. Chen, Synthesis, properties, and applications of black titanium dioxide nanomaterials, Sci Bull. 62 (2017) 431-41. https://doi.org/10.1016/j.scib.2017.01.034

[14] J. M. Herrmann, Fundamentals and misconceptions in photocatalysis, J Photochem Photobiol A: Chem. 216 (2010) 85-93. https://doi.org/10.1016/j.jphotochem.2010.05.015

[15] W. X. Tang, P. X. Gao, Nanostructured cerium oxide: preparation, characterization, and application in energy and environmental catalysis, MRS Commun. 6 (2016) 311-29. https://doi.org/10.1557/mrc.2016.52

[16] J. Kašpar, P. Fornasiero, M. Graziani, Use of CeO_2-based oxides in the three-way catalysis, Catal Today. 50 (1999) 285-98. https://doi.org/10.1016/S0920-5861(98)00510-0

[17] B. C. Steele, Fuel-cell technology: running on natural gas, Nature. 400 (1999) 619-21. https://doi.org/10.1038/23144

[18] B. C. Steele, A. Heinzel, Materials for fuel-cell technologies, Nature. 414 (2001) 345-52. https://doi.org/10.1038/35104620

[19] P. Jasinski, T. Suzuki, H. U. Anderson, Nanocrystalline undoped ceria oxygen sensor, Sens Actuators B: Chem. 95 (2003) 73-7. https://doi.org/10.1016/S0925-4005(03)00407-6

[20] D. Channei, B. Inceesungvorn, N. Wetchakun, S. Ukritnukun, A. Nattestad, J. Chen et al. Photocatalytic Degradation of Methyl Orange by CeO_2 and Fe–doped CeO_2 Films under Visible Light Irradiation, Sci Rep 4 (2014) 5757. https://doi.org/10.1038/srep05757

[21] I. S. Kim, M. Baek, S. J. Choi, Comparative cytotoxicity of Al_2O_3, CeO_2, TiO_2 and ZnO nanoparticles to human lung cells, J. Nanosci Nanotechnol. 10 (2010) 3453-8. https://doi.org/10.1166/jnn.2010.2340

[22] B. D. Johnston, T. M. Scown, J. Moger, S. A. Cumberland, M. Baalousha, K. Linge et al. Bioavailability of nanoscale metal oxides TiO_2, CeO_2, and ZnO to fish, Environ SciTechnol. 44 (2010) 1144-51. https://doi.org/10.1021/es901971a

[23] P. Periyat, F. Laffir, S. A. M. Tofail, E. A. Magner, facile aqueous sol–gel method for high surface area nanocrystalline CeO_2, RSC Adv. 1 (2011) 1794. https://doi.org/10.1039/c1ra00524c

[24] T. Zhang, T. Oyama, A. Aoshima, H. Hidaka, J. Zhao, N. Serpone, Photooxidative N-demethylation of methylene blue in aqueous TiO_2 dispersions under UV irradiation, J. Photochem. Photobiol., A: Chem. 140 (2001) 163-72. https://doi.org/10.1016/S1010-6030(01)00398-7

[25] B. Choudhury, A. Choudhury, Ce^{3+} and oxygen vacancy mediated tuning of structural and optical properties of CeO_2 nanoparticles, Mater Chem Phys. 131 (2012) 666-71. https://doi.org/10.1016/j.matchemphys.2011.10.032

[26] R. Shannon, Revised effective ionic radii and systematic studies of interatomic distances in halides and chalcogenides, Acta Crystallogr., Sect. A: Found. Crystallogr. 32 (1976) 751-67. https://doi.org/10.1107/S0567739476001551

[27] S. B. Khan, M. Faisal, M. M. Rahman, K. Akhtar, A. M. Asiri, A. Khan et al. Effect of particle size on the photocatalytic activity and sensing properties of CeO_2 nanoparticles, Int J Electrochem Sci. 8 (2013) 7284-97.

[28] F. Niu, D. Zhang, L. Shi, X. He, H. Li, H. Mai et al. Facile synthesis, characterization and low-temperature catalytic performance of Au/CeO_2 nanorods, Mater Lett. 63 (2009) 2132-5.

[29] M. Faisal, S. B. Khan, M. M. Rahman, A. Jamal, K. Akhtar, M. Abdullah, Role of ZnO-CeO_2 nanostructures as a photo-catalyst and chemi-sensor, J Mater SciTech. 27 (2011) 594-600.

[30] K. Babitha, A. Sreedevi, K. Priyanka, B. Sabu, T. Varghese. Structural characterization and optical studies of CeO_2 nanoparticles synthesized by chemical precipitation, Indian J. Pure Appl. Phys. 53 (2015) 596-603.

[31] M.H. Suzanne, S.K. Ajay, D.T. Ron, S. Nammalwar, S. Sudipta, M.R. Christopher, Anti-inflammatory Properties of Cerium Oxide Nanoparticles, small 5 (2009) 2848–2856. https://doi.org/10.1016/j.matlet.2009.07.021

[32] F-H. Chen, J-L. Her, S. Mondal, M-N. Hung, T-M. Pan, Impact of Ti doping in Sm2O3 dielectric on electrical characteristics of a-InGaZnO thin-film transistors, Appl. Phys. Lett. 102 (2013) 193515-5. https://doi.org/10.1063/1.4807014

[33] X. Chen, L. Liu, Y. Y. Peter, S.S. Mao Increasing solar absorption for photocatalysis with black hydrogenated titanium dioxide nanocrystals, Science. 331 (2011) 746-50. https://doi.org/10.1126/science.1200448

Chapter 13

Enhancing Semiconductor-Photocatalytic Organic Transformation through Interparticle Charge Transfer

C. Karunakaran[1]*, S. Karuthapandian[2]

[1]Department of Chemistry, Annamalai University, Annamalainagar 608002, India

[2]Present address: Department of Chemistry, VHNSN College, Viruthunagar 626001, India

*Email (corresponding author): karunakaranc@rediffmail.com, pandiansk2000@yahoo.com

Abstract

Diphenylamine (DPA) is photocatalytically transformed into N-phenyl-p-benzoquinonimine (PBQ) by ZnS nanoparticles and the reaction conforms to Langmuir-Hinshelwood (L-H) kinetics. The kinetic law has been deduced by studying the reaction under different experimental conditions. The kinetic parameters have been evaluated from the obtained results. Particulate ZnS with ZnO or TiO_2 or CeO_2 or CdO nanoparticles shows enhanced photocatalytic organic transformation. ZnS, ZnO, TiO_2, CeO_2 and CdO nanoparticles agglomerate in alcoholic medium. Interparticle charge transfer is the likely reason for the enhanced photocatalysis.

Keywords

ZnS, ZnO, TiO_2, CeO_2, CdO, Agglomeration

Contents

1. Introduction

While production of numerous organic chemicals and related materials for various applications, no doubt, is the foundation for modern life, its mineralization after usage, conversion into harmless naturally occurring molecules like carbon dioxide and water, is essential for sustainable living. Metal oxides and sulfides are the ores commonly occurring in the earth's crust and semiconductor oxides and sulfides act as photocatalyst. When a semiconductor is exposed to light of energy not less than that of the band gap photoexcitation occurs; electrons are promoted from the valence band (VB) to conductance band (CB) leaving holes in the VB. In presence of water or moisture, the VB hole strips an electron from hydroxide ion or water molecule adsorbed on the semiconductor-surface to produce a hydroxyl radical, the primary reactive oxidizing species (ROS), which mineralizes the organics [1]. The CB electron is picked up by molecular oxygen, adsorbed on the semiconductor-surface, to form a superoxide radical ion, which in turn, in presence of water or moisture, yields ROS. Recombination of photogenerated CB electron and VB hole suppresses the photocatalytic efficiency. Interparticle charge transfer (IPCT) reduces charge carrier recombination and thus enhances the photocatalytic activity. Enhancement of photocatalytic mineralization of phenol and carboxylic acid by particulate semiconductor mixture has been reported [2, 3]. However, such enhancement is not realized in the photocatalytic generation of iodine suggesting the hole-transfer from semiconductor to iodide ion to be faster than IPCT [4]. Conversion of one chemical into another without producing harmful or useless co-product/s is important for a sustainable environment. If the reagent required for the conversion is innocuous, it is welcome. Semiconductor-photocatalytic organic transformation is such a process. In absence of water, the organic substrate adsorbed on the semiconductor surface picks up the hole to undergo transformation. Air is the source of molecular oxygen for the reaction. Application of semiconductor-photocatalysis towards organic transformations has been reviewed by Lang et al. [5], Palmisano et al. [6] and Shiraishi and Hirai [7]. TiO_2 is widely used as photocatalyst for such transformations which include oxidation of alcohols to aldehydes [5, 7, 8] or ketones [6], amines to imines [5] and alkenes to epoxides [6] and oxidative dehydrogenation of

1,2,3,4-tetrahydroquinoline to 3,4-dihydroquinoline [5]. Imidazoles have been obtained by TiO_2-photocatalysis [6]. N-doped TiO_2 is used to photocatalytically reduce functionalized nitroarenes [6]. ZnO and Nb_2O_5 have also been reported to bring in photocatalytic organic transformations [5]. The search is on to improve the efficiency of the catalytic process [5, 7, 9, 10] and to find catalysts for new selective organic transformations [6]. Many Au-doped semiconductor nanomaterials have been synthesized to photocatalyze organic transformations [5, 8, 11]. But use of such materials as catalysts is cost prohibitive. Generally, reduction of molecular oxygen, adsorbed on the surface of the semiconductor, to superoxide radical-anion is slow and rate controlling [12]. The CB electron of ZnS is more cathodic than those of the widely employed semiconductors [13]. This large cathodic potential energetically favors reduction of molecular oxygen to superoxide radical anion. As the cathodic potential of CB electron of ZnS is larger than those of the widely employed semiconductors electron-jump from CB of ZnS to CB of widely used semiconductors is energetically favorable. To the best of our knowledge ZnS has not been employed as photocatalyst for organic transformation and mixing it with a widely employed semiconductor is to energetically enable IPCT. Here we report enhancement of ZnS-photocatalytic transformation of diphenylamine (DPA) to N-phenyl-p-benzoquinonimine (PBQ) by widely used particulate semiconductors like TiO_2, ZnO, CeO_2 and CdO. DPA is used in post-harvest treatment of apple and pear [14] and benzophenone [15] and cyanoanthracenes [16] photosensitize its oxidation; the unsensitized oxidation is slow [17] and Fe_2O_3 photocatalyzes the same [18].

2. Experimental

ZnO, TiO_2, CdO, CeO_2, and ZnS (Merck) were used as supplied. Potassium tris(oxalato)ferrate(III), $K_3[Fe(C_2O_4)_3].3H_2O$, was prepared following standard procedure. DPA (analytical reagent, Merck) was used as received. Commercial ethanol was purified through distillation over calcium oxide.

The powder X-ray diffractograms (XRD) of the samples were recorded using Cu Kα X-rays of wavelength 0.15406 nm employing a Siemens D-5000 XRD in the scan range 5-60° at a scan speed of 0.2° s^{-1} or a Bruker D8 system in a 2θ range 5-60° with a scan speed 0.050° s^{-1} or a PANalytical X'Pert PRO diffractometer in a 2θ range 15-75° at a scan rate of 0.020° s^{-1}. Nitrogen-adsorption desorption was adopted to determine the specific surface area using the Brunauer-Emmett-Teller (BET) equation. The particle sizes in methanol were measured using the Easy particle sizer M1.2, Malvern Instruments. The focal length and beam length were 100 and 2.0 mm, respectively. The UV-visible diffuse reflectance spectra were recorded using a Shimadzu UV-2600 spectrophotometer with ISR-2600 integrating sphere attachment.

The photocatalytic experiments were carried out with UV A light (predominantly at 365 nm) under the experimental set up and procedure detailed elsewhere [18]. A 6 W 254 nm low pressure mercury lamp and a 6 W 365 nm mercury lamp were also used to study the influence of radiation energy on the photocatalytic conversion. The light intensities (*I*) were determined by ferrioxalate actinometry. The experimental conditions, unless otherwise stated, were: volume of ethanolic DPA solution: 25 mL (multilamp reactor), 10 mL (microreactor), [DPA]: 5 mM, catalyst-loading: 1.0 g, airflow rate: 7.8 mL s^{-1}, *I*: 25.2 μeinstein L^{-1} s^{-1}. The UV-visible spectra were recorded using a Hitachi U-2001 UV-visible spectrophotometer. The solutions were diluted to bring down the absorbance to Beer-Lambert law limit. The PBQ formed was estimated from its absorbance at 450 nm.

3. Results and discussion

3.1 Characterization

The obtained powder XRDs of ZnO, TiO_2, CeO_2 and ZnS samples (not shown) reveal the crystalline structures as hexagonal wurtzite, tetragonal anatase, cubic fluorite and cubic zinc blende, respectively [19]. The specific surface area (*S*), determined through nitrogen adsorption-desorption using Brunauer-Emmett-Teller (BET) method, are: ZnO 12.2, TiO_2 14.7, CdO 14.5, CeO_2 11.0 and ZnS 7.7 m^2 g^{-1}. The mean particle sizes (*t*), obtained using the formula $t = 6/\rho S$, where ρ is the material density, are: ZnO 87, TiO_2 104, CdO 51, CeO_2 76 and ZnS 190 nm. The band gap energies, obtained through Kubelka-Munk plots (not given), are: ZnO 3.15, TiO_2 3.18, CeO_2 2.89 and ZnS 3.57 eV.

3.2 ZnS-photocatalysis

Fig. 1 is the UV-visible spectral time scan of ethanolic DPA solution (20 mM) illuminated with UV A light in presence of particulate ZnS. It shows the formation of PBQ (λ_{max} = 450 nm). The illuminated solution is EPR-silent revealing the absence of formation of diphenylnitroxide. Furthermore, thin layer chromatographic (TLC) experiment shows formation of a single product. The illuminated DPA solution was evaporated after the recovery of ZnS and the solid was dissolved in chloroform to develop the chromatogram on a silica gel G-coated plate employing benzene as the eluent. The PBQ formed was estimated through the measured absorbance at 450 nm using its molar extinction coefficient [20, 21]. The linear variation of [PBQ] formed with illumination time (not given) provided the initial reaction rate. The uncatalyzed photooxidation of DPA to PBQ is slow [17] and the rate of ZnS-photocatalyzed DPA oxidation was obtained from the rates of oxidation with and without ZnS under identical conditions. The rate of ZnS-photocatalyzed PBQ formation, measured at different [DPA], increases in conformity with Langmuir-Hinshelwood (L-H) kinetics (Fig. 2A). Determination of ZnS-photocatalyzed DPA oxidation rate at varied airflow rates shows enhancement of PBQ formation in accordance with L-H kinetic model (Fig. 2B). ZnS-photocatalyzed PBQ formation enhances with photon flux (Fig. 2C). Increase of ZnS-loading also increases PBQ formation, reaching a limit at high-loading (Fig. 2D). Study of ZnS-catalyzed reaction with UV A light (365 nm, 18.1 μeinstein L^{-1} s^{-1}) and UV C light (254 nm, 5.22 μeinstein L^{-1} s^{-1}), separately in a micro-reactor under identical conditions, shows PBQ formation as 16.0 and 49.1 nM s^{-1}, respectively. ZnS surface retains its activity on usage. Reuse of ZnS exhibits sustainable photocatalyzed PBQ formation. Singlet oxygen quencher azide ion (5 mM) does not suppress PBQ formation showing absence of involvement of singlet oxygen in ZnS-photocatalysis. This is in agreement with the literature [22].

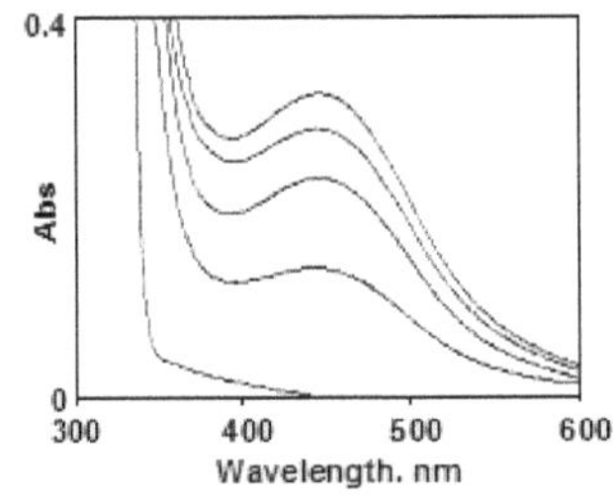

Figure 1. Action spectra showing ZnS-photocatalytic conversion of DPA into PBQ.

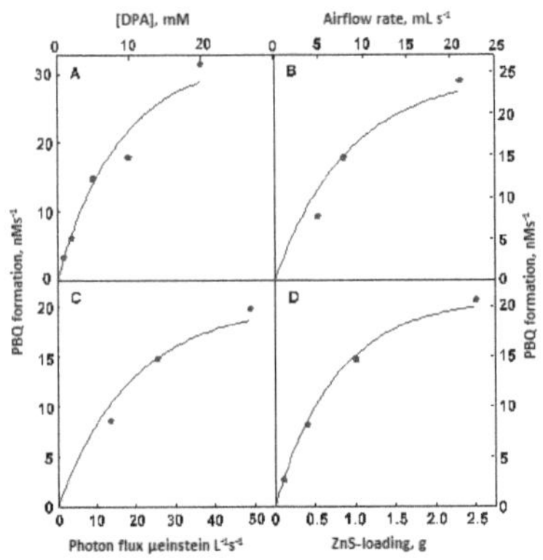

Figure 2. Variation of ZnS-photocatalytic PBQ-formation with A) [DPA], B) airflow rate, C) light intensity and D) catalyst-loading.

3.3 Kinetic law

ZnS generates charge carriers on illumination with UV light and in the absence of water the electron-hole pair brings in the photocatalytic transformation. The mechanism of ZnS-photocatalyzed conversion of DPA to PBQ is unlikely to be different from that by Fe_3O_4 [18] and the kinetic law governing the photocatalytic organic transformation is:

$$\frac{d[PBQ]_{ZnS}}{dt} = \frac{kK_1K_2SI^mC[DPA]\gamma}{(1+K_1[DPA])(1+K_2\gamma)} \qquad (1)$$

where k is the specific rate of oxidation of DPA on the ZnS surface, K_1 and K_2, respectively, are the adsorption coefficients of DPA and molecular oxygen on the illuminated surface of ZnS, γ is the airflow rate, S is the specific surface area of ZnS, C is the catalyst-loading per litre, I is the photon flux, m is an exponent with a value of unity at low photon flux but falling to 0.5 at high light intensity and $d[PBQ]_{ZnS/dt}$ is the rate of PBQ formation on the ZnS surface. The data fit to the L-H kinetic model (Figs. 2A and 2B); the fit was made using a computer program [18]. The data-fit provides the adsorption coefficients K_1 and K_2 as 65 L mol^{-1} and 0.016 mL^{-1} s, respectively, the specific reaction rate k as 1.55 μmol L m^{-2} einstein^{-1} and the value of m as 0.64; the obtained m-value is in agreement with that of Ag-doped TiO_2-photocatalyzed iodine generation [23]. Fig. 3 is the graphical presentation of the data fit (r = 0.98) employing the listed kinetic constants; the experimental rate is plotted against that predicted by the kinetic law. However, the PBQ-formation rate on ZnS surface fails to increase linearly with ZnS-loading (Fig. 2D). This is because of the high photocatalyst-loading. At high loading, the surface area of the photocatalyst exposed to light does not commensurate to the weight of the photocatalyst. The quantity of ZnS employed is beyond the critical amount corresponding to the volume of the reaction solution and the reaction vessel; the full surface of ZnS used is not exposed to light.

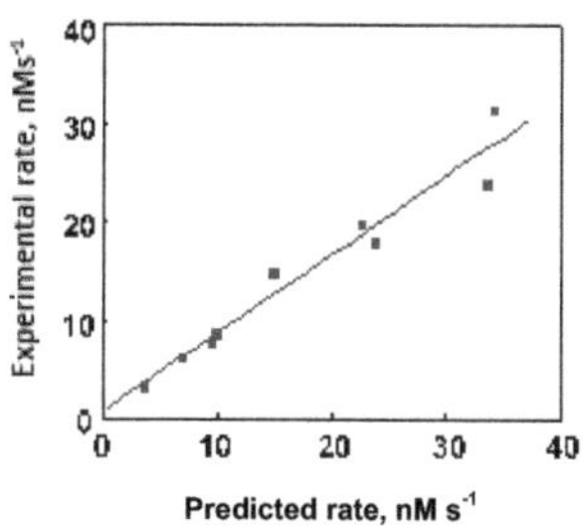

Figure 3. Graphical display of ZnS-photocatalyzed PBQ-formation data-fit.

3.4 Enhanced photocatalysis: IPCT

In coupled semiconductors, called also as composite semiconductors, vectorial transfer of photoproduced charge carriers from one semiconductor to another is possible. This charge separation enhances the photocatalytic performance [24]. In semiconductor composites, both the semiconductors exist in the same particle and the charge separation occurs within the particle. But what is observed here is enhanced photocatalytic transformation of DPA to PBQ on mixing ZnS particles with particulate ZnO or TiO_2 or CeO_2 or CdO. Fig. 4 shows the enhancement of photocatalytic formation of PBQ by ZnS mixed with ZnO or TiO_2 or CeO_2 or CdO - the two particulate semiconductors were in suspension and at continuous motion. This observed enhanced photocatalytic transformation is likely due to an interparticle charge transfer. Nanoparticles in suspension agglomerate [25].

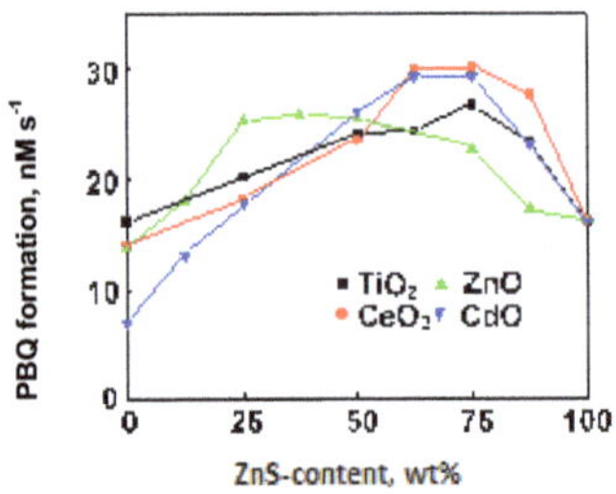

Figure 4. Enhanced PBQ-formation on mixing ZnS with ZnO or TiO_2 or CeO_2 or CdO.

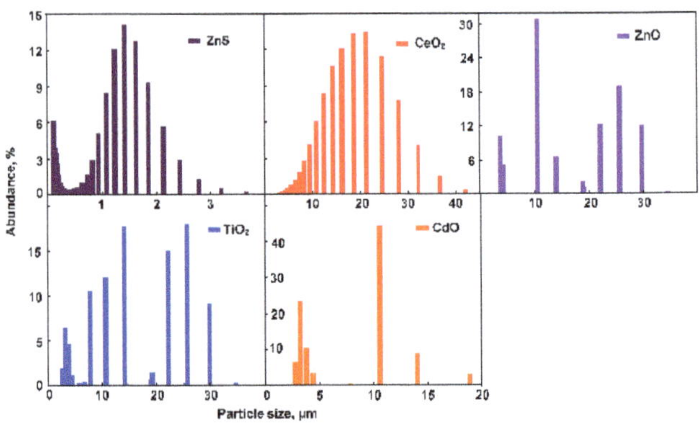

Figure 5. Particles agglomeration.

The particle size distributions of ZnS, ZnO, TiO_2, CeO_2 and CdO in ethanolic suspension are presented in Fig. 5. The mean particle sizes of ZnS, ZnO, TiO_2, CdO and CeO_2, obtained using BET results, are 190, 87, 104, 51 and 76 nm, respectively. Examination of Fig. 5 in conjunction with the obtained mean particle sizes shows agglomeration of the nanoparticles. As observed in individual semiconductor suspension, agglomeration is likely in particulate semiconductor mixtures under suspension and both the semiconductor nanoparticles are likely to be present in the agglomerates. Charge transfer between ZnS and ZnO or TiO_2 or CeO_2 or CdO nanoparticles is likely to occur when both the semiconductors are under band gap-illumination and in contact with each other. Electron from CB of a semiconductor may move to another if the latter is of lower energy and so is the hole from VB. The CB and VB edges of the investigated nanoparticulate semiconductors are presented in Fig. 6. The CB electron of ZnS is more cathodic than those of the other semiconductors. This enables transfer of CB electron of ZnS to the CB of ZnO or TiO_2 or CeO_2 or CdO. Similarly, the VB hole of ZnS is more anodic than that of CdO. This renders possible the movement of a hole from the VB of ZnS to that of CdO. But the CB of ZnS is less anodic than those of ZnO and TiO_2. This favors the migration of a VB hole of ZnO or TiO_2 to the VB of ZnS. This interparticle charge transfer enhances the photocatalytic transformation. The energy difference between the CB electrons of two semiconductors (SC) is the driving force for the interparticle electron-separation and the free energy change (ΔG) is given by [26]:

$$-\Delta G = e[E_{CB(SC1)} - E_{CB(SC2)}] \quad (2)$$

Similar equation is applicable for hole-transfer from VB of one semiconductor to another. In terms of redox chemistry, the CB and VB refer to the reduced and oxidized states in the semiconductor. In TiO_2, CeO_2, CdO and ZnO or ZnS the CB electrons refer to the reduced forms of Ti^{4+} (*i.e.*, Ti^{3+}), Ce^{4+} (*i.e.*, Ce^{3+}), Cd^{2+} (*i.e.*, Cd^{+}) and Zn^{2+} (*i.e.*, Zn^{+}), respectively. Similarly, the VB hole corresponds to the oxidized forms of the respective O^{2-} (*i.e.*, O^{-}) or S^{2-} (*i.e.*, S^{-}). The interparticle charge transfer, the transfer of electron from the CB of ZnS to CB of ZnO or TiO_2 or CeO_2 or CdO refers to the electron jump from Zn^{+} of ZnS to Zn^{2+} of ZnO or Ti^{4+} or Ce^{4+} or Cd^{2+}. On the other hand, the hole-transfer from the VB of ZnS to that of CdO refers to the electron-jump from O^{2-} of CdO to S^{-} of ZnS. Similarly, the hole-jump from the VB of TiO_2 or ZnO to the VB of ZnS implies electron movement from S^{2-} of ZnS to O^{-} of TiO_2 or ZnO. The possibility of cross-electron-hole combination, the transfer of electron from the CB of one semiconductor (SC_1) to the VB of the other (SC_2) is very remote; the very low population of the excited states renders the electron transfer between two excited states highly improbable. A possible reason for not observing the maximum photocatalytic transformation at 50% wt. composition for the semiconductor mixtures is the densities and particle sizes of the semiconductor materials and also the agglomeration.

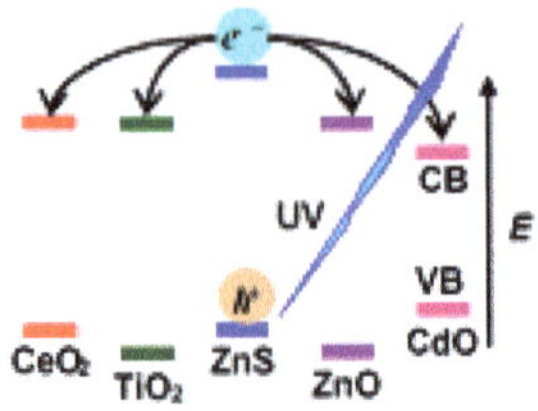

Figure 6. The interparticle charge transfer.

4. Conclusion

Charge transfer between two different band gap-illuminated particulate semiconductors enhances photocatalytic organic transformation. Particulate ZnS with high CB reduction potential is an appropriate choice to enhance photocatalytic conversion of DPA to PBQ when mixed with widely employed semiconductor photocatalysts like TiO_2, ZnO, CeO_2 and CdO. The kinetic law for the ZnS-photocatalytic conversion has been validated through experiments at different reaction conditions. The present results show mixing different particulate semiconductors is a simple and convenient strategy to obtain enhanced organic photocatalytic transformation.

Acknowledgement

Prof. C. Karunakaran is thankful to the Council of Scientific and Industrial Research (CSIR), New Delhi for the Emeritus Scientist Scheme [21(0887)/12/EMR-II].

References

[1] S. Banerjee, S.C. Pillai, P. Falaras, K.E. O'Shea, J.A. Byrne, D. Dionysiou, New insights into the mechanism of visible light photocatalysis, J. Phys. Chem. Lett. 5 (2014) 2543-2554. https://doi.org/10.1021/jz501030x

[2] C. Karunakaran, R. Dhanalakshmi, P. Gomathisankar, Semiconductor-photocatalyzed degradation of carboxylic acids: enhancement by particulate

semiconductor mixture, Int. J. Chem. Kinet. 41 (2009) 716-726. https://doi.org/10.1002/kin.20444

[3] C. Karunakaran, R. Dhanalakshmi, P. Gomathisankar, G. Manikandan, Enhanced phenol-photodegradation by particulate semiconductor mixtures: interparticle electron jump, J. Hazard. Mater. 176 (2010) 799-806. https://doi.org/10.1016/j.jhazmat.2009.11.105

[4] C. Karunakaran, P. Anilkumar, P. Vinayagamoorthy, Lack of enhanced photocatalytic formation of iodine on particulate semiconductor mixtures, Spectrochim. Acta 98 (2012) 460-465. https://doi.org/10.1016/j.saa.2012.08.079

[5] X. Lang, X. Chen, J. Zhao, Heterogeneous visible light photocatalysis for selective organic transformations, Chem. Soc. Rev. 43 (2014) 473-486. https://doi.org/10.1039/C3CS60188A

[6] G. Palmisano, E. Garcia-Lopez, G. Marci, V. Loddo, S. Yurdakal, V. Augugliaro, L. Palmisano, Advances in selective conversions by heterogeneous photocatalysis, Chem. Commun. 46 (2010) 7074-7089. https://doi.org/10.1039/c0cc02087g

[7] Y. Shiraishi, T. Hirai, Selective organic transformations on titanium oxide-based photocatalysts, J. Photochem. Photobiol. C 9 (2008) 157-170. https://doi.org/10.1016/j.jphotochemrev.2008.05.001

[8] W. Feng, G. Wu, L. Li, N. Guan, Solvent-free selective photocatalytic oxidation of benzyl alcohol over modified-TiO_2, Green Chem. 13 (2011) 3265-3272. https://doi.org/10.1039/c1gc15595d

[9] M.-Q. Yang, Y.-J. Xu, Selective photoredox using graphene-based composite photocatalysts, Phys. Chem. Chem. Phys. 15 (2013) 19102-19118. https://doi.org/10.1039/c3cp53325e

[10] N. Zhang, S. Liu, X. Fu, Y.-J. Xu, Fabrication of coenocytic Pd@CdS nanocomposite as a visible light photocatalyst for selective transformation under mild conditions, J. Mater. Chem. 22 (2012) 5042-5052. https://doi.org/10.1039/c2jm15009c

[11] C. Wang, D. Astruc, Nanogold plasmonic photocatalysis for organic synthesis and clean energy conversion, Chem. Soc. Rev. 43 (2014) 7188-7216. https://doi.org/10.1039/C4CS00145A

[12] C. Karunakaran, R. Dhanalakshmi, Photocatalytic performance of particulate semiconductors under natural sunshine – oxidation of carboxylic acids, Sol.

Energy Mater. Sol. Cells 92 (2008) 588-593. https://doi.org/10.1016/j.solmat.2007.12.009

[13] X. Chen, S. Shen, L. Guo, S.S. Mao, Semiconductor-based photocatalytic hydrogen generation, Chem. Rev. 110 (2010) 6503-6570. https://doi.org/10.1021/cr1001645

[14] A. Zanella, Control of apple superficial scald and ripening – a comparison between 1-methylcyclopropene and diphenylamine postharvest treatment, initial low oxygen stress and ultra low oxygen storage, Postharvest Biol. Technol. 27 (2003) 69-78. https://doi.org/10.1016/S0925-5214(02)00187-4

[15] T.S. Lin, J. Retsky, ESR studies of photochemical reactions of diphenylamines, phenothiazines, and phenoxazines, J. Phys. Chem. 90 (1986) 2687-2689. https://doi.org/10.1021/j100403a026

[16] Y.C. Chang, P.W. Chang, C.M. Wang, Energetic probing for the electron transfer reactions sensitized by 9,10-dicyanoanthracene and 9-cyanoanthracene and their modified zeolite particle, J. Phys. Chem. B 107 (2003) 1628-1633. https://doi.org/10.1021/jp021852j

[17] C. Karunakaran, S. Karuthapandian, Solar photooxidation of diphenylamine, Sol. Energy Mater. Sol. Cells 90 (2006) 1928-1935. https://doi.org/10.1016/j.solmat.2005.12.003

[18] C. Karunakaran, S. Karuthapandian, Oxidation of diphenylamine on illuminated Fe_2O_3 surface, Indian J. Chem. A 54 (2015) 356-360.

[19] C. Karunakaran, P. Gomathisankar, G. Manikandan, Solar photocatalytic detoxification of cyanide with bacterial disinfection by oxide ceramics, Indian J. Chem. Technol. 18 (2011) 169-176.

[20] S. Puri, W.R. Bansal, K.S. Sidhu, Benzophenone-sensitized photooxidation of diphenylamine, Indian J. Chem. 11 (1973) 828.

[21] W.R. Bansal, N. Ram, K.S. Sidhu, Reaction of singlet oxygen: Part I – Oxidation of diphenylamine with singlet oxygen ($^1\Delta_g$) produced in situ, Indian J. Chem. B 14 (1976) 123-126.

[22] M.A. Fox, C.C. Chen, Mechanistic features of the semiconductor photocatalyzed olefin-to-carbonyl oxidative cleavage, J. Am. Chem. Soc. 103 (1981) 6757-6759. https://doi.org/10.1021/ja00412a044

[23] C. Karunakaran, P. Anilkumar, P. Gomathisankar, Kinetics of Ag/TiO_2-photocatalyzed iodide ion oxidation, Monatsh. Chem. 141 (2010) 529-537. https://doi.org/10.1007/s00706-010-0288-2

[24] C. Karunakaran, S. SakthiRaadha, P. Gomathisankar, P. Vinayagamoorthy, Nanostructures and optical, electrical, magnetic and photocatalytic properties of hydrothermally and sonochemically prepared $CuFe_2O_4$/SnO_2, RSC Adv. 3 (2013) 16728-16738. https://doi.org/10.1039/c3ra41872c

[25] M. Li, M.E. Noriega-Trevino, N. Nino-Martinez, C. Marambio-Jones, J. Wang, R. Damoiseaux, F. Ruiz, E.M.V. Hoek, Synergistic bactericidal activity of Ag-TiO_2 nanoparticles in both light and dark conditions, Environ. Sci. Technol. 45 (2011) 8989-8995. https://doi.org/10.1021/es201675m

[26] R. Katoh, A. Furube, T. Yoshihara, K. Hara, G. Fujihashi, S. Takano, S. Murata, H. Arakawa, M. Tachiya, Efficiencies of electron injection from excited N3 dye into nanocrystalline semiconductor (ZrO_2, TiO_2, ZnO, Nb_2O_5, SnO_2, In_2O_3) films, J. Phys. Chem. B 108 (2004) 4818-4822. https://doi.org/10.1021/jp031260g

Chapter 14

Recent Developments in Cu_2ZnSnS_4 (CZTS) Preparation, Optimization and its Application in Solar Cell Development and Photocatalytic Applications

S.B. Patel, J.V. Gohel*

Chemical Engineering Department, S.V. National Institute of Technology, Surat – 395007, India
E-mail: sjn@ched.svnit.ac.in*

Abstract

In the present study, a topical review of recent advances in Copper Zinc Tin Sulfide - CZTS (Cu_2ZnSnS_4) preparation and its potential application related to solar cell development and as photocatalyst are discussed in detail. A rigorous review on the preparation of CZTS thin film using spin coating and spray pyrolysis methods are the main focus in the present study as these methods are easily up-scalable. The film quality controlling parameters are also discussed in detail. Recent and advanced studies on CZTS thin film preparation methods are also discussed. Lastly, some future research scopes for solar cells are explored.

Keywords

CZTS, Thin Film, Photocatalyst, Spray Pyrolysis, Spin Coating, Quality Controlling Parameters

Contents

1. Introduction

Sustainability is one of the major concerns all over the globe in recent years. The planet's conventional energy resources are depleting at a catastrophic rate owing to the human race [1]. Thus, there is a great need of research and development in solar energy conversion to meet future requirements. Solar energy can be converted into useable energy by photovoltaic (PV) cells (domestic) and solar thermal plant (large scale) [2]. Photovoltaic effect was first proposed by Becquerel in 1839 [3]; however, this phenomenon acquired the attention when Bell laboratories developed a solar cell with approximately 11% efficiency in 1954 [4]. Thereafter, tremendous amount of research is carried out on different semiconductor materials and on the development of new semiconductor materials as well [5].

Presently, crystalline or amorphous silicon solar cells are available commercially [6]. These solar cells are reported to achieve approximately 20-25% efficiency on commercial scale [7]. However, the cost of pure silicon as well as the silicon solar cell preparation cost is high. Hence, thin film solar cells have been introduced to overcome this cost issue [8]. Basically, the silicon-based solar cells require large amount of material due to their thickness (150-200 μm) requirement. Whereas, the thickness required for thin film solar cell is small (1-2 μm). So, thin film solar cells preparation process is less costly. In addition, there is flexibility in selection of low-cost substrates (e.g. glass, stainless steel, plastic plates) [9]. For an absorber layer preparation in solar cells, materials with direct band gap, such as cadmium telluride (CdTe), copper indium diselenide (CIS), and copper indium gallium diselenide (CIGS) are preferred [10]. Approximately 15% efficiency is achieved with above absorber materials based cell modules [11]. However, Cd and Se are toxic in nature. In addition, In and Te is less available in the earth crust. Thus, production of these types of solar cells is limited.

Copper Zinc Tin Sulfide (CZTS) is emerged as a great alternative absorber material recently as it requires less toxic and easily available materials. After achieving approximately 12% of Power Conversion Efficiency (PCE), CZTS has been considered as a promising contender for solar absorber material at commercial scale [12]. Owing to its high absorbance co-efficient and low direct band gap (1.4-1.6 eV), it can be used in applications, including, thermoelectric, energy harvesting as well as photocatalytic applications [13]. So far, many techniques are reported on the synthesis of CZTS nanoparticles, such as sol-gel [14], solvothermal [15], mechanochemical [16], etc. and preparation of CZTS thin films such as spray pyrolysis [17], spin coating [18], chemical vapor deposition [19], hybrid sputtering [20], pulsed lesser deposition [21], etc. to develop low-cost with high efficiency solar cells. Up to now, several review papers have been published on the synthesis and thin film preparation methods by different authors [22–26], but none have discussed thin film quality controlling parameters.

In the present review, firstly, the fundamentals of CZTS are discussed. Solar cell development requires CZTS thin films. So, two important methods of CZTS films preparation (spray pyrolysis and spin coating) and associated challenges are discussed at length, subsequently. Further, the effect of various control parameters on the band gap, morphology and structure of CZTS thin films are discussed rigorously. These two methods are much simpler and cheaper methods to produce uniform and homogeneous thin films. The effect of all major control parameters (material composition, the type of precursor, type of solvent and temperature) are discussed comprehensively in the present review. Additionally, other control parameters (type of sulfur source, the spray rate as well as spin speed) are also discussed.

2. Properties of CZTS

CZTS has great potential to replace CIS and CIGS based solar cells, which requires toxic, costly rare earth elements. CZTS is easily available at low cost and it is non-toxic. Schafer and Nitsche have fabricated CZTS for the first time in 1974. In early 2000s, the application of CZTS was boosted. In 2005, marginal efficiency (5.7%) was reported. Consequently, CZTS based solar cell became a great contender in the thin film solar cells category [27]. Cu_2ZnSnS_4 is basically invented from $CuInS_2$ by replacing In (Group III) by a mixture of Zn (Group II) and Sn (Group IV). Hence, CZTS becomes a quaternary semiconducting material, as it has a I_2-II-IV-VI_4 group. Typically, two structures are reported for CZTS, namely, stannite and kesterite. Both structures are differentiated based on location of Cu and Zn, as shown in Fig. 1. It is desirable to prepare a kesterite structure, as it is environmentally stable. The type of structure confirmation is reported by x-ray diffraction method. Basically, in the stannite structure, the (001) plane is present, while in the kesterite structure, the (112) plane is present [28].

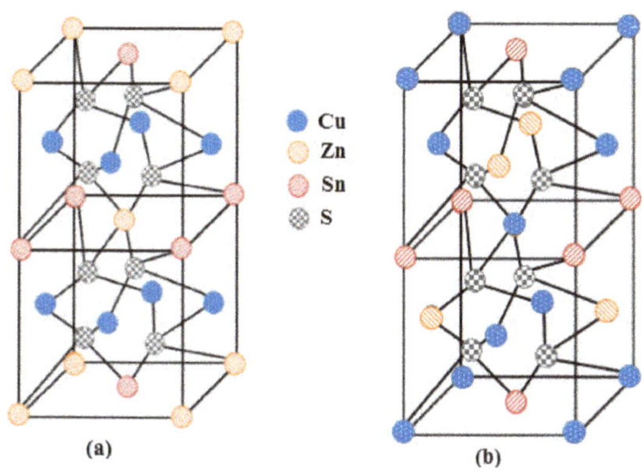

Figure 1. *CZTS Structures (a) stannite and (b) kesterite.*

Thermal stability is also reported to be high. The film prepared by spray pyrolysis [29] was reported to give only 1 wt% loss till 220 °C. This is attributed to thermal decomposition of a metal–thiourea complex. This has exceptionally high stability. Only up to 15 wt% loss is reported till 450 °C. It is attributed to oxidation of sulfides. Afterwards, there is no further weight loss observed. However, at very high temperature (above 600), it may get degrade. Essentially, CZTS is a p-type semiconducting material. It has a high absorption coefficient (more than 10^4) due to direct band gap. Hence, even a very thin film (1-2 μm) can absorb the light. CZTS has a direct band gap of 1.4-1.6 eV, which is very close to the optimum band gap (1.45 eV) value to match the solar spectrum [30]. Till date, the highest efficiency reported for CZTS based solar cells by IBM is 12.6% [7]. However, the efficiency can be further improved by different combinations of preparation methods, because the theoretical possible efficiency of CZTS is 32.2% (calculated by Shockley and Queisser) [31].

The main limiting factor for less efficiency is considered to be less V_{oc} (open-circuit voltage). Theoretically predicted V_{oc} is 1.23V [32]. However, typically reported V_{oc} is less than 0.65 V, including maximum reported value of 0.722V [33]. Low V_{oc} achievement may be attributed to: (i) presence of secondary phase or (ii) bad cell fabrication. However, CZTS is a quaternary material, so it is difficult to completely avoid the occurrence of a secondary phase. Olekseyuk et al., [34] reported detailed phase diagram preparation for a Cu_2S-ZnS-SnS_2 system. It is very clearly indicated that composition (of precursors) and temperature must be specifically controlled in order to avoid the secondary phase formation. Hence, precise control of the operating parameters is essential during preparation. Typically, various secondary phases, such as, Cu_xS_y, Cu_xO, $Cu_{2-x}Sn_xS_3$, $ZnSnO_3$, SnO_2, SnS_2, ZnS, $ZnSnO_3$ are reported. Copper containing secondary phases are more harmful to cell performance. This is attributed to the fact that highly conductive copper reduces the shunt resistance (R_{sh}), which leads to less V_{oc} and consequently less efficiency.

3. CZTS preparation

3.1 Nanoparticles synthesis

Until now, large numbers of studies are reported on CZTS nanoparticles (NPs) synthesis with various methods, such as, mechanochemical, sol-gel, solvothermal, chemical bath deposition, etc. Mainly these methods can be classified in two categories (i) mechanical synthesis methods and (ii) wet chemical methods. The prepared nano particles can be used in photocatalytic applications. The prepared nano particles can be deposited also on appropriate substrate for solar cell application.

3.1.1 Mechanical methods

The milling techniques is mainly used in the mechanical method. It is the most facile approach for bulk scale production of nanoparticles. A high energy ball mill is the most generally used equipment, in which at low temperature, impact of balls leads to fine nano-sized particles preparation. Sometimes heat treatment is given before or after milling. High homogeneity of the crystalline structures and uniform surface are the main advantages. However, a long time is required for milling. Hence, more energy is required. Additionally, there might be the presence of impurity (because of the impacts). Mokurala et al., 2014 [35] synthesized CZTS nanoparticles using the mechano-chemical method. The Copper acetate, zinc acetate, tin chloride and thiourea were mixed in ethanol. Then the mixture was charged into a pot mill. After milling, the heat treatment was done in a tubular furnace for 60 min at 250 °C (Ar atmosphere). As a result, fine particles (10-15 nm) and an appropriate band gap (1.45 eV) was achieved. Kheraj et al., 2013, [36] applied heat treatment before milling using an electric furnace at 1030 K temperature for 24 h. The bulk size powder of Cu, Zn, Sn and S were used for the synthesis.

3.1.2 Wet chemical methods

The wet chemical methods chemical reaction based technique. It is preferred for synthesis, as it offers superior control over the product's properties. It does not require any high energy equipment, thus it is a low-cost method. Two major and important wet chemical methods are sol-gel and solvo-thermal methods, which offer a wide range of selection of precursors and control over the shape and size of the NPs. In the present study, both of these methods are discussed in detail.

3.1.2.1 Sol-gel method

It is a simple and low-cost method for synthesis. The main advantage is that there is no requirement of any vacuum condition. A solution of inorganic or alkoxide is prepared in certain solvents, then it is aged (kept for definite time interval) to achieve gel formation. Finally, the product is centrifuged and washed. The main limitation is the long process time requirement. Additionally, the solvents are organic; hence it may evaporate while drying and may cause a health issue. The most important quality control parameter is the type of solvent in this method. Riha et al., 2009 [38] prepared CZTS nano particles by sol-gel method. Firstly, copper acetylacetonate, zinc acetate, and tin acetate were dissolved in oleylamine at 125 °C under inert condition. Subsequently, sulfur powder was sonicated with oleylamine. Further, both solutions were added with trioctylphosphine oxide (TOPO) at 300 °C and stirred for 75 min. CZTS nanoparticles of 13 nm size were achieved. Chen et al., 2015 [39] synthesized CZTS nano particles by sol-gel method. The solution was prepared at mild conditions. Copper chloride, zinc acetate, tin chloride and thiourea were dissolved in oleylamine in open air condition (at 250 °C). CZTS particels of 400-600 nm size were achieved with good crystallinity.

3.1.2.2 Solvothermal method

The solvothermal method has the advantage of better control of CZTS properties by using an appropriate type of solvent. The precursors are heated to 100-300 °C in a sealed autoclave for a long time. The different shapes and size of CZTS nanoparticles can be achieved by controlling process temperature and pressure. Additionally, high crystallinity can be achieved with this method. The solvothermal method gives better control over the end product than sol-gel. However, its limitation is it requires longer process time (24-48 h in autoclave). Zhou et al., 2011 [40] prepared CZTS nanoparticles by solvothermal method. Copper chloride, zinc chloride, tin chloride, thiourea and polyvinylpyrrolidone (PVP) were dissolved in ethylene glycol. Subsequently, the solution was added to a sealed autoclave. The autoclave was heated to 230 °C for 24 h and then allowed to cool to room temperature. Precipitates were collected after centrifuging. CZTS particels of 100-150 nm were achieved successfully. Yan et al., 2015 [41] synthesized CZTS nanoparticles with the solvothermal method. Firstly, copper chloride, zinc chloride, tin chloride, and thiourea precursors were dissolved in ethanol. It was then added to an autoclave operated at 200 °C for 24 h. Nano crystals of 9 nm size were achieved.

3.2 Thin film preparation

In the present study, two important methods of CZTS thin film preparation are the focus. Namely, spray pyrolysis and spin coating. Both the methods have several distinct advantages like simple equipment, easy preparation method, and no need of a vacuum at any stage. Consequently, both methods are very cost effective. Moreover, both are wet chemical based methods, so product specifications can be tuned simply by modification in chemistry. Until now, two approaches are reported for both methods: (i) single step and (ii) two step. In a single step approach, precursor solution is prepared and directly deposited on the substrate. However, in a two-step method, firstly nanoparticles (NPs) are synthesized separately and dispersed in a solvent to prepare suspension. Then the nano suspension is deposited on the substrate in a second step.

3.2.1 Single-Step method for thin film preparation

3.2.1.1 Spray Pyrolysis for thin film preparation

It is the most frequently used method for thin film preparation. The basic advantage is simple design and good reproducibility at large scale. Additionally, it doesn't require a vacuum at any stage. Thus, the method is a low-cost method. The prominent thin film quality control parameter is the substrate deposition temperature [42]. It should be optimized appropriately, as a high substrate temperature may lead to non-adherent films and low temperature may cause agglomeration. The other film quality control parameters are carrier pressure, the distance between nozzle and the substrate, precursor concentration, spray rate, spraying time and number of spray cycles.

Nakayama and Ito, 1996 [43] reported CZTS thin film preparation by spray pyrolysis method for the first-time. The precursor solution (Cu, Zn, Sn and S precursors) was sprayed on a SLG (soda lime glass) substrate. Subsequently, the substrate was annealed at 550 °C under Argon-H_2S atmosphere. The obtained structure was observed to be stannite. The band gap value was achieved to be 1.46 eV. Daranfed et al., 2012 [44] studied the effect of substrate temperature on the properties of CZTS thin films. The substrate temperature was varied from 280-360 °C. The deposition rate (66 nm/min) and deposition time (45 min) were kept constant. At low temperature (280 °C), the presence of a secondary phase ($ZnSnO_3$) was observed. Subsequently, the band gap was inappropriately high. It was also attributed to small voids and cavities, as observed in the SEM (Scanning Electron Microscope) image. However, upon increasing the temperature to 320 °C, the presence of a secondary phase was minimized. Additionally, dense homogeneous film with smallest band gap (1.41 eV) was achieved at 320 °C. Hence, it was subsequently considered to be the optimum substrate temperature. Deepa and Jampana, 2016 [45] also varied substrate temperature (from 240 to 490 °C) to optimize the opto-electrial properties. A self-fabricated automated ultrasonic spray pyrolysis unit with movement in the x-y direction (to develop different pattern in film deposition) was developed. Single phase CZTS was achieved at temperature above 400 °C.

3.2.1.2 Spin coating for thin film preparation

Spin Coating is a low-cost technique. In this method, precursor solution is prepared appropriately. Subsequently, it is deposited upon the substrate by high speed rotation. Further it is dried to get crystalline thin films. The same cycle is repeated for a definite number of times to get the appropriate thickness. Subsequently, it is annealed for a definite time interval to get a smooth thin film. Typically, precursor solution is prepared by sol-gel method, which involves the usage of low-cost precursor. Therefore, the overall process is a very low-cost technique. However, there are few limitations. Some material loss may happen due to centrifugal force. In addition, acquiring a highly pure thin film is difficult. The film quality control parameters are rotation speed; number of cycles; annealing temperature; and type of precursor.

Agawane et al., 2015 [46] prepared CZTS thin films using the sol-gel spin coating method. The solution was prepared under an inert condition to avoid the formation of a secondary phase. Copper chloride, zinc acetate, tin chloride and thiourea were dissolved in 2-methoxyethanol. The solution was deposited upon a Mo-coated soda lime glass substrate at 2000 rpm. The coating process was repeated 2 times and after each coating, the film was dried at 125 °C for 60 min. Subsequently, the film was annealed at 525 to 575 °C. XRD results revealed that as the temperature increases the sharpness of peak increases. The SEM images also suggested that the high temperature was more favorable. The solar cell was fabricated for all annealed films and cell fabrication was SLG/Mo/CZTS/CdS/i-ZnO/AZO/Al. The best solar cell efficiency of 3.01% was obtained by film annealed at 575 °C. Tunuguntla et al., 2015 [47] also adopted same procedure to prepared CZTS solution. The only change was in precursors and solvent as copper acetate, zinc chloride, tin chloride and thiourea were dissolved in DMI under a N_2 atmosphere. Annealing temperature was increased to 610 °C and as a result much improved sharp CZTS peaks in the XRD and Raman spectroscopy were achieved. The film had a particle size in the range of 700-800 nm. The solar cell was fabricated as SLG/Mo/CZTS/CdS/ZnO/ITO/Ag. 6% of PCE was achieved.

Tanaka et al., 2007 [30] prepared CZTS thin film via sol-gel spin coating under ambient condition. Cupric acetate, zinc acetate, and tin chloride were used as precursor and dimethyl alcohol as a solvent to prepare the solution. Few drops of ethanolamine were added as a stabilizer. Afterwards, the solution was spin coated on a Mo-coated glass substrate and dried at 300 °C for 5 min. Then the film was annealed for 60 min under 5% H_2S and 95% N_2 gas atmosphere at 500 °C. Good crystallinity and stoichiometry of components were achieved as the results. In 2009, Tanaka et al. [48] again improvised their work and varied the molar composition of components and also number of cycles. The cell was fabricated as Al/ZnO: Al/CdS/CZTS/Mo/SLG. A uniform thin film with 1.01% efficiency was achieved. In 2011, Tanaka et al. [49] again improvised their work and achieved 2.03% efficiency. Liu et al., 2016 [50] studied the effect of preheating of precursor on properties of CZTS films prepared by sol-gel spin coating. The preheating temperature of precursor was varied from 250 °C to 325 °C at the step rate of 25 °C rise. The films deposited at lower than 250 °C and higher than 350 °C could be peeled off easily. Then the film was annealed under sulfur atmosphere at 580 °C for 60 min. The SEM images revealed that 325 °C preheating temperature was favorable as a more uniform and dense film was observed. The film prepared using 325 °C preheating temperature exhibits high efficiency of 1.27%.

3.2.2 Two-step method for thin film preparation

In two-step method, firstly nanoparticles (NPs) are synthesized separately (various synthesis methods for nanoparticle is discussed in section 3.1). Subsequently, a thin film is prepared in a second step. Nanoparticles are dispersed in a solvent to prepare deposition nano suspension, which is further deposited on the substrate (by spin-coating or spray pyrolysis). The film properties can basically depend on the size and shape of the nanoparticles and the type of solvent. In multi-layer solar cell preparation, two-step method is generally superior, because it is a low temperature method. Hence, it can prevent degradation of layer beneath it (e.g. organic dye layer). Guo et al., 2009 [51] used copper acetylacetonate, zinc acetylacetonate, tin bis (acetylacetonate) dibromide and elemental sulfur in oleylamine for the preparation of a solution. It was heated at 225 °C for 30 min. It led to 15-25 nm CZTS nanoparticles. Subsequently, the NPs were dispersed in toluene and the film was prepared by drop-casting to achieve 0.8% efficiency. Xin et al., 2011 [52] also used the sol-gel method for CZTS nanoparticles preparation. Nanoparticles were then dispersed in toluene solvent to get nano ink, which was further deposited. The solar cell structure was fabricated as FTO/Pt/CZTS/N719/TiO_2/FTO and efficiency was measured. 3.62% efficiency was achieved. Chen et al., 2015 [53] also prepared CZTS based DSSC using the two-step method. Highly pure crystalline nanoparticles were achieved by operating an autoclave at high temperature (400 °C). The reaction time was decreased to 5 h. Nanoparticles were obtained by centrifuge and further dispersed in ethanol to prepare nano ink. The cell structure was fabricated as described by Xin et al., 2011 [52]. The fabricated cell was found to have 4.234% efficiency. Gong et al., 2016 [14] prepared it by adding precursors of Cu, Zn and Sn in the mixture of hexane, oleylamine and oleic acid. It was further heated at 65 °C. Then the sulfur solution (sulfur powder in oleylamine) was added separately and stirred at 130 °C for 30 min under N_2 atmosphere. Then the solution was cooled to 65 °C and added to the autoclave. The autoclave was operated for 24 h at 200 °C. Further, to precipitate the CZTS nanoparticles, the solution was added to toluene and isopropanol (1:1). The precipitates were collected by centrifuge and dispersed in toluene to form nano ink. The effect of EDT as solid state ligand exchange was also studied. The nanoparticles were surrounded by long chain ligands (oleylamine), which contributed to a larger inter-particle distance and led to inferior cell efficiency. Hence, to remove oleylamine and oleic acid, the NPs were treated with EDT (as solid state ligand exchanger). The EDT treated solar cell produced 5 times the efficiency compared to that prepared without EDT treatment.

4. Thin film quality control parameters

4.1 Precursor composition

The precursor composition is a very crucial control parameter for CZTS synthesis. It also affects crystallinity and optical properties. Presence of a secondary phase can be prevented by appropriate control of precursor composition. Basically, Cu/Zn+Sn ratio or S/Cu+Zn+Sn are two important parameters, which affect the optical properties. In addition, typical molar ratio of Cu:Zn:Sn:S should be 2:1:1:4. However, further optimization of composition may enhance the properties of CZTS films, hence; many studies are reported on precursor composition optimization.

4.1.1 Copper content

The Cu/Zn+Sn and Zn/Sn ratio are varied by Espindola-Rodriguez et al., 2013 [54]. With spray pyrolysis method, it was attempted to obtain Cu-poor and Zn-rich CZTS films. The optimum results were achieved with Cu-poor film (0.8 Cu/Zn+Sn). Efficiency of 0.49% was achieved. However, simultaneous Cu-poor and Zn-rich film led to secondary phase (ZnO) formation, subsequently, degraded optical properties. They further developed the work by annealing at high temperature (under sulfur atmosphere) [55, 56]. Zn-rich (1.2 Zn/Sn) CZTS films without any secondary phase formation was successfully achieved. The efficiency achieved was also high (1.4%). Rajeshmon et al., 2014 [57] studied the effect of copper content on the properties of CZTS film. Raman scattering spectra of CZTS samples prepared with increasing copper concentrations had clearly revealed that at high copper content, secondary phase (Cu_xS) was present. However, at reduced copper content, secondary phase formation could be eliminated. The optimum content was also identified with low copper content (0.98 Cu/Zn+Sn). Few other studies also reported Cu-poor films are optimum [45, 58]. Few studies have also reported an effect of Zn/Sn molar ratio. Chen et al., 2015 [39] prepared CZTS films from with Zn/Sn molar ratio from 0.75 to 1.5. Basically, Cu:Zn:Sn:S was kept to be 2:n:1:6 (with n=0.75, 1, 1.25 and 1.5). Upon increase in Zn/Sn ratio from 0.75 to 1.5, XRD peaks turn out to be sharp. It also led to decrease in Cu/(Zn+Sn) and S/Cu+Zn+Sn. Consequently, it enhanced optical properties. Optimum band gap (1.41 eV) was achieved at high Zn/Sn ratio of 1.5), without formation of any secondary phase. Few studies have reported Cu-rich films to be superior [59, 60]. Bhosale et al., 2015 [59] varied copper concentration during deposition by spray pyrolysis. The optimum results were achieved at Cu/Zn+Sn of 1.20. Kermadi et al., 2016 [60] also reported Cu-rich films to be superior with optimum Cu/Zn+Sn ratio of 1.2. However, Cu-poor film is generally reported to be superior for solar cell application, as it may decrease shunt resistance. The maximum efficiency (8.1%) is also reported to exist with low Cu/Zn+Sn ratio of 0.75 [61].

4.1.2 Sulfur content

Sulfur content also affects CZTS film properties. S-rich CZTS films are reported to be generally superior. In addition, sulfur loss can occur during heat treatment in pyrolysis or annealing. To eliminate this loss of S, a few studies report on usage of high sulfur content than the stoichiometry requirement. Swami et al., 2013 [62] used high sulfur content to prepare a solution by sol-gel method. The solution was prepared using double the S content than the stoichiometry requirement. Wei et al., 2016 [63] also studied S-rich CZTS films properties. The ratio of Cu:Zn:Sn:S was kept as 1:0.69:1.31:n, where n was kept to be 0.92, 1.84, 3.69 and 7.36. S/Cu+Zn+Sn was kept to be 0.3, 0.6, 1.25 and 2.5. CZTS films were prepared by sol-gel spin-coating method with annealing at 550 °C. The optimum results were achieved at high S/Cu+Zn+Sn ratio. Thermal stability was also observed to be good at high S/Cu+Zn+Sn ratio. Basically, the copper precursor requires more S content (than its stoichiometry requirement) to properly dissolve in the solution. Only few studies have reported less content of S (stoichiometry sulfur content) to be optimum. Seboui et al., 2014 [64] varied S/Cu+Zn+Sn ratio from 1-1.75, and got superior results at low S content. At high S content, the secondary phase formation ($Cu_4Sn_7S_{16}$) was observed. The single phase was obtained at 0.04 M sulfur content.

4.2 Type of precursor

The type of precursor affects the opto-electrial properties and ultimately solar cell performance. Rajeshmon et al., 2011 [65] reported in detail about the influence of type of precursor on opto-electrical properties. Two types of Sn salts ($SnCl_2$ and $SnCl_4$) were used to prepare different precursor solutions. Consequently, both solutions were spray coated on the substrate. The film prepared with $SnCl_4$ could achieve appropriate band gap (1.5 eV). However, the film prepared using $SnCl_2$ could only achieve 1.3 eV, which is low compared to film prepared with $SnCl_4$. The low band gap is attributed to the presence of a secondary phase (Cu_xS), as confirmed by XRD results. Consequently, solar cells prepared by a film prepared with $SnCl_4$ could achieve high V_{oc} (380 mV) and J_{sc} (2.40 mA/cm^2), which is almost four times that of $SnCl_2$ based solar cells. It is also reported that chlorine containing precursor may produce undesirable chlorides, which may deteriorate solar performance. Tanaka et al., 2014 [66] compared chlorine containing precursors ($CuCl_2$, $ZnCl_2$ and $SnCl_2$) with chlorine free precursors (copper acetate, zinc acetate and tin octylate). The SEM images and WDX (wavelength dispersive X-Ray spectroscopy) results confirmed that the film prepared with Cl-containing precursors resulted in some chloride precipitate on a few locations of the film. It further led to chlorides formation (due to high electro negativity of Cl). WDX spectra obtained from the precipitate could confirm that the composition of Cu:Zn:Sn:S (1:1.88:1:0.75) was highly different than that of the region around it (2:1.5:1:4). Consequently, efficiency achieved by Cl-free solution based solar cells were highly superior (0.35%) than that of Cl-containing solution based solar cells (0.18%). Exarhos et al., 2014 [67] also used copper, zinc and tin diethyldithiocarbamates to avoid the negative effect of chlorine. Similarly, Boshtha et al., 2016 [68] also used copper acetate and zinc acetate to prepare precursor solution (for spray pyrolysis) to avoid any chloride formation.

4.3 Type of solvent

The type of solvent may highly affect the type of secondary phase formed. Aqueous solvent may produce a Cu containing (CuS, Cu_2S, Cu_2SnS_3) secondary phase. Usage of alcohols as solvents may lead to a Zn or Sn containing ($ZnSnO_3$, SnO_2, SnS_2) secondary phase. However, secondary phase formation may be eliminated by keeping high deposition temperature. Kamoun et al., 2007 [69] used aqueous solution for CZTS film preparation by spray pyrolysis. Cu containing secondary phase (Cu_2S) was obtained at low temperature. Upon increasing the deposition temperature (>350°C), the secondary phase formation was eliminated. Kishore et al., 2009 [70] also prepared an aqueous solution for spray pyrolysis. They also reported to obtain a Cu containing secondary phase (Cu_xS and Cu_2SnS_3) at low temperature, which could be eliminated at T>370°C. Daranfed et al., 2012 [44] used alcohol (methanol) as solvent, and obtained a Zn-Sn containing secondary phase ($ZnSnO_3$) as expected. In addition, the phase could be eliminated at T>320 °C. Overall, alcohol usage thus requires low temperature. Hence, the types of solvent can affect both the initial cost (solvent cost) as well as operating cost (deposition temperature). Kannan et at., 2016 [71] also studied the effect of solvent type on the size and shape of CZTS NPs and optical properties. DI water, ethanol, ethylenediamine (EDN), diethylene glycol (DEG) and ethylene glycol (EG) were used as solvent. Morphology of NPs prepared with various solvents was studied in detail. CZTS prepared using water and ethanol as solvent led to large non-uniform particles. CZTS prepared using EDN led to rice shaped large particles. CZTS prepared using DEG and EG led to uniform spherical particles. Smaller particles were achieved with EG compared to DEG. Additionally, optical absorption and photo current were also high. Hence, it can be said that the type of solvent also affects size and shape of nanoparticles (in the film). The type of solvent is also important for preparation of nano ink (in the two-step method). Rawat et al., 2016 [72] studied CZTS NPs preparation using the sol–gel method using methanol solvent. In the second step, CZTS were further dispersed in the PEG-methanol mixture. The concentration of PEG in methanol was varied from 5-100 (%v/v). It was clearly observed that with increment in PEG content, CZTS crystallite size was decreased. The PEG content also affected the shape of nanoparticles. Appropriate nanoflakes or nanorod type particles were achieved at high PEG content.

4.4 Substrate temperature

Film morphology, composition, and grain size are highly dependent on substrate temperature for spray pyrolysis. Considerably high temperature may cause degradation or decomposition of the film. At the same time, low temperature may lead to a secondary phase (impurity) formation. Hence, the substrate temperature is crucial for solar cell performance. Vigil-Galan et al., 2013 [73] studied the effect of substrate temperature on the structural properties of CZTS. The substrate temperature was varied from 350 to 450 °C. The obtained XRD patterns are shown in Fig. 2. A single phase CZTS with preferential (112) orientation was observed for all the films. However, at low temperature, the peak intensity of the peak represent (112) orientation was low. Upon increment in temperature the peak intensity increases. The highest peak intensity was observed for the film prepared at 390 °C substrate temperature.

The product achieved was 40-60 μm in size. Pareek et al., 2015 [15] also synthesized nanoparticles from the bulk size powders of Cu, Zn, Sn and S. The powder was mixed in 1–butanol and the mixture was fed in a planetary mill. The mill was operated for a long time (approximately 30 h). Nanoparticles (133-286 nm) were achieved. Zhou et al., 2016 [37] could decrease the milling time requirement to approximately 16 h with usage of a vibration rod mill. Prepared nanoparticles were further annealed for 2 h to achieve an appropriate band gap (1.49 eV).

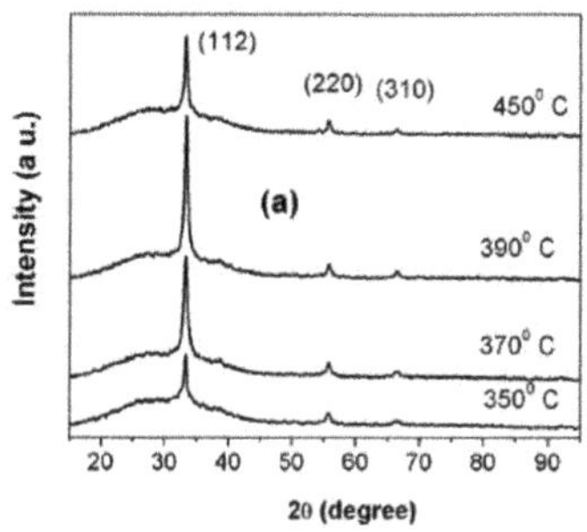

Figure 2. XRD patterns of the sprayed CZTS films as a function of the substrate temperature [73] (Reprinted from Vigil-Galan et al., 2013, *with the permission of AIP Publishing).*

Exarhos et al, 2014 [67] also studied the effect of substrate temperature on the morphological properties of CZTS film. The substrate temperature was varied from 360 to 460 °C. The SEM images are depicted in Fig. 3. At low temperature (360 °C), a uniform and smooth film was observed. However the growth rate (of the film) was low with a less than 100 nm thickness observed. Upon increase in temperature an increment in growth rate was observed. At 460 °C, a dense array of nanoplatelets were observed, also Raman spectra showed highest peak intensity at 460 °C. Hence, 460 °C substrate temperature was considered as the optimum temperature.

Swami et al., 2015 [74] also studied the effect of substrate temperature on opto-electrical properties of CZTS film (preparation by spray pyrolysis method). It was varied from 200-450 °C temperature. The film prepared at low temperature was not good (attributed to secondary phase (Cu_xS) formation), as it exhibited high resistance. Upon increase in temperature from 200 °C to 250 °C, the resistance decreased from 502.39 to 51.84 Ω-cm. Further increase (350 °C), minimum resistance (0.19 Ω-cm) was obtained. However, further increase in temperature could not further improve the resistance. The optimum temperature was thus found to be 350 °C (with band gap of 1.45 eV). Typically, same substrate temperature range (250-400 °C) is reported in various research studies [33, 61, 65, 69, 75]. Deposition at high temperature (>380 °C) may cause degradation of Sn precursor. However, few studies reported high substrate temperature to be favorable [76]. It was varied from 200–500 °C. The optimum band gap and morphology were achieved at 500 °C. High optimum temperature of 425 °C was also reported [45].

Figure 3. SEM images of the film prepared at (a) 360 °C, (b) 400 °C, (c) 440 °C and (d) 460 °C substrate temperature. [67] (Reproduced from Exarhos et al., 2014 with permission of The Royal Society of Chemistry).

4.5 Annealing temperature

For spin coating, the film quality can be controlled by the annealing temperature. After drying, the film may contain a secondary phase and defects in morphology (voids or uneven shaped, sized particles). Hence, to obtain a smooth, uniform and single phase film, a high temperature annealing treatment with specific atmosphere (H_2S, N_2, Ar or vacuum) is applied. However, at very high temperature (>600 °C), the film may get degrade. Agawane et al., 2015 [46] studied effect of annealing temperature on properties of CZTS thin films. The annealing temperature was varied from 525 to 575 °C. The SEM images revealed the effect of annealing temperature on the surface morphology of the film. The film with high annealing temperature of 575 °C depicted extra dense and uniform grains than the as-deposited film and film annealed at low temperature. Annealing temperature also affected the CZTS film thickness. The film with annealing was smooth and superior (696 nm) compared to that without annealing (1 μm). Nguyen et al., 2015 [75] studied the effect of annealing temperature and annealing time on the spray coated CZTS film. The films were annealed with sulfur powder in borosilicate glass ampoule. The annealing temperature was varied as 580 °C and 600 °C and time was varied 10 min, 30 min and 50 min. The XRD patterns and Raman spectroscopy suggested that high temperature (600 °C) was favorable. The further effect of annealing temperature and time was studied by SEM analysis. The uniform and dense film with more than 1 μm thickness was observed for as-deposited film (Fig. 4).

The annealing treatment can induce grain growth without loss in thickness. The larger grain size was observed for a high temperature. Additionally, more annealing time improved further grain size. However, extended annealing time (50 min) induced sulfurization of Mo substrate into MoS_2. Hence, 600 °C annealing temperature and 30 min annealing time was considered as the optimum. Prabeesh et al., 2016 [77] studied the effect of annealing temperature for CZTS films with sol-gel spin coating method. The annealing temperature was varied from 350-550 °C. The film was annealed under inert (N_2) condition in order to avoid the toxicity involved by annealing under a sulfur atmosphere. The optimum band gap and morphology were achieved at 500 °C temperature. Wang et al., 2017 [78] also annealed CZTS films in air to avoid the toxic approach of annealing in a sulfur atmosphere. The annealing temperature was varied from 350 to 550 °C in order to study the effect of annealing temperature. The film annealed at low temperature depicted presence of a secondary phase in Raman spectroscopy (see Fig. 5). Upon increment in annealing temperature the peaks intensity at 337 cm^{-1} and 375 cm^{-1} (correspond to CZTS) were increased and the peaks associated with the secondary phase were decreased. The solar cell performance was also improved with increment in annealing temperature, as highest efficiency of 0.25% was achieved by the film annealed at 550 °C. Further, to improve efficiency the annealing temperature was increased to 600 °C and 0.42% as thr best efficiency was obtained.

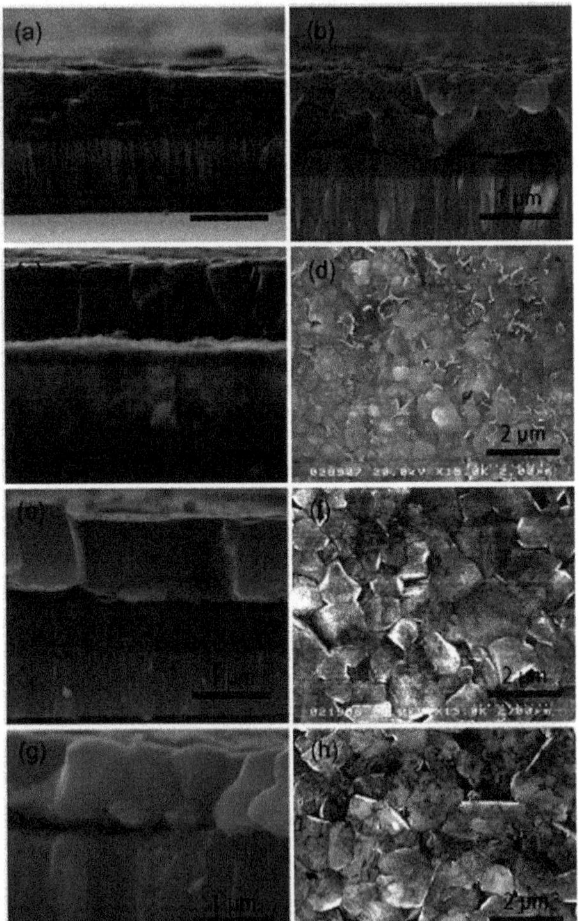

Figure 4. Cross-sectional and surface SEM images of (a) as-deposited film and CZTS films obtained by annealing the as-deposited film (b) at 580 °C for 10 min, (c and d) at 600 °C for 10 min, (e and f) at 600 °C for 30 min, and (g and h) at 600 °C for 50 min [75] (Reproduced from Nguyen et al., 2015 with permission of The Royal Society of Chemistry).

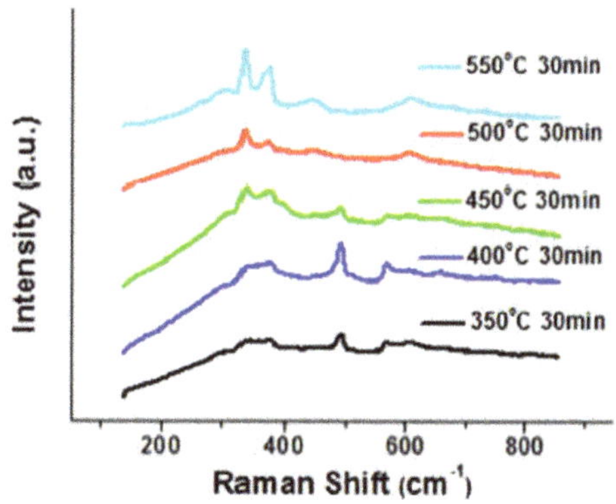

Figure 5. *Raman spectroscopy of the films annealed at different temperature [79] (Reproduced from Wang et al., 2017 with permission of The Royal Society of Chemistry).*

Park et al., 2016 [79] also studied CZTS film preparation with pre-annealing (150 to 350 °C) prior to annealing. The morphological analysis suggested that the film prepared at 350 °C led to a more uniform and dense film than that prepared at low temperature. Pre-annealing temperature was also observed to affect the composition of CZTS. As the temperature was increased from 250 °C to 350 °C the Cu/Zn+Sn ratio increased from 0.70 to 0.75. Likewise Zn/Sn ratio increased from 1.0 to 1.11. The TGA analysis suggested that at 350 °C, $SnCl_2$ get decomposed and converted into sulfide. Consequently, it depicted the maximum efficiency of 4.033%.

4.6 Sulfur source and sulfur annealing

Selection of sulfur source and sulfur annealing both affects surface morphology and stoichiometry of CZTS and ultimately affects opto-electrical properties. Selection of the sulfurization approach is crucial for the cost of the solar cell preparation as some times a vacuum is required while sulfurization which increases the cost. Wang et al., 2014 [80] proposed their work on effect of sulfur source on CZTS film morphology and size of the particles. Thiourea, thioacetamide and L-cysteine were used as the sulfur source. The effect of sulfur source was revealed by SEM and TEM images. The nanoparticles prepared using thiourea were 450 nm in size with flower like shape, and its band gap was also found to be favorable (1.43 eV). Nanoparticles prepared using L-cysteine as sulfur source exhibited non-uniform hollow spherical shape with 50 nm particle size. Whereas, thioacetamide based nanoparticles were quite smaller (2.5-3.5 nm) with spherical shape. However, the band gap was a little high (1.8 eV). Valdes et al., 2014 [81] reported the effect of thermal treatment (under sulfur vapor) on the structural, morphological and optical properties of CZTS films. Deposited films lead to secondary phase (Cu_xS) formation. After annealing (at 500 °C) with sulfur vapor, reduction in the secondary phase formation was observed. For the film deposited at 425 °C, band gap of 1.51 eV was observed. Tchognia et al., 2016 [82] also studied the effect of sulfur annealing time (with H_2S) on the properties of CZTS films prepared via sol-gel spin coating. The time was varied from 30 to 75 min. The optimum results were achieved with 75 min in terms of band gap.

4.7 Spin speed and spray rate

Spin speed and spray rate affects both the surface morphology and thickness. Low spin speed may cause voids in the films. At the same time, high spin speed may cause material loss, as coating becomes highly thin. Ultimately, both conditions may lead to degradation of solar cell performance. Hence, there is a need of identifying an appropriate optimum speed range. Optimum spin speed range depends highly on the viscosity of the solvent.

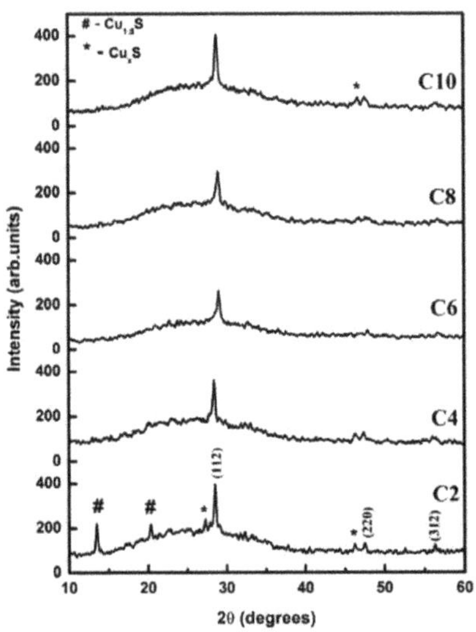

Figure 6. XRD patterns Of Sprayed Cu_2ZnSnS_4 Thin Films Prepared at Different Spray Rates [85] (Reprinted from Rajeshmon et al., 2014, *with the permission of AIP Publishing).*

Similarly, the spray rate must also be in the optimum range, as high spray rate may lead to occurrence of the secondary phase. At the same time, low spray rate may cause voids or cavities in the film. Yu et al., 2014 [83] investigated influence of spin speed on properties of the CZTS film. The spin speed was varied from 1500 to 3000 rpm. Film deposited at low speed (1500 rpm) was found to contain more voids as seen from the SEM image. Film deposited at 2000 rpm demonstrated improvement as it had less voids. Moreover, films prepared with 3000 rpm speed revealed sharp peaks, which is favorable. Additionally, grains were also observed to be large and densely packed in the SEM image. Consequently, superior efficiency (1.9%) was found with films deposited at 3000 rpm. Wu et al., 2015 [84] had prepared films at 3000 to 6000 rpm spin speed. The device fabricated with films prepared at optimum speed (4000 rpm) gave a superior result. Rajeshmon et al., 2014 [85] varied spray rate from 2 to 10 mL min^{-1} at a step of 2 mL min^{-1}in order to study the effect of spray rate on a CZTS thin film. Single phase CZTS was achieved with a film prepared at 6 mL min-1spray rate (see Fig. 6). In other cases, a secondary phase of Cu_xS was observed.

5. Recent advanced CZTS preparation and its applications

5.1 Photocatalytic applications

Very recently, CZTS nanoparticles are also studied in photocatalytic degradation applications. So far, CZTS is employed as photocatalyst for degradation of CO_2 [86], RhB [87], TNT [88], dyes [89], etc. In addition, CZTS nanoparticles are also used in water splitting to produce hydrogen [90]. Kim et al., 2016 [91] used hybrid mesoporous CZTS along with TiO_2 (instead of TiO_2) as photocatalyst to convert unwanted carbon dioxide (CO_2) to methane (CH_4). The mesoporous CZTS nanocrystals were synthesized by sol-gel method with chlorine free precursors. Subsequently, the as-synthesized nanocrystals were dried, and added to the $TiCl_4$ solution to prepare hybrid mesoporous photocatalyst, CZTS-TiO_2. Eventually, the solution was dried and finally annealed to obtain dry photocatalyst. The visible light capturing capability of hybrid photocatalyst (CZTS-TiO_2) was found to be high compared to TiO_2 photocatalyst. Consequently, a superior CO_2 to CH_4 conversion was observed. Conversion was 12 times higher than that of TiO_2. Phaltane et al., 2017 [92] were able tp degrade methylene blue (MB) dye from waste water using CZTS as photocatalyst. The CZTS nanoparticles were prepared by solvothermal method and then added to water containing MB. Subsequently, the water was exposed to visible light. Superior degradation (50%) of MB was achieved only in 45 min. Additionally, CZTS has proven its photocatalytic application in water splitting to produce hydrogen. Ha et al., 2014 [90] applied Au-CZTS film as photocatalyst for water splitting. Au-CZTS was prepared by sol-gel method and dispersed in the di water solution containing Na_2S and Na_2SO_3. The visible 1 sun (100 mW cm^{-2}) light source was applied for 1 hr and highest H_2 evolution of 102 μmol h^{-1} g^{-1} was achieved as a result. Yu et al., 2014 [93] also prepared 2 different CZTS-nobel metal (Au and Pt) hetrostructured nanoparticels as photocatalyst. The photocatalytic activity of CZTS, Au-CZTS and Pt-CZTS was studied by water splitting and RhB degradation. A 300 W Xenon lamp was used as light source. The highest hydrogen evolution of 1.02 mmol h^{-1} g^{-1} was achieved by Pt-CZTS. Additionally, the photodegradation of RhB of 95% in 2.5 h was obtained with Pt-CZTS. Whereas, a 60% degradation in 4 h was achieved with CZTS and 88% degradation in 2.5 h was achieved with Au-CZTS. Hence, in both case, Pt-CZTS was proven to be the most active photocatalyst. Table 1 shows the summary of recent studies on the photocatalytic application of CZTS.

Table 1. List of recent studies on photocatalytic application of CZTS.

Sr. No.	Photocatalyt-ic application	Photocatalyst	preparation method	Process condition	Results	Ref.
1	Degradation of CO_2 from water	CZTS/RuCE +RuCA	Magnetron Sputtering	Visible light (400-800 nm) for 3 hr	>80% degradation achieved	[87]
2	Degradation of Rhodamine B (RhB)	ZnO/CZTS and NYF:Yb, Tm/CZTS	Sol-gel	Visible light 120 min and NIR 6 hr	>90% degradation (ZnO/CZTS under visible light); >50% degradation (NYF /CZTS under NIR light.	[88]
3	Degradation of RhB and TNT	CZTS	Sol-gel	Visible light for 5 hr	98% and 86% degradation of RhB and TNT is achieved, respectively.	[88]
4	Degradation of MB dye	Au-CZTS	Sol-gel	420 nm light for 10 min	47% of MB dye reduction	[89]
5	Water splitting	Au-CZTS	Sol-gel	100 mW cm^{-2} light for 1 hr	H_2 evolution (102 μmol h^{-1} g^{-1}) and photocurrent (30 μA cm^{-2}) achieved by Au-CZTS then CZTS	[90]
6	Conversion of CO_2 to CH_4	TiO_2/CZTS	Sol-gel	Visible light 5-10 hr	12 times higher conversion than that of TiO_2.	[91]
7	Degradation of MB dye	CZTS	hydrotherm al	Visible light 45 min	50% of MB dye degradation is achieved	[92]
8	Water splitting and degradation of RhB	CZTS, Au-CZTS and Pt-CZTS	Sol-gel	300 W Xe lamp	H_2 evolution (1.02 mmol h^{-1} g^{-1} ; with Pt-CZTS) 95% of RhB degradation in 2.5 hr by Pt-CZTS.	[93]

5.2 Solar cell applications

5.2.1 Green synthesis

Synthesis can be green and environment friendly by applying low vacuum annealing or annealing under inert condition instead of the toxic approach of sulfur annealing. Wang et al., 2014 [94] prepared CZTS film using sol-gel spin coating with an environment friendly approach. Annealing under inert (Ar) atmosphere was used. It was annealed at 350 °C temperature for 1 h. Reasonable high efficiency (approximately 5%) is achieved with solar cells prepared with CZTS films. Guo et al., 2017 [95] also reported preparation of a pure CZTS film without any sulfur atmosphere while annealing. The films were dried by circulating hot air (250 °C) for 2 h. A uniform film and compact sphere shaped nanoparticles (60 nm) were achieved. Orletskyi et al., 2017 [96] also reported their work on preparation of CZTS thin film by sol-gel spin coating using environment friendly green solvent. The solution was prepared in DMSO under low vacuum. Then it was deposited on the substrate at 2000 rpm and then dried at 220 °C under ambient condition and then finally annealed at 540 °C under low vacuum. A single phase CZTS film with appropriate band gap (1.53 eV) was achieved. Boshta et al., 2016 [68] also prepared thin films with a green method via spray pyrolysis on flexible plastic substrate. The film was annealed under inert (Ar) atmosphere at 400 °C temperature for 1 h. The film depicted good crystallinity (peaks in Raman shift). Consequently prepared solar cell led to 0.15% efficiency, which can be improved by improving the surface properties.

5.2.2 Doping

Su et al., 2016 [97] studied the effect of cation substitution on performance of CZTS based solar cell. Two approaches were used. In the first approach, zinc was substituted by cadmium. In a second approach, copper was substituted by silver. Cd-doped CZTS cell depicted improvement in efficiency from 9.24% to 10.66%. Ag-doped CZTS cell showed improvement in V_{OC} from 570 to 650 mV. Hence, it was depicted that cation substitution can overcome the limitation of CZTS based solar cell (such as V_{OC} deficits, low crystallinity, low efficiency). Yang et al., 2016 [98] reported a simple method for preparation of CZTSSe thin films. The CZTS solution was prepared by dissolving elemental powder of Cu, Zn, Sn and S into a mixture of thioglycolic acid and methylamine. DI water was added to adjust the concentration and viscosity of solution. Then the solution was deposited on Mo-coated glass via spin coating and sintered at 320 °C under inert (N_2) condition. Then after, the film was selenized by annealing with Se powder in a graphite box for 10 min at 510 °C. The cell was fabricated as glass/Mo/CZTSSe/CdS/i-ZnO/ITO/Ag and 5.64% efficiency was achieved. Further, Na-doped CZTSSe film was prepared. Further, similar solar cells were fabricated. Positive effect of Na-doping was observed and 23% enhancement in efficiency was achieved (6.96% efficiency).

5.3 Recent developments in CZTS films

Wu et al., 2015 [86] used the spin coating method to prepare a CZTS film on a perovskite layer. In which, the rotation speed was varied from 3000 to 6000 rpm also annealing time was varied from 5, 10 and 15 min. Optimum results were obtained at 4000 rpm and 10 min annealing as it gave approximately 20 nm sized nanoparticles and a 200 nm thick CZTS film. It was depicted for the first time that CZTS films can be utilized as hole transporting material in perovskite based tandem solar cells. The cell was fabricated as FTO/TiO_2/Pervoskite/CZTS/Au and achieved 12.75% PCE, which is very promising. Khanzada et al., 2016 [99] also prepared perovskite based solar cells in which CZTS was applied as HTM and achieved 15.4 % cell efficiency. Cheng et al., 2016 [100] prepared CZTS nanoparticles via a solvothermal method and then dispersed it with P3HT and PCBM in dichlorobenzene to prepare a solution of active layer. The concentration of CZTS nanoparticles was varied from 0 to 2 mg mL^{-1}. The solution of active layer was deposited on the substrate using a spray pyrolysis method. The cell was fabricated as ITO/ZnO/Active layer/MoO_3+Ag. The concentration of CZTS NPs have an effect on the carrier generation rate (which lead to high or low current density and efficiency). It was observed that, small concentration of NPs (0.5 to 1 mg mL^{-1}) in the active layer showed improvement in J_{SC} due to high carrier generation compared to the layer without NPs. However, an active layer containing more CZTS NPs (1.5 to 2 mg mL^{-1}) showed a decrement in J_{SC} (10.21 and 9.67) due to an increment in carrier recombination rate. Hence, 3.65% of highest cell efficiency was achieved by cell containing 1 mg mL^{-1} concentration. Park et al., 2016 [101] prepared CZTS based DSSC using the single step method by sol-gel spin coating instead of the two-step method, previously reported [52, 53]. Enhanced solar cell efficiency (4.494%) was achieved. Additionally, after adding platinum (Pt) as counter electrode, further enhanced efficiency (5.749%) was achieved.

6. Summary and future scope

The present study outlined CZTS preparation, optimization and its applications. CZTS is a great absorbing material with direct band gap of 1.4-1.6 eV, which is close to the theoretical optimum (1.45 eV). The main applications are solar cell development and photocatalytic degradation. Firstly, properties of CZTS materials are discussed. Further, CZTS thin film solar cell development is discussed at length. Two decisive thin film preparation methods (spin coating and spray pyrolysis) are discussed in detail along with their advantages and limitation. The thin film quality controlling parameters are discussed in detail for both the methods and studies on these parameters have been discussed comprehensively. By going through the entire review, it can be concluded that CZTS based solar cell can be considered as next generation thin film solar cells. Till now, 12.6% of maximum efficiency is registered with CZTS based solar cell. However, theoretical calculated efficiency is 32%. Hence, there is a wide scope available for improvement of efficiency. The reason behind present (low) efficiency is low Voc, which may be attributed to inappropriate cell combination or to the presence of a secondary phase. Hence, improvement in Voc can be a future research scope.

Some attempts have been done by different researchers, although still improvement in CZTS based solar cell is needed to acquire it on commercial level. Until now, doping of Na, Se, Ge, Ag and Cd in CZTS layer have been reported in order to improve efficiency or Voc (or both). CZTS can work as a p-type conductor, so that it can be applied as HTM in other cells. Up to now, it is applied as HTM in DSSCs. However, only two studies have been reported to use CZTS as hole transport material (HTM) in place of conventional spiro-MeOTAD in perovskite based solar cell. The cost of perovskite based solar cells is reduced drastically as the cost of CZTS is much less than Spiro-MeOTAD. In addition, CZTS is highly stable and its preparation does not require any specific atmospheric condition as required with conventional spiro-MeOTAD. So, there is a great need of devoted research to CZTS film preparation, optimization and its application, especially in solar cell applications.

In Addition, CZTS was proven to be a potential photocatalyst for degradation in the recent years. However, only a few limited degradation applications are reported till now. It includes photocatalytic degradation of CO_2, RhB, TNT, dyes etc. Because, highly superior degradation can be achieved by CZTS, there is a huge research opportunity for further studies on photocatalytic degradation by CZTS.

References

[1] J.F. Mercure, P. Salas, An assessement of global energy resource economic potentials, Energy 46 (2012) 322–336. https://doi.org/10.1016/j.energy.2012.08.018

[2] F. Bayrak, N. Abu-Hamdeh, K.A. Alnefaie, H.F. Oztop, A review on exergy analysis of solar electricity production, Renew. Sustainable Energy Rev. 74 (2017) 755–770. https://doi.org/10.1016/j.rser.2017.03.012

[3] M. Pehnt, Dynamic life cycle assessment (LCA) of renewable energy technologies, Renew. Energy 31 (2006) 55–71. https://doi.org/10.1016/j.renene.2005.03.002

[4] Z.M. Salameh, F. Dagher, W.A. Lynch, Step-down maximum power point tracker for photovoltaic systems, Sol. Energy 46 (1991) 5 279–282. https://doi.org/10.1016/0038-092X(91)90095-E

[5] M.A. Green, Third generation photovoltaics: solar cells for 2020 and beyond, Physica E 14 (2002) 65–70. https://doi.org/10.1016/S1386-9477(02)00361-2

[6] R.W. Miles, G. Zoppi, I. Forbes, Inorganic photovoltaic cells, Mater. Today 10, 11 (2007) 20–27. https://doi.org/10.1016/S1369-7021(07)70275-4

[7] M.A. Green, Solar cell efficiency tables (version 49): Solar cell efficiency tables (version 49), Prog Photovoltaics Res Appl. 25 (2017) 3–13. https://doi.org/10.1002/pip.2855

[8] A.C. Mayer, S.R. Scully, B.E. Hardin, M.W. Rowell, M.D. McGehee, Polymer-based solar cells, Mater. Today, vol. 10 (2007) 28–33. https://doi.org/10.1016/S1369-7021(07)70276-6

[9] J.H. Heo, S.H. Im, J.H. Noh, T.N. Mandal, C.S. Lim, J.A. Chang, Y.H. Lee, H.J. Kim, A. Sarkar, M.K. Nazeeruddin, M. Gratzel, Efficient inorganic–organic hybrid heterojunction solar cells containing perovskite compound and polymeric hole conductors, Nat. Photonics 7 (2013) 486–491. https://doi.org/10.1038/nphoton.2013.80

[10] G. Conibeer, M. Green, R. Corkish, Y. Cho, E.C. Cho, C.W. Jiang, T. Fangsuwannarak, E. Pink, Y. Huang, T. Puzzer, T. Trupke, Silicon nanostructures for third generation photovoltaic solar cells, Thin Solid Films 511 (2006) 654–662. https://doi.org/10.1016/j.tsf.2005.12.119

[11] N. Ali, A. Hussain, R. Ahmed, M.K. Wang, C. Zhao, B.U. Haq, Y.Q. Fu, Advances in nanostructured thin film materials for solar cell applications, Renew. Sustainable Energy Rev. 59 (2016) 726–737. https://doi.org/10.1016/j.rser.2015.12.268

[12] L. E. Valle Rios, K. Neldner, G. Gurieva, S. Schorr, Existence of off-stoichiometric single phase kesterite, J. Alloys Compd. (2016) 408–413. https://doi.org/10.1016/j.jallcom.2015.09.198

[13] P. Kush, K. Deori, A. Kumar, S. Deka, Efficient hydrogen/oxygen evolution and photocatalytic dye degradation and reduction of aqueous Cr (vi) by surfactant free hydrophilic Cu_2ZnSnS_4 nanoparticles, J. Mater. Chem. A 3 (2015) 8098-8106. https://doi.org/10.1039/C4TA06551D

[14] S. Kumar, B. Kasubosula, M. Loorits, J. Raudoja, V. Mikli, M. Altosaar, M. Grossberg, Synthesis of Cu_2ZnSnS_4 Solar Cell Absorber Material by Sol-gel

Method, Energy Procedia 102 (2016) 102–109. https://doi.org/10.1016/j.egypro.2016.11.324

[15] Z. Gong, Q. Han, J. Li, L. Hou, A. Bukhtiar, S. Yang, B. Zou, A solvothermal route to synthesize kesterite Cu_2ZnSnS_4 nanocrystals for solution-processed solar cells, J. Alloys Compd. 663 (2016) 617–623. https://doi.org/10.1016/j.jallcom.2015.12.181

[16] D. Pareek, K. R. Balasubramaniam, P. Sharma, Synthesis and characterization of bulk Cu_2ZnSnX_4 (X: S, Se) via thermodynamically supported mechano-chemical process, Mater. Charact. 103 (2015) 42–49. https://doi.org/10.1016/j.matchar.2015.03.014

[17] M. Courel, J. A. Andrade-Arvizu, A. Guillen-Cervantes, M. M. Nicolas-Marin, F. A. Pulgarin-Agudelo, O. Vigil-Galan, Optimization of physical properties of spray-deposited Cu_2ZnSnS_4 thin films for solar cell applications, Mater. Des. 114 (2017) 515–520. https://doi.org/10.1016/j.matdes.2016.10.068

[18] A. Safdar, M. Islam, M.A. Akram, M. Mujahid, Y. Khalid, S.I. Shah, Reaction Time and Film Thickness Effects on Phase Formation and Optical Properties of Solution Processed Cu_2ZnSnS_4 Thin Films, J. Mater. Eng. Perform. 25 (2016) 457–465. https://doi.org/10.1007/s11665-015-1874-6

[19] T. Washio, T. Shinji, S. Tajima, T. Fukano, T. Motohiro, K. Jimbo, H. Katagiri, 6% Efficiency Cu_2ZnSnS_4-based thin film solar cells using oxide precursors by open atmosphere type CVD, J. Mater. Chem. 22 (2012) 4021-4024. https://doi.org/10.1039/c2jm16454j

[20] T. Tanaka, T. Nagatomo, D. Kawasaki, M. Nishio, Q. Guo, A. Wakahara, A. Yoshida, H. Ogawa, Preparation of Cu_2ZnSnS_4 thin films by hybrid sputtering, J. Phys. Chem. Solids 66 (2005) 1978–1981. https://doi.org/10.1016/j.jpcs.2005.09.037

[21] S.A. Vanalakar, G.L. Agawane, S.W. Shin, M.P. Suryawanshi, K.V. Gurav, K.S. Jeon, P.S. Patil, C.W. Jeong, J.Y. Kim, J.H. Kim, A review on pulsed laser deposited CZTS thin films for solar cell applications, J. Alloys Compd. 619 (2015) 109–121. https://doi.org/10.1016/j.jallcom.2014.09.018

[22] K. Ramasamy, M. A. Malik, P. O'Brien, Routes to copper zinc tin sulfide Cu_2ZnSnS_4 a potential material for solar cells, Chem. Commun. 48 (2012) 5703-5714. https://doi.org/10.1039/c2cc30792h

[23] H. Zhou, W.C. Hsu, H.S. Duan, B. Bob, W. Yang, T. B. Song, C.J. Hsu, Y. Yang, CZTS nanocrystals: a promising approach for next generation thin film

photovoltaics, Energy Environ. Sci. 6 (2013) 2822-2838. https://doi.org/10.1039/c3ee41627e

[24] S. Abermann, Non-vacuum processed next generation thin film photovoltaics: Towards marketable efficiency and production of CZTS based solar cells, Sol. Energy 94 (2013) 37–70. https://doi.org/10.1016/j.solener.2013.04.017

[25] M.P. Suryawanshi, G.L. Agawane, S.M. Bhosale, S.W. Shin, P.S. Patil, J.H. Kim, A.V. Moholkar, CZTS based thin film solar cells: a status review, Mater. Technol. 28 (2013) 98–109. https://doi.org/10.1179/1753555712Y.0000000038

[26] X. Song, X. Ji, M. Li, W. Lin, X. Luo, H. Zhang, A Review on Development Prospect of CZTS Based Thin Film Solar Cells, Int. J. Photoenergy2014 (2014) 1–11. https://doi.org/10.1155/2014/613173

[27] H. Katagiri, K. Jimbo, W.S. Maw, K. Oishi, M. Yamazaki, H. Araki, A. Takeuchi, Development of CZTS-based thin film solar cells, Thin Solid Films 517 (2009) 2455–2460. https://doi.org/10.1016/j.tsf.2008.11.002

[28] A. Polizzotti, I. L. Repins, R. Noufi, S. H. Wei, D. B. Mitzi, The state and future prospects of kesterite photovoltaics, Energy Environ. Sci. 6 (2013) 3171-3182. https://doi.org/10.1039/c3ee41781f

[29] M. A. Majeed Khan, S. Kumar, M. Alhoshan, A. S. Al Dwayyan, Spray pyrolysed Cu_2ZnSnS_4 absorbing layer: A potential candidate for photovoltaic applications, Opt. Laser Technol. 49 (2013) 196–201. https://doi.org/10.1016/j.optlastec.2012.12.012

[30] K. Tanaka, N. Moritake, H. Uchiki, Preparation of Cu_2ZnSnS_4 thin films by sulfurizing sol–gel deposited precursors, Sol. Energy Mater. Sol. Cells 91 (2007) 1199–1201. https://doi.org/10.1016/j.solmat.2007.04.012

[31] W. Shockley, H. J. Queisser, Detailed Balance Limit of Efficiency of p-n Junction Solar Cells, J. Appl. Phys. 32 (1961) 510-519. https://doi.org/10.1063/1.1736034

[32] H. Sugimoto, H. Hiroi, N. Sakai, S. Muraoka, T. Katou, Over 8% efficiency Cu_2ZnSnS_4 submodules with ultra-thin absorber, presented at the IEEE Photovoltaic Spec. Conf. 2012 (2012) 002997–003000.

[33] M. Yang, X. Ma, Z. Jiang, Z. Li, S. Liu, Y. Lu, S. Wang, The Cu_2ZnSnS_4 solar cell with high open circuit voltage, Phys. B 509 (2017) 50–54. https://doi.org/10.1016/j.physb.2017.01.001

[34] I. D. Olekseyuk, I. V. Dudchak, L. V. Piskach, Phase equilibria in the Cu_2S–ZnS–SnS_2 system, J. Alloys Compd.. 368 (2004) 135–143. https://doi.org/10.1016/j.jallcom.2003.08.084

[35] K. Mokurala, S. Mallick, P. Bhargava, Low Temperature Synthesis and Characterization of Cu_2ZnSnS_4 (CZTS) Nanoparticle by Solution Based Solid State Reaction Method, Energy Procedia, vol. 57 (2014) 73–78. https://doi.org/10.1016/j.egypro.2014.10.010

[36] V. Kheraj, K. K. Patel, S. J. Patel, D. V. Shah, Synthesis and characterisation of Copper Zinc Tin Sulphide (CZTS) compound for absorber material in solar-cells, J. Cryst. Growth 362 (2013) 174–177. https://doi.org/10.1016/j.jcrysgro.2011.10.034

[37] Y. Zhou, S. Xi, C. Sun, H. Wu, Facile synthesis of Cu_2ZnSnS_4 powders by mechanical alloying and annealing, Mater. Lett. 169 (2016) 176–179. https://doi.org/10.1016/j.matlet.2016.01.116

[38] S.C. Riha, B.A. Parkinson, A.L Prieto, Solution-based synthesis and characterization of Cu_2ZnSnS_4 nanocrystals, J. Am. Chem. Soc. 131 (2009) 12054-12055. https://doi.org/10.1021/ja9044168

[39] J. Chen, Q. Chen, Y. Ni, Y. Yamaguchi, T. Wang, Z. Jia, X. Dou, S. Zhuang, The synthesis of Cu_2ZnSnS_4 nanoparticles via an open-air solution route: influences of Zn precursor content, J. Sol-Gel Sci. Technol., 75 (2015) 25–30. https://doi.org/10.1007/s10971-015-3670-z

[40] Y.L. Zhou, W.H. Zhou, Y.F. Du, M. Li, S.X. Wu, Sphere-like kesterite Cu_2ZnSnS_4 nanoparticles synthesized by a facile solvothermal method, Materials Letters 65 (2011) 1535-1537. https://doi.org/10.1016/j.matlet.2011.03.013

[41] X. Yan, X. Hu, S. Komarneni, Solvothermal synthesis of CZTS nanoparticles in ethanol: Preparation and characterization, J. Korean Phys. Soc. 66 (2015) 1511–1515. https://doi.org/10.3938/jkps.66.1511

[42] A. Arunachalam, S. Dhanapandian, C. Manoharan, M. Bououdina, G. Ramalingam, M. Rajasekaran, M. Radhakrishnan, A.M. Ibraheem, Influence of sprayed nanocrystalline Zn-doped TiO_2 photoelectrode with the dye extracted from Hibiscus Surattensis as sensitizer in dye-sensitized solar cell, Ceram. Int. 42 (2016) 11136–11149. https://doi.org/10.1016/j.ceramint.2016.04.020

[43] N. Nakayama, K. Ito, Sprayed films of stannite Cu_2ZnSnS_4, Applied Surface Science 92 (1996) 171–175. https://doi.org/10.1016/0169-4332(95)00225-1

[44] W. Daranfed, M. S. Aida, N. Attaf, J. Bougdira, H. Rinnert, Cu_2ZnSnS_4 thin films deposition by ultrasonic spray pyrolysis, J. Alloys Compd. 542 (2012) 22–27. https://doi.org/10.1016/j.jallcom.2012.07.063

[45] D. K.G., N. Jampana, Development of an automated ultrasonic spray pyrolysis system and the growth of Cu_2ZnSnS_4 thin films, J. Anal. Appl. Pyrolysis 117 (2016) 141–146. https://doi.org/10.1016/j.jaap.2015.12.004

[46] G.L. Agawane, A.S. Kamble, S.A. Vanalakar, S.W. Shin, M.G. Gang, J.H. Yun, J. Gwak, A.V. Moholkar, J.H. Kim, Fabrication of 3.01% power conversion efficient high-quality CZTS thin film solar cells by a green and simple sol–gel technique, Mater. Lett. 158 (2015) 58–61. https://doi.org/10.1016/j.matlet.2015.05.036

[47] V. Tunuguntla, W.C. Chen, P.H. Shih, I. Shown, Y.R. Lin, J.S. Hwang, C.H. Lee, L.C. Chen, K.H. Chen, A nontoxic solvent based sol–gel Cu_2ZnSnS_4 thin film for high efficiency and scalable low-cost photovoltaic cells, J. Mater. Chem. A 3 (2015) 15324–15330. https://doi.org/10.1039/C5TA02833G

[48] K. Tanaka, M. Oonuki, N. Moritake, and H. Uchiki, "Cu_2ZnSnS_4 thin film solar cells prepared by non-vacuum processing," Sol. Energy Mater. Sol. Cells, vol. 93, no. 5, pp. 583–587, May 2009. https://doi.org/10.1016/j.solmat.2008.12.009

[49] K. Tanaka, Y. Fukui, N. Moritake, and H. Uchiki, "Chemical composition dependence of morphological and optical properties of Cu_2ZnSnS_4 thin films deposited by sol–gel sulfurization and Cu_2ZnSnS_4 thin film solar cell efficiency," Sol. Energy Mater. Sol. Cells, vol. 95, no. 3, pp. 838–842, Mar. 2011. https://doi.org/10.1016/j.solmat.2010.10.031

[50] R. Liu, M. Tan, X. Zhang, J. Chen, S. Song, and W. Zhang, "Impact of sol–gel precursor treatment with preheating temperature on properties of Cu_2ZnSnS_4 thin film and its photovoltaic solar cell," J. Alloys Compd., vol. 655, pp. 124–129, Jan. 2016. https://doi.org/10.1016/j.jallcom.2015.09.028

[51] Q. Guo, H. W. Hillhouse, R. Agrawal, Synthesis of Cu_2ZnSnS_4 Nanocrystal Ink and Its Use for Solar Cells, J. Am. Chem. Soc.131 (2009) 11672–11673. https://doi.org/10.1021/ja904981r

[52] X. Xin, M. He, W. Han, J. Jung, Z. Lin, Low-Cost Copper Zinc Tin Sulfide Counter Electrodes for High-Efficiency Dye-Sensitized Solar Cells, Angew. Chem., Int. Ed. 50 (2011) 11739–11742. https://doi.org/10.1002/anie.201104786

[53] S. Chen, H. Tao, Y. Shen, L. Zhu, X. Zeng, J. Tao, T. Wang, Facile synthesis of single crystalline sub-micron Cu_2ZnSnS_4 (CZTS) powders using solvothermal treatment, RSC Adv. 5 (2015) 6682–6686. https://doi.org/10.1039/C4RA12815J

[54] M. Espindola-Rodriguez, M. Placidi, O. Vigil-Galan, V. Izquierdo-Roca, X. Fontane, A. Fairbrother, D. Sylla, E. Saucedo, A. Perez-Rodriguez, Compositional optimization of photovoltaic grade Cu_2ZnSnS_4 films grown by pneumatic spray pyrolysis, Thin Solid Films 535 (2013) 67–72. https://doi.org/10.1016/j.tsf.2012.12.082

[55] O. Vigil-Galan, M. Espíndola-Rodriguez, M. Courel, X. Fontane, D. Sylla, V. Izquierdo-Roca, A. Fairbrother, E. Saucedo, A. Perez-Rodriguez, Secondary phases dependence on composition ratio in sprayed Cu_2ZnSnS_4 thin films and its impact on the high power conversion efficiency, Sol. Energy Mater. Sol. Cells 117 (2013) 246–250. https://doi.org/10.1016/j.solmat.2013.06.008

[56] O. Vigil-Galan, M. Courel, M. Espindola-Rodriguez, D. Jimenez-Olarte, M. Aguilar-Frutis, E. Saucedo, Electrical properties of sprayed Cu_2ZnSnS_4 thin films and its relation with secondary phase formation and solar cell performance, Sol. Energy Mater. Sol. Cells, 132 (2015) 557–562.

[57] V. G. Rajeshmon, M. R. R. Menon, C. S. Kartha, K. P. Vijayakumar, Effect of copper concentration and spray rate on the properties Cu_2ZnSnS_4 thin films deposited using spray pyrolysis, J. Anal. Appl. Pyrolysis, 110 (2014) 448–454. https://doi.org/10.1016/j.jaap.2014.10.014

[58] D. K.G., G. Chandrabose, N. Jampana, "Tailoring the properties of Cu_2ZnSnS_4 thin films grown by ultrasonic spray pyrolysis," J. Anal. Appl. Pyrolysis 120 (2016) 356–360. https://doi.org/10.1016/j.jaap.2016.06.003

[59] S. M. Bhosale, M. P. Suryawanshi, J. H. Kim, A. V. Moholkar, Influence of copper concentration on sprayed CZTS thin films deposited at high temperature, Ceram. Int. 41 (2015) 8299–8304. https://doi.org/10.1016/j.ceramint.2015.02.124

[60] S. Kermadi, S. Sali, F.A. Ameur, L. Zougar, M. Boumaour, A. Toumiat, N.N Melnik, D.W. Hewak, A. Duta, Effect of copper content and sulfurization process on optical, structural and electrical properties of ultrasonic spray pyrolysed Cu_2ZnSnS_4 thin films, Mater. Chem. Phys. 169 (2016) 96–104. https://doi.org/10.1016/j.matchemphys.2015.11.035

[61] T.H. Nguyen, S. Fujikawa, T. Harada, J. Chantana, T. Minemoto, S. Nakanishi, S. Ikeda, Impact of Precursor Compositions on the Structural and Photovoltaic

Properties of Spray-Deposited Cu_2ZnSnS_4 Thin Films, ChemSusChem, 9 (2016) 2414–2420. https://doi.org/10.1002/cssc.201600641

[62] S. K. Swami, A. Kumar, V. Dutta, Deposition of Kesterite Cu_2ZnSnS_4 (CZTS) Thin Films by Spin Coating Technique for Solar Cell Application, Energy Procedia, 33 (2013) 198–202. https://doi.org/10.1016/j.egypro.2013.05.058

[63] Z. Wei, M. Zhu, J. D. McGettrick, G. P. Kissling, L. M. Peter, T. M. Watson, The effect of additional sulfur on solution-processed pure sulfide Cu_2ZnSnS_4 solar cell absorber layers, MRS Adv. 1 (2016) 2815–2820. https://doi.org/10.1557/adv.2016.425

[64] Z. Seboui, A. Gassoumi, N. Kamoun-Turki, Evolution of sprayed Cu_2ZnSnS_4, Mater. Sci. Semicond. Process. 26 (2014) 360–366. https://doi.org/10.1016/j.mssp.2014.05.004

[65] V. G. Rajeshmon, C. S. Kartha, K. P. Vijayakumar, C. Sanjeeviraja, T. Abe, Y. Kashiwaba, Role of precursor solution in controlling the opto-electronic properties of spray pyrolysed Cu_2ZnSnS_4 thin films, Sol. Energy 85 (2011) 249–255. https://doi.org/10.1016/j.solener.2010.12.005

[66] K. Tanaka, M. Kato, H. Uchiki, Effects of chlorine and carbon on Cu_2ZnSnS_4 thin film solar cells prepared by spray pyrolysis deposition, J. Alloys Compd. 616 (2014) 492–497. https://doi.org/10.1016/j.jallcom.2014.07.101

[67] S. Exarhos, K. N. Bozhilov, L. Mangolini, Spray pyrolysis of CZTS nanoplatelets, Chem. Commun. 50 (2014) 11366-11369. https://doi.org/10.1039/C4CC05162A

[68] M. Boshta, S. Binetti, A. Le Donne, M. Gomaa, M. Acciarri, A chemical deposition process for low-cost CZTS solar cell on flexible substrates, Mater. Technol. 32 (2016) 251-255. https://doi.org/10.1080/10667857.2016.1200837

[69] N. Kamoun, H. Bouzouita, B. Rezig, Fabrication and characterization of Cu_2ZnSnS_4 thin films deposited by spray pyrolysis technique, Thin Solid Films 515 (2007) 5949–5952. https://doi.org/10.1016/j.tsf.2006.12.144

[70] Y. B. Kishore Kumar, G. Suresh Babu, P. Uday Bhaskar, V. Sundara Raja, Preparation and characterization of spray-deposited Cu_2ZnSnS_4 thin films, Sol. Energy Mater. Sol. Cells 93 (2009) 1230–1237. https://doi.org/10.1016/j.solmat.2009.01.011

[71] A. G. Kannan, T. E. Manjulavalli, J. Chandrasekaran, Influence of Solvent on the Properties of CZTS Nanoparticles, Procedia Eng. 141 (2016) 15–22. https://doi.org/10.1016/j.proeng.2015.08.1112

[72] K. Rawat, H. J. Kim, P. K. Shishodia, Synthesis of Cu_2ZnSnS_4 nanoparticles and controlling the morphology with polyethylene glycol, Mater. Res. Bull., 77 (2016) 84–90. https://doi.org/10.1016/j.materresbull.2016.01.012

[73] O. Vigil-Galan, M. Courel, M. Espindola-Rodriguez, V. Izquierdo-Roca, E. Saucedo, and A. Fairbrother, Toward a high Cu_2ZnSnS_4 solar cell efficiency processed by spray pyrolysis method, *J. Renew. Sustain. Energy* 5 (2013) 0–14.

[74] S. K. Swami, N. Chaturvedi, A. Kumar, V. Dutta, Effect of deposition temperature on the structural and electrical properties of spray deposited kesterite (Cu_2ZnSnS_4) films, Sol. Energy 122 (2015) 508–516. https://doi.org/10.1016/j.solener.2015.09.027

[75] T. H. Nguyen, W. Septina, S. Fujikawa, F. Jiang, T. Harada, S. Ikeda, Cu_2ZnSnS_4 thin film solar cells with 5.8% conversion efficiency obtained by a facile spray pyrolysis technique, RSC Adv. vol. 5 (2015) 77565–77571. https://doi.org/10.1039/C5RA13000J

[76] S. M. Bhosale, M. P. Suryawanshi, M. A. Gaikwad, P. N. Bhosale, J. H. Kim, A. V. Moholkar, Influence of growth temperatures on the properties of photoactive CZTS thin films using a spray pyrolysis technique, Mater. Lett. 129 (2014) 153–155. https://doi.org/10.1016/j.matlet.2014.04.131

[77] P. Prabeesh, I. P. Selvam, S. N. Potty, Effect of annealing temperature on a single step processed Cu_2ZnSnS_4 thin film via solution method, Thin Solid Films 606 (2016) 94–98. https://doi.org/10.1016/j.tsf.2016.03.037

[78] Z. Wang, R. Gauvin, and G. P. Demopoulos, Nanostructural and photo-electrochemical properties of solution spin-coated Cu_2ZnSnS_4 –TiO_2 nanorod forest films with an improved photovoltaic performance, *Nanoscale* 9 (2017) 7650–7665. https://doi.org/10.1039/C7NR01422H

[79] S. N. Park, S. J. Sung, K. J. Yang, D. H. Kim, K. Y. Cho, J. K. Kang, Effect of Morphology of Solution-Processed Precursor Films on Cu_2ZnSnS_4 Thin Film Solar Cells, J. Nanosci. Nanotechnol. 16 (2016) 10758–10762. https://doi.org/10.1166/jnn.2016.13234

[80] W. Wang, H. Shen, X. He, J. Li, Effects of sulfur sources on properties of Cu_2ZnSnS_4 nanoparticles, J. Nanopart. Res. 16 (2014) 2437. https://doi.org/10.1007/s11051-014-2437-8

[81] M. Valdes, G. Santoro, M. Vazquez, Spray deposition of Cu_2ZnSnS_4 thin films, J. Alloys Compd. 585 (2014) 776–782. https://doi.org/10.1016/j.jallcom.2013.10.009

[82] J. H. N. Tchognia, Y. Arba, B. Hartiti, A. Ridah, J. M. Ndjaka, P. Thevenin, Effect of sulfurization time on properties of Cu_2ZnSnS_4 thin films obtained by sol–gel deposited precursors, Opt. Quantum Electron. 48 (2016) 1-7. https://doi.org/10.1007/s11082-016-0424-2

[83] Y. Yu, J. Ge, T. Prabhakar, Y. Yan, Effects of spin speed on the properties of spin-coated Cu_2ZnSnS_4 thin films and solar cells based on DMSO solution, in IEEE Photovoltaic Spec. Conf. 2014 (2014) 0448–0451.

[84] Q. Wu, C. Xue, Y. Li, P. Zhou, W. Liu, J. Zhu, S. Dai, C. Zhu, S. Yang, Kesterite Cu_2ZnSnS_4 as a Low-Cost Inorganic Hole-Transporting Material for High-Efficiency Perovskite Solar Cells, ACS Appl. Mater. Interfaces, 7 (2015) 28466–28473. https://doi.org/10.1021/acsami.5b09572

[85] V. G. Rajeshmon, C. S. Kartha, and K. P. Vijayakumar, Modification of optoelectronic properties of sprayed CZTS thin films through spray rate variation, *AIP Conf. Proc.* 1591 (2014) 1686–1688. https://doi.org/10.1063/1.4873077

[86] T. Arai, S. Tajima, S. Sato, K. Uemura, T. Morikawa, T. Kajino, Selective CO_2 conversion to formate in water using a CZTS photocathode modified with a ruthenium complex polymer, Chem. Commun. 47 (2011) 12664-12666. https://doi.org/10.1039/c1cc16160a

[87] Y. Yang, W. Que, X. Zhang, X. Yin, Y. Xing, M. Que, H. Zhao, Y. Du, High-quality Cu_2ZnSnS_4 and $Cu_2ZnSnSe_4$ nanocrystals hybrid with ZnO and $NaYF_4$:Yb, Tm as efficient photocatalytic sensitizers, Appl. Catal., B 200 (2017) 402–411. https://doi.org/10.1016/j.apcatb.2016.07.022

[88] S. S. Shinde, Photocatalytic degradation of RhB and TNT and photocatalytic water splitting with CZTS microparticles, J. Semicond. 36 (2015) 073003. https://doi.org/10.1088/1674-4926/36/7/073003

[89] P. S. Dilsaver, M. D. Reichert, B. L. Hallmark, M. J. Thompson, J. Vela, Cu_2ZnSnS_4-Au Heterostructures: Toward Greener Chalcogenide-Based Photocatalysts, J. Phys. Chem. C 118 (2014) 21226–21234. https://doi.org/10.1021/jp5062336

[90] E. Ha, L. Y. S. Lee, J. Wang, F. Li, K. Y. Wong, S. C. E. Tsang, Significant Enhancement in Photocatalytic Reduction of Water to Hydrogen by Au/Cu_2ZnSnS_4 Nanostructure, Adv. Mater. 26 (2014) 3496–3500. https://doi.org/10.1002/adma.201400243

[91] K. Kim, A. Razzaq, S. Sorcar, Y. Park, C. A. Grimes, S. I. In, Hybrid mesoporous Cu_2ZnSnS_4 (CZTS) – TiO_2 photocatalyst for efficient photocatalytic conversion of CO_2 into CH_4 under solar irradiation, RSC Adv. 6 (2016) 38964–38971. https://doi.org/10.1039/C6RA02763F

[92] S. A. Phaltane, S. A. Vanalakar, T. S. Bhat, P. S. Patil, S. D. Sartale, L. D. Kadam, Photocatalytic degradation of methylene blue by hydrothermally synthesized CZTS nanoparticles, J. Mater. Sci.: Mater. Electron. 28 (2017) 8186–8191. https://doi.org/10.1007/s10854-017-6527-0

[93] X. Yu, A. Shavel, X. An, Z. Luo, M. Ibanez, and A. Cabot, Cu_2ZnSnS_4-Pt and Cu_2ZnSnS_4-Au Heterostructured Nanoparticles for Photocatalytic Water Splitting and Pollutant Degradation, J. Am. Chem. Soc. 136 (2014) 9236–9239. https://doi.org/10.1021/ja502076b

[94] J. Wang, P. Zhang, X. Song, L. Gao, Cu_2ZnSnS_4 thin films: spin coating synthesis and photoelectrochemistry, RSC Adv. 4 (2014) 21318-21324. https://doi.org/10.1039/C4RA01139B

[95] X. Guo, J. Han, H. Zhang, X. Lv, J. Yan, R. Sun, Sol-gel synthesis of Cu_2ZnSnS_4 thin films under mild conditions, J. Alloys Compd., 697 (2017) 361–366. https://doi.org/10.1016/j.jallcom.2016.12.122

[96] I.G. Orletskyi, M.M Solovan, V.V Brus, F. Pinna, G. Cicero, P.D. Maryanchuk, E.V. Maistruk, M.I. Ilashchuk, T.I Boichuk, E. Tresso, Structural, optical and electrical properties of Cu_2ZnSnS_4 films prepared from a non-toxic DMSO-based sol-gel and synthesized in low vacuum, J. Phys. Chem. Solids 100 (2017) 154–160. https://doi.org/10.1016/j.jpcs.2016.09.015

[97] Z. Su, W. Li, G. Asim, T. Y. Fan, L. H. Wong, Cation substitution of CZTS solar cell with > 10% efficiency, in IEEE Photovoltaic Spec. Conf. 43rd 2016 (2016) 0534–0538.

[98] Y. Yang, X. Kang, L. Huang, S. Wei, D. Pan, A general water-based precursor solution approach to deposit earth abundant $Cu_2ZnSn(S,Se)_4$ thin film solar cells, J. Power Sources 313 (2016) 15–20. https://doi.org/10.1016/j.jpowsour.2016.01.085

[99] L.S. Khanzada, I. Levchuk, Y. Hou, H. Azimi, A. Osvet, R. Ahmad, M. Brandl, P. Herre, M. Distaso, R. Hock, W. Peukert, Effective Ligand Engineering of the Cu_2ZnSnS_4 Nanocrystal Surface for Increasing Hole Transport Efficiency in Perovskite Solar Cells, Adv. Funct. Mater. 26 (2016) 8300–8306. https://doi.org/10.1002/adfm.201603441

[100] J. Cheng, Z. Dai, B. Chen, R. Ji, X. Yang, R. Hu, J. Zhu, L. Li, Well-Dispersed Cu_2ZnSnS_4 Nanocrystals Synthesized from Alcohols and Their Applications for Polymer Photovoltaics, Nanoscale Res. Lett. 11 (2016) 550. https://doi.org/10.1186/s11671-016-1761-6

[101] S. N. Park, S. J. Sung, K. J. Yang, J. K. Kang, D. H. Kim, Low-Cost Nanoporous Cu_2ZnSnS_4 Thin-Film Counter Electrode for Dye-Sensitized Solar Cells, J. Nanosci. Nanotechnol. 16 (2016) 10490–10494. https://doi.org/10.1166/jnn.2016.13182

Chapter 15

Modeling and Optimization of Photocatalytic Degradation Process of 4-Chlorophenol using Response Surface Methodology (RSM) and Artificial Neural Network (ANN)

Pritesh S. Patel[1], Vimal Gandhi [1*], Mihir P. Shah[1], Thillai Sivakumar Natarajan[2], K. Natarajan[2], R. J. Tayade[2*]

[1]Department of Chemical Engineering, Dharmsinh Desai University, College Road, Nadiad- 387 001, Gujarat, India

[2]Inorganic Materials and Catalysis Division, Central Salt and Marine Chemicals Research Institute (CSMCRI),), G. B. Marg, Bhavnagar-364 021, Gujarat, India.

E-mail address: rtayade@gmail.com,vggandhi.ch@ddu.ac.in

Abstract

Present study focuses on modeling and optimization of photocatalytic degradation of 4-Chlorophenol (4-CP) using Response Surface Methodology (RSM) and Artificial Neural Network (ANN). Titanium nanotube synthesized by hydrothermal method were used and characterized using various physico-chemical and electronic techniques. The influence of operational parameters was investigated by employing face centred experimental design.. Response surface model was developed and its significance was evaluated by an ANOVA study. This process was also modeled by a novel approach of ANN. A two layer feed forward neural network with back propagation algorithm was employed. Architecture of ANN was optimised based on fractional factorial design. Optimum parameters were found to be eight hidden layer neurons, *tansig* transfer function in the hidden as well as output layer. Optimum conditions were found using the response surface model with initial concentration of 40 mg/L, catalyst dose of 184.84 mg/L, and initial pH of 3.94. From the ANN model, the optimum conditions were found to be with initial concentration of 42 mg/L, catalyst dose of 212.51 mg/L, and initial pH of 3.38. These optimum conditions were experimentally verified and reasonably good agreement was found between predicted and experimental conditions.

Keywords

Photocatalytic Degradation, 4-Chlorophenol (4-CP), Response Surface Methodology (RSM), Artificial Neural Network (ANN)

Contents

1. Introduction

Heterogeneous photocatalysis is considered as multivariate process as many operational parameters such as catalyst dose, initial concentration, degradation time, pH of solution, light intensity, reactor configuration, and temperature etc., can affect the degradation of the organic compound [1-3]. Difficulty in solving the equation of radiant energy balance, spatial distribution of absorbed light, mass transfer and complicated mechanism of photocatalytic processes compel us to think about modeling techniques based on statistical regression and artificial intelligence [3, 4]. From optimization and reactor design viewpoints, it is essential to evaluate the effect of such operational parameters. Using a conventional approach of one-factor-at-a-time (OFAT) is a complex method to evaluate the effects of different parameters on an experimental outcome. This approach assesses one factor at a time while keeping others at hold values instead of all simultaneously. This method is time-consuming, expensive and often leads to misinterpretation of results when interactions between different parameters are present [5-8]. Another approach is Response Surface Methodology (RSM) based on statistical design of experiments in which all parameters are varied simultaneously in a systematic manner [7-9]. This approach was adopted by many researchers to model and optimize the photocatalytic degradation of broad range of organic compounds [4, 7-13].

Artificial Neural Networks (ANNs), the fastest growing facet of artificial intelligence [14-15], are widely used in the field of engineering and science for modeling, prediction, classification, pattern-recognition, and process control [16-23]. ANNs are computer based methods to simulate learning processes of the human nervous system. The learning ability of ANNs is different from other modeling techniques. Unique advantages such as no requirement of prior phenomenological knowledge about the system being modeled and any mathematical relationship favor their usage [3-4, 14-17, 24-25]. Many authors have shown application of ANN as modeling tool for various Advanced Oxidation Processes (AOPs) [3-4, 26-33].

In this study, a photocatalytic degradation process of 4-CP was modeled and optimized by RSM as well as ANN. 4-CP has been reported as environmental persistent pollutant in various literature for its acute toxicity, non-biodegradability, and carcinogenic properties [34-39]. First of all titanium dioxide nanotubes (TNTs) were fabricated by hydrothermal method and characterised by X-ray Diffraction (XRD), N_2-Sorption study, Scanning Electron Microscopy (SEM) and Transmission Electron Microscopy (TEM). The influence of various operational parameters was investigated by conducting statistical experimental design. Catalyst dose, initial concentration, and initial pH were taken as parameters of interest. A three factor three level face centred design (FCD) was employed. Experimental results were further used by RSM and ANN for model development. Parametric optimization was carried out using both models. A reasonably well agreement was found between experimental and predicted optimum conditions.

2. Experimental methodology

2.1 Chemicals and materials

Titanium dioxide (pure anatase) was procured from Aldrich Chemical Company, Inc. Sodium hydroxide (98%) pellets were purchased from Qualigens Chemicals Limited, Mumbai. 4-CP was procured from Ranbaxy Fine Chemicals Limited, New Delhi. Ultrapure-water (18.5 MΩcm) was used for preparation of solution obtained from Milli-Q water purification system. HPMV UV Lamp 125 W was obtained from Crompton Greaves, New Delhi. Irradiated samples were analysed by Cary 500 UV-Vis Spectrophotometer (Varian, Palo Alto, CA). Absorbance wavelengths of 225 nm was used for analysis for 4-CP.

2.2 Synthesis of catalyst

Mixture of 10 M NaOH (50 ml) and TiO_2 powder (1.2 g) was placed into a 250 ml beaker and stirred for 30 min using a magnetic stirrer, followed by 5 min of sonication for ensuring well dispersed suspension of particles. After that the mixture was transferred to a teflon lined stainless steel autoclave and put into a hot oil bath preheated at 130 °C for hydrothermal heating. Reaction was carried out at a speed of 250 rpm for a time duration of 48 hr. After the hydrothermal process, white-colored precipitate was obtained and the precipitate was repetitively washed with deionized water until the solution pH became 7. The removal of Na^+ ions on the surface of and inside the obtained precipitated particles was carried out by an ion-exchanging process in 0.1 N HCl solution for 12 hours. After the ion-exchanging process, the precipitate was washed again with deionised water until pH came to 7, followed by drying at 70 °C for 12 h in an oven. Then, the dried precipitates were calcined at 250 °C for 2 h under air atmosphere in a tubular furnace [40]. Finally prepared TNTs were characterised by PXRD, N_2-Sorption Study, SEM and TEM analysis.

2.3 Characterisation of catalyst

Powder X-ray diffraction (XRD) was used for crystal phase identification. Powder X-ray diffraction patterns were recorded with Phillips X'pert MPD system using $CuK\alpha_1$ radiation (λ= 0.154056 nm). Diffraction patterns were measured in 2θ ranges 5-80° at a scan rate of 0.1° sec^{-1}. The crystallite size was estimated from characteristic peak of 2θ = 25.3° by the Scherrer formula (Eq. (1)), with shape factor (K) of 0.9.

$$\text{Crystallite size} = K\lambda / (W \cos \theta) \quad (1)$$

Where, $W = W_b - W_s$, W_b = broadened profile width of experimental catalyst, W_s = standard profile width of reference silicon sample, and λ is the wavelength of X-ray radiation ($CuK\alpha_1 = 0.15405$ nm).

A quantitative study of the specific surface area, pore volume, and pore size distribution was performed using a Micromeritics ASAP 2010, USA static volumetric adsorption system by nitrogen adsorption-desorption isotherms at 77.4 K. Prior to nitrogen adsorption, the samples were degassed at 373 K under vacuum (0.005 mmHg). Surface area and pore size distribution were determined using the BET equation and BJH method respectively. The surface morphology of the catalysts was determined with scanning electron microscope (SEM) Leo Series 1430 VP equipped with INCA, Energy Dispersive System (EDX). The catalysts were supported on aluminium stubs using silver paint and then sputter coated with gold (by using Polaris Sputter Coater, Model Polaron SC7620, Quantum Technologies). The surface morphology was further confirmed by transmission electron microscope (TEM) using JEOL JEM-2010 Electron microscope. Samples were prepared by dispersing the catalyst powders in spectrosol and allowing a drop of the resultant suspension to dry on a carbon support film covering a standard copper grid.

2.4 Experimental set up

In this study an annular-slurry batch reactor was employed. Main components of the reactor are outer glass vessel, quartz jacket (double walled quartz tube), UV lamp, wooden box, cooling system, magnetic stirrer. Various components of the reactor are shown in Fig. 1. UV lamp is kept inside a jacketed glass shield. Cooling water is being circulated through the jacket to maintain a temperature of 20 °C. The void annular space is utilised for reaction contents. A magnetic stirrer is used to keep the catalyst suspended in the medium.

2.5 Experimental procedure

Prior to commencing illumination, a suspension containing required amount of the catalyst and 300 ml aqueous solution of 4-CP was put into the glass vessel of the reactor as shown in Fig. 1 and sonicated for 5 min. 0.05 N HCl or NaOH was used to adjust the pH of the solution. The prepared suspension was continuously stirred for 1 h in dark condition for establishing adsorption desorption equilibrium of organic compound on the catalyst. At regular intervals of time 5 ml of reaction aliquots were withdrawn by syringe from the irradiated suspension. Total course of reaction was 4 hours. For the first 1 hour, the sampling time interval was 10 minute and for the remaining 3 hours samples were collected at interval of 1 hr. The catalyst was separated by centrifuge from the reaction samples prior to analysis. The concentration of residual 4-CP in the reaction mixture was determined by Spectrophotometry. A wavelength of 225 nm was used as characteristic wavelength for UV-Vis Spectrophotometry. Percentage degradation was calculated using Eq. (2).

$$\% \text{ Degradation} = \left(\frac{C_0 - C}{C_0}\right) \cdot 100 \quad (2)$$

Where, C_0 is the initial concentration and C is the concentration at any time.

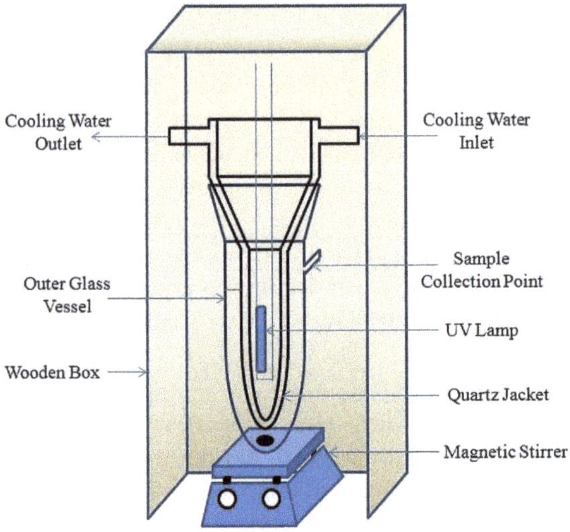

Figure 1. Annular-slurry photoreactor.

3. Modeling of photocatalytic degradation of 4-CP

3.1 Response surface model development

The RSM involves the following steps: (1) design of experiment (2) performing statistically designed experiments; (3) estimating the coefficients of a mathematical model using regression analysis technique; (4) predicting the responses and checking the adequacy of the model, and (5) optimization of the process using a mathematical model.

In this study, the central composite design (CCD) was used for the response surface modeling and optimization of the photodegradation process. The independent variables taken into consideration were as follows: initial concentration of 4-CP solution, amount of TNT, and initial pH of the solution. Percentage degradation of 4-CP after 4 h of irradiation time was taken as dependent variable. A three factor-three level face centred design (FCD), having eight cubic points, six axial points (star points)(α=1), and four replicates at centre point, comprising of a set of eighteen experiments, was adopted for obtaining response surface second order model to optimize photocatalytic degradation of 4-CP using TNTs. Three levels of each factor corresponded to their low, middle or centre, and high values. The experimental range and levels of independent variables are listed in Table 1. Due to the large difference in the order of magnitude of the value of independent factors, each factor was coded in between -1 and 1 referring to its low and high value respectively.

$$X_c = 2\left(\frac{x - x_{low}}{x_{high} - x_{low}}\right) - 1 \qquad (3)$$

Table 1. Experimental Ranges and Levels of Independent Variables.

	Independent Variable	Level		
		Low (-1)	Centre (0)	High (1)
X1	Initial Concentration (mg/L)	40	70	100
X2	Catalyst Dose (mg/L)	33.33	133.33	233.33
X3	Initial pH	2	6	10

Eq. (3) was used for coding and decoding of factors, where X is the actual factor value and X_c is the coded factor value. Table 2 shows a set of experiments that has been designed using FCD. Design of experiments was done using Minitab 16.0.1 (Minitab Inc., State College, PA). Each factor was shown in its coded value. The experiments were performed in a random manner in order to avoid any systematic bias in the outcome.

After completion of all experiments, the second order polynomial empirical relationship as given by Eq. (4) was fitted to the experimental data.

$$y = \beta_0 + \sum_{i=1}^{n} \beta_i x_i + \sum_{\substack{i=1 \\ j=i+1}}^{n} \beta_{ij} x_i x_j + \sum_{i=1}^{n} \beta_{ii} x_i x_i \qquad (4)$$

Where, n is number of levels, y is response variable, x_i indicates independent variables, and β_0, β_i, β_{ii}, β_{ij} are coefficients of intercept, linear, interaction, and quadratic terms respectively. The coefficients of the quadratic polynomial model were calculated by a multiple regression analysis on the experimental data. The statistical significance and adequacy of model and its coefficients were analyzed using the analysis of variance (ANOVA) using Minitab 16. ANOVA was followed by a Fischer F-test to check significance of model and its coefficients. A significant level of $P < 0.05$ (confidence level - 95 %) was used in all tests. After that insignificant terms were removed from the quadratic model and surface plots were drawn to visualize the effect of the operational parameters.

Table 2. Design Matrix for RSM.

Runs	Point Type	Independent Variable: Initial Concentration (mg/L)	Catalyst Dose (mg/L)	Initial pH
1	Cubic Point	-1	-1	-1
2		1	-1	-1
3		-1	1	-1
4		1	1	-1
5		-1	-1	1
6		1	-1	1
7		-1	1	1
8		1	1	1
9	Axial or Star Point	+α	0	0
10		- α	0	0
11		0	+α	0
12		0	- α	0
13		0	0	+α
14		0	0	- α
15	Centre Point	0	0	0
16		0	0	0
17		0	0	0
18		0	0	0
-1= low value, 0= centre value, 1= high value, α= star or axial point value(1)				

3.2 Artificial neural network model development

The ANN modelling involves following steps: assembling and pre-processing of data, selection of ANN architecture, optimisation of ANN architecture, and external validation of ANN model. In this study, MATLAB 7.10.0 (R2010a) (The MathWorks Inc.) was used as platform for necessary calculations and data processing.

ANN models are purely data driven models, they require a representative data set consisting of independent and dependent variables of the system under investigation. Initial concentration of reaction solution, catalyst dose, initial pH of the solution, and time were used as inputs (independent variables) to ANN and percentage degradation was taken as output (dependent variable) of ANN. The results of the experiments that were performed during response surface modeling were used in the ANN modeling. Here the only difference is that time is one of the independent variables. To enhance the reliability of the ANN models, pre-processing should be applied on the ANN input and output data in prior before using them in the design of the ANN models. Due to the large difference in the order of magnitude of the value, the available datasets were normalised into 0.2–0.8 interval using MATLAB intrinsic normalisation function: *mapminmax* in order to avoid solution divergence [3, 16]. So scaling of data value (x_i) to new normalised value (x_{ni}) follows

$$X_{ni} = 0.6\left(\frac{X_i - X_{min}}{X_{max} - X_{min}}\right) + 0.2 \quad (5)$$

Correspondingly, ANN model outputs were also denormalised to actual values by using Eq. (5) inversely. In addition to more accurately evaluate the generalization and robustness of ANN models in the present work, all input-output data samples were divided into three sets as training set, validation set, and testing set using MATLAB intrinsic function: *dividerand* with the size ratio of 70:15:15%.

Network architecture includes defining numbers of inputs and outputs, numbers of hidden layers, numbers of hidden neurons in each hidden layer, activation transfer function in each layer, and suitable training function. However, according to the universal approximation theory [15], an ANN with a single hidden layer with proper number of hidden neurons can map any nonlinear function to any desired accuracy. In this work, the ANN architecture employed was a two layer feed forward neural network with back propagation (BPNN). In the ANN model key parameters, including number of neurons in the hidden layer, activation transfer function in the hidden as well as output layer play a crucial role in establishing a good ANN regression model with high precision accuracy and stability. Optimization of ANN is a process of finding the right combination of these parameters that yields best prediction performance. It involves creation of neural network objects by taking appropriate value of parameters, training using suitable training function, and evaluation of performance of neural networks. In this work, Levenberg-Marquardt training algorithm (trainlm) was used due to its advantage of faster convergence [16]. Normalised mean square errors as of training and validation steps, given by Eq. (6), were taken as evaluation parameters to measure the prediction performance ANN model.

$$MSE = \frac{1}{n}\sum_{i=1}^{n}(y_i - y_i^{NN})^2 \quad (6)$$

Where y_i is actual normalized output and y_i^{NN} is neural network predicted normalized output. So above selected architecture was optimized by conducting fractional factorial design. Table 3 depicts all parameters that need to be optimized and their range. All possible combinations of ANN architecture were tabulated in Table 4.

Table 3. ANN Optimization: Parameters and Their Levels.

Independent parameter	Level
Number of hidden layer neurons	2, 4, 6, 8, 10,12
Transfer function in hidden layer	tansig, logsig
Transfer function in output layer	tansig, purelin

Table 4. ANN Optimization: Fractional Factorial Design – Design Matrix.

Run	Hidden layer Neurons	Hidden layer Transfer function	Output layer Transfer function
1	Tansig	tansig	2
2	Tansig	tansig	4
3	Tansig	tansig	6
4	Tansig	tansig	8
5	Tansig	tansig	10
6	Tansig	tansig	12
7	Tansig	purelin	2
8	Tansig	purelin	4
9	Tansig	purelin	6
10	Tansig	purelin	8
11	Tansig	purelin	10
12	Tansig	purelin	12
13	Logsig	purelin	2
14	Logsig	purelin	4
15	Logsig	purelin	6
16	Logsig	purelin	8
17	Logsig	purelin	10
18	Logsig	purelin	12

4. Optimization of photocatalytic degradation of 4-CP

Both response surface model and ANN model of photocatalytic degradation of 4-CP were optimised subsequently. Here the primary goal was to find the optimum values of independent variables occurring in quadratic equation that would maximize the response. Minitab inbuilt response optimizer was used to optimize the response surface model while the genetic algorithm toolbox in MATLAB was used to optimize the ANN model. Genetic algorithm always perform a minimization task so to maximize the percentage degradation the objective function was slightly modified as,

$$\max f(x) = -\min(-f(x)) \quad (7)$$

5. Results and discussion

5.1 X-ray powder diffraction and N_2 sorption study

The photocatalytic activity of TNTs depends on the crystallinity and crystal structure. Fig. 2a shows XRD pattern of synthesized TNTs. It can be seen that TNTs are highly crystalline in the 2θ range of 20-80^0. Characteristic peak of anatase phase at $2\theta = 25.3^0$ was observed which was giving indication of the presence of an anatase phase in TNTs. The crystallite size of TNTs was found to be 32 nm. The N_2 sorption study demonstrates TNTs follow type II isotherm as shown in Fig. 2b. The BET surface area was found to be 185.24 m^2/g.

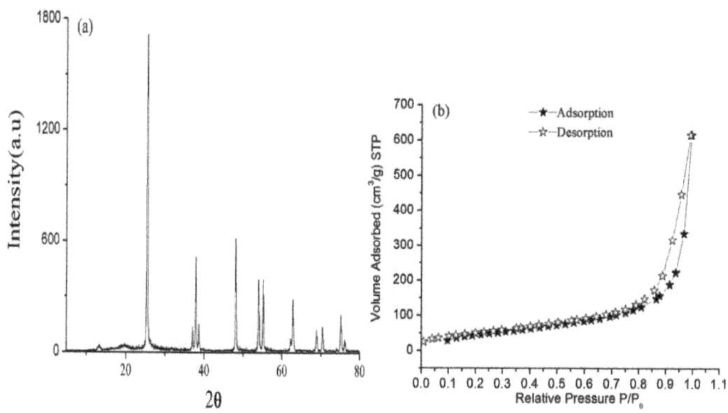

Figure 2. (a) XRD pattern of TNTs (b) N_2 adsorption-desorption isotherm.

5.2 SEM and TEM analysis

SEM and TEM images are depicted in Fig. 3. Both have confirmed nanotubular morphology of synthesized material. Diameter and wall thickness of TNT were found to be 8.3 nm and 2 nm respectively from Fig. 3c. Fig. 3d confirms that ends of TNTs are open.

5.3 Photocatalytic activity of synthesized TNTs

Photocatalytic activity of TNTs was evaluated by carrying out degradation of 4-CP solution (40 mg/L) under UV irradiation by taking 133.33 mg/L of TNTs. Degradation reactions were carried out under the following conditions: (i) 4-CP solution in absence of photocatalyst under UV irradiation (Photolysis) (ii) reaction mixture of 4-CP solution and

TNTs under UV irradiation (Photocatalysis) (iii) reaction mixture of 4-CP solution and TNTs in absence of UV irradiation (Dark Reaction). In above all cases, reactions were carried out at natural pH of 6.5. Fig. 4 shows percentage degradation of 4-CP as function of time under different conditions mentioned above. From Fig. 4, it can be said that a catalyst alone can remove only around 15% of 4-CP by physical adsorption while UV radiation alone can remove only 45% of 4-CP by photolysis. But using catalyst in presence of UV light was capable of removing 90% of 4-CP by photocatalysis in same time of irradiation. An increase in degradation of 4-CP can be attributed to photoactivity of synthesized TNTs.

Figure 3. (a) SEM image of TNTs (b), (c), (d) TEM images of TNTs.

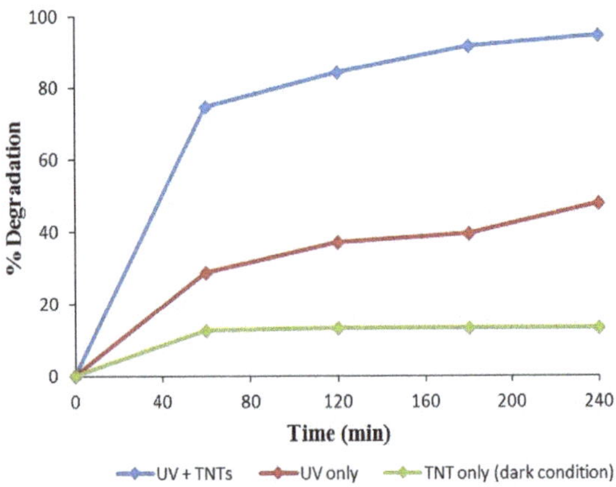

Figure 4. Photocatalytic Activity of Synthesized TNT; 40 mg/L of initial concentration of 4-Chlorophenol, 133.33 mg/L of catalyst dose, 6 pH.

5.4 Response surface model development and optimization

5.4.1 Fitting quadratic model and its ANOVA study

A quadratic polynomial model was fitted to experimental data using multiple linear regressions. An adequacy and statistical significance of developed model and its coefficients were evaluated by conducting analysis of variance (ANOVA). The fitted quadratic model to experimental data as coded unit was given by Eq. (8).

$$y = 84.8044 - 7.6810x_1 + 1.3787x_2 - 1.9736x_3 - 0.6995x_1x_1 - 5.4124x_2x_2 - 2.5769x_3x_3 - 4.0076x_1x_2 + 0.4483x_1x_3 - 0.4939x_2x_3 \quad (8)$$

Where, y is percentage degradation of 4-CP, and x_1, x_2, x_3 are initial concentration in mg/L, catalyst dose in mg/L, and initial pH in coded units respectively. A comparison between experimental and predicted values (by response surface model) of percentage degradation has been shown in Table 5. Analysis of variance (ANOVA) showed that the quadratic model predictions were reasonably correlated with experimental values in factor space analyzed in this study as shown in Table 5 with the coefficient of multiple determination (R^2-value) of 0.99, which meant that the quadratic model was able to manifest 99% of total variability. Here adjusted R^2-value of 0.9788 was also in good agreement with the R^2-value, which was indicative of good predictability of the model. A significant level of $P < 0.05$ (confidence level - 95 %) was used as evaluating criterion. For any of the terms in the models, a large F-value (small P -value) would indicate a more significant effect on the respective response variables. The Model F-value of 88.36 implies the model is significant. There is only a 0.01% chance that a "Model F-Value" this large could occur due to noise. Numerical values of ANOVA for quadratic model were tabulated in Table 6. All terms appearing in this quadratic model have P- value less than 0.05, hence it can be said that model doesn't carry unnecessary terms. The F-Value of Lack of Fit of model was found to be 7.19, which implied that Lack of Fit of model was insignificant relative to pure error. The P- value of Lack of Fit was 0.068, which was greater than 0.05, which also proves adequate fit of the model. Table 7 shows results of ANOVA for regression coefficients for quadratic model. ANOVA shows that all linear terms, initial concentration, catalyst dose, and initial pH of solution have higher significant effect (for all linear terms P- values < 0.05) on percentage degradation of 4-CP. In addition to that quadratic terms of catalyst dose and initial pH as well as interaction terms of initial concentration and catalyst dose are found to be having significant influence on the response. Insignificant terms are removed from the quadratic polynomial model and used for further analysis.

Table 5. Numerical Values of Experimental and Predicted Response by Fitted Quadratic Model.

Runs	Independent variable			Response variable	
	Initial concentration (mg/L)	Catalyst Dose (mg/L)	Initial pH	% Degradation	
				Experimental	Predicted
1	-1	-1	-1	80.1637	80.33831
2	1	-1	-1	72.9985	72.09488
3	-1	1	-1	92.2475	92.09853
4	1	1	-1	68.1698	67.8248
5	-1	-1	1	76.4523	76.48236
6	1	-1	1	70.1981	70.03213
7	-1	1	1	85.6784	86.26708
8	1	1	1	64.2761	63.78655
9	+α	0	0	92.4303	91.78592
10	- α	0	0	74.5198	76.42394
11	0	+α	0	77.1484	78.01332
12	0	- α	0	80.3758	80.77064
13	0	0	+α	82.9781	84.20108
14	0	0	- α	80.2172	80.25398
15	0	0	0	86.1548	84.80442
16	0	0	0	85.4278	84.80442
17	0	0	0	85.1984	84.80442
18	0	0	0	84.9562	84.80442
R2 = 0.99 , Adj. R2 = 0.9788 , 1= low value, 0 = centre value, 1= high value, α = star or axial point value (=1)					

Table 6. ANOVA and Lack of Fit for Fitted Quadratic Model.

Sources	DF	Seq. SS	MS	F- Value	P-Value	Remark
Model	9	1037.37	115.26	88.36	0.0001	Significant
Linear Terms	3	647.93	215.98	165.58	0.0001	Significant
Quadratic Terms	3	257.39	85.80	65.78	0.0001	Significant
Interaction Terms	3	132.04	44.02	33.74	0.0001	Significant
Res. Error	8	10.44	1.30			
Lack of Fit	5	9.63	1.93	7.19	0.068	Insignificant
Pure Error	3	0.80	0.27			
Total Error	17	1047.81				
Significant terms are indentified by ANOVA at 95% confidence interval DF = degree of freedom, Seq. SS= sequential sum of square of errors, MS= mean square errors						

Table 7. ANOVA of Regression Coefficients of Fitted Quadratic Equation.

Variable	Regression Coefficient	F-Value	P- Value	Remark
Intercept	84.8044	-	-	
Linear				
X1	-7.6810	452.30	0.0001	Significant
X2	1.3787	14.57	0.005	Significant
X3	-1.9736	29.86	0.001	Significant
Quadratic				
X11	-0.6995	1.02	0.343	
X22	-5.4124	60.85	0.0001	Significant
X33	-2.5769	13.79	0.006	Significant
Interaction				
X12	-4.0076	98.50	0.0001	Significant
X13	0.4483	1.23	0.299	
X23	-0.4939	1.50	0.256	
Significant factors are indentified by ANOVA at 95% confidence interval				

5.4.2 Influence of key operational parameters

In this study, effect of key operational parameters like initial concentration of 4-CP, catalyst dose, and initial pH of solution were investigated. For visualization the main effect of individual factors and two factor interaction, main effect plots and surface plots were drawn respectively. The main effect plot of operational parameters has been depicted in Fig. 5.

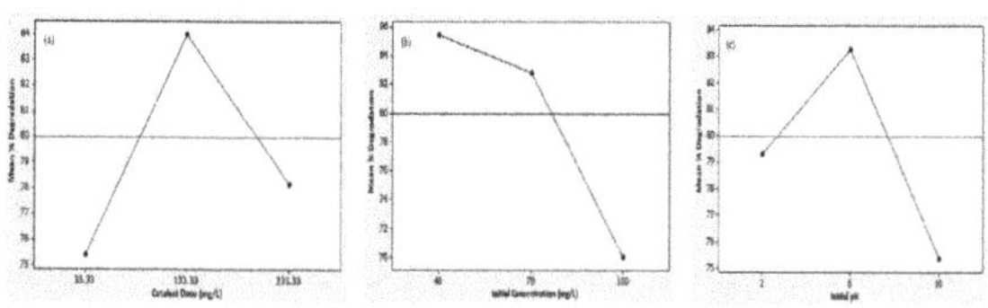

Figure 5. Main Effects of Operational Parameters : % Degradation Vs (a) Effect of Catalyst Dose (b) Effect of Initial Concentration (c) Effect of Initial pH.

The slope of the main effect curve is proportional to the size of the effect and the direction of the curve specifies a positive or negative influence of the effect [13]. It was observed that the photocatalytic degradation rate initially increases with catalyst loading and then decreases at high values because of light scattering and screening effects. This study revealed that increasing catalyst concentration from 33.33 mg/L to 133.33 mg/L increased degradation performance from 75% to 84% initially but further increase in catalyst dose from 133.33 mg/L to 233.33 mg/L decreased mean percentage degradation from 84% to 78% (Fig. 5a). Although the number of active sites in solution will increase with increase in catalyst loading, a point appears to be reached where light penetration is compromised because of an increased turbidity of mixture. A further increase in catalyst loading beyond the optimum will result in non-uniform light intensity distribution, so that the degradation rate would indeed be lower with increased catalyst dosage. Another reason for lower degradation rate at high catalyst dose could be the tendency towards agglomeration (particle – particle interaction) at high solids concentration, resulting in a reduction in surface area available for light absorption. Initial concentration of substrate always plays an important role in the photocatalytic process. It was noted that the degradation performance decreases with an increase in substrate concentration. As shown in Fig. 5b, by increasing 4-CP concentration from 40 mg/L to 70 mg/L, the mean percentage degradation slightly reduced from 85% to 82%. But sudden reduction in mean percentage degradation from 82% to 70% was found when initial concentration was increased from 70 mg/L to 100 mg/L. This can be attributed as an increase in initial concentration leads to more and more absorption of UV light by organic compound molecules and hence the net flux of light reaching to photocatalyst surface reduces. In turn this effect decreases numbers of electron-hole pairs responsible for the photocatalytic reaction. Another reason can be given as an increase in initial concentration of substrate results into more and more adsorption on the catalyst surface and requirement of $OH^{\cdot}$ and $O_2^{\cdot-}$ radicals needed for degradation also increases. Due to increased adsorption, the surface available for photon absorption decreases and results into a lesser number of $OH^{\cdot}$ and $O_2^{\cdot-}$ radicals than the actual requirement and hence degradation performance decreases. Photocatalytic degradation of 4- CP is accompanied by generation of several intermediates. An increase in substrate concentration leads to generation of some of the intermediates which may be adsorbed onto the surface of the photocatalyst. Slow diffusion of these intermediates from catalyst surface can result in deactivation of the active sites of photocatalyst and consequently reduction in degradation rate. The pH of an aqueous solution determines the surface charge of the photocatalyst and the size of aggregates it forms. In this study we found that the degradation rate of 4-CP was higher in acidic medium than alkaline medium. As pH of the solution of 4-CP was increased from 2 to 6, its degradation performance increases form 79% to 83%. While an increase in pH from 6 to 10 decreased mean percentage degradation from 83% to 76 % (Fig. 5c). While in basic medium 4-CP molecules dissociates and forms anion and hence due to negative surface charge, electrostatic repulsion decreases adsorption of 4-CP on photocatalyst which resulted into decreases in degradation of 4-CP in basic condition. In acidic condition 4-CP molecule does not dissociate. But in hydroxyl functional group of 4-CP, a partial negative charge is developed on the Oxygen atom because of its high electronegativity. The electrostatic attraction between partial negatively charged 4-CP molecules and positively charged catalyst surface results into more and more adsorption onto the surface of photocatalyst and subsequently increase degradation rates.

To visualize the effect of the interaction of independent variables on the dependent ones, using quadric polynomial model surface and corresponding contour plots were drawn by varying two of the independent variables within the experimental range while holding the other factors at their central values. Fig. 6 shows interaction effect of initial concentration and catalyst dose on percentage degradation at hold value of pH of 6. Curvature can be seen in the surface plot which implies significant interaction between initial concentration and catalyst dose. This revealed that an increase in catalyst dose at lower concentration favours the degradation performance while an increase in catalyst dose at higher concentration is detrimental to degradation performance in given factor space. From Fig. 6, it can be said that the optimum condition should lie in the rage of 120-220 mg/L for catalyst dose and 40-50 mg/L for initial concentration of solution. Fig. 7 shows interaction effect of catalyst dose and pH on percentage degradation at hold value of initial concentration of 70 mg/L. The surface can be seen curvy which indicates strong interaction between catalyst dose and pH. This shows that maximum degradation performance can be obtained at intermediate catalyst dose in an acidic medium. From Fig. 7, it can be said that optimum condition should lie in the rage of 100-200 mg/L for catalyst dose and 2-5 for initial pH of the solution. Fig. 8 shows interaction effect of pH and initial concentration on percentage degradation at hold value of catalyst dose of 133.33 mg/L. The surface can be seen less curvy which indicates weak interaction between pH and initial concentration. It implies that lower initial concentration and acidic condition favour the degradation performance. From Fig. 8, it can be said that optimum condition should lie in the rage of 40-50 mg/L for initial concentration and 2-4 for initial pH of the solution.

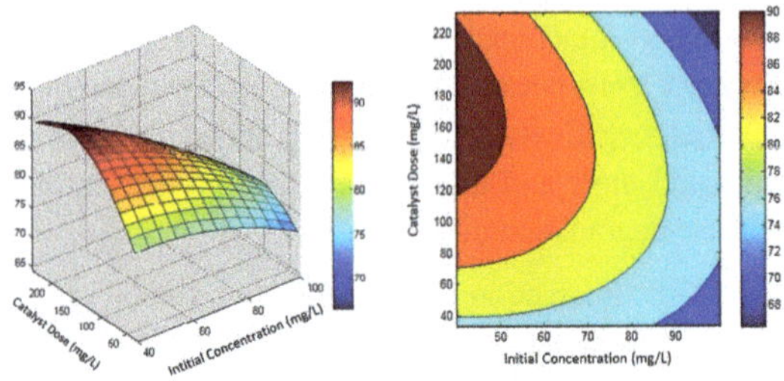

Figure 6. Interaction Effect of Initial Concentration and Catalyst Dose at Initial pH 6, (a) Surface (b) Contour Plot.

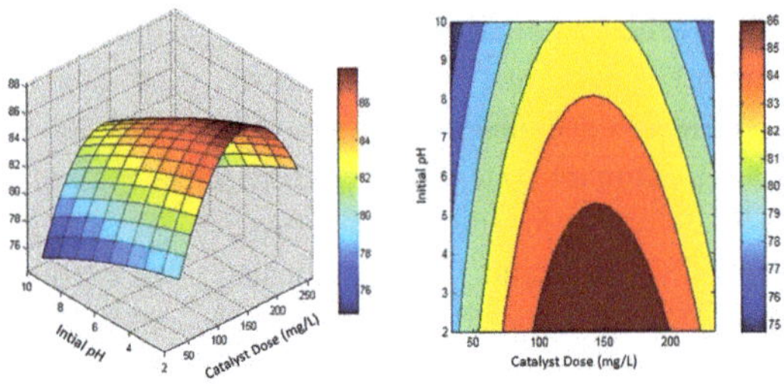

Figure 7. Interaction Effect of Initial ph and Catalyst Dose at Initial Concentration 70 mg/L, (a) Surface (b) Contour Plot.

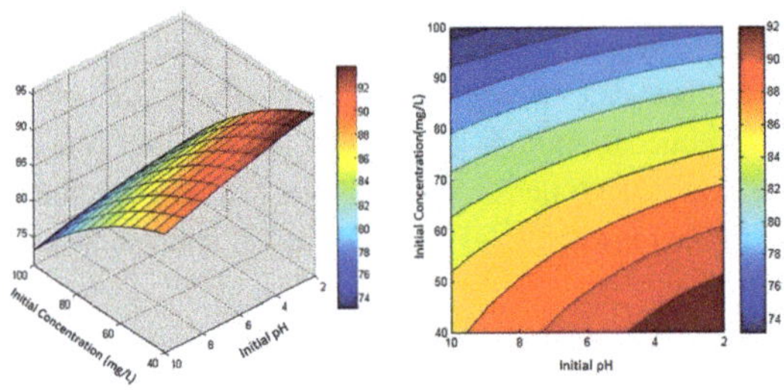

Figure 8. Interaction Effect of Initial ph and Initial Concentration at Catalyst Dose of 133.33 mg/L, (a) Surface (b) Contour Plot.

5.4.3 Optimization using response surface model

Here the primary goal was to find out the optimum values of independent variables occurring in quadratic equation that would maximize the response. The response surface model (Eq. (8)) was optimized for being used in a built response optimizer in Minitab 16. The results of optimization were tabulated in Table 8. Optimum parameters were found in coded unit so later on it is decoded using Eq. (3). These optimum parameters were also verified experimentally. A reasonably good agreement was found between experimental and predicted response.

Table 8. Response Surface Model: Optimum Parameters.

Parameter	Coded value	Decoded value	Experimental Verification
Initial Concentration (mg/L)	-1	40	40
Catalyst Dose (mg/L)	0.5151	184.84	185
Initial pH	-0.5150	3.94	4
% Degradation (Pred.)	93.82	93.82	92.23

5.5 Artificial neural network model development and optimization

5.5.1 Optimization of ANN architecture

As per fractional factorial design matrix as shown in Table 4, neural network objects having each possible combination of parameters were created, trained, and evaluated in terms of prediction performance. Above procedure for each neural network object was repeated several times to avoid random correlation due to the random initialization of the weights. Numerical values of normalised training error and validation error were tabulated in Table 9. The same values have been plotted in Fig. 9. From Fig. 9 it can be seen that a minimum training error occurred in three cases i) *tansig-tansig*, 8 hidden layer neurons, ii) *tansig-tansig*, 10 hidden layer neurons, iii) *logsig-purelin*, 10 hidden layer neurons. But when comparing all of them in term of validation error ANN object having *tansig-tansig*, 8 hidden layer neurons was found to be having minimum validation error and hence this architecture was taken as optimum architecture. Hence optimum parameters of selected ANN architecture are tabulated in Table 10. Schematic representation of optimized ANN architecture was shown in Fig. 10. Performance of optimum neural network during training was depicted in Fig. 11. Both training error curve and validation error curve have same characteristic which manifests that no significant over fitting has occurred. The correlation between actual values and predicted values for training, validation and test sets were shown in Fig. 12. Correlation coefficient (R-value) of each of them is above 0.985 that confirms good agreement between actual and predicted values.

Table 9. ANN Optimization: Fractional Factorial Design – Numerical Values of Training and Validation Errors.

Run	Hidden layer Neurons	Hidden layer Transfer function	Output layer Transfer function	Normalised MSE x 104 (training)	Normalised MSE x 104 (Validation)
1	tansig	tansig	2	6.04825	9.21185
2	tansig	tansig	4	3.97	5.947595
3	tansig	tansig	6	3.2125	5.16368
4	tansig	tansig	8	2	4.7708
5	tansig	tansig	10	1.945	5.935075
6	tansig	tansig	12	2.87	9.748933
7	tansig	purelin	2	5.8225	7.006175
8	tansig	purelin	4	4.073333	8.813333
9	tansig	purelin	6	3.09	4.146753
10	tansig	purelin	8	3.073333	4.3534
11	tansig	purelin	10	2.34	5.66145
12	tansig	purelin	12	3.736667	5.922725
13	logsig	purelin	2	6.75	5.460267
14	logsig	purelin	4	3.243333	4.558267
15	logsig	purelin	6	3.195	4.561
16	logsig	purelin	8	3.69575	8.673733
17	logsig	purelin	10	1.865	6.558333
18	logsig	purelin	12	2.296667	4.569467

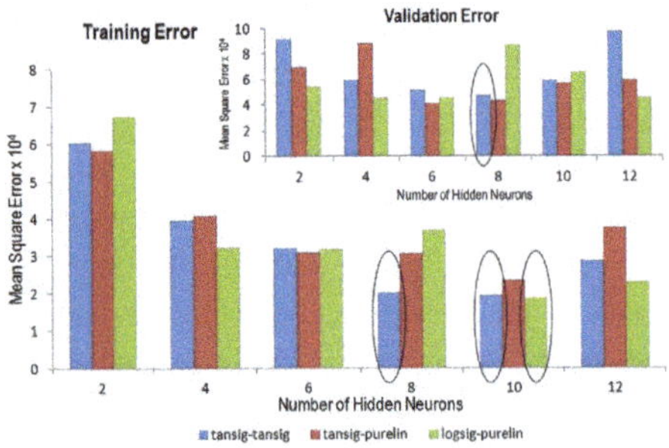

Figure 9. Training and Validation Error as Function of Number of Hidden Neurons and Transfer Function in Hidden Layer and Output Layer.

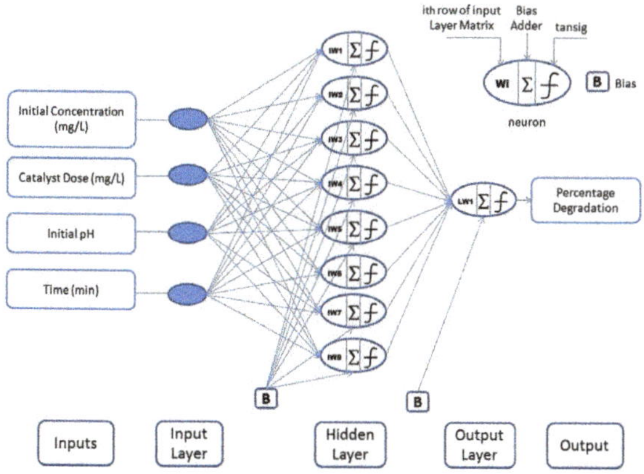

Figure 10. Schematic Representation of Optimized ANN Architecture.

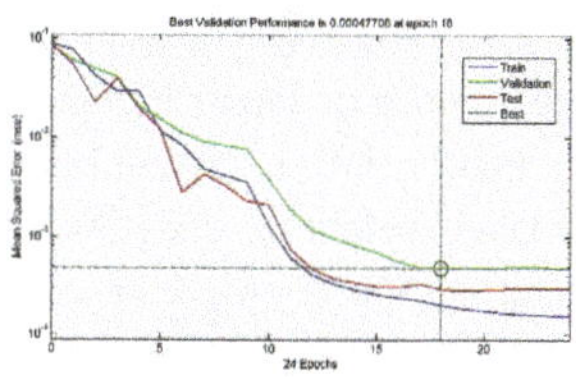

Figure 11. ANN Performance During Training.

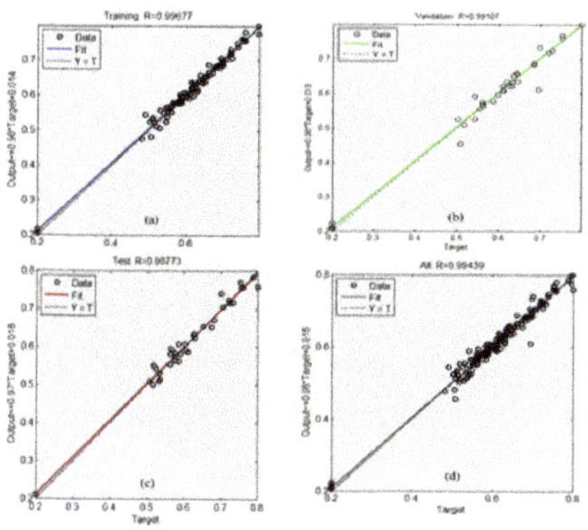

Figure 12. Correlation Analysis between Experimental and Predicted values, (a) Training (b) Validation (c) Testing (d) All.

Table 10. Optimum Parameters of ANN.

Parameter	Level
Number of hidden layer neurons	8
Transfer function in hidden layer	*Tansig*
Transfer function in output layer	*Tansig*

5.5.2 Relative importance of input variables

As proposed by Garson, the neural net weight matrix can be used to assess the relative importance of the various input variables on the output variables using Eq. (9) [4].

$$I_{ik} = \sum_{j=1}^{Nh}\left(\left(|W_{ij}^{IW}| / \sum_{i=1}^{Ni}|W_{ij}^{IW}|\right)|W_{jk}^{LW}|\right) / \sum\left\{\sum_{j=1}^{Nh}\left(\left(|W_{ij}^{IW}| / \sum_{i=1}^{Ni}|W_{ij}^{IW}|\right)|W_{jk}^{LW}|\right)\right\} \quad (9)$$

Where, Nh is the number of hidden neurons, Ni is the number of inputs, Wij stands for weight of connection from i to j, IW and LW corresponds to the input and output layer weight matrix, and i, j, k indicate nodes of input, hidden, and output layer respectively. Table 11 lists input layer weight matrix (IW{1,1}) and bias matrix to hidden layer (b{1}), While Table 12 lists the output layer weight matrix (LW{2,1}) and bias matrix to output layer (b{2}). Using Garson equation relative importance of various parameters affecting percentage degradation were found and represented in Fig. 13. All parameters studied here were found to have significant effect on percentage degradation efficiency. Among all the parameters initial concentration and time were the strongest influential parameters having relative importance of 35% and 31% respectively.

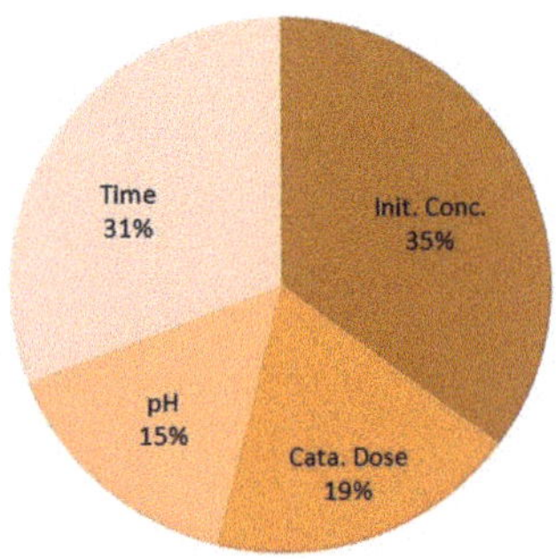

Figure 13. Relative Importance of Parameters.

Table 11. Weights and Biases to Hidden Layer.

Input Layer Weight Matrix (IW{1,1})					Bias to Hidden layer (b{1})
Hidden layer nodes	Input 1	Input 2	Input 3	Input 4	
Node 1	-1.68449	2.393884	-0.50042	1.160354	1.525866
Node 2	0.671192	-0.80857	-2.31335	0.742723	1.679522
Node 3	-1.6527	0.412398	-2.04942	-2.09816	-1.42328
Node 4	2.297944	-0.07207	0.022974	-0.74497	1.481131
Node 5	-0.07145	0.024184	-0.04904	-15.4929	-16.2222
Node 6	4.045752	-0.91124	0.36974	-0.80486	2.341204
Node 7	-1.52337	-0.76097	0.39456	0.603191	-2.00309
Node 8	1.190547	-2.09609	0.4863	1.081481	-2.1816

Table 12. Weights and Biases to Output Layer.

Layer Weight Matrix (LW{2,1})									Bias to output layer (b{2})
Output node	Hidden layer node								
	Node 1	Node 2	Node 3	Node 4	Node 5	Node 6	Node 7	Node 8	
Node 1	0.2020	0.1595	-0.1463	-2.5566	-5.6374	1.3702	-1.2500	0.2389	-5.4160

5.5.3 Optimization using ANN model

Here the primary goal was to find out optimum values of independent variables that would maximize the response. Here, the genetic algorithm was used to find out global optimum in the entire factor space that was being explored. Here, ANN model was optimized in terms of initial concentration, catalyst dose and initial pH for fixed length of time of 240 min to compare the optimum conditions with the same obtained from RSM. Optimum conditions found by the ANN model were verified experimentally. Experimental results were in accordance with predicted values as shown in Table 13.

Table 13. Comparison of Predicted Optimum Parameters and Experimental Validation.

Parameter	Predicted Optimum parameters	Experimental Verification
Initial Concentration (mg/L)	42	40
Catalyst Dose (mg/L)	212.51	210
Initial pH	3.38	3.4
% Degradation (Pred.)	91.27	90.67

6. Conclusion

In this study, RSM and ANN were used to model and subsequently optimize photocatalytic degradation process of 4-CP. The influence of operational parameters including catalyst dose, initial concentration, and initial pH were investigated by employing three level face centred design of experiment. In RSM, a quadratic model was developed by fitting experimental data through the regression method. An ANOVA study was carried out to check significance and adequacy of the model and its coefficients. The results demonstrated that the model had no unnecessary terms and lack of fit was also significant. The study revealed that all linear effects, quadratic effects of catalyst dose and initial pH, and interaction effect between initial concentration and catalyst dose were strongly influential towards degradation performance. From the main effects, it was found that an increase in catalyst dose had positive effect up to a certain limit afterwards it was detrimental to the degradation performance. An increase in initial concentration was found to have negative influence on degradation efficiency. In case of the initial pH of the solution, an acidic condition was favouring the degradation performance. Two layer feed forward neural network with back propagation was employed where optimum parameters were chosen by conducting experimental design. An optimum ANN architecture was found to be eight hidden layer neurons and *tansig* transfer function in the hidden as well as output layer. Relative importance of inputs to ANN was determined using Garson equation based on weights of connection between neurons. Among all parameters, initial concentration and time were strongest influential parameters having relative importance of 35% and 31% respectively. Both models exhibited reasonably well correlation between experimental and predicted values. Optimum conditions found by each of the models were experimentally verified and good harmony was observed between predicted and experimental conditions.

References

[1] S. Ahmed, M.G. Rasul, W.N. Martens, R. Brown, M.A. Hashib, Heterogeneous photocatalytic degradation of phenols in wastewater: A review on current status and developments, Desalination 261 (2010) 3-18. https://doi.org/10.1016/j.desal.2010.04.062

[2] K. Natarajan, P. Singh, H.C. Bajaj, R.J. Tayade, Facile synthesis of $TiO_2/ZnFe_2O_4$ nanocomposite by sol-gel auto combustion method for superior visible light photocatalytic efficiency, Korean Journal of Chemical Engineering, 33 (2016) 1788-1798. https://doi.org/10.1007/s11814-016-0051-4

[3] W.K. Jo, G.T. Park, R. J. Tayade, Synergetic effect of adsorption on degradation of malachite green dye under blue LED irradiation using spiral-shaped photocatalytic reactor, J. Chem. Technol. Biotechnol. 90 (2015) 2280-2289. https://doi.org/10.1002/jctb.4547

[4] M.B. Kasiri, H. Aleboyeh, A. Aleboyeh, Modeling and optimization of heterogeneous photo-fenton process with response surface methodology and artificial neural network, Environ. Sci. Technol. 42 (2008) 7970–7975. https://doi.org/10.1021/es801372q

[5] G.E.P. Box, K.B. Wilson, On the experimental attainment of optimum conditions, Journal of the Royal Statistical Society 13 (1951) 1–45.

[6] D.C. Montgomery, Design and analysis of experiments, 5th ed., New York: Wiley (2001).

[7] R.H. Myers, D.C. Montgomery, Response surface methodology: Process and product optimization using designed experiments, 2nd ed., New York: Wiley (2002).

[8] Y.R. Smith, A. Kar, V.R. Subramaniam, Investigation of physicochemical parameters that influence photocatalytic degradation of Methyl Orange over TiO_2 nanotubes, Ind. Eng. Chem. Res. 48 (2007) 10268-10276. https://doi.org/10.1021/ie801851p

[9] V.A. Sakkas, A.Md. Islam, C. Stalikas, T.A. Albanis, Photocatalytic degradation using design of experiments: A review and example of the Congo red degradation, J. Haz. Mat. 175 (2010) 33-44. https://doi.org/10.1016/j.jhazmat.2009.10.050

[10] M.S. Secula, G.D. Suditu, I. Poulios, C. Cojocaru, I. Cretescu, Response surface optimization of the photocatalytic decolorization of a simulated dyestuff effluent, Chem. Eng. J. 141 (2008) 18–26. https://doi.org/10.1016/j.cej.2007.10.003

[11] C. Liyana-Pathirana, F. Shahidi, Optimization of extraction of phenolic compounds from wheat using response surface methodology, Food Chemistry 93 (2005) 47-56. https://doi.org/10.1016/j.foodchem.2004.08.050

[12] Aleboych, N. Daneshvar, M.B. Kasiri, Optimization of C. I. Acid Red 14 azo dye removal by electrocoagulation batch process with response surface methodology, Chem. Eng. Process. 47 (2008) 827–832. https://doi.org/10.1016/j.cep.2007.01.033

[13] M. Gfrerer, E. Lankmayr, Screening, optimization and validation of microwave-assisted extraction for the determination of persistent organochlorine pesticides, Analytica Chimica. Acta. 533 (2005) 203–211. https://doi.org/10.1016/j.aca.2004.11.016

[14] D.M. Himmelblau, Accounts of experiences in the application of artificial neural networks in chemical engineering, Ind. Eng. Chem. Res. 47 (2008) 5782-5796. https://doi.org/10.1021/ie800076s

[15] L. Jin, P. Nikiforuk, M. Gupta, Direct adaptive output tracking control using multilayered neural networks, IEEE Proc, 140 (1993) 393–398.

[16] B. Zhao, Y. Su, Artificial neural network-based modelling of pressure drop coefficient for cyclone separators, Chemical Engineering Research and Design 88 (2010) 606–613. https://doi.org/10.1016/j.cherd.2009.11.010

[17] K. Elsayed, C. Lacor, Modeling analysis and optimization of aircyclones using artificial neural network, response surface methodology and CFD simulation approaches, Powder Technology 212 (2011) 115–133. https://doi.org/10.1016/j.powtec.2011.05.002

[18] J.A. Burns, G.M. Whitesides, Feed-forward neural networks in chemistry: mathematical systems for classification and pattern recognition, Chem. Rev. 93 (1993) 2583-2601. https://doi.org/10.1021/cr00024a001

[19] H.M. Cartwright, Applications of artificial intelligence in chemistry, Oxford University Press: Oxford, U.K., 1993.

[20] T. Aoyama, Y. Suzuki, H. Ichikawa, Neural networks applied to structure-activity relationships, J. Med. Chem. 33 (1990) 905-908. https://doi.org/10.1021/jm00165a004

[21] N. Quis, T. Sejnowski, Predicting the secondary structure of globular proteins using neural networks models. J. Mol. Biol. 202 (1998) 865-884.

[22] R. Baratti, G. Vacca, A. Servida, Neural networks modeling of distillation column. Hydrocarbon Process. (1995) 35-38.

[23] M.L. Thompson, M.A. Kramer, Modeling chemical process using prior knowledge and neural networks. AIChE J. 40 (1994) 1328-1340. https://doi.org/10.1002/aic.690400806

[24] J.W. Prasad, S.S. Bhagwat, Simple neural network models for prediction of physical properties of organic compounds. Chem. Eng. Technol. 25 (2002) 1041-1046. https://doi.org/10.1002/1521-4125(20021105)25:11%3C1041::AID-CEAT1041%3E3.0.CO;2-5

[25] D. Howard, M. Beale, M. Hagan, Neural network toolbox for use with MATLAB® user's guide, Version 5, Mathworks Inc, 2006.

[26] A.R. Khataee, Photocatalytic removal of C.I. Basic Red 46 on immobilized TiO_2 nanoparticles: Artificial neural network modelling, Environ. Techno. 8 (2009) 1155-1168. https://doi.org/10.1080/09593330903133911

[27] A.R. Khataee, O. Mirzajani, UV/peroxydisulfate oxidation of C. I. Basic Blue 3: Modeling of key factors by artificial neural network, Desalination 251 (2010) 64–69. https://doi.org/10.1016/j.desal.2009.09.142

[28] D. Salari, N. Daneshvar, F. Aghazadeh, A.R. Khataee, Application of artificial neural networks for modeling of the treatment of wastewater contaminated with methyl tert-butyl ether (MTBE) by UV/H_2O_2 process, J. Haz. Mat. B. 125 (2005) 205-210. https://doi.org/10.1016/j.jhazmat.2005.05.030

[29] S. Gob, E. Oliveros, S.H. Bossmann, A.M. Braun, R. Guardani, C.A.O. Nascimento, Modeling the kinetics of a photochemical water treatment process by means of artifical neual networks, Chem. Eng. Process. 38 (1999) 373–382. https://doi.org/10.1016/S0255-2701(99)00028-8

[30] J.E.F. Moraes, F.H. Quina, C.A.O. Nascimeto, D.N. Silva, O. Chiavone-Filho, Treatment of saline wastewater contaminated with hydrocarbons by the photo-fenton process, Environ. Sci. Technol. 38 (2004) 1183–1187. https://doi.org/10.1021/es034217f

[31] V.K. Pareek, M.P. Brungs, A.A. Adesina, R. Sharma, Artificial neural network modeling of a multiphase photodegradation system, J. Photochem. Photobiol. A 149 (2002) 139–146. https://doi.org/10.1016/S1010-6030(01)00640-2

[32] Aleboyeh, M.B. Kasiri, M.E. Olya, H. Aleboyeh, Prediction of azo dye decolorization by UV/H_2O_2 using artificial neural networks, Dyes and Pigments 77 (2008) 288–294. https://doi.org/10.1016/j.dyepig.2007.05.014

[33] Duran, J.M. Monteagudo, M. Mohedano, Neural networks simulation of photo-Fenton degradation of Reactive Blue 4. Appl. Catal., B: Environ. 65 (2006) 127–134. https://doi.org/10.1016/j.apcatb.2006.01.004

[34] G. Benoit, Degradation of chlorophenols by ozone and light, Fresenius Environ. Bull. 3 (1994) 331–338.

[35] M.S. Goswami, R.P. Singh, Kinetics of chlorophenol degradation by benzoate-induced culture of Rhodococcus erythropolis M1, World J. Microbiol. Biotechnol. 18 (2002) 779–783. https://doi.org/10.1023/A:1020429109524

[36] M. Hugul, I.Boz, R. Apak, Photocatalytic decomposition of 4-CP over oxide catalysts, J. Haz. Mat. B 64 (1999) 313–322. https://doi.org/10.1016/S0304-3894(98)00272-6

[37] T. Poznyak, J. Vivero, Degradation of aqueous phenol and chlorinated phenols by ozone, Ozone: Sci. and Eng. (2005) 447-458. https://doi.org/10.1080/01919510500351529

[38] M.H. Priya, G. Madras, Kinetics of photocatalytic degradation of chlorophenol, nitrophenol, and their mixtures, Ind. Eng. Chem. Res. 45 (2006) 482-486. https://doi.org/10.1021/ie050846h

[39] N. Venkatachalam, M. Palanichamy, V. Murugesan, Sol–gel preparation and characterization of alkaline earth metal doped nano TiO_2: Efficient photocatalytic degradation of 4-chlorophenol, J. Mol. Cat. A: Chem. 273 (2007) 177–185. https://doi.org/10.1016/j.molcata.2007.03.077

[40] R.J. Tayade, D.L. Key, Synthesis and characterization of titanium dioxide nanotubes for photocatalytic degradation of aqueous nitrobenzene in the presence of sunlight, Materials Science Forum 657 (2010). 62-74 https://doi.org/10.4028/www.scientific.net/MSF.657.62

Chapter 16

Role of Ultrasound in the Synthesis of Nanoparticles and Remediation of Environmental Pollutants

Pankaj[1*], Srimayee Sahu[1], Shubhi Misra[1] and Hemlata Srivastava[2]

[1]Department of Chemistry, Dayalbagh Educational Institute, Agra 282 005, India
[2]Departnemt of English, Agra College, Dr. BRA University, Agra 282 003, India
[*]Email: profpankaj99@gmail.com

Abstract

The present review briefly discusses two important aspects of the application of power ultrasound. Firstly, the application of ultrasound in the synthesis of nanoparticles, using various state-of-the-art sonochemical methods for the synthesis of nanoparticles of metals of s-, p-, d- and f- blocks and their compounds using ultrasound alone or in combination with other techniques. The advantage of using ultrasound lies in controlling the size, morphology and physical state (amorphous / crystalline) of nanoparticles through the variation in frequency, power and duration of sonication. Secondly, the use of ultrasound in the remediation of pollutants in aqueous effluents, such as, metal ions, organic acids, dyes, pesticides, pharmaceuticals, preservatives etc., has been discussed. Ultrasound is undoubtedly a very promising futuristic tool for both these technologies.

Keywords

Ultrasound, Cavitation, Sonophotocatalyst, Nanoparticles, Sonochemical

Contents

1. Introduction

Ultrasound has a tremendous role in chemical sciences in almost all fields of technology and manufacturing. This review is dedicated to highlight the effects of ultrasound propagation in solutions and then its role in the field of synthesis of nanoparticles and finally in the quantization. The review has been, therefore, broadly divided into two major sections. The first segment deals with the synthesis of nanoparticles using different techniques in combination with ultrasound whereas, in the second segment a discussion on the effects of the propagation of ultrasound in solutions and its role in the remediation and quantization of pollutants has been discussed.

2. Part I: Synthesis of nano particles involving ultrasound

A nanoparticle is a particle having at least one of its dimensions in the nano range[1]. The ultrafine particle range between 1-100 nm in size, fine particles are 100-2500 nm in size and the coarse particles are 2500-10000 nm. Nanoparticles possess unique characteristic properties due to its small size, since at nano level quantum behavior comes into existence therefore, gold and platinum nanoparticles behave as magnetic particles[2]. The increased interest in nanotechnology is due to its various unique properties. The advancement in nanotechnology produced various materials and devices with new functionalities in the chemical, physical and biological field. The different forms of nanomaterials are nanocomposite, nanocapsules, nanofibers, fullerenes etc.[3] Nanotechnology is now an independent and emerging field of science, with lots of future innovations. Fabrication of nanoparticles can be done in two ways: Top-down & Bottom-up. Top- down synthesis is possible through milling / attrition whereas the bottom – up is through the pyrolysis, solvothermal reaction and sol- gel method. Similar other methods, such as, biomimetic synthesis and microwave assisted synthesis are also important[4]. Ultrasound has been found to be a useful tool in the synthesis of nanoparticles. These are the sound waves whose frequency is higher than the upper audible limit of human hearing (<16 kHz). Ultrasound devices operate at different frequencies from 20 kHz to several gigahertz[5]. Ultrasound, ranging from 20 to 100 kHz, is used in chemically important systems[6].

Generally cavitation and nebulization are the two phenomenon associated with ultrasound. The formation, growth and the collapse of bubbles in the liquid takes place in acoustic cavitation, which is responsible for the sonochemical process, while, when the ultrasound radiation passes through the liquid formation mist occurs which is called the nebulization[7]. Ultrasonic spray pyrolysis (USP) is based on the principle of nebulization and helps in preventing agglomeration which is often formed during the traditional pyrolysis technique. Sonochemistry is, thus, the application of ultrasound to a chemical process or technology. Based on the nature of cavitation, sonochemistry is broadly divided into two classes namely; homogenous sonochemistry of liquids and heterogeneous sonochemistry of liquid – liquid or liquid - solid systems[5]. The application of power ultrasound in the production of nanomaterials has manifold effects. The first area of application is particle synthesis and precipitation, the second is dispersing of materials in liquids in order to break particle agglomeration[8]. Sonication yield and the rate depend upon the various factors like temperature, sonication time, sample volume and concentration[9]. Application of power ultrasound for producing nanomaterials is presently a quickly developing and promising area of scientific research. Amongst a variety of approaches, the utilization of ultrasound for material synthesis has been extensively attempted over the years, and is now established as one of the most powerful tools in nanostructured materials synthesis[7]. The prepared nanostructures are characterized by using XRD, SEM, FT-IR, TEM, STEM, and AFM. Debye-Scherrer's equation is used to determine the size of nanoparticle, using following equation:

$$D = \frac{K\lambda}{\beta Cos\theta}$$

Where, D =diameter size particle; K= shape factor (0.9), its value is close to unity ; λ=X-ray wavelength; β= full width at half maximum (FWHM) ; θ=Bragg angle; The FWHM can be calculated as;

$$\beta_{total}^2 = \beta_{instrument}^2 + \beta_{sample}^2 .$$

Since, λ, the wavelength of X-ray, is already known[10], the size of particles can be easily calculated. Ultrasound at high frequencies is a versatile tool used to achieve finer nanoparticles, which is simple and cost effective[11].

2.1 Sonochemical and combinatorial techniques used in the nano-patricle synthesis

2.1.1 Sonochemical methods

The sonochemical method has been used extensively to synthesize novel materials with unusual properties[12]. Many methods have been developed to make nanoparticles using ultrasound through sonochemical reduction, sonochemical deposition and ultrasonic spray pyrolysis methods[8].

2.1.1 Sonochemical reduction

a) Metals: Nanoparticle of various metals may be produced via the sonochemical route. Sonochemical reduction method is widely used for the synthesis of metallic nanoparticles. One of the interesting observations is that the size of the particle is dependent upon the concentration of metal and frequency of ultrasound. A new synthetic approach termed as sonochemical assisted metal displacement reduction is introduced for the synthesis of gold and platinum nanoparticles. In this a precursor of gold and platinum is reduced on the surface of copper or iron in the ultrasonic bath and the metal displacement reduction occurs[7].

b) Metaloxides: The monometallic and bimetallic nanoparticles are of great interest due to the catalytic and optoelectronic applications. Metal oxides are easily produced via the sonochemical technique. The sonication of the solution, in ambient conditions, enhances the rate of formation of nanoparticles, besides controlling the uniform size and distribution. Such nanoparticles possess higher surface area with improved phase and physico—chemical nature. The metal oxide powder has fascinating properties, such as amorphous Fe_2O_3 nanoparticles and $Mn_2(CO)_{10}$ in decahydronaphthalene show an unusual resistance to the phase transition γ-Fe_2O_3 to more stable α-Fe_2O_3 during high temperature heat-treatment[13].

c) Metal chalcogenides: Metal chalcogenide nanoparticles are synthesized by irradiating the aqueous solution of metal salt and a chalcogen source. The first approach for synthesizing the metal chalcogenides was introduced by Suslick and his coworkers[14], by sonicating the organometallic precursors in the presence of chalcogen source, such as, nanostructured MoS_2 prepared sonochemically[7], which is superior to the conventional MoS_2.

2.1.1.1 Sonochemical deposition:

Ultrasound may also be used to forcefully deposit nanoparticles on the surface of substrates. This process reduces the reaction time. In addition to the deposition, a sonochemical route for coating a metal oxide with metal sulfide or silica can also be done. This method is used to produce the hetero- structured core or shell nanocomposites. The hollow nanostructures may be prepared using the inorganic nanoparticles by depositing them on solid substrates.

2.1.1.2 Ultrasonic spray pyrolysis:

The spray pyrolysis is generally used for the synthesis of ultrafine nanoparticle formation. A fine powder and film of metals, metal alloys etc. are prepared by decomposition of metal salts (aqueous). In general, metal nitrates, chlorates, and acetates are employed as precursors; in case of metal chalcogenide synthesis, chalcogen sources are added to the precursor solution. As long as suitable precursors are available, fine powders of nearly any metal oxide or chalcogenide can be simply prepared using the USP method[15,16].

2.1.2 Combination methods

2.1.2.1 Sol-gel and sonochemical method

TiO_2 nanoparticle may be synthesized using titanium isopropoxide and iso-propyl alcohol. Titanium (IV) isopropoxide is dissolved in iso-propyl alcohol and mixed with the mixture of HCl and propanol dropwise with continuous stirring. The resultant solution is sonicated and a gel is obtained, which is centrifuged, dried and heated. TiO_2 nanoparticles are obtained. The result shows that the acidity promotes amorphous nature. The crystallinity of the nanoparticle increased[17] when the molar ratio of acid and alkoxide was between 2 and 3. ZnO nanoparticle is prepared by dissolving zinc acetate dihydrate in ethanol and adding to this triethylamine (TEA) drop wise to stabilize the solution. One sample is stirred for half an hour. Another solution is placed in an ultrasonic bath. Both are deposited on a glass substrate using the spin coating method. The film is dried and then calcinated. The result shows that the size of the nanoparticles is increased on ultrasonic irradiation[18].

2.1.2.2 Sonoelectrochemical method

Aluminum nanoparticles may be synthesized via the sonoelectrochemical method. The solution of $AlCl_3$ and $LiAlH_4$ in THF is sonicated with an ultrasonic horn, used both as the cathode and the ultrasound emitter. The sonoelectrode produces a sonic pulse followed by the current pulse. The synthesis is performed using 3 electrode cells (counter electrode, reference electrode and working electrode). A current pulse is passed through the solution of $AlCl_3$ and $LiAlH_4$ in THF. The resultant product is washed and dried under vacuum. The result reveals that the reduction of Al ions can occur only through electrolytic methods and nanoparticles with sizes ranging from 10 to 20 nm may be obtained[19]. Similarly, tungsten nanoparticle may be synthesized using an electrolyte of tri-sodium citrate (complex), sodium tungstate, ferrous sulphate and citric acid as a buffer. The horn is immersed in electrolyte which acts as cathode and platinum as anode. The pulse of ultrasound is passed. The obtained nanoparticles are body centered[20].

2.1.2.3 Sonochemical and solvothermal method

CdS nanorods are synthesized using cadmium nitrate, ethylenediamine, and sodium sulfide. Solution of cadmium nitrate in ethylenediamine is mixed with sodium sulfide. Mixture is then sonicated and washed with ultra-distilled water. The result shows that the uniform hexagonal CdS nanoparticles are obtained. Nanoparticles synthesized by combined method are more amorphous than the sonochemical method[21].

2.1.2.4 Sonochemical and microwave assisted method

Bismuth vanadate nanoparticles can be prepared by using the reagents $Bi(NO_3)_3.5H_2O$ (Bismuth nitrate pentahydrate), $NaHCO_3$ (sodium bicarbonate) and Rhodamine B. NH_4VO_3(Ammonium metavanadate), diethylene glycol, ethylene glycol, glycerol and mannitol. Bismuth nitrate pentahydrate, ammonium metavanadate are dissolved in diethylene glycol and distilled water respectively. Milky yellow precipitate is heated and placed in an ultrasonic and microwave combined reactor. The solution is centrifuged, washed and dried. The dried product is calcined and nanoparticles are obtained. Various morphologies such as nut-like, potato-like and broccoli- like $BiVO_4$ can be obtained by using different solvents. The nanoparticles, thus obtained, possess different optical properties and photo catalytic activities[22].

ZnO nanoparticles can also be synthesized via combined microwave and ultrasonic method. For the preparation $Zn(CH_3COO)_2$ (Zinc acetate dehydrate), NaOH (Sodium hydroxide), $Zn(acac)_2$ (Zinc acetylacetonate hydrate), $ZnCl_2$ (zinc chloride), HEPES (2-[4-(2- hydroxyethyl)-1-piperazinyl] ethanesulfonic acid) and methylene blue are required. Two solutions of zinc acetate solution and HEPES are mixed and irradiated with ultrasound followed by microwave heating. On cooling the product is centrifuged, washed and dried. The shape and size depends upon the molar ratio of Zn/HEPES, pH value of HEPES solution and the precursor used for obtaining zinc. The nanostructure exhibits greater photocatalytic activities[23].

From the above discussion it can be seen that ultrasound can be used in many ways to synthesize nanoparticles and one has to apply the best combination for optimum results. In the underlying sections, a brief summary of nanoparticles of various elements of the periodic table is discussed and only important references related to metals using sonochemical technique have been mentioned to economic the space and emphasize the role of ultrasound in the preparation of nanoparticles of almost all classes of metal.

2.2 Synthesis of Nano-particles of metals and their compounds

Sonochemistry and ultrasonic spray pyrolysis (USP) are the most successful examples of the utilization of ultrasound in synthesis. In particular, both methods have a superior ability to create nano composites compared to other synthetic methods. USP has been found to be an excellent method to produce organic-inorganic hybrid materials[24].

2.2.1 s-block elements

a) Barium nanoparticles are studied in detail due to their tunable microwave applications such as phase shifter, antennas and other application in the ceramic industry.

$BaZr_{0.3}Ti_{0.7}O_3$ (Barium zirconate titanate) nanoparticles are produced using solution of barium chloride dihydrate, zirconium oxychloride octahydrate, and titanium chloride in de- ionized water. The solution is sonicated for 30 minutes in sonication chamber containing sodium hydroxide solution in the atmosphere of argon gas. The solution is cooled and washed with de- ionized water until neutralized and finally dried in oven at 100°C for 24 hours. The nanoparticles thus obtained, are cubic perovskite in nature and their morphology depends upon the concentration of NaOH[25]. $BaTiO_3$ (Barium titanate) nanoparticles may be prepared using the barium chloride dihydrate and titanium chloride. Concentrated $TiCl_4$ solution is added to de-ionized water with continuous stirring along with barium chloride and excess of NaOH. The precipitate is centrifuged and dried to get a white product, which itself is the Barium titanate nanoparticle. A good crystalline and narrow diffraction peaks appears which confirms the formation of barium titante nanoparticles[26].

b) Lithium nanoparticle has drawn special attention because of their potential application in lithium-ion batteries and luminescence. $LiCoO_2$ nanoparticle has been synthesized under sonochemical conditions. The treatment of NaOH with $CoCl_2$ (or $CoSO_4$) produces Co_3O_4 nanoparticles, followed by a coating of LiOH on Co_3O_4 nanoparticles. Calcination brings about the formation of the homogenous $LiCoO_2$ nanoparticles. The result shows homogeneity in shape and size[27]. $Li_4Ti_5O_{12}$ nanoparticle preparation via sonochemical process includes a coating of LiOH on TiO_2 which is then calcined to yield a homogenous $Li_4Ti_5O_{12}$ nanoparticle. The resulting product is characterized as a nanoparticle of tetragonal shape which is responsible for the existence of a transient supercritical state[28].

c) Magnesium nanoparticle is widely used in flame-retardant composite materials and in the paper industry. $Mg(OH)_2$ (magnesium hydroxide) nanoparticles is prepared from the magnesium itself, without any requirement of additive. Mixture of magnesium powder, water and ethanol is irradiated with ultrasound. NaCl is used to speed up the reaction. Sample obtained is centrifuged and dried. Magnesium hydroxide is formed by the interaction of metal ion with the hydroxyl ions. The reaction occur in the following manner;

$$H_2O \rightarrow H^+ + OH^-$$

$$Mg^{2+} + 2OH^- \rightarrow Mg(OH)_2$$

At different sonochemical condition various structure may be formed such as platelet-shaped, nanometric and de-agglomerated powders[29].

2.2.2 p- blockelements

a) Antimony nanoparticles have been used in several nanodevices, besides it can be used as a catalyst. SbSeI (Antimony seleno-iodide) is produced using the stoichiometric ratio of the chemicals: Sb, Se, I2and carbon nanotubes in ethanol. A dark sol, obtained under sonication condition, is washed with ethanol to remove the impurity of the resultant product and dried at room temperature as a dark purple substance. The nanoparticle of antimony seleno-iodide in multiwalled carbon nanotube is highly anisotropic 1D structure30. SbS1-xSexI may be sonically prepared. Antimony and selenium solution in ethanol were mixed and placed in a sonication bath. The slurry produced changes the colour gradually. After drying a SbS1-xSexI xerogel is created. The 1D nanostructure is produced with the highly anisotropic bonding, found normally in crystallographic structure. There was also a possibility of obtaining these nanowire solutions with definite width of forbidden energy gap[31].

b) Indium nanoparticles are widely applied in the field of dry cells, manufacturing picture tubes for television. In_2S_3 (Indium Sulfide) exists in 3 different forms: α- In_2S_3, β-In_2S_3 and γ-In_2S_3. The various methods used to prepare indium sulfide nanoparticle are (a) wet chemical method where In_2S_3 is prepared using $InCl_3$ and H_2S and (b) sonically by the reaction of indium chloride and thioacetamide. This reaction carried out at different temperature yield different kinds of nanoparticle; at zero degree amorphous product is obtained while at room temperature crystalline product is formed[32].

c) Lead nanoparticles have wide application in the field of catalytic activity, electrical conductivity. Coordination compounds of lead are of great interest due to their sorption property. $[Pb_2(4, 4'\text{- bipy})(NO_3)_4]_n$ Lead (II) coordination polymer nanoparticles can be prepared using lead (II) nitrate and 4, 4'- bipy in MeOH. Upon mixing, the precipitate obtained was filtered, washed with acetone and ether and finally dried in air. The geometry of compound formed was 3- Dimensional polymer. This polymer is used as the precursor for many nanomaterials[33]. PbE (Lead chalcogenides) nanoparticles, where E can be S, Se, Te, have been synthesized using $Pb(CH_3COO)_2.3H_2O$, Se (or S or Te) and ethylenediamine. A three step mechanism is given for the preparation. $Pb(CH_3COO)_2.3H_2O$ is dissolved in ethylenediamine and mixed with Se (or S or Te) in the solution before placing it in the sonication bath for 3-8 hours. The precipitate obtained is black or greyish black in colour which is washed with distilled water & anhydrous ethanol and dried with vacuum pump. Result shows that the nanoparticles formed are in cubic form with no impurity. The ultrasound irradiation is responsible for the acceleration of the reaction and prevention of the aggregation of nanoparticles[34].

$PbMoO_4$ nanoparticle is prepared by dissolving Na_2MoO_4 and $Pb(NO_3)_2$ separately in distilled water. The two solutions are mixed and then sonicated. Precipitate obtained is centrifuged, washed and dried in air. pH plays an important role in creating different morphology of $PbMoO_4$ crystal. Shape varies from spherical to pagoda[35].

$$Pb^{2+} + MoO_4^{2-} \rightarrow PbMoO_4$$

d) Tin nanoparticles has fascinating role in sensing molecule, possess storage properties and various electrochemical application. SnS (Tin sulfide) nanoparticle[36] preparation can be done using the sodium sulfide hydrate, triethanolamine (TEA), tin chloride dihydrate, polyvinyl alcohol and ethanol. Tin chloride powder is dissolved in TEA, whereas sodium sulfide is dissolved in distilled water. The two solutions were mixed and the pH of the solution was maintained at 12. Upon sonication a black precipitate was obtained, which was cooled, washed and dried.

$$SnCl_2.2H_2O + Na_2S.xH_2O \rightarrow SnS + 2NaCl + xH_2O$$

The nanoparticles were spherical in shape. The SnS crystal gives a blue shift and the band gap energy of 1.74 eV. Carbon supported Sn nanoparticles has been reportedly synthesized sonochemically using the reduction reaction *in situ.* The required reagents include tin chloride, carbon, Nafion solution, potassium chloride and sulfuric acid. The carbon supported Sn nanoparticles were prepared by mixing $SnCl_2$ and carbon in ethylene glycol. Sonicated mixture yield a solid sample which is washed with absolute ethanol and dried under vacuum. The particle size was small[37].

2.2.4 d-block elements

a) Manganese Nanoparticles are extensively used as separators, nanoelectronic, optical devices and fuel cells. Mn_3O_4 (manganese oxide) nanoparticle is prepared by the ultrasonic assisted approach. The materials used in synthesis are manganese chloride and ethanol amine. They are mixed and placed in the ultrasonic cleaner, until a clear brown solution is obtained. Distilled water is added to the resulting dark brown solution which is centrifuged and washed with excess absolute ethanol before being dried. A highly pure and crystallized sample is obtained[38]. In another method Mn_3O_4 nanoparticle is prepared by using anhydrous ethanol instead of ethanol amine. Manganese chloride was dissolved in a mixture of anhydrous ethanol and distilled water (20/80 ratio) and sonicated for 1 hour at 500°C. The precipitate of manganese hydroxide is separated by centrifugal process and washed with deionized water. The resultant nanoparticles possess cubic morphology. Nanoparticles are paramagnetic in nature[39]. Mn_3O_4 nanoparticles can also be formed using the manganese acetate as precursor. It is a one-step mechanism. A solution of $Mn(CH_3COO)_2$, tetra methyl ammonium hydroxide (or NaOH), ethanol (or water) is sonicated without requirement of calcination process. Sphere like nanocrystal of spinel Mn_3O_4 are formed[40].

$$Mn(CH_3COO)_2 + NaOH \text{ (or TMAH)} + H_2O\text{(or ethanol)} \rightarrow Mn_3O_4$$

b) Nickel Nanoparticle is considered to be a promising material in the manufacturing of super paramagnetic material, gas sensors and in dye sensitized solar cell. NiO (Nickel oxide) nanoparticle is synthesized in three steps:

- Precipitation of nickel hydroxide as precursor through the drop wise addition of NaOH to nickel nitrate with stirring.
- Under ultrasonic irradiation nickel hydroxide changes to nickel oxide.
- Precursor is filtered and washed with distilled water. The product is dried with the help of an oven.

The nanoparticles are highly crystalline in nature with cubic geometry[41,42]. Another method for the preparation of NiO nanoparticle is through the addition of NaOH to the solution of $Ni(CH_3COO)_2{\cdot}2H_2O$ in ethanol (or water). The mixture is centrifuged, washed using ethanol and calcinated.

$$Ni(CH_3COO)_2{\cdot}2H_2O + NaOH{,}HOH \rightarrow NP\text{-}NiO$$

The NiO crystals are spherical in nature and can be used in developing NiO – NP based sensors[43]. α - Ni(OH) (α-nickel hydroxide) nanoparticle is synthesized sonochemically using nickel nitrate and urea. Aqueous solution of nickel nitrate is mixed with urea and sonicated. The precipitate obtained is separated by centrifugation, washed and dried. This sample of nanoparticles is found to be stable in alkaline medium[44].

c) Platinum nanoparticle is extensively used as chemo and bio sensors, as data storage and fluorescent sensors. Pt (Platinum) nanoparticle is sonically prepared using reduction method. Reduction of H_2PtCl_6 with methanol (ethanol or propanol) occurs in the presence of a capping polymer. The different polymer as a capping agent are (CHI) chitosan, (PEG) polyethylene glycol and (PAmHU) poly(amidehydroxyurethane). The aqueous solution of H_2PtCl_6 (containing capping polymer) is mixed with alcohol and sonicated. Refluxing of metal salts/ complex is done to yield a nanocrystalline metal powder. The nanoparticles formed are found capped with polymer[45].

d) Molybdenum nanoparticle is widely used as lubricant additive and in optical sensors. MoS_2 (Molybdenum sulfide) nanoparticle is prepared by ultrasonically irradiating a slurry of molybdenum hexacarbonyl and sulfur in 1,2,3,5- tetramethylbenzene in inert medium (Ar). A black powder is obtained which is filtered and washed with pentane. The resultant powder is heated to remove excess molybdenum and sulfur. The MoS_2 nanoparticles are spherical in shape and behave as high temperature lubricant[46].

e) Cobalt nanoparticle is used in electrochemistry, gas sensors, energy storage and catalysis. Co_3O_4 (Cobalt oxide) nanoparticle is prepared via single step. The precursor used for preparing cobalt nanoparticle is cobalt acetate with NaOH (or tetra methyl ammonium hydroxide) in solvents. The reaction occurs in ultrasonic bath. Calcination is not needed. The sphere like cobalt oxide nanoparticles shows cubic spinel geometry[47].

$$Co\,(CH_3COO)_2 + NaOH\ (\text{or TMAH}) + HOH \rightarrow Co_3O_4$$

$(CoL_1).0.5DMF.1.5CH_3OH$ (Cobalt (II) coordination polymer) nanoparticle is synthesized by a sonochemical process. To the solution of cobalt (II) acetate in methanol, ligand [H_2L_1 = 5- (4-Carboxy phenyl azo) salicyilic acid] in DMF is added dropwise in ultrasonic bath. The precipitate obtained is filtered and washed with the double distilled methanol and then dried. This polymer nanoparticle serves as a precursor for the synthesis of Co_3O_4. The calcination of the coordination compound in air results in the cubic structure of Co_3O_4 nanoparticles[48].

$CoAl_2O_4$ (Cobalt aluminate) nanoparticle is prepared using the starting reagents $Co(NO_3)_2.6H_2O$, $Al(NO_3)_3.9H_2O$ and a precipitating agent (urea). A stoichiometric solution of these reagents is irradiated with ultrasound radiation in the environment of Ar. pH changes gradually till the precipitation ends. The precipitate is washed with water and ethanol and dried in vacuum. This dried product act as the precursor for $CoAl_2O_4$ when subjected to calcination. Urea is used as precipitating agent because of two reasons: it can be removed easily and also doesn't enter into the matrix of oxide. Monophasic cobalt aluminate crystals are formed[49].

f) Zirconium nanoparticles are used as catalyst in degradation of recalcitrant pollutant and in ceramic industry. $ZrFe_2O_3$ (Zirconium ferrate) nanoparticle shows dual activity (photo and phanton). Chemicals required for the synthesis are zirconium oxychloride, iron acetate and ammonium solution. Aqueous solution of Zirconium oxychloride and iron acetate in water at pH 6.5 is sonicated until the mixture turns brown. The sample is centrifuged, dried in an oven and calcinated. $ZrFe_2O_3$ nanoparticle is used in degradation of azo / non azo dyes and has a lower band gap[50].

g) Gold nanoparticles have tunable optical properties, used in biomedical diagnostic and analytics. For the synthesis of Au nanoparticle, solution of $HAuCl_4$ is dissolved in distilled water and mixed with different concentration of sodium dodecylsulphate (SDS) and poly(N-vinyl-2 pyrrolidene), a capping reagent and sonicated until the solution changes to pink from light yellow. The shape of anisotropic gold nanoparticle depends upon the ratio of [$HAuCl_4$]/ [SDS][51]. In another method gold nanoparticle is prepared by ultrasonically irradiating a solution of $HAuCl_4$ in 1-propanol with PVP to prevent clustering until the color of the Au (III) solution changes from pale yellow to reddish purple. The size of the particle is dependent on the ultrasound frequency[52].

h) Cadmium nanoparticles attracted much attention due to their unique semiconducting, optical and photocatalytic properties and used in photovoltaic devices. Cadmium telluride nanoparticle is prepared by two different methods: (a) A solution of elementary tellurium and NaOH as well as a precursor $CdSO_4.8/3H_2O$ in EDTA is mixed and irradiated using ultrasonic probe under nitrogen atmosphere. The slurry form is centrifuged, washed by distilled water and then a dark gray product is dried. (b) A solution of elementary tellurium and NaOH and the precursor, used in this synthesis is $CdCl_2$ in EDTA is irradiated by ultrasound for 2 hours. Formation of the product in both case possess amorphous nature. The reaction is exothermic and morphology of CdTe nanoparticles are cubic in shape[53].

i) Silver nanoparticles are responsible for the photocatalytic activity, optical and electronic properties. Silver nanoparticle is prepared through the process of ultrasound irradiation via reduction method. The starting material is silver nitrate ($AgNO_3$). Silver nitrate solution is mixed with sodium borohydride dropwise. The grey coloured reaction mixture is irradiated and the slurry is separated using centrifugation. The product obtained is washed with double distilled water and dried in oven (The silver nanoparticles can also be prepared by the same process using sodium citrate as the reducing agent). Results show that the silver nanoparticles possess high degree of crystallinity. A surface plasmon resonance band is also observed. The shape and size depends upon the reducing agent used. Strong reducing agents yield larger nanoparticle compared to weak reducing agents[54].

j) Zinc nanoparticles are most popular because of their wide band gap (3.37 eV). The properties like photoluminescence and photocatalytic activity arises due to the doping in zinc nanoparticles. ZnS (zinc sulfide) nanoparticles is characterized using the following ionic liquids: 1-butyl-3- methylimidazolium bis (trifluoromethylsulfonyl) imide ([C4mim][NTf2]), 1-hexyl-3- methylimidazolium bis(trifluoromethylsulfonyl) imide ([C6mim][NTf2]), and 1-octyl-3- methylimidazolium bis (trifluoromethylsulfonyl) imide ([C8mim][NTf2]). Zinc acetate ($Zn(CH_3COO)_2.2H_2O$) is dissolved in deionized water and mixed with ionic liquid and thioacetamide. The reaction mixture is treated with ultrasonic radiation. A white colour precipitate is centrifuged and washed with ethanol and deionized water and finally dried in vacuum. The yield of nanoparticle is high and quantum confinement effect is observed. The wide band gap semiconductor is used in the preparation of blue laser and photodetectors[55].

An iron nanoparticle possesses superparamagnetic nature and is widely used in the MRI and in the manufacturing of spintronic devices. Iron oxide nanoparticle is prepared using the ferrous chloride and ferric chloride. The iron chloride is dissolved in water and irradiate in the presence of ammonia (maintained pH). Slurry obtained is washed with ethanol and water and dried in a vacuum. The iron oxide is round in shape and has monodispersity[56].

2.2.4 f-block elements

a) Lanthanum nanoparticles are intensively used due to their wide application in sensors fuel cells, catalysts and as monitoring devices. $LaFeO_3$ nanoparticles is prepared using starting materials as lanthanum carbonate and iron pentacarbonyl. The use of ultrasonic radiation affects (a) in the synthesis of the rods of $La_2O\ (CO_3)_2.1.4.H_2O$ and (b) obtaining Fe_2O_3 from $Fe(CO)_5$ and (c) in situ dispersion of the amorphous Fe_2O_3 on the prepared $La_2O(CO_3)_2.1.4.H_2O$. The fine powder of $LaFeO_3$ is obtained by calcined precursor formed by dispersing the Fe_2O_3 on the prepared $La_2O(CO_3)_2.1.4.H_2O$. Nanoparticles obtained are of high purity and homogeneity[57]. $La_{1-x}Sr_xMnO_3$ nanoparticles is synthesized using a sonochemical method and involving reagents; $KMnO_4$, $MnCl_2.4H_2O$, $La(NO_3)_3.6H_2O$, $Sr(NO_3)_2$. These chemicals are first dissolved in de-ionized water by stirring and then KOH is added to maintain the alkaline nature. Then the solution is ready for the sonication process. The solution is cooled, centrifuged, dried and then post-annealed in air. The nanoparticles, thus formed, possess a single perovskite structure[58].

b) Cerium nanoparticle is used for producing highly sensitive and selective sensing probe for determining lead ions in the solution. CeO_2 (cerium dioxide) nanopowder is prepared using different precursors $Ce_2O(CO_3)_2H_2O$, $Ce(NO_3)_3.9H_2O_3$, and $Ce(NO_3)_3.6H_2O$. Cerium nitrate hexahydrate is used to produce ceria powder. Both Cerium nitrate hexahydrate and sodium hydroxide are dissolved separately in water and sodium hydroxide is added dropwise in Cerium nitrate hexahydrate under sonication bath, at constant temperature of the reaction mixture. Finally, the solution is centrifuged, dried up and then cooled. The nanoparticles, thus obtained, possess greater energy efficacy[59].

$CePO_4$ nanoparticles is prepared using chemicals such as $NH_4H_2PO_4$ and $Ce(NO_3)_3{\cdot}6H_2O$. Firstly the $NH_4H_2PO_4$ aqueous solution is irradiated and then $Ce(NO_3)_3{\cdot}6H_2O$ is added dropwise. Using ammonia to adjust the pH, the white precipitate is obtained, which is then centrifuged and dried after washing with ethanol. Various morphologies are obtained for $CePO_4$ nanoparticles such as coral reef, nanowire, nanorods at different conditions. Resultant nanostructure such as coral reef is used in electrochemical nanosensors and it is easy to use these nanoparticles as inorganic fluorescent probe[60].

2.3 Bimetallic nanoparticles

i) Platinum – Ruthenium bimetallic nanoparticles has potential application in optoelectronic and catalysis field. It is used as the catalyst in methanol oxidation which occur at methanol fuel cell. For the preparation of Pt – Ru, a solution of ruthenium chloride, potassium tetrachloroplatinate, K_2PtCl_4, propanol and sodium dodecyl sulfate (or polyvinyl -2- pyyrolidone) in perchloric acid is sonicated. Perchloric acid is used to dissolve the $RuCl_3$. The mixture is sonicated in inert medium of argon. The reaction is slow but the reduction of ruthenium occurs faster in the presence of platinum[61].

ii) Gold – Silver core – shell bimetallic nanoparticle is widely used in electronic devices, data storage etc. The required chemicals are chloroauric acid trihydrate, silver nitrate, polyethylene glycol and ethylene glycol. Au – Ag bimetallic nanoparticle is synthesized by co-reduction method. Aqueous solution of chloroauric acid trihydrate and silver nitrate, that contains polyethyleneglycol and ethyleneglycol, is sonicated under an inert atmosphere of argon. The colour of these bimetallic nanoparticles is dark purple. Reduction of $AuCl_4^-$to metallic gold and Ag^+ to metallic silver occur in presence of alcohols. The crystal formed contains more gold nanoparticles than the silver[62].

iii) Mg(II) doped ZnO nanoparticle is prepared using Zinc acetate ($Zn(CH_3COO)_2.2H_2O$), magnesium acetate ($Mg(CH_3COO)_2.4H_2O$, diethylene glycol (DEG), lithium hydroxide ($Li(OH).H_2O$). Zinc acetate and lithium hydroxide are dissolved in diethylene glycol. The resultant solution, subjected to sonication, turns transparent. Magnesium acetate is added, followed by sonication. White precipitate is centrifuged and washed. The magnesium doped zinc oxide has smooth surface and the ability to act as the photocatalyst[63].

iv) Erbium (III) doped zinc oxide nanoparticle is synthesized using the starting material $Zn(NO_3)_2.6H_2O$, $Er(NO_3)_3.5H_2O$, TEA (triethanolamine) and PEG-20000. $Zn(NO_3)_2.6H_2O$ was dissolved in water then $Er(NO_3)_3.5H_2O$, TEA and PEG were introduced sequentially with stirring to form a clear solution. The transparent mixture yields a white precipitate upon sonication, which is centrifuged, washed and dried in air. TEA is precipitating agent and PEG-20000 is used to control the shaping size of zinc oxide nanoparticles. The luminescence intensity is greater for Erbium doped zinc oxide then the pure zinc oxide. The shape of nanoparticles is spherical and the Erbium is incorporated inside the ZnO grains[64].

3. Part-II: Applications of ultrasound for the remediation & quantisation of environmental pollutants

In this section of the review, we have tried to focus on some applications of ultrasound with emphasis on environmental remediation. Nevertheless without going in the specific details and applications of ultrasound, it is now well established that the introduction of ultrasound in any aqueous solution, sewage waste or industrial effluents brings about few characteristic changes, such as, degassing, agglomeration of particulate matter, fragmentation of particulate material, precipitation of metallic ions and respective hydroxides, oxides or metallic particles, besides the sterilization i.e. killing of bacterial and fungal colonies. These effects are mainly due to cavitation produced as a result of propagation of ultrasound. Therefore, a simple propagation of ultrasound in a system triggers a series of physical and chemical effects, which may be discussed, as under, individually and exclusively.

3.1 Degassing

Degassing[65-67] is the first effect of ultrasonic propagation, which removes dissolved gases but also increases the nucleation temperature and thus reducing ice formation compared to unsonicated water and its solutions[65]. Ultrasound may just not delay nucleation in the crystallization process but can also postpone further, if sonication continues for a longer duration[66].

3.2 Sterilisation process

Another physical effect of ultrasound propagation is on the sterilization processes[68 - 70], through the bacterial and fungal cell bodies under implosive collapse of cavitating bubbles. Pretreatment of sewage sludge, with low frequency ultrasound, ruptures cell wall. The intercellular substance, thus released, has been found to act as bio-available substrate in the anaerobic digestion process[71]. Methane production from the municipal solid waste has been reported to increase by about 24% after 45 days of anaerobic codigesion[72]. A biological study involving the effect of ultrasound on five microbes, namely Enterobacter *aero genes*, Bacillus *subtilis* and three species of Staphylococcus showed lethal effect of ultrasound on Enterobacter *aero genes* and Bacillus *subtilis*. However, three species of Staphylococcus were not affected[69]. Authors thus concluded that microbes with a thicker and soft capsule were not destroyed by the ultrasound, but those with relatively thin but hard capsules were broken down[69]. Similarly antibiotics are reported to be more effective in killing planktons bacteria, bacterial bio-films, Chlamydia and bacteria in plant, when delivered in the presence of low intensity ultrasound[70]. Anodisation of titanium and ultrasonication of natural water controlled bio films formation very effectively. Studies revealed that 30 min of sonication (24 kHz, 400 W) could bring the bacterial density of tap-water, cooling tower water and reservoir water to about three order decrease and two order decrease in sea water[73] for Gram-negative, Gram-positive, Pseudomonas *sp*, Flavobacterium *sp*, Bascillus *sp* and Micrococcus *sp*[73]. Mechanical force associated with ultrasound has been found to control the fouling and enhance membrane filtration in bio-reactors used for digestion of waste activated sludge[74].

3.3 Photocatalytic activity

Ultrasound has been used in the photocatalytic process in more than one ways. Firstly, sonochemical synthesis of photo-sensitive materials led in several instances to better catalytic composites. Secondly, ultrasound also improves photocatalytic activity.

Sonochemical fabrication of mesoporous TiO_2 inside diatom frustules for photocatalyst yielded photochemical results better than P-25 Degussa. This was attributed to larger number of accessible active sites[75]. Iron doped TiO_2 (anatase nanotubes) were found to reduce band gap energy from 3.2 to 2.75 eV, which showed photocatalytic activity 2 – 4 times higher than the value measured for pure TiO_2 nanotubes[76]. Efficiency of the photocatalytic degradation of Cresol Red[77] was 98.7 % in 60 min of duel frequency ultrasound (20/40 kHz) using active carbon fiber loaded Fe^{3+} - TiO_2. However, in a comparative study of photocatalytic activity of TiO_2 and TiO_2/Pt, synthesized by sol-precipitation, sonochemical and hydrothermal methods respectively, the degradation of Methylene Blue was found to be maximum with TiO_2 nano-powder and TiO_2/Pt nano-composite, synthesized through hydrothermal method[78]. In another comparative study of the photocatalytic activity of nano-particles, synthesized through ultrasound assisted sol-gel method (precursor $Ti[OC_4H_9]_4$) on the photodegradation Methyl Orange have been reported. The average particle dimension of bare TiO_2 and composites doped with rare earths (La & Ce) and co-doped with La-Ce were found to be in the range of 14-21 nm. The optical absorption edges for La-TiO_2, Ce-TiO_2 and La-Ce-TiO_2 shifted to the visible light region by 21 nm, 29 nm and 35 nm respectively. The photocatalytic performance of rare earth doped titania was much higher than the pure titania[79]. Low temperature sonochemical synthesis of boron doped TiO_2 with an average crystallite size of anatase B-TiO_2 in the range of about 6.2 nm and band gap of 2.74 eV, exhibited excellent photcatalytic degradation of Methylene Blue dye[80]. Fluorine doped TiO_2 [F-TiO_2] showed better photocatalytic activity in the degradation of Methyl Orange besides a potent antibacterial activity for Staphylococcus *aureus* (12 mm) and Pseudomonas *aernginosa* (5 mm)[81].

The photocatalytic activity[82] of visible light responsive photocatalyst $ZrTiO_4/Bi_2O_3$ (band gap 2.8 eV) on the decomposition of 4-Chlorophenol was reported to be better than commercially available Degussa P 25. KS Suslick [83] synthesized $BiVO_4$ using ultrasonic spray pyrolysis and found significantly superior photocatalytic activity compared to commercially available $BiVO_4$ and WO_3, likely due to difference in particle morphology. Ultrasound assisted synthesis of heterogeneous g-C_3N_4/$BiVO_4$ composite exhibited an interaction between $BiVO_4$ and g-C_3N_4 at a hetero-junction on g-C_3N_4/ $BiVO_4$ composite, which enhance the degradation and transfer of photo generated electron – hole pairs, thus converting NO to NO_2^- and NO_3^- to about four times larger extent than that of the pure composite[84]. Sonochemical synthesis of nano-crystallite Bi_2O_3, in the presence of surfactant polyvinylpyrrolidone, produced crystallites in the size range of 40–100 nm, with photoabsorption properties shifting from UV region to visible light and a band gap of 2.85 eV. Photocatalytic degradation of Methyl Orange by nano-crystallite Bi_2O_3 was found to be about 86 % with 100 min under visible light illumination ($\lambda > 400$ nm)[85]. α-Bi_2O_3 nano sheet, synthesized by graphite oxide assisted sonochemical route, has been reported to be a better photocatalyst with narrower band gap (2.25 eV) compared to other two forms of α -Bi_2O_3 porous and nanosheet covered microrods, for degrading organic pollutants under visible light[86]. Nano laminar Bi_2WO_6 has been found to be a stronger photocatalyst for the degradation of Rhodamine B compared to bulk Bi_2WO_6 nanoparticles[87]. As an extension of this work, Bi_{2-x}SbxWO$_6$ nanorodswith 2-5 % antimony doping, has been reported to greatly improve the degradation of Rhodamine B[88]. Further Ag/AgCl hybrid photocatalyst has been reported to be an excellent photocatalyst for the degradation of various dye molecules under visible light[89]. Degradation of Indigo Carmine has been reported in the presence of Ag^0–$PbMoO_4$ photocatalyst and the results indicate that the presence of Ag^0 nanoparticles on the surface of $PbMoO_4$ significantly increase the photocatalysis due to surface Plasmon resonance effect[90]. Ag/ZnO nanocomposites produced via ultrasonic spray pyrolysis method, has been found to be of considerable interest for the degradation of pollutants from textile wastewaters[91]. AbBr/Ag_3PO_4 composite quasi microcubes, synthesized sonochemically exhibited much higher photocatalytic performance and stability on the degradation of Rhodamine B dye[92]. Few ternary sulphides, such as $CdIn_2S_4$; $ZnIn_2S_4$ and $AgIn_5S_8$, prepared through ultrasonic pyrolysis method, have also been found to be good visible light photocatalysts[93]. Photocatalytic and bactericidal activity of sonochemically prepared Fe_2O_3-SnO_2 nanoparticles has been reported for the degradation of Rhodamine B and E.*coli* respectively and reported that the hydrothermally synthesized nanocomposites were better than those synthesized sonochemically[94].

3.4. Quantinization processes

For quantization of micro quantities of pollutants by dedicated techniques, such as, gas chromatography – tandem mass spectrometry, High performance liquid chromatography, gas chromatography – mass spectrometry, gas chromatography with μ - electron capture etc. require additional techniques of pre-concentration, such as, leaching, solvent extraction, emulsification and micro-extraction, without which quantization of pollutants could be erroneous. Fortunately, under these circumstances too, ultrasonication provides an answer. Few representative studies, which involve these steps are detailed below.

3.4.1 Ultrasound–leaching

Polybrominated diphenyl ethers play a great role in the polymers but during usage time these are released into the environment. Solid floating organic droplets, used for analysis for BDE-209 in sediment samples were small and thus the determination of PBDE was possible only after using ultrasound leaching. Use of ultrasound leaching technique also needed minimum organic solvent consumption in the determination of PBDE[95].

3.4.2 Solvent – extraction method

Ultrasound irradiation and vortex agitation enhanced the efficiency of solvent micro-extraction for ultrasound assisted emulsification microextraction, ultrasound assisted surfactant enhanced microextraction, vortex assisted liquid – liquid microextraction, vortex assisted surfactant enhanced emulsification liquid-liquid microextraction and ionic liquid based ultrasound assisted dispersive liquid-liquid microextraction for pre-concentration of pollutants, such as, rosmarinic acid, luteolin-7-*O*-glucoside, caffeic acid, carnosic acid and carnosol, all state-of-the-art analytical techniques[96]. Optimization of antioxidant activity of Marjoram (*Orifanummajorana L.)*, extracted using ultrasound, maximized the values of extract of total phenol content, at 15 ^{0}C in 15 min[97].

3.4.3 Emulsification and micro-extraction

Ionic liquid based ultrasound assisted emulsification microextraction combined with high performance liquid chromatography – ultraviolet has been reported for the pre-concentration and detection of organic ultraviolet filters in the environmental water samples, over the concentration range of 5-1000 ng/ml, with a limit of 0.5-1 ng/ml[98]. In the European Union the maximum admissible concentration of organo-chlorine pesticides (OCP) in the environment and drinking water is 0.1μg/L and 0.5μg/L respectively for total concentration of all OCPs. The analysis of OPC could be done by applying polyethylene Pasteur pipette-based ultrasound-assisted emulsification extraction, with a good enrichment factor in the range between 128 and 328 and depending on the analyses. The initial detection limit was between 0.8 and 10 ng/L but when organo-chlorine pesticides are extracted by ultrasound-assisted emulsification micro-extraction with GC-μECD, the detection limit ranges improved from 0.002 to 0.016μg/L[99,100]. Similarly for extraction of organo-phosphorous pesticides from water and soil samples[101], through toluene – water emulsion, the pre-concentration factors of 2390 and 1390 were reported for diazinon and chloropyrifos respectively, with a detection limit between 0.01 to 0.1 µg L^{-1}. An in-syringe ultrasound assisted emulsification micro-extraction[102] of organo-phosphorus pesticides from water samples for gas chromatographic analysis using micro-electron capture detector, improved the detection limit between 1-2 ng L^{-1}.

Five phenyl urea pesticides (i.e, diuron, diflubenzuron, teflubenzuron, flufenoxuronn and chlorfluazauron) in the environmental water samples, have been determined using a new micro-extraction technique i.e, in - situ metathesis reaction combined with ultrasound - assisted emulsification micro-extraction[103]. The optimized technique provided good repeatability (RSD 2.4 to 3.5%), linearity (0.5 - 500 µg L^{-1}), low LODs (0.06 - 0.08 µg L^{-1}) and great enrichment factor (244- 268). Carbamate pesticides in the environmental water samples are reported to have been analysed using low – density solvent based ultrasound - assisted emulsification micro-extraction for the analysis by GC-MS[104] with a lower detection limit between 0.01 – 0.1 µg L^{-1}. Similarly, ultra-pre-concentration and determination of multiple pesticide residue in water sample[105], using ultrasound assisted dispersive liquid – liquid micro-extration and its subsequent quantization by GC-FID for multiple pesticides such as triazine herbicides, organophosphorous pesticides miticides and pyrethroid pesticides has been reported to be between 0.09- 0.57 µg L^{-1}.

Concentration of two fungicides Pyrimethanil and Imazalil in river water have been determined by ultrasound assisted emulsification micro-extraction and reverse phase high performance liquid chromatography, with an enrichment factor of 169 and 172 for Pyrimethanil and Imazaliil respectively106. Estimation of eight toxic volatile organic compounds (Benzene, Chlorobenzene, 1,3-dichlorobenzene, ethylbenzene, o-xylenes and Styrene) in air, water and soil has been reported by ultrasound – assisted emulsification microextraction followed by gas chromatography with an enrichment factor of 96 – 284 for lake water samples[107]. Different types of phenols such as endocrine disrupting phenols, chlorophenols and nonylphenols can also be extracted by ultrasound-assisted emulsification micro-extraction method. However, Endocrine disrupting phenols in sea water samples and detergent samples, such as, 4-tert-butylphenol; 4-cumylphenol; 4-octylphenol; 2,4-di-tert-butylphenol, and 4-nonylphenol. For sea water samples, the limit of detection was 0.5 – 2.8 ng L^{-1} and the limit of quantification was 1.8 – 9.3 ng L^{-1}, however, for detergent samples, the limit of detection and quantification was reported to be 0.4 - 2.4 ng L^{-1} and 1.6 - 8.2 ng L^{-1} respectively[108]. A similar ultrasound assisted emulsification micro-extraction and quantization by near-IR spectroscopy, for trace amounts of nonylphenol in water samples has also been reported[109]. Several chlorophenols, such as, 2-chlorophenol, 3-chlorophenol, 2,6-dichlorophenol and 3,4-

dichlorophenol could be estimated by ultrasound-assisted liquid-liquid micro-extraction using ultrasonication probe under which chlorophenols could easily diffused through donor phase and acceptor phase. Ultrasonication has been reported to accelerate the mass transfer rate and minimize the fluid loss with increasing extraction process[110] with detection limit up to 0.03- 0.05 μg L^{-1}. Chlorinated anilines, which are antifungal and toxic to mankind and sea life, such as, 4-chloroaniline, 2,6-dichloroaniline and dichloron could be determined using ultrasonication assisted emulsification micro-extraction coupled with high performance liquid chromatography – ultraviolet detection[111]. Polybrominated flame retardants are organobromine compounds such as polybrominated diphenyl ethers: BDE-47, BDE-99, BDE-100 and BDE-153, which could be extracted from water samples by ultrasound-assisted emulsification micro-extraction technique and estimated through gas chromatography – mass spectrometry[112].

Polycyclic aromatic hydrocarbons (PAHs) are dangerous environmental pollutants of both natural and anthropogenic origin. These PAHs from soil, food, sediments and sewage sludge need detection and removal, but the impurities present in the sorbents pose serious limitation in the quantification of PAHs. These impurities may be flushed out through sonication of sorbent and subsequent detection of PAHs though liquid chromatography with fluorescence[113]. 100% recovery of many persistent organic pollutants such as, poly-aromatic hydrocarbons, poly-chlorinated biphenyls and some organo-chlorinated pesticides have been reported for many semipermeable devices used for dialysis through ultrasonic assistance[114]. Similarly one-step ultrasound extraction and purification for gas chromatographic analysis of hydrocarbons from marine sediments has been found to be a more convenient and accurate method[115]. Using the same method of extraction, petroleum oils could also be estimated by IR spectroscopy under optimized analytical condition[116]. Triclosan, an antifungal and antibacterial agent used in tooth paste, hand cleanser, air fresheners and deodorants etc. could also be extracted by in-tube ultrasonication assisted emulsification microextraction and subsequent detection with gas chromatography-electron capture detection[117]. Preservatives such as sodium benzoate, p-hydroxybenzoic acid, methyl paraben, ethyl paraben and propyl paraben are the organic acids and their esters, which are widely used as antimicrobial agents in many cosmetics, foods and pharmaceutical products. Ultrasonic assisted emulsification micro-extraction of these organics may be lead to almost total recovery with pre-concentration factor of 109 - 490, leading to their estimation in μg L^{-1} range[118]. Phenolic preservatives, such as, parabens and triclosan and related phenols in water could be extracted only in 20 min through ultrasound assisted emulsification micro-extraction[119]. Simultaneous analysis of nine pharmaceuticals[120], such as, paracetamol, metoprolol, bisoprolol, betaxolol, ketoprofen, naproxen, ibuprofen, flufenamic acid and tolfenamic acid, in waste water have been determined using high performance liquid chromatography with a hybrid triple quodrupole- linear ion trap mass spectrometry, with enrichment factors between 255-340 and recoveries between 88-111%. Analytical methods for the determination of four chlorobenzens, namely, 1,2,3-trichlorobenzene; 1,2,3,4-tetrachlorobenzene; hexachlorobenzene and 1-chloro-4-nitrobenzene using ultrasonic assisted water extraction and solvent micro-extraction and subsequent analysis by gas chromatography-ion trap mass spectrometry in soil samples, with limits of quantification range of 0.7-27.3 ng g^{-1} have been reported[121]. Results for the determination of ten perfluronated compounds[122] in sludge amended soil through ultrasonic extraction and analysis by liquid chromatography – tandem mass spectrometry showed lowest level of detection limits ranging from 0.13 to 18.3 ng g^{-1}. Determination of eight fluoroquinolones[123] in ground water samples using ultrasonic assisted ionic liquid dispersive liquid-liquid micro-extraction prior to high performance liquid chromatography and fluorescence detection has been reported to have a detection limit between 0.8 -1.3 ng L^{-1} and the enrichment factors between 122- 205.

Ultrasound assisted emulsification micro-extraction of organolead and organomanganese compounds from sea water and their determination by GC-MS, with a detection limits range from 7.0-41 ng L^{-1} and recoveries from 84-118 % has been reported[124]. Ultrasound assisted emulsification dispersive liquid-liquid micro-extraction of trace silver in 3 minutes with a detection limit of 0.45 μg $L^{-1}$and enhancement factor of 35 has been reported[125]. Ultrasonic assisted emulsification micro-extraction and pre-concentration of trace quantities of metals, such as, aluminum, bismuth, cadmium, cobalt, copper, iron, gallium, indium, nickel, lead, thallium and zinc, in water, with limits of detection as 0.13, 0.48, 0.19, 0.28, 0.29, 0.27, 0.27, 0.38 0.44, 0.47, 0.52 and 0.17 μg L^{-1} respectively have been reported[126]. Ultrasound assisted emulsification micro-extraction of total amount of iron in water and tea samples[127], has been reported with a detection limit of 7.4 μg L^{-1}. Ultrasound assisted cloud point extraction for speciation and indirect spectrophotometric determination of Cr(III) and (VI) in water samples has been reported[128] with a detection limit of 12 ng L^{-1} and average 100% recovery. Optimized ultrasound assisted emulsification – micro-extraction followed by ICP – OES determination of Lanthanum and Cerium in urine and water samples[129] have shown a detection limit of 0.012 μg L^{-1} for Lanthanum and 0.61 μg L^{-1} for Cerium respectively. As inorganic Se (IV) is more toxic than the organic selenium and hazardous to aquatic organisms as compared to Se(VI), the extraction and speciation of inorganic selenium in environmental water sample is important which is reported by ultrasonication-assisted emulsification micro-extraction using extraction solvents denser than water samples[130].

3.4.4 Ultrasound bleaching

A novel and environmental friendly technique is developed to reduce the processing cost of cotton bleaching consisting in the laccase bleaching of cotton associated with Sonochemical and hydrothermal cavitation. During the chemical bleaching process, the whiteness of the cotton fabrics increases and the amount of hydrogen peroxide required (reduction of 50%) and energy consumption in terms of temperature (reduction of 40 °C and also optimum time (reduction of 90 min.) is reduced[131].

3.4.5 Sonochemical degradation processes

There are numerous reports of degradation processes for inorganic and organic compounds, involving ultrasound. Interestingly, this effect becomes synergetic when ultrasound is coupled with anyone of the other techniques, such as oxidative, catalytic, electrolytic, microwave assisted, use of Fenton's reagent etc. The inclusion of all such studies would make this review too exhaustive and hence only a few references of the work from this laboratory are being added for further reading for the interested workers in this field[132 - 144].

3.4.6 Ultrasonically improved galvanochemical technology

The conventional method of decomposition, using flocculation and coagulation techniques for the removal of oils and metal ions from industrial waste, requires the addition of chemicals or electricity in case of electro-coagulation. Contrary to this, the Sono-Galveno-Chemical technology uses the application of a galveno-chemical reaction between iron and coke, both of which are economically viable. The ultrasonic processing reduces the particle size and thus provides an enormously large active surface area to make the cleaning process / removal of pollutants more effective[145].

4. Conclusion

From the above discussion, covering a wide range of metals and their compounds from the four blocks of elements in the periodic table, it is clear that the application of ultrasound has been tried universally in the best laboratories of the world and the results have been overwhelming. Changes in the concentration of precursors used, frequency, duration and power of ultrasound, besides the combination of other techniques with ultrasound have produced nanoparticles of desired size, texture and morphology for different application for which these are required to be produced. Even the crystallinity or amorphous character of the product could be controlled. There are immense advantages of using the sonochemical method over other methods, as under:

1. Sonochemical method is the faster method, therefore more productive in terms of time and energy consumption.
2. Products synthesized by sonochemical methods have thicker walls than the conventional methods, hence greater stability.
3. The control of particle size is easier in the sonochemical process, which is otherwise not possible by any other technique.
4. Preparation of amorphous products is possible, therefore, enhanced characteristics such as adsorption, catalytic activity etc. may be expected.

Thus we may finally say that although ultrasound may by itself not a complete analytical technique but is surely is a very important analytical tool, which has a tremendous application in many techniques, both in the synthesis of nano-composites and their detection, where the sensitivity and detection limits are surprisingly improved from μg L^{-1} to ng L^{-1}.

References

[1] Saba Hasan, A review on Nanoparticles: Their Synthesis and Types, Research Journal of Recent Sciences. 4(ISC-2014), 1-3 (2015)

[2] Information on https://shellzero.wordpress.com/2012/05/14/ nano-materials – and – its - properties/

[3] Mai Mostafa Khalaf, Hikmet Gamal Ibrahimov and Etibar Hummat Ismailov, Nanostructured Materials: Importance, Synthesis and Characterization-A Review, Chemistry Journal, 2(3)(2012)118-125.

[4] René Overney, Nanothermodynamics and Nanoparticle Synthesis (2010) NME 498A/A.

[5] Kenneth S. Suslick sonochemistry, Kirk-Othmer Encyclopedia of Chemical Technology, fourth edition, John Wiley & Sons, Inc.: New York, 26(1998)516-541.

[6] Ta Yeong Wu, Ningqun Guo, Chee Yang Teh, Jacqueline Xiao and Wen Hay "Applications of Ultrasound Technology in Environmental Remediation" in: Advances in Ultrasound Technology for Environmental Remediation, Springer Briefs in Molecular Science Green Chemistry for Sustainability, 2012, 13 – 93.

[7] Jin Ho Bang and Kenneth S. Suslick, Applications of Ultrasound to the Synthesis of Nanostructured Materials, Adv. Mater. 22(2010), 1039–1059. https://doi.org/10.1002/adma.200904093

[8] A.A. Novik, Applying Of Ultrasound for Production of Nanomaterials XXII Session of the Russian Acoustical Society Moscow, Session of the Scientific Council of Russian, Academy of Science on Acoustics, 17 (2010)276-278

[9] Gedanken, Sonochemistry and its application to nanochemistry, Current Science, 85(12)(2003)1720-1722

[10] Ahmad Monshi, Mohammad Reza Foroughi and Mohammad Reza Monshi, Modified Scherrer Equation to Estimate More Accurately Nano-Crystallite Size Using XRD, World Journal of Nano Science and Engineering, 2(2012)154-160. https://doi.org/10.4236/wjnse.2012.23020

[11] Irudayaraj Johnson, H. Joy Prabu and S. Arulappan, Ultrasound tailoring of industrially important nanoparticles, International, Journal of Scientific Research Engineering & Technology (IJSRET), 2 (2) (2013) 056-059.

[12] M. Mohapatra and S. Anand, Synthesis and applications of nano-structured iron oxides/ hydroxide, a review, International Journal of Engineering, Science and Technology, 2 (8) (2010) 127-146.

[13] Sambandam Anandan and Muthupandian Ashokkumar, Sonochemical synthesis of noble monometallic and bimetallic nanoparticles for catalytic applications in: Cavitation, 2014, 55-88.

[14] B. C. Gates, Catalytic chemistry, Wiley& Sons., New York, (1992)406.

[15] S. J. Doktycz and K. S. Suslick, Interparticle Collisions Driven by Ultrasound, 247(1990)1067-1069.

[16] M. I. Díez-Garcíaa, V. Manzi-Orezzolia, M. Jan kulovskaa, S. Anandanb, P. Bonetea, R.Gómez and T. Lana-Villarreal, Effects of ultrasound irradiation on the synthesis of metal oxide nanostructures, Physics Procedia, 63 (2015) 85 – 90. https://doi.org/10.1016/j.phpro.2015.03.014

[17] Sabriye Piskin, Sibel Kasap and Muge Sari Yilmaz, Synthesis of TiO_2 nanoparticles by Sol- Gel and sonochemical combination, Chemical and Molecular Engineering, 10(1)(2016) 117 – 119.

[18] M. Heshmat, H. Abdizadeh and M. R. Golobostanfard, Sonochemical assisted synthesis of ZnO nanostructured thin films prepared by sol-gel method, Procedia Materials Science 11(2015), 486-490. https://doi.org/10.1016/j.mspro.2015.11.070

[19] Chinnathambi Mahendiran, Raman Ganesan and Aharon Gedanken, Sonoelectrochemical Synthesis of Metallic Aluminum Nanoparticles, Eur. J. Inorg. Chem. 2009(14) (2009) 2050–2053. https://doi.org/10.1002/ejic.200900097

[20] H. Lei, Y.-J. Tang, J.-J. Wei, J. Li, X.-B. Li and H.-L. Shi, Synthesis of tungsten nanoparticles by sonoelectrochemistry, Ultrasonics Sonochemistry, 14(2007)81–83. https://doi.org/10.1016/j.ultsonch.2006.01.008

[21] Pinar Acar Bozkurt and Burak Derkus, Synthesis and characterization of CdS nanorods by combined Sonochemical - solvothermal method, Materials Science-Poland, 34(3)(2016)684–690. https://doi.org/10.1016/j.jallcom.2012.11.017

[22] Yafang Zhang, Guangfang Li, Xiaohui Yang, Hao Yang, Zhong Lu and Rong Chen, Monoclinic $BiVO_4$ micro/nano structures: Microwave and ultrasonic wave combined synthesis and their visible-light photocatalytic activities, Journal of Alloys and Compounds, 551(2013)544–550.

[23] Qin Li, Hui Li, Runming Wang, Guangfang Li, Hao Yang and Rong Chen, Controllable microwave and ultrasonic wave combined synthesis of ZnO micro-/nanostructures in HEPES solution and their shape-dependent photocatalytic activities, Journal of Alloys and Compounds, 567 (2013)1–9. https://doi.org/10.1016/j.jallcom.2013.03.077

[24] Veronica Sáez and Timothy J. Mason, Sonoelectrochemical Synthesis of Nanoparticles, Molecules, 14 (2009) 4284-4299. https://doi.org/10.3390/molecules14104284

[25] Panpailin Seeharaj, Banjong Boonchom, Piyanut Charoonsuk, Pattaraporn Kim-Lohsoontorn and Naratip Vittayakorna, Barium zirconatetitanate nanoparticles synthesized by the sonochemical method, Ceramics International, 39 (2013) S559–S562. https://doi.org/10.1016/j.ceramint.2012.10.135

[26] Sirajit Vuttivong, Surasak Niemcharoen, Panpailin Seeharaj, Wanwilai C. Vittaya kornand and Naratip Vittayakorn, Sonochemical Synthesis of Spherical $BaTiO_3$ Nanoparticles, Ferroelectrics, 457(4)(2013) 44–52. https://doi.org/10.1080/00150193.2013.847332

[27] Jong Pil Park, Jea Young Park, Cha Hwan Hwang, Myung-ho Choi, Jee Eon Kim, Kang Min Ok, and Il-Wun Shim, Synthesis of $LiCoO_2$ Nanoparticles by a Sonochemical Method under the Multibubble Sonoluminescence Conditions, Bull. Korean Chem. Soc. 31(2)(2010)327-330. https://doi.org/10.5012/bkcs.2010.31.02.327

[28] Seung Soo Lee, Ki-Taek Byun, Jong Pil Park, Sin Kyu Kim, Ho-Young Kwak and Il-Wun Shim Preparation of $Li_4Ti_5O_{12}$ nanoparticles by a simple sonochemical method, Dalton Trans. (37)(2007)4182–4184. https://doi.org/10.1039/b707164g

[29] Olga Baidu kova and Ekaterina V. Skorb, Ultrasound-assisted synthesis of magnesium hydroxide nanoparticles from magnesium, Ultrasonics Sonochemistry, 31 (2016) 423– 428. https://doi.org/10.1016/j.ultsonch.2016.01.034

[30] M. Jesionek, M. Nowak, P. Szperlich, D. Stroz J. Szala, K. Jesionek and T. Rzychon, Sonochemical growth of antimony selenoiodide in multiwalled carbon nanotube, Ultrasonics Sonochemistry, 19 (2012) 179–185. https://doi.org/10.1016/j.ultsonch.2011.06.006

[31] M. Nowak, B. Kauch, P. Szperlich, D. Stro_z, J. Szala, T. Rzychon, Ł. Bober, B.Toron and A. Nowrot, Sonochemical preparation of $SbS_{1-x}Se_xI$ nanowires, Ultrasonics Sonochemistry, 17 (2010)487–493. https://doi.org/10.1016/j.ultsonch.2009.10.006

[32] S. Avivi, O. Palchik, V. Palchik, M. A. Slifkin, A. M. Weiss and A. Gedanken Sonochemical Synthesis of Nanophase Indium Sulfide, Chem. Mater, 13(2001) 2195- 2200. https://doi.org/10.1021/cm010162+

[33] Akbar Abdollahi, Gholam Hossein Shahverdizadeh, Samad Motameni-Tabatabaii and Sara Hosseinchi-Gareaghaji, Sonochemical synthesis and characterization of nanoparticles lead(II) coordination polymer, National Conference on Nanotechnology and Green Chemistry, 936(1–3)(2009) 206–212.

[34] Qing Li,Yi Ding, Mingwang Shao,Ji, Wu,Guihua Yu, Yitai Quan, Sonochemical synthesis of nano crystalline lead chalcogenides, Materials Research Bulletin, 38(2003)549 -543.

[35] Linlin YANG, Yonggang WANG, Yujiang WANG, Xiaofeng WANG and Gaorong HAN, Morphology-controlled synthesis of $PbMoO_4$ crystals by a simple sonochemical method, Journal of the Ceramic Society of Japan, 120 (12) (2012)609-612.

[36] Yashar Azizian - Kalandaragh, Ali Khodayari, Zaiping Zeng, Christos S. Garoufalis, Sotirios Baskoutas and Lionel Cervera Gontard, Strong quantum confinement effects in SnS nanocrystals produced by ultrasound-assisted method, J Nanopart Res. 15(2013)1 – 9.

[37] Sambandam Anandan and Jerry J. Wu, Sonochemical synthesis of carbon supported Sn nanoparticles and its electrochemical application, Ultrasonics Sonochemistry, 21(2014)1954–1957. https://doi.org/10.1016/j.ultsonch.2014.02.018

[38] Shuijin Lei, Kaibin Tang, Zhen Fang, and Huagui Zheng Ultrasonic-Assisted Synthesis of Colloidal Mn_3O_4 Nanoparticles at Normal Temperature and Pressure, Crystal Growth & Design, 6(8)(2006)1757-1760. https://doi.org/10.1021/cg050402o

[39] Abdulhadi Baykal, Hüseyin Kavas, Zehra Durmuş, Mine Demir, Sinan Kazan, RamazanTopkaya and Muhammet S. Toprak Sonochemical synthesis and chracterization of Mn_3O_4 nanoparticles, Cent. Eur. J. Chem. 8(3)(2010), 633–638.

[40] Azadeh Askarinejad and Ali Morsali, Direct ultrasonic-assisted synthesis of sphere-like nanocrystals of spinel Co_3O_4 and Mn_3O_4, Ultrasonics Sonochemistry, 16 (2009) 124– 131. https://doi.org/10.1016/j.ultsonch.2008.05.015

[41] S. Mohseni Meybodi, S.A. Hosseini, M. Rezaee, S.K. Sadrnezhaad and D. Mohammadyani, Synthesis of wide band gap nanocrystalline NiO powder via a sonochemical method, Ultrasonics Sonochemistry, 19 (2012)841–845. https://doi.org/10.1016/j.ultsonch.2011.11.017

[42] D. Mohammadyani, S. A. Hosseini and S. K. Sadrnezhaad, Sonochemical Synthesis of Nickel Oxide Nano-particle-Proceeding of the 3rdconference on nanostructures, March, 2010, 722 – 725.

[43] Alireza Aslani, Vahid Oroojpour, Mehrab Fallahi, Sonochemical synthesis, size controlling and gas sensing properties of NiO nanoparticles, Applied Surface Science. 257(2011)4056–406. https://doi.org/10.1016/j.apsusc.2010.11.174

[44] P. Jeevanandam, Yu. Koltypin, and A. Gedanken, Synthesis of Nanosized alpha-Nickel Hydroxide by a Sonochemical Method, Nano Letters, 1 (5)(2001)263-266. https://doi.org/10.1021/nl010003p

[45] Laura Obreja, N. Foca, M. I. Popa and V. Melnig, Alcoholic reduction platinum nanoparticles synthesis by Sonochemical Method, Tomul I, s. Biomaterials in Biophysics, Medical Physics and Ecology, (2008) 31-36.

[46] Millan M. Mdleleni, Taeghwan Hyeon, and Kenneth S. Suslick, Sonochemical synthesis of nanostructured Molybdenum Sulfide, J. AM. Chem. Soc. 120(1998). 6189-6190. https://doi.org/10.1021/ja9800333

[47] Azadeh Askarinejad and Ali Morsali, Direct ultrasonic-assisted synthesis of sphere-like nanocrystals of spinel Co_3O_4 and Mn_3O_4, UltrasonicsSonochemistry, 16 (2009) 124– 131. https://doi.org/10.1016/j.ultsonch.2008.05.015

[48] Shiva Hojaghani, Moayed Hosseyni Sadr and Ali Morsali Sonochemical Synthesis of a Cobalt (II) Coordination Polymer Nano-structure with azo ligand: A New

Precursor for Preparation Pure Phase of Co_3O_4 Nanostructure, JNS. 3(2013)109-114.

[49] Weizhong Lv, Qi Qiu, Fang Wang, Shaohui Wei and Bo Liu, Zhongkuan Luo Sonochemical synthesis of cobalt aluminate nanoparticles under various preparation parameters, Ultrasonics Sonochemistry, 17 (2010)793–801. https://doi.org/10.1016/j.ultsonch.2010.01.018

[50] Sankar Chakma, Vijayanand S. Moholkar, Sonochemical synthesis of mesoporous $ZrFe_2O_5$ and its application for degradation of recalcitrant pollutants, RSC Adv. 5(2015) 53529-53542. https://doi.org/10.1039/C5RA06148B

[51] Jong-Eun Park, Mahito Atobe and Toshio Fuchigam, Synthesis of multiple shapes of gold nanoparticles with controlled sizes in aqueous solution using ultrasound, Ultrasonics Sonochemistry, 13 (2006)237–241. https://doi.org/10.1016/j.ultsonch.2005.04.003

[52] Kenji Okitsu, Muthupandian Ashokkumar, and Franz Grieser Sonochemical Synthesis of Gold Nanoparticles: Effects of Ultrasound Frequency, J. Phys. Chem. B. 109(44) (2005)20673-20675. https://doi.org/10.1021/jp0549374

[53] Irena Ban, Matjaž Kristl, Valerija Danč, Anita Danč and Miha Drofenik, Preparation of cadmium telluride nanoparticles from aqueous solutions by sonochemical method, Materials Letters, 67 (2012)56–59. https://doi.org/10.1016/j.matlet.2011.09.001

[54] Irshad A. Wani, Aparna Ganguly, Jahangeer Ahmed and Tokeer Ahmad, Silver nanoparticles: Ultrasonic wave assisted synthesis, optical characterization and surface area studies, Materials Letters, 65 (2011)520–522.

[55] Elaheh K. Goharshadi, Sayyed Hashem Sajjadi, Roya Mehrkhah and Paul Nancarrow, Sonochemical synthesis and measurement of optical properties of zinc sulfide quantum dots, Chemical Engineering Journal, 209 (2012)113–117. https://doi.org/10.1016/j.cej.2012.07.131

[56] Md. Nazrul Islam, Le Van Phong, Jong-Ryul Jeong and Cheol Gi Kim, A facile route to sonochemical synthesis of magnetic iron oxide (Fe_3O_4) nanoparticles, Thin Solid Films, 519 (2011)8277–8279. https://doi.org/10.1016/j.tsf.2011.03.108

[57] M. Siva kumar, A. Gedan ken, W. Zhong, Y. H. Jiang, Y. W. Du, I. Bru kental, D. Bhattacharya, Y. Yeshurun and I. Nowik, Sonochemical synthesis of nanocrystalline $LaFeO_3$, J. Mater. Chem. 14(2004) 764 –769.

[58] Phan Van Cuong, Nguyen Hong Quang, Soo-HO Kim, Saket Asthana, Joonghoe Dho and Do- Hyung Kim, $La_{1-x}Sr_xMnO_3$ perovskite nanoparticles-sonochemical synthesis and their properties, proceedings of IWNA , 17(2007)184-186.

[59] Dipak Vitthal, Pinjari Aniruddha and Bhalchandra Pandit, Room temperature synthesis of crystalline CeO_2 nanopowder: Advantage of sonochemical method over conventional method, Ultrasonics Sonochemistry, 18 (2011) 1118–1123. https://doi.org/10.1016/j.ultsonch.2011.01.008

[60] Amin Shiralizadeh Dezfuli, Mohammad Reza Ganjali and Parviz Norouzi, Facile sonochemical synthesis and morphology control of $CePO_4$ nanostructures via an oriented attachment mechanism: Application as luminescent probe for selective sensing of Pb2+ ion in aqueous solution, Materials Science and Engineering C. 42 (2014) 774– 781. https://doi.org/10.1016/j.msec.2014.06.012

[61] K.Vinodgopal, Yuanhua He, Muthupandian Ashokkumar and Franz Griser, Sonochemically prepared platinum-ruthenium bimetallic nanoparticles, The Journal of Physical Chemistry B. 110(9) (2006)3849-3852. https://doi.org/10.1021/jp060203v

[62] Sambandam Anandan, Franz Grieser, and Muthupandian Ashokkumar, Sonochemical Synthesis of Au-Ag Core-Shell Bimetallic Nanoparticles, J. Phys. Chem. C. 112(39) (2008)15102–15105. https://doi.org/10.1021/jp806960r

[63] Xianyong Lu, Zhaoyue Liu, Ying Zhu and Lei Jiang Sonochemical synthesis and photocatalytic property of zinc oxide nanoparticles doped with magnesium(II), Materials Research Bulletin,46 (2011) 1638–1641.

[64] Jun Geng, Guang-Hui Song and Jun-Jie Zhu, Sonochemical Synthesis of Er^{3+}-Doped ZnO Nanospheres with Enhanced up-conversion Photoluminescence, Hindawi Publishing Corporation, Journal of Nanomaterials, 2012 (2012) 1 – 5. https://doi.org/10.1155/2012/317857

[65] Deyang Yu, Baolin Liu and Bochun Wanga, The effect of ultrasonic waves on the nucleation of pure water and degassed water, Ultrasonics Sonochemistry, 19 (2012) 459–463. https://doi.org/10.1016/j.ultsonch.2011.08.005

[66] Hossein Kiani, Da-Wen Sun, Adriana Delgado and Zhihang Zhang, Investigation of the effect of power ultrasound on the nucleation of water during freezing of agar gel samples in tubing vials, Ultrasonics Sonochemistry, 19 (2012) 576–581.

[67] Liyan Liu, Yang Yang, Penghong Liu and Wei Tan, The influence of air content in water on ultrasonic cavitation field –Ultrasonics Sonochemistry 21 (2014) 566–571. https://doi.org/10.1016/j.ultsonch.2013.10.007

[68] Pankaj and Manju Chauhan, Sonochemical removal of hardness and sterilization of underground potable water, J Ind Council Chem. 23(1)(2006)38–40

[69] Shengpu Gao , Gillian D. Lewis , Muthupandian Ashokkumar , Yacine Hemar, Inactivation of microorganisms by low-frequency high-power ultrasound: 1. Effect of growth phase and capsule properties of the bacteria, Ultrasonics Sonochemistry Jan;21(1)(2014)446-453

[70] Hao Yu, Siping Chen and Ping Cao, Synergistic bactericidal effects and mechanisms of low intensity ultrasound and antibiotics against bacteria: A review, Ultrasonics Sonochemistry, 19(2012) 377–382. https://doi.org/10.1016/j.ultsonch.2011.11.010

[71] Basak Savun , Uwe Neis , Nilsun H. Ince and Orhan Yenigün , Pretreatment of Sewage Sludge by Low-frequency Ultrasoun, J. Adv. Oxid. Technol. 15(2)(2012)374-379.

[72] Alessandra Cesaro, Vincenzo Naddeo, Valeria Amodio and Vincenzo Belgiorno, Enhanced biogas production from anaerobic codigestion of solid waste by sonolysis, Ultrasonics Sonochemistry, 19 (2012) 596–600. https://doi.org/10.1016/j.ultsonch.2011.09.002

[73] SD Ruth Nithila, B Anandkumar, SC Vanithakumari, RP George, U Kamachi and RK Dayal, Studies to control biofilm formation by coupling ultrasonication of natural waters and anodisation of titanium, Ultrasonic Sonochemistry, 21(2014) 189 – 199. https://doi.org/10.1016/j.ultsonch.2013.06.010

[74] Meilan Xu, Xianghua Wen, Xia Huang, Zhiyong Yu and Min Zhu, Mechanisms of membrane fouling controlled by online ultrasound in an anaerobic membrane bioreactor for digestion of waste activated sludge, Journal of Membrane Science, 445 (2013) 119–126. https://doi.org/10.1016/j.memsci.2013.06.006

[75] Lin Maoa, Jia Liu, Shenmin Zhu, Di Zhang, Zhixin Chen and Chenxin Chen, Sonochemical fabrication of mesoporous TiO2 inside diat-om frustules for photocatalyst, Ultrasonics Sonochemistry, 21 (2014) 527–534. https://doi.org/10.1016/j.ultsonch.2013.09.001

[76] XiuzhenWeia, Huijuan Wanga, Guangfeng Zhua and Liping Zhub, Iron-doped TiO_2 nanotubes with high photocatalytic activity under visible light synthesized by

an ultrasonic-assisted sol-hydrothermal method, Ceramics International, 39 (2013) 4009–4016. https://doi.org/10.1016/j.ceramint.2012.10.251

[77] Shaofeng Xiong, Zhoulan Yin, Yuanjin Zhou, Xianzhong Peng, Wenbin Yan, Zhixiong Liu, and Xiangyu Zhang ,The Dual-frequency (20/40 kHz) Ultrasound Assisted Photocatalysis with the Active Carbon Fiber-loaded Fe^{3+}-TiO_2 as Photocatalyst for Degradation of Organic Dye, Bull. Korean Chem. Soc. 34(10)(2013)3039 – 3045. https://doi.org/10.5012/bkcs.2013.34.10.3039

[78] Vojka Zunicˇ, Marija Vukomanovic, Srecˇo D. Škapin, Danilo Suvorov and Janez Kovacˇ, Photocatalytic properties of TiO_2 and TiO_2/Pt: A sol-precipitation, sonochemical and hydrothermal approach, Ultrasonics Sonochemistry, 21(1)(2014)367-375. https://doi.org/10.1016/j.ultsonch.2013.05.018

[79] Hongtao Gao, Wenchao Liu , Bing Lu and Fangfang Liu, Photocatalytic Activity of La, Y Co-Doped TiO2 Nanoparticles Synthesized by Ultrasonic Assisted Sol–Gel Method, Nanoscience and Nanotechnology, 12(2012) 3959–3965. https://doi.org/10.1166/jnn.2012.5859

[80] X.-K. Wang, C. Wang and D. Zhang, Low temperature sonochemical synthesis of boron doped TiO_2 nanoparticles with enhanced visible light induced photocatalytic activity, Materials Research Innovations, 16(6)(2012)395-399. https://doi.org/10.1179/1433075X11Y.0000000052

[81] M. Umadevi, M. Sangari, R. Parimaladevi, A. Sivanantham and J. Mayandi, Enhanced photocatalytic, antimicrobial activity and photovoltaic characteristics of fluorine doped TiO_2 synthesized under ultrasound irradiation, Journal of Fluorine Chemistry, 156 (2013) 209–213. https://doi.org/10.1016/j.jfluchem.2013.10.011

[82] Bernaurdshaw Neppolian, Youngae Kim, Muthupandian Ashokkumar, Hiromi Yamashita and Heechul Choi, Preparation and properties of visible light responsive $ZrTiO_4$/Bi_2O_3 photocatalysts for 4-chlorophenol decomposition, Journal of Hazardous Materials, 182 (2010) 557–562. https://doi.org/10.1016/j.jhazmat.2010.06.069

[83] Scott S.Dunkle, Richard J. Helmich, and Kenneth S. Suslick, $BiVO_4$ as a Visible-Light Photocatalyst Prepared by Ultrasonic Spray Pyrolysis, The physical chemistry C letters, 113(2009)11980–11983.

[84] Man Ou, Qin Zhong, Shule Zhang and Lemeng Yu, Ultrasound assisted synthesis of heterogeneous g-C_3N_4/$BiVO_4$ composites and their visible-light-induced photocatalytic oxidation of NO in gas phase, Journal of Alloys and Compounds, 626 (2015) 401–409. https://doi.org/10.1016/j.jallcom.2014.11.148

[85] Lisha Zhang, Wenzhong Wang, Jiong Yang a, Zhigang Chen, Wenqing Zhang, Lin Zhou and Shengwei Liu, Sonochemical synthesis of nanocrystallite Bi_2O_3 as a visible-light-driven photocatalyst, Applied Catalysis A: General, 308 (2006) 105–110. https://doi.org/10.1016/j.apcata.2006.04.016

[86] Dong Xie, Qingmei Su, Jun Zhang, Gaohui Du and Bingshe Xu, Graphite oxide-assisted sonochemical preparation of a-Bi_2O_3nanosheets and their high-efficiency visible light photocatalytic activity, J Mater Sci. 49 (2014) 218–224. https://doi.org/10.1007/s10853-013-7695-9

[87] Feng-Jun Zhang , Fa-Zhi Xie , Jin Liu , Wei Zhao and Kan Zhang , Rapid sonochemical synthesis of irregular nanolaminar-like Bi_2WO_6 as efficient visible-light-active photocatalysts, Ultrasonics Sonochemistry, 20 (2013) 209–215. https://doi.org/10.1007/s10853-013-7900-x

[88] Anamika Singh, Dimple P. Dutta, Mainak Roy, A.K. Tyagi and M. H. Fulekar, Sonochemical synthesis, characterization, and photocatalytic properties of Bi2-xSbxWO6 nanorods, J Mater Sci. 49 (2014) 2085–2097. https://doi.org/10.1007/s10853-013-7900-x

[89] Deliang Chen, Seung Hwa Yoo, Qingsong Huang, Ghafar Ali and Sung Oh Cho , Sonochemical Synthesis of Ag/AgCl Nanocubes and Their Efficient Visible-Light-Driven Photocatalytic Performance, Chemistry of a European journal, 18(17)(2012) 5192-5200. https://doi.org/10.1002/chem.201103787

[90] Gobinda Gyawali, Rajesh Adhikari, Bhupendra Joshi, Tae Ho Kim, Vicente Rodríguez-González and Soo Wohn Lee, Sonochemical synthesis of solar-light-driven Ag°-$PbMoO_4$ photocatalyst, Journal of Hazardous Materials, 263P (2013) 45– 51. https://doi.org/10.1016/j.jhazmat.2013.03.065

[91] Kamil Burak Dermenci, Bora Genc, Burcak Ebin, Tugba Olmez-Hanci and Sebahattin Gurmen, Photocatalytic studies of Ag/ZnO nanocomposite particles produced via ultrasonic spray pyrolysis method, Journals of Alloys and Compounds, 586(2014) 267-273. https://doi.org/10.1016/j.jallcom.2013.10.004

[92] Jinsong Xie, Difang Zhao, Changan Tian, Chengliang Han and Hui Zhang, Simple ultrasonic construction of AgBr/Ag_3PO_4 hybrid quasi-microcube with improved visible – driven photocatalytic property, Micro & Nano Letters, 8(7)(2013)353–356. https://doi.org/10.1049/mnl.2013.0141

[93] Jianhui Huang,Wahkit Cheuk, Yifan Wu, Lee, Frank S. C. and Wingkei Ho, Template-free synthesis of ternary sulfides submicrospheres as visible light

photocatalysts by ultrasonic spray pyrolysis, Catalysis Science & Technology, 2(9)(2012)1825-1827. https://doi.org/10.1039/c2cy20053h

[94] C. Karunakaran, S.Sakthi Raadha and P.Gomathisankar, Photocatalytic and bactericidal activities of hydrothermally and sonochemically prepared Fe_2O_3–SnO_2 nanoparticles, Materials Science in Semiconductor Processing, 16(177) (2013)818-824. https://doi.org/10.1016/j.mssp.2012.12.030

[95] Nerina B. Lana, Paula Berton, Adrian Covaci, Adrián G. Atencio, Néstor F. Cioccoe and Jorgelina C. Altamiranoa, Ultrasound leaching–dispersive liquid–liquid microextraction based on solidification of floating organic droplet for determination of poly brominated diethyl ethers in sediment samples by gas chromatography-tandem mass spectrometry, Journal of Chromatography A, 1285(2013) 15-21. https://doi.org/10.1016/j.chroma.2013.02.027

[96] Vasil Andruch, Ioseph S. Balogh, Martin Burdel, Lívia Kocúrová and Jana Šandrejová, Application of ultrasonic irradiation and vortex agitation in solvent microextraction, TrAC Trends in Analytical Chemistry, 49(2013)1-19.

[97] Mohammad Hossain, Nigel Brunton, Ankit Patras, Brijesh Tiwari, C. O'Donnell, Optimization of Ultrasound Assisted Extraction of Antioxidant Compounds from Marjoram (*Origanum majorana L.*) Using Response Surface Methodology, Ultrasonics Sonochemistry, 19(3)(2012)582-590. https://doi.org/10.1016/j.ultsonch.2011.11.001

[98] Dandan Ge and Hian Kee Lee, A new 1-hexyl-3-methyl imidazoliumtris (pentafluoro ethyl) tri fluoro phosphate ionic liquid based ultrasound-assisted emulsification microextraction for the determination of organic ultraviolet filters in environmental water samples, Journal of Chromatography A, 1251 (2012) 27–32. https://doi.org/10.1016/j.chroma.2012.06.048

[99] Yufeng Zhang and Hian Kee Lee, Application of ultrasound-assisted emulsification microextraction based on applying low-density organic solvent for the determination of organochlorine pesticides in water samples, Journal of Chromatography A, 1252 (2012) 67– 73. https://doi.org/10.1016/j.chroma.2012.06.065

[100] Senar Ozcan, Ali Tor and Mehmet Emin Aydin, Application of ultrasound-assisted emulsification-microextraction for the analysis of organochlorine pesticides in waters, water research, 43(2009) 4269-4277.

[101] Yaser Abdollahzadeh, Yadollah Yamini, Ali Jabbari, Ali Esrafili and Mohammad Rezaeec, Application of ultrasound-assisted emulsification microextraction

followed by gas chromatography for determination of organophosphorus pesticides in water and soil samples, Anal. Methods, 4, (2012) 830 - 837. https://doi.org/10.1039/c2ay05653d

[102] Yi-Song Su and Jen-Fon Jen, Determination of organophosphorous pesticides in water using in-syringe ultrasound-assisted emulsification and gas chromatography with electron-capture detection, Journal of Chromatography A, 1217 (2010) 5043–5049. https://doi.org/10.1016/j.chroma.2010.06.006

[103] Jiaheng Zhang, Zhe Liang, Songqing Li, Yubo Li, Bing Peng, Wenfeng Zhou and Haixiang Gao, In-situ metathesis reaction combined with ultrasound-assisted ionic liquid dispersive liquid–liquid microextraction method for the determination of phenyl urea pesticides in water samples, Talanta, 98 (2012) 145–151. https://doi.org/10.1016/j.talanta.2012.06.062

[104] Liang Guo and Hian Kee Lee, Low-density solvent based ultrasound – assisted emulsification microextraction and on – column derivatization combined with gas chromatography – mass spectrometry for the determination of carbamate pesticides in environmental water samples, Journal of Chromatography A, 1235(2009)1-9. https://doi.org/10.1016/j.chroma.2012.02.045

[105] Shumin Cui, Qianxia Chen, Weiping Wang, Jigen Miao, Aijun Wang, and Jianrong Chen Ultra-Preconcentration and Determination of Multiple Pesticide Residues in Water Samples Using Ultrasound-Assisted Dispersive Liquid–Liquid Microextraction and GC-FID, Chromatographia, 76(2013) 671–678. https://doi.org/10.1007/s10337-013-2441-7

[106] Zhixi Gao, Yanhong Wu, Huajun, Zhao, Fangying Ji, Qiangz He and Si Li, Concentration determination of new fungicide in river water by ultrasound-assisted emulsification micro-extraction and reversed-phase high performance liquid chromatography, Anal. Methods, 4(2012) 2365–2368. https://doi.org/10.1039/c2ay25372k

[107] Mei-I. Leong and Shang-Da Huang, Determination of volatile organic compounds in water using ultrasound assisted emulsification microextraction followed by gas chromatography, J. Sep. Sci. 35(2012),688–694. https://doi.org/10.1002/jssc.201100610

[108] Ming-Wei Shu, Mei-I Leong, Ming-Ren Fuh and Shang-Da Huang, Determination of endocrine-disrupting phenols in water samples by a new manual shaking-enhanced, ultrasound-assisted emulsification microextraction method, Analyst, 137(2012)2143-2150. https://doi.org/10.1039/c2an16117f

[109] Wenbo Yuan, Bingren Xiang, Liyan Yu and Liangzhen Zhu, Feasibility study on ultrasound-assisted emulsification microextraction-near infrared spectroscopy technique for the determination of traces of nonylphenol in water samples, J. Near Infrared Spectrosc. 20(2012) 675-685. https://doi.org/10.1255/jnirs.1023

[110] Yu-Ying Chao, Yi-Ming Tu, Zhi-Xuan Jian , Hsaio-Wen Wang and Yeou-Lih Huang, Direct determination of chlorophenols in water samples through ultrasound-assisted hollow fiber liquid–liquid–liquid microextraction on-line coupled with high-performance liquid chromatography, Journal of Chromatography A, 1271 (2013) 41– 49. https://doi.org/10.1016/j.chroma.2012.11.039

[111] Abilasha Ramkumar, Vinoth Kumar Ponnusamy, Jen-Fon and Jen n, Rapid analysis of chlorinated anilines in environmental water samples using ultrasound assisted emulsification microextraction with solidification of floating organic droplet followed by HPLC-UV detection, Talanta, 97 (2012) 279–284. https://doi.org/10.1016/j.talanta.2012.04.031

[112] Ariel R. Fontana, Rodolfo G.Wuillouda, Luis D. Martínez and Jorgelina C. Altamiranoa, Simple approach based on ultrasound - assisted emulsification - microextraction for determination of polibrominated flame retardants in water samples by gas chromatography–mass spectrometry, Journal of Chromatography A, 1216 (2009) 147–153. https://doi.org/10.1016/j.chroma.2008.11.034

[113] Y. Moliner-Martínez, R.A. González-Fuenzalida, R. Herráez-Hernández, P. Campíns –Falcó and J. Verdú-Andrés, Cleaning sorbents used in matrix solid-phase dispersion with sonication: Application to the estimation of polycyclic aromatic hydrocarbons at ng/g levels in marine sediments, Journal of Chromatography A, 1263 (2012) 43– 50. https://doi.org/10.1016/j.chroma.2012.09.034

[114] Julen Bustamante, Patricia Navarro, Gorka Arana, Albertode Diego, Juan and Manuel Madariaga, Ultrasound assisted dialysis of semi-permeable membrane devices for the simultaneous analysis of a wide number of persistent organic pollutants, Talanta, 114(2013) 32–37. https://doi.org/10.1016/j.talanta.2013.03.076

[115] Marco Pietroletti, Serena Mattiello, Francesca Moscato, Federico Oteri and Mauro Mecozzi, One Step Ultrasound Extraction and Purification Method for the Gas Chromatographic Analysis of Hydrocarbons from Marine Sediments: Application to the Monitoring of Italian Coasts, Chromatographia, 75(2012) 961–971. https://doi.org/10.1007/s10337-011-2172-6

[116] Huidong Qiu and Guobing Luo, A simple and rapid method for determination of petroleum oils in sewage sludge samples with ultrasonic solvent extraction by infrared spectrophotometry under optimized analytical conditions, Anal. Methods, (2012) 4, 3891-3896. https://doi.org/10.1039/c2ay25910a

[117] Hou-Kung Shih, Chiao-Wen Lin, Vinoth Kumar Ponnusamy, Abilasha Ramkumaraand Jen-Fon Jen, Rapid analysis of triclosan in water samples using an in-tube ultrasonication assisted emulsification microextraction coupled with gas chromatography-electron capture detection, Anal. Methods, 5(2013), 2352-2359. https://doi.org/10.1039/c3ay40104a

[118] Yadollah Yamini, Abolfazl Saleh, Mohammad Rezaee, Leila Ranjbar and Morteza Moradi, Ultrasound- assisted emulsification microextraction of various preservatives from cosmetics, beverages, and water samples, Journal of Liquid Chromatography & Related Technologies, 35(2012) 2623–2642.

[119] Jorge Regueiro, Maria Llompart, Elefteria Psillakis , Juan C. Garcia-Monteagudo and Carmen Garcia-Jaresa,, Ultrasound-assisted emulsification–microextraction of phenolic preservatives in water, Talanta, 79 (2009) 1387–1397. https://doi.org/10.1016/j.talanta.2009.06.015

[120] M.M. Parrill Vázquez, P. Parrill Vázquez, M. Martínez Galer, M.D. Gil Garcí and A. Uclés, Ultrasound-assisted ionic liquid dispersive liquid–liquid microextraction coupled with liquid chromatography-quadrupole-linear ion trap-mass spectrometry for simultaneous analysis of pharmaceuticals in wastewaters, Journal of Chromatography A, 1291 (2013) 19– 26. https://doi.org/10.1016/j.chroma.2013.03.066

[121] Li Wanga, Linling Wang, Jing Chen, Wenjun Du, Guoliang Fan and Xiaohua Lu, Ultrasonic-assisted water extraction and solvent bar microextraction followed by gas chromatography–ion trap mass spectrometry for determination of chlorobenzenes in soil samples, Journal of Chromatography A, 1256 (2012) 9– 14. https://doi.org/10.1016/j.chroma.2012.07.044

[122] Ana I. Garcıa-Valcarcel, Esther Miguel and Jose L. Tadeo, Determination of ten perfluorinated compounds in sludge amended soil by ultrasonic extraction and liquid chromatography-tandem mass spectrometry, Anal. Methods, 4(2012)2462-2468. https://doi.org/10.1039/c2ay25387a

[123] M.M. Parrilla Vázquez, P. Parrilla Vázquez, M. Martínez Galera, M.D. Gil García, Determination of eight fluoroquinolones in groundwater samples with ultrasound-assisted ionic liquid dispersive liquid–liquid microextraction prior to high-

performance liquid chromatography and fluorescence detection, Analytica Chimica Acta, 748 (2012) 20– 27. https://doi.org/10.1016/j.aca.2012.08.042

[124] Natalia Campillo, Juan Ignacio Cacho, Javier Marín, PilarViñas and Manuel Hernández-Córdoba, Ultrasound-assisted emulsification microextraction of organolead and organomanganese compounds from seawater, and their determination by GC-MS, Micro chim Acta, 181(2014)97–104.

[125] Xiaodong Wen, Lamei Kong, Meihui Chen, Qingwen Deng, Xia Zhao and Jie Guo, A new coupling of spectrophotometric determination with ultrasound-assisted emulsification dispersive liquid–liquid microextraction of trace silver, Spectrochimica Acta Part A: Molecular and Biomolecular Spectroscopy, 97 (2012) 782–787. https://doi.org/10.1016/j.saa.2012.07.078

[126] Hassan Sereshti, Yeganeh Entezari, and Heravi, Soheila Samadi, Optimized ultrasound-assisted emulsification microextraction for simultaneous trace multielement determination of heavy metals in real water samples by ICP-OES, Talanta, 97 (2012) 235–241. https://doi.org/10.1016/j.talanta.2012.04.024

[127] Gholamreza Khayatian and Shahed Hassanpoor, Ultrasound Assisted Emulsification Microextraction Based on dimethyl (E)-2-[(Z)-1-acetyl)-2-hydroxy-1-propenyl]-2-butenedioate for Determination of Total Amount of Iron in Water and Tea Samples, J. Chin. Chem. Soc. 59(5) (2012), 659-666. https://doi.org/10.1002/jccs.201100447

[128] Mahdi Hashemia, Seyed and Mosayeb Daryanavarda, Ultrasound-assisted cloud point extraction for speciation and indirect spectrophotometric determination of chromium (III) and (VI) in water samples, Spectrochimica Acta Part A, 92 (2012) 189– 193. https://doi.org/10.1016/j.saa.2012.02.073

[129] Hassan Sereshti, Ahmad Rohani Far and Soheila Samad, Optimized ultrasound-assisted emulsification- microextraction followed by ICP-OES for simultaneous determination of Lanthanum and Cerium in urine and water samples, Analytical Letters, 45(2012) 1426–1439. https://doi.org/10.1080/00032719.2012.675490

[130] Nahid Mashkouri Najafia, Hamed Tavakoli, Yaser Abdollahzadeh and Reza Alizadeh, Comparison of ultrasound-assisted emulsification and dispersive liquid–liquid microextraction methods for the speciation of inorganic selenium in environmental water samples using low density extraction solvents, Analytica Chimica Acta, 714 (2012) 82– 88. https://doi.org/10.1016/j.aca.2011.11.063

[131] Idalina Gonçalves, Madalena Martins, Ana Loureiro, Andreia Gomes, Artur Cavaco-Paulo and Carla Silva, Sonochemical and hydrodynamic cavitation

reactors for laccase/hydrogen peroxide cotton bleaching, Ultrasonics Sonochemistry, 21(2) (2014) 774-781. https://doi.org/10.1016/j.ultsonch.2013.08.006

[132] Pankaj, Shikha Goyal and Prem Kishore Patnala, Degradation of Reactive, Acid and Basic textile dyes in the presence of Ultrasound and rare Earths [La and Pr], Ultrasonics Sonochemistry, 21(6)(2014) 1994-2009.

[133] Pankaj Srivastava, Prem Kishore Patnala and Shikha Goyal, Sonolytic decolourisation of Reactive Orange 107 dye in the presence of Titanium dioxide and Rare Earths, International Journal of Innovative Research in Science & Engineering, 2(4) (2014) 140-148.

[134] Pankaj Srivastava, Shikha Goyal and Rajesh Tayade, Ultrasound-assisted adsorption of Reactive Blue 21 dye on TiO2 in the presence of some rare earths (La, Ce, Pr & Gd), Canadian Journal Chem Engg. 92(1)(2014) 41-51. https://doi.org/10.1002/cjce.21799

[135] Pankaj Srivastava, Prem Kishore Patnala and Shikha Goyal, Sonolytic decolourisation of Acid Red 88 dye in the presence of Titanium dioxide and Rare Earths, Journal of Applicable Chemistry, 2 (2013) 66-72.

[136] Pankaj, Shikha Goyal and Prem Kishore Patnala, Role of Ceric ion (Ce^{4+}) in the Sonosorption of Acid Red 114, Reactive Blue 21 and Basic Violet 16 dyes on TiO_2, J. Pure & Appl.Ultrasonics, 35(2013)129-132.

[137] Pankaj and Shikha Goyal, Sonochemical decolourisation of Reactive Blue 21 and Acid Red 114 in the presence of TiO_2 and Rare Earths, Material Science Forum, Switzerland, 734 (2013) 237-247.

[138] Pankaj, Theoretical and Experimental Sonochemistry Involving Inorganic Systems in: "Aqueous Inorganic Sonochemistry", Pankaj and M. Ashokkumar (Eds.), Springer, UK. Chapter 9, 2010, pp. 213 - 271.

[139] Pankaj, Manju Chauhan. "Sonochemical Study on Multivalent Cations (Fr, Cr & Mn)" in: Theoretical and Experimental Sonochemistry Involving Inorganic Systems, Pankaj and M. Ashokkumar (Eds.), Springer, UK, Chapter 10, 2010, pp. 273 - 285.

[140] Pankaj and Mayank Verma, "Sonochemical degradation of phenol in the presence of inorganic catalytic materials" in: Theoretical and Experimental Sonochemistry Involving Inorganic Systems, Pankaj and M. Ashokkumar (Eds.), Springer, UK. Chapter 11, 2010, pp.287 – 313.

[141] Mayank Verma and Pankaj, “Sono-photo-catalytic degradation of Amines In water” in: Theoretical and Experimental Sonochemistry Involving Inorganic Systems, Pankaj and M. Ashokkumar (Eds), Springer, UK, Chapter 12, 2010, pp.315 – 336.

[142] Pankaj and Mayank Verma, “Sonophotocatalytic behavior of cerium doped salts of Cu(II), Co(II) and Mn(II) in the degradation of phenol”, Indian J. Chem. 48A(2009)367-371.

[143] Pankaj,Mayank Verma and Himanshi Rikhy. “Sono-photo-catalytic behavior of Cerium in the Degradation of Potassium Iodide”, J. Pure & Appl. Ultrasonics, 31(3) (2009)105-109.

[144] Manisha V. Bagal and Parag R Gogate, Waste water treatment using hybrid treatment schemes based on cavitation and Fenton chemistry : A Review, Ultrasonics Sonochemistry, 21 (2014) 1 – 14. https://doi.org/10.1016/j.ultsonch.2013.07.009

[145] Vladimir O. Abramov, Anna V. Abramova, Petr P. Keremetin, Marat S. Mullakaev, Georgiy B. Vexler and Timothy J. Mason, Ultrasonically improved galvanochemical technology for the remediation of industrial wastewater, Ultrasonics Sonochemistry, 21 (2014) 812–818. https://doi.org/10.1016/j.ultsonch.2013.08.013

Keyword Index

About the Editors

Dr. Rajesh J. Tayade

Dr. Rajesh Tayade is currently working in the Inorganic Materials and Catalysis Division, CSIR-Central Salt & Marine Chemicals Research Institute, (CSIR-CSMCRI), Bhavangar, India since 1993. He had received Master degree in Electronic Science from Savitribai Phule Pune University, Pune, India and Ph. D. from Saurashtra University, Rajkot, India. During his doctoral study he has synthesised tuned bandgap, and easy to separate photocatalytic semiconductor nanomaterials for photocatalytic degradation of organic compounds present in water. He has received Fast Track Young Scientist Award from Department of Science and Technology, India to work in the area of development of Light Emitting Diode (LEDs) based photocatalytic reactor for purification of water. He has published 47 research articles in international journals of repute and 3 book chapters in knowledge-based editions published by renowned international publisher. He has published 5 edited book with Trans Tech Publications Ltd., Switzerland. He had worked as a visiting scientist for one year at South African Institute for Advanced Materials Chemistry, University of Western Cape, Bellville, Cape Town, South Africa. He also worked as Brain Pool Professor at Kyungpook National University, Daegu during 2013-2014 in the field of photocatalysis. His current research interest includes the development of photocatalytic materials and surfaces, application of photocatalytic processes for environmental protection and remediation purposes, hydrogen production as well as design and development of light emitting diode (LEDs) based photocatalytic reactors.

Dr. Vimal Gandhi

Dr. Vimal Gandhi is, presently working as Associate Professor, with Department of Chemical Engineering, Dharmsinh Desai University since last 17 years. He received his Bachelor (1999) and Master degree (2005) in Chemical Engineering from Dharmsinh Desai University and received Ph.D. in the area of Application of nanomaterials for waste water treatment from the same University in 2011.

In the field of research and consultancy, He has guided several graduates and undergraduate students for their project work. He has more than 20 publications/presentation in International/National reputed journals and conferences to his credit. His research interest is synthesis, characterization and application of nanomaterials in the field of Environment Engineering. He has also organized various training programme for chemical industries in Gujarat including GNFC, PI Industries, Huntsman, Transpek Silox etc.

He is a member of several professional bodies, including Indian Institute of Chemical Engineers (IIChE), International Congress of Chemistry and Environment (ICCE) and Indian Society of Technical Education (ISTE). He is additionally functioning as Secretary of Indian Institute of Chemical Engineers- Nadiad Regional Center. He is also giving his professional services to Bharuch Environ Infrastructure Limited and Environ Technology Limited, Ankleshwar as an Independent Director.

www.ingramcontent.com/pod-product-compliance
Ingram Content Group UK Ltd.
Pitfield, Milton Keynes, MK11 3LW, UK
UKHW020659041026
15590UKWH00001B/365

9 781945 291586